PREPARING A WORKFORCE FOR THE NEW BLUE ECONOMY

PREPARING A WORKFORCE FOR THE NEW BLUE ECONOMY

People, Products and Policies

Edited by

LIESL HOTALING

RICHARD W. SPINRAD

ELSEVIER

Elsevier
Radarweg 29, PO Box 211, 1000 AE Amsterdam, Netherlands
The Boulevard, Langford Lane, Kidlington, Oxford OX5 1GB, United Kingdom
50 Hampshire Street, 5th Floor, Cambridge, MA 02139, United States

Library of Congress Cataloging-in-Publication Data
A catalog record for this book is available from the Library of Congress

British Library Cataloguing-in-Publication Data
A catalogue record for this book is available from the British Library

ISBN: 978-0-12-821431-2

For information on all Elsevier publications visit our website at https://www.elsevier.com/books-and-journals

Publisher: Candice Janco
Acquisitions Editor: Louisa Munro
Editorial Project Manager: Grace Lander
Production Project Manager: Sujatha Thirugnana Sambandam
Cover Designer: Mark Rogers

Typeset by TNQ Technologies

Contents

Contributors

Norm Augustine
Lockheed Martin Corporation, Bethesda, MD, United States

Jesse H. Ausubel
The Rockefeller University, New York, NY, United States

Christine Bassett
United States National Oceanic and Atmospheric Administration (NOAA), Washington, DC, United States

Anna Kate Baygents
Mississippi Enterprise for Technology, Stennis Space Center, MS, United States

Nick Beaird
Rutgers, The State University of New Jersey, New Brunswick, NJ, United States

Matthew Breece
School of Marine Science and Policy, University of Delaware, Lewes, DE, United States

William Yancey Brown
Bureau of Ocean Energy Management, Department of the Interior, Washington, DC, United States

Eric P. Chassignet
Center for Ocean-Atmospheric Prediction Studies (COAPS), Florida State University, Tallahassee, FL, United States

Rita Colwell
Distinguished University Professor at the University of Maryland at College Park and Johns Hopkins Bloomberg School of Public Health and the Founder and Chairman of CosmosID, Inc, Rockville, MD, United States

Bob Coniglione
Contract Deputy Program Manager, Mission Control Manager, and IT Manager at the National Buoy Center, Stennis Space Center, MS, United States

Thomas Cuff
United States National Oceanic and Atmospheric Administration (NOAA), Washington, DC, United States

Benjamin Cuker
Department of Marine and Environmental Science, Hampton University, Hampton, VA, United States

Alan Curry
The Rockefeller University, New York, NY, United States

Thomas G. Drake
United States Navy Office of Naval Research, Arlington, VA, United States

John Englander
Rising Seas Institute, The Explorers Club, Boca Raton, FL, United States

Peer Fietzek
Manager Science & Research, Kongsberg Maritime Contros GmbH, Hamburg, Germany

Darin Figurskey
United States National Oceanic and Atmospheric Administration (NOAA), Washington, DC, United States

Melissa Fischel
Director of Operations and Workforce Development at TMA BlueTech, San Diego, CA, United States

Nicholas I. Flocken
United States National Oceanic and Atmospheric Administration, Washington, DC, United States

Ben Freeston
Founder, Magicseaweed, Plymouth, United Kingdom

Oscar Garcia-Pineda
Founder, Water Mapping, Gulf Breeze, FL, United States

Tim Gallaudet
United States National Oceanic and Atmospheric Administration, Washington, DC, United States

Simon Geerlofs
Pacific Northwest National Laboratory, Seattle, WA, United States

Deidre Gibson
Department of Marine and Environmental Science, Hampton University, Hampton, VA, United States

Alfredo Giron
Center for Ocean Solutions, Stanford University, Monterey, CA, United States

Scott Glenn
Rutgers, The State University of New Jersey, New Brunswick, NJ, United States

Carl Gouldman
United States Integrated Ocean Observing System, NOAA Integrated Ocean Observing System Office, Silver Spring, MD, United States

Steve Hall
Pembroke Coastal Forum and Marine Energy Wales Pembrokeshire, United Kingdom

Jim Hanlon
Former CEO, Centre for Ocean Ventures and Entrepreneurship (COVE), Nova Scotia, Canada

Liesl Hotaling
Eidos Education, Highlands, NJ, United States

Michael Jones
President of TMA BlueTech (TMA), San Diego, CA, United States

Laurie Jugan
Mississippi Enterprise for Technology, Stennis Space Center, MS, United States

Paula Keener
Center for Oceans and Human Health and Climate Change Interactions, Arnold School of Public Health, University of South Carolina, Columbia, SC, United States; Global Ocean Visions, LLC, Charleston, SC, United States

Judith T. Kildow
Middlebury Institute of International Studies, Monterey, CA, United States; Bigelow Laboratory for Ocean Sciences, East Boothbay, ME, United States

Barb Kirkpatrick
Gulf of Mexico Coastal Ocean Observing System (GCOOS), Department of Oceanography, Texas A&M University, College Station, TX, United States

Louisa Koch
National Oceanic and Atmospheric Administration (NOAA), Washington, DC, United States

Randy Kochevar
Oceans of Data Institute, Education Development Center (EDC), Waltham, MA, United States

Martin Koehring
Head of The Economist Group's World Ocean Initiative, London, United Kingdom

Josh Kohut
Rutgers, The State University of New Jersey, New Brunswick, NJ, United States

Bev MacKenzie
Director of Policy, Institute of Marine Engineering, Science & Technology (IMarEST), London, United Kingdom

Justin Manley
Founder and Principal Consultant, Just Innovation, Kingston, MA, United States

Craig McLean
National Oceanic and Atmospheric Administration (NOAA), Washington, DC United States

Travis Miles
Rutgers, The State University of New Jersey, New Brunswick, NJ, United States

David Millar
Government Accounts Director, Americas at Fugro, Frederick, MD, United States

S. Bradley Moran
College of Fisheries and Ocean Sciences, University of Alaska, Fairbanks, AK, United States

Laura Nazzaro
Rutgers, The State University of New Jersey, New Brunswick, NJ, United States

Matthew J. Oliver
School of Marine Science and Policy, University of Delaware, Lewes, DE, United States

Irene Penesis
Blue Economy Cooperative Research Centre, Newnham, TAS, Australia; Australian Maritime College, University of Tasmania, Launceston, TAS, Australia

Dwayne E. Porter
Center for Oceans and Human Health and Climate Change Interactions, Arnold School of Public Health, University of South Carolina, Columbia, SC, United States

Noelani Puniwai
Hawai'inuiākea School of Hawaiian Knowledge, Honolulu, HI, United States

Perry Raso
Matunuck Oyster Farm, South Kingstown, RI, United States

Ralph Rayner
London School of Economics, London, United Kingdom; National Oceanic and Atmospheric Administration, Washington, DC, United States

A.J. Reiss
United States National Oceanic and Atmospheric Administration (NOAA), Washington, DC, United States

Mitch Roffer
President of Fishing Oceanography, Inc., Melbourne, FL, United States

Vladimir Ryabinin
Intergovernmental Oceanographic Commission of UNESCO, Paris, Île-de-France, France

Grace Saba
Rutgers, The State University of New Jersey, New Brunswick, NJ, United States

Paul A. Sandifer
Center for Coastal Environmental and Human Health, College of Charleston, Charleston, SC, United States; Center for Oceans and Human Health and Climate Change Interactions, Arnold School of Public Health, University of South Carolina, Columbia, SC, United States

Oscar Schofield
Rutgers, The State University of New Jersey, New Brunswick, NJ, United States

Geoffrey I. Scott
Center for Oceans and Human Health and Climate Change Interactions, Arnold School of Public Health, University of South Carolina, Columbia, SC, United States

Sherry Scully
Executive Director of the Centre for Ocean Ventures and Entrepreneurship (COVE) Workforce Initiative, Nova Scotia, Canada

Alexandra Skrivanek
United States National Oceanic and Atmospheric Administration, Washington, DC, United States

Emily Smail
United States National Oceanic and Atmospheric Administration, Center for Weather and Climate Prediction, College Park, MD, United States

Paul Snelgrove
Departments of Ocean Sciences and Biology, Memorial University of Newfoundland, St John's, NL, Canada

Richard W. Spinrad
Oregon State University, Corvallis, Oregon, United States

Margaret Spring
Monterey Bay Aquarium, Monterey, California, United States

Kathryn Sullivan
Potomac Institute for Policy Studies, Former Administrator of the National Oceanic and Atmospheric Administration (NOAA), Washington, DC, United States

Vasily Titov
Pacific Marine Environmental Laboratory, United States National Oceanic and Atmospheric Administration, Seattle, WA, United States

Jan van Smirren
Ocean Sierra, LLC, Sugar Land, TX, United States

Hans Van Sumeren
Northwestern Michigan College, Traverse City, MI, United States

Samuel Walker
Group Marine Science Expert for BP International, Houston, TX, United States

Don Walsh
International Maritime Inc., Myrtle Point, and Oregon State University, Corvallis, OR, United States

Kevin Wark
FV Dana Christine II, Barnegat, NJ, United States

Wendy Watson-Wright
7 Mile Bay Consulting, Albany, PE, Canada

Sheldon Whitehouse
United States Senate, Washington, DC, United States

John Whittington
Blue Economy Cooperative Research Centre, Newnham, TAS, Australia

Author biographies

Norman R. Augustine (page 553–556)

Norman R. Augustine was raised in Colorado and attended Princeton University where he graduated magna cum laude with a BSE and MSE in Aeronautical Engineering. He has served as US Undersecretary of the Army and Acting Secretary of the Army, as Chairman and Chief Executive Officer of Lockheed Martin Corporation, and as Lecturer with the Rank of Professor at Princeton University. Mr. Augustine was Chairman of the American Red Cross, the Council of the US National Academy of Engineering, the Association of the United States Army, the Aerospace Industries Association, and the US Defense Science Board. He is a former President of the American Institute of Aeronautics and Astronautics and the Boy Scouts of America. He was a Regent of the University System of Maryland, Trustee of Princeton and MIT, and is a Trustee Emeritus of The Johns Hopkins University. He served on advisory boards to the US National Aeronautics and Space Administration (NASA), the US Departments of Homeland Security, Energy, Defense, Commerce, Transportation, and Health and Human Services (NIH), and served for 16 years on the US President's Science Advisory Board. He is a member of the US National Academy of Sciences, the US National Academy of Engineering, the American Philosophical Society, the American Association of Arts and Sciences, the Council on Foreign Relations, and a fellow of the Explorers Club. Mr. Augustine has been presented the National Medal of Technology by the President of the United States, the Joint Chiefs of Staff Distinguished Public Service Award, and the NASA Distinguished Public Service Medal. He has five times received the US Department of Defense's highest civilian decoration, the Distinguished Service Medal, and holds honorary degrees from 35 universities. Since retiring he has chaired or cochaired 39 committees and commissions, mostly on behalf of the government. He has authored or coauthored four books and has traveled in 129 countries and stood on both the North and South poles of the Earth.

Jesse H. Ausubel (page 249–258)

Jesse Ausubel began his career in 1977 as a resident fellow with the Climate Research Board of the US National Academy of Sciences in Washington, DC. He helped organize the first UN World Climate Conference (1979) and authored much of the original documentation of the World Climate

Program and the International Geosphere-Biosphere (Global Change) Program. Since 1989, Mr. Ausubel has served at Rockefeller University, where he leads a program to elaborate the technical vision of a large, prosperous society that emits little harmful and spares large amounts of land and sea for nature. Mr. Ausubel initiated and helped lead the Census of Marine Life, Barcode of Life Initiative, and Encyclopedia of Life, and conceived the International Quiet Ocean Experiment. An adjunct scientist of the Woods Hole Oceanographic Institution, University Fellow of Resources for the Future, and author or editor of 160 publications, Mr. Ausubel and his group have pioneered studies of marine eDNA.

Christine Bassett (page 63–84)

Christine Bassett is a 2020 John A. Knauss Marine Policy Fellow hosted by the US National Weather Service. Christine supports the Office of Observations Director in his role as Chair of the World Meteorological Organization (WMO)/Intergovernmental Oceanographic Commission Joint Collaborative Board and the WMO Standing Committee on Marine Meteorological and Oceanographic Services (SC-MMO). She works with the SC-MMO's Secretariat and Chair to develop and implement a revised framework for global marine weather services, disaster risk reduction in coastal environments, response to maritime environmental disasters, and international efforts to coordinate sea ice analysis and forecasts. She is also a PhD Candidate in Geological Sciences at the University of Alabama where she develops and applies novel proxies of paleoclimatology and paleoceanography in the Aleutian Island chain and wider North Pacific Basin.

Anna Kate Baygents (page 357–364)

Anna Kate Baygents served as the Project Manager at the Mississippi Enterprise for Technology (MSET), supporting the Marine Industries Science and Technology (MIST) Cluster to seek and assist companies interested in doing business with or locating at Stennis Space Center and along the Mississippi Gulf Coast. She served as Project Manager for the Governor Bryant's Ocean Task Force, focusing on the Blue Economy. She assisted in a number of documents describing and quantifying Blue Economy efforts along the Mississippi coast and in the state. Currently, Anna Kate works as a Program Manager for the Center for International Education at The University of Southern Mississippi.

Nicholas Beaird (page 417–424)

Nicholas Beaird is a Physical Oceanographer and an Assistant Teaching Professor at Rutgers University. He completed his PhD at the University of

Washington and postdoctoral fellowships and Woods Hole Oceanographic Institution and Oregon State University before joining Rutgers. Along with the other coauthors, he has led the development of the Masters in Operational Oceanography. The authors comprise the Center for Ocean Observing Leadership (RUCOOL) at Rutgers University. RUCOOL creates knowledge of our ocean planet by pushing the limits of science and new technologies while inspiring and teaching future generations of ocean explorers.

Matthew Breece (page 151–159)

Matthew Breece graduated with a BA in Biology and Economics from St. Mary's College of Maryland in 2005, received his MS in Natural Resource from Delaware State University in 2012, and earned his PhD in Oceanography from the University of Delaware in 2017. Currently he is working as a Post-doctoral Researcher for the University of Delaware, continuing his research on anadromous fish movement and habitat using advance tagging and tracking techniques to reduce human impacts to populations. He also plays a critical role in Project Recover searching for MIAs throughout the world.

William Yancey Brown (page 179–191)

William Yancey Brown is the Chief Environmental Officer of the Bureau of Ocean Energy Management in the US Department of the Interior, where he oversees science and regulation for protection of the environment in energy and nonenergy minerals development on the Outer Continental Shelf of the United States. He is a graduate of the Escola Americana do Recife (High School), University of Virginia (BA Biology, with highest distinction), The Johns Hopkins University (MAT), University of Hawaii (PhD, Zoology), and Harvard Law School (JD). He was born in Artesia, California.

Eric Chassignet (page 47–61)

Eric Chassignet is Distinguished Professor in the Department of Earth, Ocean, and Atmospheric Science and the Director of the Center for Ocean-Atmospheric Prediction Studies (COAPS) at Florida State University, a center of excellence that promotes interdisciplinary research in ocean–atmosphere–land–ice interactions to increase our understanding of the physical, social, and economic consequences of climate variability. For the past 20 years, Professor Chassignet has been the coordinator of a broad partnership of institutions (academic, government, and industry) to develop and demonstrate the performance and application of high-resolution, real-time global and basin-scale ocean prediction systems using the HYbrid Coordinate Ocean Model (HYCOM). These systems are currently used operationally by the US Navy at the Naval Oceanographic Office (NAVOCEANO), Stennis Space Center, MS, and by the National Oceanic and Atmospheric Administration (NOAA) at the National Centers for Environmental Prediction (NCEP), Washington, D.C. This project received the 2008 National Ocean Partnership Program (NOPP) Excellence in Partnering Award. Dr. Chassignet is also codirector of the Florida Climate Institute and cochair of OceanPredict.

Rita Colwell (page 237–244)

Rita Colwell is Distinguished University Professor at the University of Maryland at College Park and Johns Hopkins Bloomberg School of Public Health and President of CosmosID, Inc. Her interests are focused on global infectious diseases, water, and health. Dr. Colwell served as 11th Director of the US National Science Foundation and Cochair of the US Committee on Science, National Science and Technology Council. Dr. Colwell holds advisory positions in the US Government, nonprofit science policy organizations, and private foundations. She has authored or coauthored 20 books and more than 800 scientific publications. Dr. Colwell served as Chairman of the Board of Governors of the American Academy of Microbiology, President of the American Association for the Advancement of Science, Washington Academy of Sciences, American Society for Microbiology, Sigma Xi National Science Honorary Society, International Union of Microbiological Societies, and American Institute of Biological Sciences. Dr. Colwell is a member of the National Academy of Sciences, Royal Swedish Academy of Sciences, Stockholm, Royal Society of Canada, and Royal Irish Academy, American Academy of Arts and Sciences, and American Philosophical Society. Dr. Colwell has been awarded 63 honorary degrees from institutions of higher education and is the recipient of the Order of the Rising Sun, Gold and Silver Star, and the International Prize for Biology, bestowed by the Emperor of Japan, the 2006 National Medal of Science awarded by the President of the United States, the 2010 Stockholm Water Prize awarded by the King of Sweden, and the Lee Kuan Yew Water Prize of 2018. A geological site in Antarctica, Colwell Massif, has been named in recognition of her work in the Polar Regions.

Bob Coniglione (page 483–487)

Bob Coniglione is a retired US Navy Master Chief who served 21 years in the field of Meteorology and Oceanography. He holds an AA in general science from the University of Phoenix, is a graduate of the US Navy's Senior Enlisted Academy, and has over 1800 hours of Navy meteorology and oceanography education and training. His Navy career includes tours on the Aircraft Carriers NIMITZ and MIDWAY, special operations on five US Navy and one Canadian ship as well as tours as Enlisted Advisor to the Oceanographer of the Navy and as Command Master Chief of the Naval Meteorology and Oceanography Command. Following his retirement from the Navy, Bob joined General Dynamics Information Technology where he spent 13 years and departed as a Director of Operations. He joined PAE as a contract manager for the Mission Control and Information Technology branches of the National Data Buoy Center at the Stennis Space Center in Mississippi.

Thomas Cuff (page 63–84)

Thomas Cuff is the Director, Office of Observations for the US National Weather Service (NWS) and is responsible for the collection of space, weather, water, and climate observational data to support forecasts and impact-based decision support services for the nation. Previously, Tom served as the Director for the Ocean Prediction Center, and prior to joining NWS, he had a naval career that spanned over 30 years as a naval officer retiring at the rank of Captain and for 22 years as a navy civilian that included work on the US Oceanographer of the Navy's staff and Technical Director of the navy's largest operational oceanography center. Tom holds a Bachelor's degree from the US Naval Academy and a Master's in Meteorology from the University of Maryland. He chairs the World Meteorological Organization's (WMO) Standing Committee on Marine Meteorological and Oceanographic Services and is a member of the Joint WMO/Intergovernmental Oceanographic Commission Collaborative Board.

Benjamin Cuker (page 469–477)

Benjamin Cuker is a Professor of Marine and Environmental Science at Hampton University, having taught at Historically Black Universities since 1981. Dr. Cuker created several programs to promote human diversity in the aquatic sciences and has been recognized for this work with several awards, including the Association for the Sciences of Limnology and Oceanography (ASLO) award for excellence in education, the ASLO service award, and a Pew Fellowship in Marine Conservation. His work extends from Arctic lakes to the Chesapeake Bay. He is the editor and lead author of the book *Diet for a Sustainable Ecosystem: The Science for Recovering the Health of the Chesapeake Bay and its People*. He. has numerous publications in various scientific journals.

Alan Curry (page 249–258)

Alan Curry joined The Rockefeller University's Program for the Human Environment in 2012. He has researched diffusion of technologies in energy, transport, materials, and human performance and helped manage national and international projects in geoscience, ocean exploration, ocean acoustics, and environmental DNA, and development of software for analysis of growth processes.

Thomas G. Drake (page 261–268)

Tom Drake is Head of the US Office of Naval Research (ONR) Ocean Battlespace and Expeditionary Access Department. The department is responsible for a vertically integrated program of Navy and Marine Corps basic and applied research in the environmental sciences, including physical oceanography, marine meteorology, coastal geosciences, ocean acoustics, remote sensing, marine mammals, and Arctic and global prediction. The department also supports applied research and advanced technology development in the fields of ocean engineering, unmanned undersea vehicles, and all aspects of undersea warfare including antisubmarine and mine warfare. The department manages six oceanographic research vessels that are part of the University-National Oceanographic Laboratory System. Drake serves as cochair of the Subcommittee on Ocean Science and Technology under the US National Science and Technology Council to represent US Navy science and technology at the national level. Dr. Drake joined the US Federal Civilian Service in 2003 and the Senior Executive Service in 2013. Prior to joining the Federal Civilian Service, Drake was a tenured professor at North Carolina State University from 1995 to 2003. From 1990 to 1995, he was a research oceanographer at Scripps Institution of Oceanography, University of California, San Diego. He holds a BS degree from the Massachusetts Institute of Technology and a PhD from the University of California, Los Angeles.

John Englander (page 305–317)

John Englander is an oceanographer, consultant, and leading expert on sea level rise. He has a BSc degree in Geology and Economics from Dickinson College. He works with organizations to help them understand the risks of increased flooding, advocating for "intelligent adaptation." He is President and cofounder of the nonprofit Rising Seas Institute. Previously he served as CEO of The International SeaKeepers Society and The Cousteau Society. He is a Research Fellow at University of California-Santa Cruz and a Fellow of IMarEST (Institute of Marine Engineering Science and Technology) and The Explorers Club.

Peer Fietzek (page 377–384)

Peer Fietzek works as Business Development Manager Ocean Science at Kongsberg Maritime. He is a convinced advocate for science–industry collaboration and currently focuses on innovative application projects for quantitative acoustic sensors. After having graduated in physics in 2007 from the Technische Universität Darmstadt, Germany, including studies at the Istanbul Technical University (İTÜ), Turkey, he has been spending his entire career in the maritime sector working for a research institute, a start-up, an SME (small/medium enterprise), and a global corporation. He worked at GEOMAR Helmholtz Centre for Ocean Research Kiel and at CONTROS Systems & Solutions GmbH as a sensor developer for chemical oceanographic parameters followed by several technical management positions in the company until its acquisition in 2015.

Darin Figurskey (page 63–84)

Darin Figurskey has been with the US National Weather Service (NWS) since January 1990. Darin has worked at NWS offices in Flint, Michigan, Detroit, Michigan, Lubbock, Texas, White Lake, Michigan, Buffalo, New York, and Raleigh, North Carolina, prior to him becoming the Operations Branch Chief at the Ocean Prediction Center in December, 2015. Darin is responsible for the operational forecast and warning services for the Ocean Prediction Center's area of responsibility that covers the high seas of the north Atlantic and north Pacific. The Ocean Prediction Center also produces forecast information for part of the Arctic Ocean. Darin became Chair of the World Meterological Organization/Intergovernmental Oceanographic Commission Ship Observations Team in October, 2017.

Melissa Fischel (page 349–355)

Melissa Fischel is Director of Operations and Workforce Development at TMA BlueTech. Ms. Fischel supports the growth of the organization through programs and initiatives. Melissa has an MBA from University of California, Irvine, and a BA in Economics from University of California, San Diego. She brings 20+ years of industry experience including a Fortune 100 company, start-ups, and nonprofits.

Nicholas I. Flocken (page 271–284)

Nicholas I. Flocken is Senior Policy Advisor to the US Undersecretary of Commerce for Oceans and Atmosphere. He provides advice and strategic recommendations for how to advance administration initiatives to support the American blue economy including reducing the seafood deficit, preventing and reducing marine debris, promoting the sustainable use of ocean resources and engagements in the Arctic. He represents the US National Oceanic and Atmospheric Administration (NOAA) in several high-level interagency groups including the informal water subcabinet working group and is the Deputy Representative to the Maritime Domain Awareness Executive Steering Committee. Before coming to NOAA, he engaged with Congress on international conservation issues and has political leadership experience on local, state, and federal campaigns. He has a Bachelor's Degree from Denison University in Economics and Political Science and a Master's Degree from the University of London, Birkbeck, in Management and Corporate Governance with Business Ethics.

Ben Freeston (page 319–321)

Ben is responsible for product development and innovation at Surfline.com. As the founder of Magicseaweed, Ben spent almost two decades playing at the intersection of data and product in surfing, becoming an evangelist for the power of new data science techniques to help deliver the next generation of surf zone modeling. He lives in California with his wife and three children and still tries to surf everyday, regardless of the conditions.

Tim Gallaudet (page 271–284)

Tim Gallaudet, PhD, Rear Admiral, US Navy (retired) is the Assistant Secretary of Commerce for Oceans and Atmosphere and Deputy Administrator of the National Oceanic and Atmospheric Administration (NOAA). From 2017 to 2019, he served as Acting Undersecretary of Commerce for Oceans and Atmosphere and NOAA Administrator. Previously, he served for 32 years in the US Navy, completing his service in 2017 as the Oceanographer of the Navy. Today, he leads NOAA's Blue Economy activities advancing marine transportation, sustainable seafood, ocean exploration and mapping, marine tourism and recreation, and coastal resilience. He directs NOAA's support to the Administration's Indo-Pacific Strategy, oversees NOAA's Arctic research, operations, and engagement, and leads implementation of NOAA's science and technology strategies for Artificial Intelligence, Unmanned Systems, Omics, Cloud, Data, and Citizen Science. RDML Gallaudet has a Bachelor's Degree from the US Naval Academy and a Master's and Doctorate Degree from Scripps Institution of Oceanography, all in oceanography.

Oscar Garcia (page 289–291)

Oscar Garcia earned his PhD in Coastal and Marine System Science Program from Texas A&M University-Corpus Christi, Texas. He currently owns and operates Water Mapping, a consultant group focused on geological exploration and coastal and marine environmental monitoring using state-of-the-art technologies including satellite imagery and geophysical data followed by aerial and on-site observations to provide spatial and temporal analysis for coastal monitoring, oil exploration, and oil spills.

Simon Geerlofs (page 171–178)

Simon Geerlofs is an Advisor for the Blue Economy and Coastal Science at Pacific Northwest National Laboratory, specializing in the intersection of maritime technology, coastal economies, and renewable energy. In 2018 and 2019, Mr. Geerlofs served on assignment to the Department of Energy, Water Power Technologies Office to develop the Powering the Blue Economy initiative. Previous to his employment with PNNL, Mr. Geerlofs gained extensive experience with local jurisdictions and diverse stakeholder groups engaged in coastal management issues as the Marine Programs Manager for the Northwest Straits Commission. Before that, Mr. Geerlofs served as a Legislative Assistant for ocean policy to US Senator Maria Cantwell, working closely with Congress, constituents, agencies, and others to enact effective marine policy. Simon has a Master's Degree from the University of Washington School of Marine Affairs and served as a John A. Knauss Marine Policy Fellow in 2006.

Deidre Gibson (page 469–477)

Deidre Gibson is the Chair of the Department of Marine and Environmental Science at Hampton University. She earned her BS in Oceanography from the University of Washington, PhD in Marine Science from the University of Georgia/Skidaway Institute of Oceanography. Dr. Gibson is a broadly trained biological oceanographer with research interests centered on the trophic ecology, reproductive biology, and population dynamics of zooplankton, but more specifically, gelatinous zooplankton, and currently oyster restoration. While at Hampton University, she has served as Principal Investigator on several US National Science Foundation and National Oceanic and Atmospheric Administration grants that continue to train the next generation of African American marine scientists.

Alfredo Giron-Nava (page 457–460)

Alfredo Giron-Nava is a Mexican postdoctoral scholar at the National Center for Ecological Analysis and Synthesis at UC Santa Barbara in partnership with Future Earth. In 2019, he obtained his PhD from Scripps Institution of Oceanography, where he developed quantitative methods to develop dynamic management strategies for fisheries that are heavily affected by fishing pressure and environmental variability. In May 2019, Alfredo participated in the first Global Planning Meeting for the UN Decade of Ocean Sciences for Sustainable Development, where, along with a group of Early Career Ocean Professionals (ECOP), he started an ongoing effort to consider ECOP contributions to every aspect of the Decade. Since then, Alfredo has been a strong advocate to include ECOP in global, regional, and national committees and events related to the Decade and broader ocean sustainability issues, often serving as a point of contact to hear and synthesize the views of ECOP around the world. Alfredo is also the first recipient of the Walter Munk Scholar Award, awarded by the Marine Technology Society (MTS) and the Walter Munk Foundation for the Oceans to a young scientist pioneering activities of exploration and discovery.

Scott Glenn (page 417–424)

After graduating from the Massachusetts Institute of Technology and Woods Hole Oceanographic Institution Joint Program in 1983 with a Sc.D. in Ocean Engineering, Dr. Glenn began an over 30-year research career of implementing sustained real-time ocean observation and forecast systems first for offshore oil exploration at Shell Development Company (1983-1986), then for the Naval Oceanography Command supporting fleet operations while at Harvard University (1986-1990), and, since 1990, for a wide range of scientific and societal applications at Rutgers University.

Carl Gouldman (page 33–45)

Carl Gouldman has been Director of the US Integrated Ocean Observing System (IOOS) Office in the National Oceanic and Atmospheric Administration (NOAA), since February 2017. Carl served as the Deputy Director of the program from June 2014 to February 2017 and has been in NOAA since 2000. US IOOS is a coordinated network of people and technology that work together to generate and disseminate continuous data on our coastal waters, Great Lakes, and oceans. Before NOAA, Carl spent 3 years in the education department at the Chesapeake Bay Foundation where he led field programs teaching students about bay ecology and conservation. He holds a BS in Political Science from Duke University and a Master's (MEM) in Coastal Environmental Management from the Nicholas School of the Environment and Earth Sciences at Duke.

Steve Hall (page 495–502)

Steve Hall is Chief Executive Officer of the Society for Underwater Technology (SUT), an international learned society for marine engineering, science, technology, and policy headquartered in London, with branches in 10 countries and members in 40. Before his SUT role, he served 27 years in the UK Natural Environment Research Council, based mainly at the National Oceanography Centre, initially conducting tracer chemistry observations at sea, administering the UK contribution to the World Ocean Circulation Experiment, and managing the early "Autosub" autonomous underwater vehicle science missions. After 2006, Steve moved into a policy role, advising governments and stakeholders on a wide range of ocean policy issues, eventually becoming Head of the UK delegation to the Intergovernmental Oceanographic Commission of UNESCO, where he was elected as Vice Chair 2015–17. He was also responsible for tsunami warning systems for UK overseas territories. Steve has a long-standing interest in education and training for marine scientists and was one of the first cohorts of Chartered Marine Scientists. He is a Fellow of IMarEST & SUT and a member of the Marine Technology Society.

Jim Hanlon (page 365–370)

Jim Hanlon is a 40-year veteran of the ocean tech industry, having worked in product design, marketing, and management for companies in Atlantic Canada, Ontario, and New England. His career has spanned the aerospace and defense sectors as well as the marine environmental monitoring field. He has been an owner in two separate high-tech companies that have successfully grown and been purchased by multinationals. Jim holds an electrical engineering degree from Dalhousie University's Faculty of Engineering and an MBA in marketing from Saint Mary's University in Halifax and is a registered Professional Engineer. In June of 2020, Jim was elected a Fellow of the Canadian Academy of Engineering. He serves on many advisory boards for ocean research and commercialization ventures.

Liesl Hotaling (page 387–405)

Liesl Hotaling is President of Eidos Education. She currently also holds a position at the University of South Florida. Previously she served as the

Chief Education Officer for the Beacon Institute for Rivers and Estuaries, the Assistant Director for the Center for Innovation in Engineering and Science Education at Stevens Institute of Technology, and the Partnerships and Collaborations Coordinator for the Centers of Ocean Sciences Education Excellence. She holds a BS in Marine Science, an MAT in Science Teaching, and an MS in Maritime Systems (ocean engineering). She currently serves as the Vice President of Communications for the Marine Technology Society and specializes in real-time data education projects and hands-on STEM educational projects supporting environmental observing networks.

Michael Jones (page 347–355)

Michael Jones is President of TMA BlueTech (TMA), which is an organized nonprofit industry association founded in 2007. It is active across the "Triple Helix"—academia/education, industry, and policy. Its Mission Statement is "Promoting Sustainable, Science-Based Ocean & Water Industries." Originally focused on San Diego—the largest US BlueTech (ocean and water tech) cluster—it has members across the United States and abroad. TMA is a Strategic Partner with the US Department of Commerce and helped found the BlueTech Cluster Alliance, which includes 10 leading BlueTech clusters from eight countries. Michael has a broad background as a Board member, entrepreneur, investment banker, and Founder/President of TMA. He received his undergraduate degree from the University of Arizona and an MA from The Johns Hopkins University School of Advanced International Studies (SAIS). His graduate work included a year in Bologna, Italy; a year in Lima, Peru; and a 5-month internship at the EU headquarters in Brussels.

Laurie A. Jugan (page 357–364)

Laurie A. Jugan is a consultant supporting the Mississippi Enterprise for Technology (MSET) and the Marine Industries Science and Technology (MIST) Cluster, formalized under a Small Business Administration contract with the University of Southern Mississippi. For these two efforts, she serves as the Program Director, creating and implementing programs supporting small businesses and business interactions with organizations with promising partnership opportunities. Prior to this, she supported a technology-based small company, providing technical and administrative services under US Navy and National Oceanic and Atmospheric Administration contracts. Ms. Jugan holds Master's degrees in Oceanography and Business Management and is an active member of the Marine Technology Society.

Paula Keener (page 213–236)

Paula Keener is a marine biologist and Founder of Global Ocean Visions, LLC. She spent 18 years with the US National Oceanic and Atmospheric Administration's (NOAA's) Office of Ocean Exploration and Research. She served on national and international committees and advisory boards focused on building ocean capacities across multiple sectors. She participated in United Nations Intergovernmental Oceanographic Commission—UNESCO initiatives to build capacity for sustainable development and ocean literacy. She served on the US President's Panel on Ocean Exploration and was the US Colead for the trilateral Atlantic Ocean Research Alliance Ocean Literacy Working Group and Chair of the European Union–funded Horizon 2020 Sea Change International Advisory Group representing nine countries. She is past President of the National Marine Educators Association. She is a member of the South Carolina Department of Natural Resources Marine Advisory Committee and the South Atlantic Fisheries Management Council Habitat and Ecosystem-based Management Advisory Panel.

Judith T. Kildow (page 135–146)

Judith T. Kildow is Founding Director of the National Ocean Economics Program (NOEP) (www.OceanEconomics.org), retired, 2018, from the Center for the Blue Economy (CBE) at the Middlebury Institute of International Studies, and now CBE Advisory Council member, and editorial board member for their *Journal of Ocean and Coastal Economics*. She is a Board of Trustees member for the Bigelow Laboratory of Ocean Science. Dr. Kildow has an AB in Political Science from Grinnell College and a PhD in International Relations and Science Policy from The Fletcher School of Law and Diplomacy, Tufts University. She began her career at the Scripps Institution of Oceanography as Associate Director of the Center for Marine Affairs. For the next 40+ years she dedicated her life teaching and doing research to provide evidence that oceans and coasts are valuable and need protection and conservation. She spent 25+ years on the faculty, tenured, at Massachusetts Institute of Technology, was Senior Social Scientist at the Monterey Bay Aquarium Research Institute, James W. Rote Distinguished Professor, California State University Monterey Bay, Research Faculty/Senior Fellow at Harvard University, University of Southern California, and University of Vermont. She was a member of The National Advisory Committee on Oceans and Atmospheres, has served on numerous US National Academy of Sciences boards and committees, and has published and lectured throughout the world in the fields of coastal and ocean science policy and economics about the value of oceans and coasts and the high risks from climate change impacts to our shores.

Barbara Kirkpatrick (page 245–248)

Barbara Kirkpatrick is the Executive Director for the Gulf of Mexico Coastal Ocean Observation System (GCOOS). She has over 35 years of experience in human and environmental epidemiology and started her career as a Respiratory Care Supervisor at Duke University Medical Center before receiving a Master's Degree in Health Occupations Education at North Carolina State University and a Doctorate in Educational Leadership from the University of Sarasota. Kirkpatrick joined Mote Marine Laboratory and shifted her research to environmental human health, particularly the respiratory effects linked to harmful algal blooms. While there she led an 11-year study to scientifically document the impacts of red tide on humans—particularly those who have chronic respiratory diseases. Now, as Executive Director of GCOOS, she has been instrumental in broadening the scope of the ocean monitoring organization to include biological aspects of ocean monitoring—particularly monitoring for toxic algal blooms and marine animal movements.

Louisa Koch (page 451–455)

Louisa Koch, Director of Education at the US National Oceanic and Atmospheric Administration (NOAA), educates and inspires the public and future workforce about the Earth System. She works with NOAA's amazing array of people, partners, places, and information. Louisa also serves as lead for NOAA's National Team for Regional Collaboration, helping to address complex, place-based challenges that require interdisciplinary approaches and regionally tailored solutions. Ms. Koch served as NOAA's acting Deputy Undersecretary and Deputy Assistant Administrator for Research. Prior to joining NOAA, Ms. Koch worked at the US Office of Management and Budget, the US Department of Defense, and the Joint Economic Committee of the US Congress. Ms. Koch earned a Master's in Electrical Engineering from the Massachusetts Institute of Technology and a Bachelor's in Physics from Middlebury College.

Randy Kochevar (page 445–449)

Randy Kochevar trained as a deep-sea biologist at the University of California, Santa Barbara, studying the physiology of hydrothermal vent and hydrocarbon seep animals. The discovery of similar animal communities in the Monterey Submarine Canyon brought Randy up the coast, initially to the Monterey Bay Aquarium Research Institute (MBARI), and then to the Monterey Bay Aquarium, where he was involved in exhibit development, website development and print publishing, and media relations. From 2008 to 2015, Randy worked at Stanford University's Hopkins Marine Station, collaborating with Dr. Barbara Block on electronic tagging of marine apex predators including tunas, billfish, and sharks. In 2015, Randy was invited to take over as director of the Oceans of Data Institute at the Education Development Center (EDC), where he oversees a portfolio of projects to build data literacy skills in K-16 students and to help build pathways for data science and analytics careers. In addition to his role at EDC, Randy teaches Environmental Science and Environmental Science Lab at Monterey Peninsula College.

Martin Koehring (page li–liv)

Martin Koehring is the head of The Economist Group's World Ocean Initiative, inspiring bold thinking, new partnerships and the most effective action to build a sustainable ocean economy. He also leads The Economist Intelligence Unit's policy and thought leadership work on sustainability, climate change and natural resources in the EMEA region. He has led major projects in areas such as food sustainability, sustainable infrastructure and ocean sustainability. He is also a member of the Advisory Committee for the UN Environment Programme's Global Environment Outlook for Business.

His previous roles at the EIU, where he has been since 2011, include managing editor, global health lead and Europe editor.

He earned a bachelor of economic and social studies in international relations from Aberystwyth University and a master's degree in international relations from the College of Europe.

Josh Kohut (page 151–159)

Josh Kohut graduated cum laude in 1997 with a BS in Physics from the College of Charleston and earned his PhD in Physical Oceanography from Rutgers University in 2002. Since then, he has been a central player in the Rutgers University Center for Ocean Observing Leadership serving in several capacities including Technical Director and now as a member of the Faculty Leadership. Using networks of ocean observing technologies, his research and extension programs focus on the physical ocean processes that structure marine ecosystems. He has developed relationships and partnerships across stakeholder groups to ensure his work is collaborative, relevant, and informs applications, decision-making, and management of ocean resources. His projects range from coastal processes impacting storm intensity, beach water quality, and fisheries habitat off the coast of the Mid-Atlantic to regional-scale questions centered on marine ecosystem dynamics within the coastal seas around Antarctica.

Bev Mackenzie (page 489–494)

Bev Mackenzie has a degree in chemical oceanography and a PhD in Physical Oceanography from the University of Liverpool, UK. She learned the ropes in volunteer engagement at the University of Oxford and moved to the Institute of Marine Engineering, Science & Technology (IMarEST) as a Technical Coordinator in 2004. She is now the Director of Policy with her role looking at strategies that provide opportunities for the exchange of ideas and practices and opportunities to influence the development of marine policy and legislation at local, regional, and global levels in support of the United Nations Sustainable Development Goals. Dr. Mackenzie's specialist knowledge is in operational oceanography and ocean observing and issues relating to maritime industries and the environment. Dr. Mackenzie oversees the IMarEST contributions to the work of a number of intergovernmental organizations and is the IMarEST lead on representation at the International Maritime Organization, the Intergovernmental Oceanographic Commission of UNESCO, the Intergovernmental Panel on Climate Change, the Informal Consultative Process on Oceans and the Law of the Sea, and the London Convention/London Protocol. She is a Fellow of the IMarEST and of the Marine Biological Association and is a Chartered Marine Scientist and Chartered Scientist. She has been a member of the Board of the Science Council (UK) and currently sits on the Board of Plymouth Marine Laboratory.

Justin Manley (page 371–376)

Justin Manley is an innovative technologist and executive with experience in start-up, public corporation, academic, and public sectors. He attended the Massachusetts Institute of Technology (MIT) and earned a BS in History and a BS and an MS in Ocean Engineering.

Craig McLean (page 513–525)

Craig McLean is the Assistant Administrator for Research at the US National Oceanic and Atmospheric Administration (NOAA), responsible for ocean, weather, and climate research. He was commissioned in the NOAA Corps, one of the Nation's seven uniformed services, in 1981, and retired from Active Duty with NOAA in 2006, then becoming deputy and later director in NOAA's research Line Organization. McLean has commanded NOAA research ships, dived in deep-sea submersibles to hydrothermal vents and the wreck of the RMS Titanic, and worked as an attorney and program manager within NOAA. At the federal interagency level, he is a cochair of the US National Science and Technology Council, Subcommittee on Ocean Science and Technology. He has served for 10 years as the US Representative to the UN Intergovernmental Oceanographic Commission and was among the original members contributing to the UN Decade of Ocean Science for Sustainable Development, and now part of the 20-person global team constituting the Executive Planning Group for the UN Decade of Ocean Science. McLean was recognized as a 250th Anniversary Fellow from his alma mater, Rutgers University, and holds a Juris Doctor degree from Quinnipiac University.

Travis Miles (page 417–424)

Travis Miles is a physical oceanographer interested in understanding how the atmosphere, cryosphere, earth, and ecosystems connect through the ocean across a broad range of temporal and spatial scales by using and developing ocean observation networks that sample across spatial and temporal scales in combination with process oriented numerical modeling. He received a B.S. in Marine science and Meteorology from North Carolina State University (NCSU); Master's in Physical Oceanography from NCSU; and a PhD from Rutgers University studying sediment resuspension and transport during Nor'easters and Hurricanes, specifically Sandy in 2012. He is currently an Assistant Research Professor in the Rutgers University Center for Ocean Observing Leadership (RUCOOL).

David Millar (page 323–332)

David Millar is Fugro's Government Accounts Director for the Americas region. Based in the Washington, D.C. area, Mr. Millar serves as Fugro's account manager for the US government, other national governments in the Americas, the United Nations, the World Bank, and other multilateral development banks. He is responsible for overseeing the development and execution of Fugro's partnerships with public sector clients and overseeing Fugro's collaborative science activities (with government, academia, and NGOs) within the Americas Region. Mr. Millar also directs Fugro's global support of Seabed 2030 and the United Nations Decade of Ocean Science for Sustainable Development. He is a member of the General Bathymetric Chart of the Oceans (GEBCO) Guiding Committee, a member of the US National Oceanic and Atmospheric Administration's (NOAA) Ocean Exploration Advisory Board and has more than 30 years of ocean mapping, marine geophysical, and hydrographic survey experience. Over his professional career, Mr. Millar has been involved in various aspects of ocean mapping, spanning field operations, operations management, software development management, business development, and executive management. He holds a BSc in Math and Physics from Mount Allison University and a BScE in Survey Engineering from the University of New Brunswick.

S. Bradley Moran (page 407–416)

S. Bradley Moran is Dean of the University of Alaska Fairbanks College of Fisheries and Ocean Sciences. He previously served as Acting Director of the Obama Administration's National Ocean Council, Assistant Director for Ocean Sciences in the White House Office of Science and Technology Policy, and Program Director in the Chemical Oceanography Program at the US National Science Foundation (NSF). In his prior academic position as Professor of Oceanography at the University of Rhode Island Graduate School of Oceanography, he served as Codirector of the Rhode Island NSF Experimental Program to Stimulate Competitive Research and as Assistant Vice President for Research. He received a BSc in Chemistry from Concordia University, a PhD in Oceanography from Dalhousie University, and conducted his postdoctoral research at Woods Hole Oceanographic Institution.

Laura Nazzaro (page 151–159)

Laura Nazzaro graduated with a BA in Biology from The College of New Jersey in 2008 and earned her MS in Oceanography from Rutgers University in 2011. Since then she has continued working for the Rutgers Center for Ocean Observing Leadership, providing software support for various projects ranging from quality control of remotely sensed data streams to coupling of habitat models to physical ocean models for various Mid-Atlantic Bight species.

Matthew Oliver (page 151–159)

Matthew Oliver is the Patricia and Charles Robertson Professor of Marine Science and Policy at the University of Delaware. He earned a Master's in Biology from Cal Poly, San Luis Obispo, and a Doctorate in Oceanography from Rutgers University. Dr. Oliver operates the Rutgers Ocean Exploration,

Remote Sensing and Biogeography Lab, and the Robotics Discovery. Dr. Oliver won the Presidential Early Career Award for Scientists and Engineers, is an Alfred P. Sloan Research Fellow, and has earned the Outstanding Doctoral Graduate Advising and Mentoring Award. He uses ocean observing to understand coupled physical–biological systems in the ocean.

Irene Penesis (page 335–348)

Irene Penesis is the Research Director for the Blue Economy CRC-Co Ltd. Irene is on secondment from the University of Tasmania where she led a multidisciplinary team working in the field of marine renewable energy and contributes to educating maritime engineering students, at the Australian Maritime College—National Centre for Maritime Engineering and Hydrodynamics. Professor Penesis has established strong industry, government, and research partnerships (nationally and internationally) and is leading an Australian Renewable Energy Agency (ARENA)-funded project, "Tidal Energy in Australia—Assessing Resource and Feasibility to Australia's Future Energy Mix." She also contributes to international-impactful activities including Member of the Marine Energy Research and Innovation Centre (MERIC) Scientific International Committee since 2018, and Chair of Marine Renewable Energy Specialist Committee of the International Towing Tank Conference (ITTC) between 2011 and 2017, a peak international body developing technical guidelines and procedures relevant to the hydrodynamic testing of wave energy converters, marine current/tidal turbines, and offshore wind turbines.

Dwayne E. Porter (page 213–236)

Dwayne E. Porter (MS, PhD) holds a joint administration/faculty appointment with the Arnold School of Public Health (ASPH) and the Baruch Institute for Marine and Coastal Sciences at the University of South Carolina (UofSC). Within ASPH, Dr. Porter is the Associate Chair and Director of Graduate Studies for the Department of Environmental Health Sciences. Dr. Porter is also a faculty member in the Environment and Sustainability Program and the School of the Earth, Ocean and the Environment, and directs the Community Engagement Core of the US National Institute of Environmental Health Sciences (NIEHS) Center for Oceans and Human Health and Climate Change Interactions (OHHC2I) at the UofSC. In addition, Dr. Porter directs the US National Oceanic and Atmospheric Administration (NOAA) National Estuarine Research Reserves System's Centralized Data Management Office. Dr. Porter's research interests include the roles that technology and technological innovations play in monitoring, assessing, modeling, and managing coastal environmental resources and associated environmental and public health issues, with a specific focus on underserved communities.

Noelani Puniwai (page 461–468)

Noelani Puniwai is passionate about cultivating the next generation of students to mālama ʻāina. She has been trained academically (PhD in Natural Resources and Environmental Management, University of Hawaii Mānoa; MSc in Environmental Science, Washington State University; BA in Marine Science, UH Hilo) to practice malama ʻāina/kai. Yet through her experiences in her culture, science training, and student mentorship, she's recognized that feeling aloha ʻāina, he alo a he alo, face-to-face must also be practiced, moving beyond exclusively intellectual pursuits. Her research interests include coastal ecosystems, cultural geography, knowledge coproduction, and seascapes. She believes that we can use the rigor and methodologies of pono science, the foundational wisdom of our kūpuna, and our experiential daily practice of aloha ʻāina to awaken responsible action for the future of our Hawai'i.

Perry Raso (page 165–168)

Perry Raso, owner of Matunuck Oyster Bar and an oyster farmer, holds Bachelor's and Master's degrees from the University of Rhode Island (URI) in Aquaculture and Fisheries Technology. He founded Matunuck Oyster Farm in 2002, in East Matunuck (South Kingstown), Rhode Island, and the Pond to Plate vision was formed. The restaurant and oyster farm have received national acclaim in publications including *The New York Times*, *National Geographic*, *USA Today*, *Yankee Magazine*, *The Boston Globe*, and *Chicago Tribune*. His business was featured on the PBS show "Start Up." In 2016, he gave a TedX talk in Providence sharing his story and outlining the "Importance of Sustainable Aquaculture in Our Future."

Ralph Rayner (page 113–134)

Ralph Rayner is an ocean science and technology specialist whose career has encompassed numerous commercial, academic, and governmental roles associated with making and using ocean observations, measurements, and models in support of delivering benefits across a wide range of end uses and markets. Dr. Rayner is a Professorial Research Fellow at the London School of Economics, is responsible for industry outreach in the Integrated

Ocean Observing System program office of the US National Oceanic and Atmospheric Administration, and serves as the President of the Society for Underwater Technology. He holds editorial roles for a number of publications concerned with ocean science and technology and is Editor in Chief of the *Journal of Operational Oceanography*. Ralph is a former chair of the Intergovernmental Oceanographic Commission Global Ocean Observing System Steering Committee and has served in a wide variety of roles in designing and implementing national, regional, and global ocean observation and forecasting capabilities. He holds a first degree in Biology, a Master's in Underwater Science and Technology, and a Doctorate in Physical Oceanography.

A.J. Reiss (page 63–84)

A.J. Reiss is the Director of the US National Weather Service's Ocean Prediction Center and is responsible for its mission to sustain more than $2 trillion American marine economy through life and cargo saving forecasts, warnings, and decision support to mariners in the North Atlantic and Pacific Oceans. He is a retired US Navy Captain who served on ships and then as a Naval Oceanographer, being privileged to command a maritime weather center and major operational oceanography center. A.J. is a 1990 graduate of the US Naval Academy and later earned his Master's degree in Air & Ocean Sciences from the US Naval Postgraduate School. He is from the suburbs of Akron, Ohio, but confides that most of his growing up was on a Lake Erie fishing boat. Lately, he finds dignity and wonder in all, but really enjoys spending time with his family, ideally outdoors and anywhere near the ocean.

Mitchell Roffer (page 161–164)

Mitchell Roffer is President of Fishing Oceanography, Inc., a scientific consulting company that focuses on fisheries oceanography and remote sensing along with environmental education and advocacy, as well as sustainable alternative energy, waste disposal, and bioremediation solutions and was founded in 2019 in Melbourne Beach, Florida. Prior to this, he was founder (1989) and President of Roffer's Ocean Fishing Forecasting Service, Inc. (ROFFS). He received his doctorate in Biological Oceanography—fisheries oceanography at the University of Miami's Rosenstiel School for Marine and Atmospheric Science (RSMAS). While many in the fisheries world know Dr. Roffer for his pioneering and continued work with operational fishing forecasting analyses and habitat classification, he has received significant national and international acclaim for the innovative oil oceanographic analyses products after the British Petroleum DeepWater Horizon explosion and oil spill, as well as other pollution monitoring products after hurricanes, Katrina, Rita, and Harvey. Roffer has been an active member of the US National Aeronautics and Space Administration's (NASA) prestigious Biodiversity and Ecological Forecasting Science Team since 2003.

Vladimir Ryabinin (page 527–532)

Vladimir Ryabinin serves as the Executive Secretary of the Intergovernmental Oceanographic Commission (IOC) of UNESCO and Assistant Director-General of UNESCO. Dr. Ryabinin is an oceanographer, marine engineer, climatologist, and emeritus meteorologist of The Russian Federation. His research has led to a number of achievements in numerical weather prediction, marine meteorological services, offshore engineering, ocean and climate science. Since the early 1980s, Dr. Ryabinin has been involved as an expert in coordination activities of the United Nations. He has contributed to core design of the Global Ocean Observing System, International Polar Year 2007/2008, and is now leading the design of the UN Decade of Ocean Science for Sustainable Development (2021–30).

Grace Saba (page 417–424)

Grace Saba is an Assistant Professor at Rutgers University. Her broad research interests are in the fields of coastal marine organismal ecology and physiology, with emphasis on how organisms interact with their environment (physical-biological coupling) and other organisms (food web dynamics and predator-prey interactions), how physiological processes impact biogeochemistry (nutrient cycling and carbon sequestration), and how climate change (i.e., ocean acidification, warming) impacts these processes. Dr. Saba earned a Bachelor of Science degree in Aquatic Biology from the University of California Santa Barbara and her Ph.D. in Marine Science from the College of William & Mary at the Virginia Institute of Marine Science.

Paul A. Sandifer (page 213–236)

Paul A. Sandifer (BS, PhD) is Director of the Center for Coastal Environmental and Human Health at the College of Charleston, South Carolina, and Deputy Director for the Center for Oceans and Human Health and Climate Change Interactions at the University of South Carolina. He is experienced in ecological and aquaculture research, natural resource management, science policy, and environmental health science. Previously he worked in the National Oceanic and Atmospheric Administration (NOAA), overseeing the agency's Oceans and Human Health Program, as Senior Science Advisor to the NOAA Administrator, and Chief Science Advisor for the National Ocean Service. Before NOAA, Paul was a scientist, manager, and agency Director with the South Carolina Department of Natural Resources. He served on the US Commission on Ocean Policy and is an Honorary Life Member of the World Aquaculture Society and a Fellow of the American Association for the Advancement of Science and the Ecological Society of America.

Oscar Schofield (page 417–424)

Oscar Schofield is a biological oceanographer, Distinguished Professor and Department Chair at Rutgers University. He is interested how the physics and chemistry regulates ocean ecosystems, with a primary research focus on the physiology and ecology of phytoplankton. His research is conducted in a range of oceans from the rapidly warming/melting along the West Antarctic Peninsula to sustained studies along the northeast United States. He earned a degree in biology and marine science and attended the University of California at Santa Barbara for his undergraduate and PhD where he conducted research on photosynthesis and environmental impacts of the Antarctic ozone hole.

Geoffrey I. Scott (page 213–236)

Geoffrey I. Scott is currently a Clinical Professor and Chair of the Department of Environmental Health Sciences in the Arnold School of Public Health at the University of South Carolina. He is also the Director of the US National Institutes of Environmental Health Sciences' Center for Oceans and Human Health and Climate Change Interactions, housed in the Arnold School of Public Health, with five collaborating universities. Throughout his career he has worked for the US Environmental Protection Agency (EPA) on the ecotoxicology of Water Disinfection By-products and Oil Spills; private industry (oil spills and vibrio cholerae outbreaks), and the US National Oceanic and Atmospheric Administration (NOAA) where he was the Director of the Center for Coastal and Environmental Health and Biomolecular Research from 2001 to 2014. Dr. Scott has served on over 100 graduate student research committees and has directed the research of >55 Master's and PhD students. He has published over 162 peer-reviewed scientific publications and has authored/coauthored more than 23 national reports.

Sherry Scully (page 365–370)

Sherry Scully is the Executive Director of the Centre for Ocean Ventures and Entrepreneurship (COVE) Workforce Initiative, where she leads a research project relating to workforce development across the national marine and ocean industries, to attract, train, and retain highly qualified people to sustain the competitiveness and growth of the Canadian shipbuilding and marine industries. With this focus on current and future workforce, Dr. Scully also conducts projects and programs and develops educational resources for teachers and youth in the k-12 system and leads an annual ocean institute for educators in several regions across the country. Sherry holds an interdisciplinary PhD in Social Psychology and Management from Dalhousie University, as well as an MBA with a concentration in Change Strategy, and two degrees in Education.

Alexandra Skrivanek (page 271–284)

Alexandra Skrivanek is a Policy Analyst and 2020 National Oceanic and Atmospheric Administration (NOAA) Sea Grant Knauss Marine Policy Fellow in the Office of the Assistant Secretary of Commerce for Oceans and Atmosphere. Currently, she supports NOAA leadership in the areas of coral reef health and marine-protected areas and serves as the Executive Secretary of the NOAA Blue Economy Executive Committee. Previously, as a National Science Foundation Graduate Research Fellow, she assimilated evidence preserved in ancient Caribbean coral reefs for the response of sea level and ice sheets to a past warm climate to improve projections of future sea-level changes and developed outreach activities to support evidence-based discussions on coastal resilience and marine chemistry. Dr. Skrivanek earned a Bachelor's degree in Earth and Environmental Sciences with minors in Museum Studies and Oceanography at the University of Michigan (2014) and a PhD in Geology at the University of Florida (2019).

Emily Smail (page 285–288)

Emily Smail is the Executive Director of the GEO Blue Planet Initiative and a Senior Faculty Specialist at the US National Oceanic and Atmospheric Administration (NOAA)—University of Maryland Cooperative Institute for

Climate and Satellites. She also serves as the cochair of the GEO AquaWatch Initiative's outreach and user engagement working group and supports outreach and education efforts for the NOAA CoastWatch/OceanWatch program. Previously, Dr. Smail worked in informal science education at the Waikiki Aquarium and policy in the US Senate through the Knauss Marine Policy Fellowship Program. She received a BS in Biology from the Pennsylvania State University and a PhD in Biology from the University of Southern California where her research focused on water quality and marine biogeochemistry.

Paul Snelgrove (page 17–32)

Paul Snelgrove is a Professor of Ocean Sciences and Biology at Memorial University of Newfoundland. Since 2008, he has led the Natural Sciences and Engineering Research Council of Canada (NSERC) Canadian Healthy Oceans Network, a national research network working on all three of Canada's oceans to develop new tools and approaches to support sustainable oceans. He also currently plays the role of Associate Scientific Director of The Ocean Frontier Institute, which brings together researchers in Atlantic Canada and beyond to support research on safe and sustainable oceans. From 2003 to 2013, Dr. Snelgrove held a Canada Research Chair in Boreal and Cold Ocean Systems, and prior to that an NSERC Industrial Chair in Fisheries Conservation. He led the synthesis of the International Census of Marine Life research program, where he was a member of the program's Scientific Steering Committee. Dr. Snelgrove published the book *Discoveries of the Census of Marine Life: Making Ocean Life Count* with Cambridge University Press in 2010, and his TED Global talk in 2011 has been viewed by over 330,000 people. In 2013, he was awarded the Timothy Parsons Medal for Excellence in Marine Sciences in Canada, and in 2020, he took up a new part-time role as Departmental Science Advisor for Fisheries and Oceans Canada.

Richard W. (Rick) Spinrad (page 87–111)

Rick W. Spinrad is a Professor of Oceanography at Oregon State University (OSU), Member of the Ocean Studies Board of the US National Academies of Science, Engineering, and Medicine, and past President of the Marine Technology Society and The Oceanography Society. In 2016, he retired as Chief Scientist of the National Oceanic and Atmospheric Administration (NOAA), as appointed by President Barack Obama. Dr. Spinrad was the Vice President for Research at OSU and previously was the head of NOAA's Research Office and the US National Ocean Service. He coled the White House Committee developing the nation's first set of ocean research priorities. Dr. Spinrad was a Senior Executive with the US Navy, where he was awarded the Distinguished Civilian Service Award (highest civilian award in the Navy). He held faculty appointments at three universities, served as Executive Director at the Consortium for Oceanographic Research and Education, and was President of Sea Tech, Inc. He also created the National Ocean Sciences Bowl. Dr. Spinrad was the US permanent representative to the United Nations' Intergovernmental Oceanographic Commission. Spinrad received Presidential Rank Awards from Presidents Bush and Obama and is a Fellow of four professional societies. Dr. Spinrad holds degrees from The Johns Hopkins University and OSU.

Margaret Spring (page 533–551)

Margaret Spring is Chief Conservation and Science Officer at the Monterey Bay Aquarium, where she oversees ocean conservation policy initiatives, the Seafood Watch sustainable seafood program and research of California's ocean ecosystems. From 2009 to 2013, Margaret held top leadership roles at the US National Oceanic and Atmospheric Administration. Prior to that, she led The Nature Conservancy's California coastal and marine program. From 1999 to 2007, she served on Capitol Hill as senior counsel, then general counsel, to the US Senate Committee on Commerce, Science, and Transportation, where she played a key role in advising members of Congress on ocean and climate issues and developing legislation on major science and policy topics. Margaret is a state appointee to the Board of Trustees of the California Ocean Science Trust and a founding board member of the Monterey Bay Fisheries Trust. Margaret previously served on the Ocean Studies Board of the US National Research Council. She is a graduate of Duke University Law School and Dartmouth College.

Kathryn Sullivan (page 479–482)

Kathryn Sullivan has a long career as a distinguished scientist, astronaut, and executive. She was one of the first six women to join the US National Aeronautics and Space Administration (NASA) astronaut corps in 1978 and holds the distinction of being the first American woman to walk in space. Her submersible dive to the Challenger Deep in June of 2020 made her the first person to both orbit the planet and reach its deepest point, as well as the first woman to dive to full ocean depth. Dr. Sullivan has held a variety of senior executive positions since leaving NASA, including Presidential appointments to the National Science Board and as Chief Scientist, Deputy Administrator and Administrator of the National Oceanic and Atmospheric Administration (NOAA). She currently serves on the boards of International Paper, Accenture Federal Services, the National Audubon Society and Terra Alpha Investments and is a Senior Fellow at the Potomac Institute for Policy Studies. Dr. Sullivan earned a Bachelor of Science degree in Earth Sciences from University of California at Santa Cruz and a PhD in Geology from Dalhousie University, Nova Scotia. Dr. Sullivan is a member of the National Academy of Engineering and the American Academy of Arts and Sciences and has been elected into the Astronaut Hall of Fame, Women Aviators Hall of Fame, Women Divers Hall of Fame, and Ohio Veterans Hall of Fame. She was recognized as one of the 46 distinguished First Women by *Time Magazine* (2017), the 15 Women Changing the World by the World Economic Forum (2015), and Time's 100 Most Influential People (2014). She has been awarded the Explorers Club Medal, the Rachel Carson Award, an Emmy, and nine honorary degrees. She is the author of the children's book *To the Stars!* (Charlesbridge Press, 2016) and *Handprints on Hubble, An Astronaut's Story of Invention* (MIT Press, 2019).

Vasily V. Titov (page 293–304)

Vasily V. Titov is the Lead Scientist of the National Oceanic and Atmospheric Administration's (NOAA) Center for Tsunami Research, the joint operation of NOAA's Pacific Marine Environmental Laboratory and the University of Washington. He is also an Affiliate Assistant Professor of the Earth and Space Sciences of the University of Washington. He received his PhD in Coastal and Ocean Engineering in 1997 at the University of Southern California. His undergraduate degree is in Mathematics from the Novosibirsk State University. He has developed the MOST model to simulate tsunami dynamics that is now being used for operational NOAA tsunami forecast and for tsunami studies and forecasts in many countries.

Jan van Smirren (page 201–209)

Jan van Smirren, BSc, MSc, has gained over 35 years of experience in commercial oceanography. Mr. van Smirren is a Consultant Oceanographer with Ocean Sierra LLC, prior to this he was Partnership Managing Director for the Energy Sector at British Maritime Technology, and previously occupied global and regional senior management positions in the United Kingdom, Singapore, and United States with Fugro. His oceanographic experience has seen him undertake studies in the North and South Atlantic, South China Sea, Indian Ocean, Mediterranean Sea, Pacific Ocean, Gulf of Mexico, and Caribbean. This has included both field measurement and monitoring system campaigns and the detailed analysis and interpretation of data. Mr. Van Smirren is the Treasurer of the Gulf of Mexico Coastal Ocean Observing System Regional Association and a founding member and Honorary Secretary of the Houston Branch of the Society of Underwater Technology.

Hans Van Sumeren (page 433–443)

Hans Van Sumeren is the Director of the Great Lakes Water Studies Institute at Northwestern Michigan College and has over 28 years of experience across multiple sectors of the blue economy. He has led the development and delivery of numerous courses, trainings, and degree pathways focused on technical skills required for careers in the marine industry. Prior to Northwestern Michigan College, Hans worked at the University of Michigan in the Ocean Engineering Laboratory and the Marine Hydrodynamics Laboratories where he served at various research and administrative positions. He holds both a BSE and MSE in Naval Architecture and Marine Engineering from the University of Michigan.

Samuel Walker (page 193–200)

Samuel Walker currently serves as the Group Marine Science Expert for BP International, working on the Environment Team within BP's Safety and Operational Risk function. His primary responsibilities include establishing group-wide requirements, identifying risks, and leading the overall strategy for the marine science discipline. Prior to joining BP, Dr. Walker held technical and strategic project development positions in the private, government, academic, and nonprofit sectors. He has been directly involved in the ocean observing community for the past decade, including technical positions at the regional and national levels of the US Integrated Ocean Observing System (IOOS). During his time with the IOOS Program Office, he supported the Deepwater Horizon spill response as the lead scientist for the Subsurface Monitoring Unit within the Unified Area Command. Dr. Walker has an undergraduate degree in Geography and holds a Master's of Science in Public Health (MSPH) and PhD in Environmental Health Sciences with a focus on coastal ecosystem health and modeling.

Don Walsh (page 3–16)

Don Walsh, Captain, US Navy (retired), PhD, is an oceanographer, explorer, and ocean engineer. During his 24-year naval career, he served in submarines at sea and in ocean-related research while ashore. In 1960, Lieutenant Walsh and Jacques Piccard dove to the deepest place in the ocean in the Navy's Bathyscaph Trieste. Life Magazine said this was one of the greatest explorations of the 20th century. In addition to deep ocean exploration, he has made over 80 expeditions to the Polar Regions including the South and North poles. The Walsh Spur in the Antarctic is named for him. After 24 years of naval service, Dr. Walsh became a dean and faculty member at the University of Southern California. Until 2018, he served as Honorary President of the Explorers Club. In 2001, he was elected a member of the National Academy of Engineering (NAE) in recognition of his work in undersea engineering.

Kevin Wark (page 151–159)

Kevin Wark is the owner-operator of the Fishing Vessel Dana Christine II. Captain Wark is a third-generation Long Beach Island resident from Barnegat Light N.J. with 40 years of commercial fishing experience in the Mid-Atlantic region and two decades of collaborative research work with academic partners including the US National Oceanic and Atmospheric Administration (NOAA) science center at Wood Hole, Massachusetts. Captain Wark has served on numerous state and federal advisory panels and currently serves on the New Jersey Marine Fisheries council as a voting member. He has worked with NOAA as a Take Reduction Team member on Large Whale, Harbor Porpoise and Bottle Nose Dolphin, received a national award from *National Fisherman* magazine in 2012, and continues to work collaboratively on fisheries science and management.

Wendy Watson-Wright (page 17–32)

Wendy Watson-Wright served as the Executive Secretary and Assistant Director General of UNESCO's Intergovernmental Oceanographic Commission (IOC-UNESCO) in Paris from 2010 to 2015 and retired at the end of 2019 as the inaugural CEO of the Atlantic Canadian-led Ocean Frontier Institute. For most of her career, she held various senior positions within Fisheries and Oceans Canada, including 8 years as Assistant Deputy Minister of Science. She is currently Founder and CEO of 7 Mile Bay in Prince Edward Island, Canada. Dr. Watson-Wright is a member of the UN

Joint Group of Experts on the Scientific Aspects of Marine environmental Protection (GESAMP) as well as of the Boards of the International Ocean Institute (IOI) and the Fondation du Prince Albert II de Monaco (Canada). She has been on a number of other boards and panels including the European Marine Board External Review panel (Chair), the Belmont Forum, Expert Panel Review, Collaborative Research Action on Trans-disciplinary Research for Ocean Sustainability (Chair), and the Scientific Advisory Board of the Wendy Schmidt Ocean Acidification XPRIZE. A Killam scholar, she holds a PhD in Physiology from Dalhousie University in Halifax, Nova Scotia.

Sheldon Whitehouse (page 505–511)

US Senator Sheldon Whitehouse, a member of the Senate Environment and Public Works Committee (EPW), plays a key role in crafting policies addressing environmental protection and climate change. In 2011, he joined with Democrats and Republicans to form the Senate Oceans Caucus to increase awareness of and find common ground on issues facing the oceans and coasts. The Caucus helped gain Senate approval of four international fisheries treaties and passage of the IUU Fishing Enforcement Act that will prevent illegal, unreported, and unregulated fishing. Whitehouse has worked to boost federal support for fisheries science and cooperative fisheries research as well as efforts to improve transparency and efficiency in the commercial and recreational fisheries management process. Senator Whitehouse has long advocated for a dedicated fund to support ocean and coastal research, restoration, and conservation. In 2015, the National Oceans and Coastal Security Fund (now the National Coastal Resilience Fund) was created to provide grants that support work for the oceans, coasts, and Great Lakes. It received funding for the first time as part of the fiscal year 2018 spending bill. He also led the successful bipartisan effort to reauthorize the US Environmental Protection Agency's National Estuaries Program to protect and study coastal habitats. Whitehouse has led bipartisan efforts to address the growing threat of marine plastic and other marine debris. He authored the Save Our Seas Act, a bipartisan law to reauthorize the US National Oceanic and Atmospheric Administration's marine debris program and to strengthen America's role in combating global marine debris. Whitehouse's second ambitious bill to clean up the oceans, the bipartisan Save Our Seas 2.0 Act, has passed the Senate. A graduate of Yale University and the University of Virginia School of Law, Whitehouse served as US

Attorney and Attorney General of Rhode Island before being elected to the Senate in 2006. In addition to EPW, he is a member of the Budget Committee; the Judiciary Committee; and the Finance Committee.

John Whittington (page 335–348)

John Whittington is the Chief Executive of the Blue Economy CRC-Co Ltd. Prior to joining the Blue Economy CRC, John was the Secretary (CEO) of the Tasmanian Department of Primary Industries, Parks, Water and Environment where he was responsible for the sustainable planning and management of Tasmania's wild fisheries, aquaculture, agriculture, animal welfare, and biosecurity. Dr. Whittington has also been the Director of Parks and Wildlife and the Director General of Lands, responsible for the conservation and management of Tasmania's 3.2 million hectare reserve estate. He has previously chaired the Tasmanian Institute of Agriculture Advisory Board and was a Director of the Institute for Marine and Antarctic Studies, and through his career has led major research, development, and extension programs and portfolios in fisheries, animal and plant health, and conservation science. John is also a Director of Tasmanian Irrigation Pty Ltd, a company established to design, construct, and manage large-scale irrigation schemes.

Foreword

When Rick Spinrad and Liesl Hotaling approached me to write a foreword for a book on developing the workforce for the New Blue Economy, I did not hesitate. This is a topic that is very close to my heart. I lead The Economist Group's World Ocean Initiative, which focuses on inspiring bold thinking, new partnerships, and the most effective action to build a sustainable ocean economy (WOI, 2020a). The ocean is already the world's biggest employer, supporting the livelihoods of more than 3 billion people and providing food for more than 2.6 billion (Figueres et al., 2017). The transition from the "old" blue economy—one that unfortunately still relies too heavily on oil and gas, high-carbon shipping, overfishing, and unsustainable tourism—to a new economy, one that is both sustainable and built on data, information, and knowledge, requires a major transition for technology, talent, and skills too.

The year 2020 was supposed to be the "ocean super year," with major progress expected for ocean and climate conferences as well as key targets under Sustainable Development Goal 14. And then the coronavirus pandemic hit the world economy. A lot has been written about the impact of the global pandemic on the ocean economy: from the collapse in global trade, shipping, and tourism to the delay of offshore wind projects and the rise in marine plastic pollution as a result of rising demand for single-use plastics used in personal protective equipment.

As the global economy emerges from the pandemic-induced slump, attention is turning to how to "build back better"—and how to "build back bluer" in the ocean economy. Building a sustainable ocean economy during the 2020s and beyond is not only desirable—it has become an imperative if the world is to meet ambitious global climate targets and harness the ocean's full potential to contribute to the global recovery. At the World Ocean Initiative, we have been highlighting both the challenges and opportunities of creating a sustainable ocean economy by 2030 (WOI, 2020a).

What is the New Blue Economy? It is an economy founded on emerging capabilities for acquiring data and developing knowledge that harnesses ocean resources for economic growth while protecting ocean health and ensuring social equity. The ocean's contribution to global economic growth has been significant and growing. A seminal OECD study estimated the size of the ocean economy at around US$1.5T in 2010 (around 3% of global GDP) and predicted the ocean's contribution to double in size from 2010 levels to US$3T by 2030 (OECD, 2017). However, for that growth to be sustainable and create sustainable jobs, several transitions are required, nurturing opportunities in renewable ocean energy, decarbonization of shipping, sustainable aquaculture, ecotourism, and ocean data.

As the blue economy continues to expand, energy demand will continue to rise. New energy innovations are emerging in areas such as battery technologies, wind turbines and foundations, tidal stream technologies, wave energy conversion, and thermal energy conversion. These have huge potential to create both sustainable value and jobs (WOI, 2020b). The International Energy Agency expects global offshore wind capacity to increase 15-fold and attract around US$1T of cumulative investment by 2040. This growth will create lots of jobs: the Global Wind Energy Council projects that, by 2024 already, more than 77,000 trained workers will be required to deliver new offshore wind-energy projects in North America, China, Taiwan, Japan, Vietnam, and South Korea.

The energy transition will also have major impacts on shipping. According to targets adopted by the International Maritime Organisation, the shipping industry is set to reduce its greenhouse gas emissions by at least 50% by 2050, compared with 2008 levels, with the ambition for the sector to achieve net-zero emissions by 2050. Meeting these targets implies not only improvements in operational and energy efficiency in ships, but also a major transition away from heavy fuel oil to electrification and alternative fuels, such as ammonia and hydrogen, and enhanced understanding of environmental factors. The US national Energy Storage Association (ESA) expects that new energy-storage goals could lead to the creation of around 200,000 jobs in the United States alone by 2030. New jobs and skills will also be required for the production of these new fuels and the associated fuel infrastructure as well as building new low-carbon vessels.

Aquaculture is another important area full of opportunity for building a sustainable ocean economy and creating new sustainable jobs. One-third of wild fisheries are currently overfished, with harmful fisheries subsidies contributing to overcapacity and overfishing, as well as to illegal, unreported, and unregulated fishing. Against this backdrop, aquaculture is the fastest-growing animal-protein production sector in the world, according to the UN's Food and Agriculture Organisation (FAO). Already more than 20M people are employed in aquaculture, and this number is set to increase significantly. A huge opportunity lies in making aquaculture more sustainable, by accelerating the shift away from wild-caught fish for feed, and using environmental knowledge more effectively. New innovation and technologies in fishmeal replacements are emerging, from microalgae and insect-based feed to genetically modified plants and gas fermentation (WOI, 2020c).

Tourism is among the sectors most affected by the coronavirus pandemic and the travel restrictions that have come with it. However, the pandemic may accelerate a trend toward more conscientious traveling that has supported the rise in ecotourism. Tourism will play a crucial role within the New Blue Economy. The 2016 OECD study had predicted that maritime and coastal tourism would replace offshore oil and gas as the largest ocean-based industry in terms of gross value added by 2030. The need to understand the tourism sector's impact on the environment, notably the ocean, is creating demand for new data, technologies and skills—central to the concept of the New Blue Economy. This shift has already sparked innovation in better data gathering to support environmental impact assessments and marine spatial planning. For example, eco-conscientious resorts are investing in coastal management services, such as marine rangers.

As the UN Decade of Ocean Science for Sustainable Development (2021–30) commences, the need to harness ocean science and innovation will accelerate—from satellite imaging and remote sensors to big data and artificial intelligence. The need to monitor ocean degradation, and reverse it, will continue to spark technological innovation and progress. For example, the Seabed 2030 Project endeavors to map the entire ocean floor by 2030. Meanwhile, new surveillance and product-tracing technologies will help to tackle illegal fishing and human-rights abuses at sea. Data overload will present a growing problem. Hence, artificial intelligence and machine learning will play a growing role in making sense of the growing volume of ocean data.

This book highlights some of the most exciting areas in which ocean technology disruption and the data revolution of the New Blue Economy are paving the way for job creation. From harnessing ocean renewables and sustainable aquaculture to shipping decarbonization and blue tourism, the New Blue Economy offers a myriad of challenges and opportunities for new talent, skills, and jobs.

As the world emerges from the coronavirus pandemic and lays the foundations for building back better, unlocking the potential of the New Blue Economy holds enormous potential. Rick Spinrad and Liesl Hotaling have brought together an exceptional group of experts and leaders for this book: oceanographers, marine policy specialists, policymakers, marine engineers, data literacy leaders, educators, economists, geologists, technologists, and entrepreneurs. Their insights make a significant contribution to advancing our understanding of the New Blue Economy and how to build the workforce that harnesses its potential.

Martin Koehring Head,
World Ocean Initiative,
The Economist Group

References

Figueres, J.M., Lamy, P., Podesta, J.D., February 17, 2017. Our Oceans Need Protecting. This is How We Can Help. World Economic Forum. https://www.weforum.org/agenda/2017/02/the-worlds-biggest-employer-and-seventh-largest-economy-needs-help-this-is-how-to-save-our-oceans.

OECD, The Ocean Economy in 2030. https://www.oecd.org/environment/the-ocean-economy-in-2030-9789264251724-en.htm.

World Ocean Initiative, 2020a. A Sustainable Ocean Economy in 2030: Opportunities and Challenges. https://woi.economist.com/sustainable-ocean-economy-2030/.

World Ocean Initiative, 2020b. Accelerating Energy Innovation for the Blue Economy. https://ocean.economist.com/energyinnovation.

World Ocean Initiative, 2020c. Fish Feed of the Future. https://ocean.economist.com/innovation/articles/fish-feed-of-the-future.

Further reading

World Ocean Initiative. https://ocean.economist.com/.

Acknowledgments

This book emerged from discussions with innumerable friends and colleagues, many of whom have contributed within. In addition to those authors included here, who shared their time, talents, and willingness to participate, we express special gratitude to our many colleagues at the Marine Technology Society, the US National Oceanic and Atmospheric Administration, Oregon State University, and the US Navy, for sharing their perspectives and insights over many years of discussion. One of us (RS) also owes an enormous debt of gratitude to Alanna Spinrad for her patience, tolerance, and extraordinary good humor as this project regularly displaced many previously scheduled activities. And one of us (LH) appreciates both Rick and Alanna Spinrad for dealing with my intrusions upon their schedules.

SECTION I

Exploration, observation, and prediction

The economic value of the ocean has developed over time through identification of resources and development of new capabilities. These activities have evolved with the applications of exploration and research to provide new understanding. **Don Walsh** provides an important historical perspective, framed in terms of exploration and engineering, with specific examples of market development that have provided the foundation for an ocean economy. **Wendy Watson-Wright and Paul Snelgrove** discuss the specific advances in generating relevant and reliable data from the ocean. These improvements in observational capabilities serve as the basis for building a knowledge-based economy around ocean information. **Carl Gouldman** describes the development and operationalization of the US Integrated Ocean Observing System, a critical component of the Global Ocean Observing System. Modeling and prediction efforts are the focal point for the New Blue Economy. **Eric Chassignet** provides an historical summary of developments in ocean modeling, around which predictive tools have emerged to feed the "currency of the realm" of the New Blue Economy: data/information/knowledge. Finally, in this section, we are introduced to one component of the public sector provision of ocean products and services, in the chapter by **A.J. Reiss et al.**, describing the diversity and value of ocean products provided by the US National Oceanic and Atmospheric Administration (NOAA) Ocean Prediction Center.

CHAPTER 1

Exploration and technology—key building blocks for the new blue economy

Don Walsh
International Maritime Inc., Myrtle Point, and Oregon State University, Corvallis, OR, United States

Sea power—sum of all national uses of ocean space

"Sea Power" is a term people often associate solely with a nation's military activities in the ocean. This is not correct. Sea power is the total sum of a nation's uses of the sea. It includes military, maritime commerce, fisheries, and resource development as well as conservation activities.

All the world's coastal nations are inherently sea powers even though they may only make limited use of the ocean.

There are four broad activity areas that comprise a nation's sea power:

Ocean sciences

Not a pure scientific discipline, oceanography is an interdisciplinary field involving the application of several different sciences to the marine environment. The primary product of oceanographic work is predictive information. That is, knowledge about the oceans that can be applied to beneficial uses of the sea. Examples are location and quality of a fish stock; knowledge of subsea floor geology leading to gas and oil finds; location of seafloor mining prospects, the best routing of seafloor cables, and forecasting marine weather (see, for example, Programs that Deliver, Section III). Also, more extensive knowledge of the seas leads to better conservation measures.

Ocean engineering

Ocean engineering is the application of relevant engineering arts and sciences to ocean-related problems. It's also an interdisciplinary field and not a specific engineering discipline such as aerospace, electrical,

Preparing a Workforce for the New Blue Economy
ISBN 978-0-12-821431-2
https://doi.org/10.1016/B978-0-12-821431-2.00011-1

mechanical, or civil engineering. The principal goal of ocean engineering is to build machines and equipment to do work in the sea. However, this does not define whether that work is *useful*.

Economics test and related public policy

The economics test determines if the work being done is useful. That is, will what you are doing in, on, or under the sea produce a viable commercial opportunity?

The public policy test is more imprecise. Consider the role of navies. Governments define that spending on navies is an important national use of the sea. While the cost of having a navy does fail the economic test, there is no question that it is vital for national security.

A more specific example is when a government may establish and support a national fishing industry because there is insufficient private capital to do this. It may operate that industry as a monopoly, or eventually sell it to the people who operate it as it becomes self-sustaining. Even though the enterprise may operate at a loss, public policy decisions of the central government will determine whether this work at sea is useful. That is, is it in the national self-interest to do this?

Man-made constraints

The fourth level of sea power development is outside of the rather precise scientific and technological determinism of ocean-related, science, engineering, and economics. Here we get into the often-interrelated questions of public policy, law of the sea, politics, and historic cultural and ethnic problems. Sometimes these issues will trump the rational scientific, technological, and economics verities.

Exploration—discovery leads to scientific knowledge

Exploration begins with curiosity and lots of it. A simple definition could be, "Exploration is the activation of curiosity and then acting upon it." It is the very first stage of the scientific method with the observation of something or some place and the organized defining (a hypothesis) of what is seen. Scientific knowledge has its origins here.

It is human nature to see something and wonder "why" though few may act on finding out. It is the explorers who act on their curiosity and take action to create new knowledge for humankind. Exploration is the first step in creating new knowledge for the benefit of all.

Today, it's a popular theme that "It has all been found; there is nothing left to explore." This is ignorance in action. Certainly, with many of the "big ones" such as the highest mountains, going to the moon and the deepest place in the ocean has been done. However, a visit, or even multiple visits, does not mean the end of exploration at those places. For every high-visibility "big one" there are thousands of exciting places and things to be discovered on our planet. In this respect, the largely unexplored world ocean, our planet's largest geographic feature, will remain a very compelling destination for the explorer.

Exploration of the world ocean—the evolution of oceanography

Despite thousands of years of human history, exploration, and discovery, we have only begun to fully explore the world ocean. It is a journey that will be a long one.

The earliest uses of the seas were as a highway to distant shores for conquest, resource development, and later, commercial activities. Scientific knowledge was primarily anecdotal. Knowledge that was based on mariners' experiences with sea conditions, weather, and navigational safety. That is, things seen and experienced.

Most historians of ocean science agree that the organized, interdisciplinary field now known as oceanography dates from the beginning of the 19th century. It was not born of some singular event. Rather, it evolved in response to an increasing curiosity by naturalists about our planet's largest geographic feature.

It is not surprising that the earliest marine scientists were from the naturalist community. In the beginning, their primary interests in the sea were in the plant and animal life of the coastal oceans. By the second half of the 19th century in England and Europe, there was an evolving group of "citizen naturalists" among the wealthier classes (Adler, 2019). Despite this growing interest in the sea, it was primarily for living things and did not include much in the earth and physical sciences. That would come much later.

By the mid-1800s, the population of naturalists was increasing in England, Europe, and North America. At first, most did their work as individuals, but they soon realized that they needed dedicated facilities and supporting equipment to do more than just investigate "tide pools." This led to the creation of shoreside marine laboratories and field stations where groups of scientists could collaborate using shared facilities (Adler, 2019). At first, they were primarily populated with scientists from the host country. However, it was not long before they realized that ocean-related investigations were transnational.

As the early oceanographic work progressed from individual investigators to organized groups, governments began to become involved with supportive policies and funding. Leaders began to understand that the better you understood the sea, the better this knowledge could serve national interests. This was a slowly evolving self-interest on the part of the major maritime nations. Therefore, much of the significant scientific progress at sea and onshore was due to increasing active participation of enlightened government leaders and also wealthy private sponsors.

Most historians of ocean science agree that oceanography's foundation as a scientific field began with the Royal Society of London's organization and its execution of the government-sponsored Challenger Expedition, 1872—76. The Royal Navy corvette *HMS Challenger* was specially fitted out with laboratories and instrumentation for the 70,000-mile, globe-circling expedition. Many new instruments and sampling devices had to be developed at a time when little off-the-shelf equipment existed.

The observations made and samples collected resulted in a 50-volume, 29,500-page report (Thompson et al., 1880; Linklater, 1972). These volumes and the samples collected are still referred to today. While the expedition involved several branches of science, in biology alone some 4700 new species were discovered.

In Europe, a preeminent example of an influential, wealthy patron was Prince Albert I of Monaco (1848—1922). He was an active ocean explorer during the last years of the 19thcentury and into the early 20th century. Though he was head of state, Prince Albert I devoted much of his life to oceanography, using his wealth, royal yachts, and active participation to greatly increase contemporary knowledge of the sea. His high-visibility position as a sovereign ruler greatly helped bring the importance of the world ocean to public attention.

International marine sciences cooperation began slowly in the mid-1800s. But it was not until the early part of the 20th century that it began to be widely adopted when the inherent economies and efficiencies from pooling knowledge and physical facilities were recognized.

In the late 19th century, there were British and European concerns about commercial overfishing in the North Atlantic and its bordering seas (Adler, 2019; Scearce, 2009). This was facilitated by better scientific knowledge of the stocks and more efficient catch methods such as the use of steam-powered fishing vessels to replace sailing ones.

Fishery resources exploitation, its regulation, and conservation depended on knowledge of the stocks and their temporal variability. Governments, commercial groups, and regional associations asked scientists to help figure out what could be done to remedy it. This meant that data had to be collected and organized in order to produce information on which regulatory actions could be taken. A continuing need for physical samples and measurements required a concurrent effort to develop the tools needed for this work. This was an early beginning of supporting marine technologies.

Meanwhile, among an increasing scientifically literate public, the high-visibility fisheries problems also increased their interest in marine life. Fisheries research was perhaps the first major ocean use activity in the evolution of oceanography as we know it today.

Marine science progress in the late 1800s benefited from increasing governmental interests in developing knowledge of the seas. Specifically, knowledge that could be an instrument of the state's power. Political entities slowly began to understand the nature and use of sea power through the lens of national self-interest.

For the curious, increased abilities to capture live and dead creatures fueled public imagination and spurred development of private and public aquaria. As a result, there began a popular development of citizen naturalists. Their interests were fueled by an increasing volume of ocean-themed reporting in the press as well as more extensive literature in the form of books and journals.

As the 19th century gave way to the 20th century, private philanthropy continued to play a major role in the creation of the larger research entities. There are two examples from the United States. In 1903, the creation of the Scripps Oceanographic Institution was largely made possible by support from that family. And in 1930, it was a $3 million grant from the Rockefeller Foundation that was instrumental in the creation of the Woods Hole Oceanographic Institution.

In the last two decades of the 19th century, as sail power gave way to steam power, the use of dedicated seagoing vessels for marine science began. By the beginning of the 20th century, the combination of marine biological stations on land and oceangoing ships stimulated a rapid increase in knowledge about a three-dimensional world ocean.

During the two world wars, there was limited international cooperation that had taken place in the prewar years. However, due to the specialized ocean science demands of warfare, great progress was made in the United States and other major naval powers. Superior knowledge of the marine environment was absolutely vital for prosecution of war at sea and was especially urgent for antisubmarine warfare (URI, 2012). Record funding was invested in wartime oceanographic research.

After WWII, governments' support of large-scale marine science and engineering programs has continued to increase. By every metric, postwar oceanography became "big science." For most large programs, funding and facilities requirements were beyond the purse of private and philanthropic institutions. Despite this, there was no diminution of significant support from privately funded institutions. And indeed, the numbers have increased. A few of the most prominent in the United States are: Mote Marine Laboratory Sarasota, Florida (1967); Harbor Branch Oceanographic Institution Fort Pierce, Florida (1971), Monterey Bay Aquarium Research Institute (MBARI) Moss Landing, California (1987); Waitt Institute at La Jolla, California (2008); Schmidt Oceanographic Institute Palo Alto, California (2009), and the Dalio Philanthropies Ocean X Initiative, Connecticut (2018). The special advantages of private philanthropy are that they are often more nimble to take actions and can focus on "far horizon" issues that government funding is not addressing. In the future, this area will be an increasing partner with government support for joint private—public programs to the benefit of both sectors.

In the 20th century, technological advancements permitted seagoing scientists to greatly increase their productivity and access to the global ocean. More efficient seagoing research vessels and development of other platforms from spacecraft to deep diving submersibles expanded the seagoing scientists' capabilities. Concurrently, technological advancements made possible new sensing and sampling capabilities. There was also a recognition that modern oceanography was indeed "big science" that involved extensive logistical support both on shore and at sea. It could not be done on the cheap.

Also, governments of the major maritime powers started to invest more in ocean studies as a matter of national self-interest (IOC, 2017). As a result, an increasing number of scientists elected to follow the money and become oceanographers. Driven by national security and commerce concerns, the result has been an accelerated increase in knowledge about the global seas.

Ocean engineering—providing the tools to do work in the sea

The origins of ocean engineering probably began thousands of years ago in the Classical Era (8 BCE—CE 6) when the first port and coastal engineering infrastructure works were undertaken. In the millennia since then, the simple extension of terrestrial engineering skills from shore into the sea was sufficient to meet requirements in that "wet space."

Operationally, ocean engineering developments have produced a remarkable array of devices and systems to make uses of the sea more effective. Since its beginnings in the 19th century, oceanography has been done primarily at the microscale level. Here the scientist undertakes close-up observations at discrete sites. This work was done remotely by lowering sensors, imaging devices, and physical samplers from a surface ship platform. Direct viewing of the site was not possible.

In the 1950s, technological advances made it possible to put the investigator in situ (at the actual worksite). The change began slowly in the post WWII years with the development of the self-contained underwater breathing apparatus (SCUBA). While the major development and use of this equipment were intended for the recreational diving market, it was also adopted by some marine scientists. For a few hundred dollars investment, they could now do work inside the ocean down at the work to depths as great as 120 feet. Now microscale research could be done with the investigator up close at the site. But to go deeper than the modest depths permitted by SCUBA, the diving scientist would need manned or unmanned submersibles.

By the late 1950s, the first practical submersibles were put into use as research platforms (Busby, 1976; Forman, 2000; Geyer, 1977). Initially, manned vehicles were the only choice as the development of suitable electronics, power systems (batteries), and instrumentation was limited by the technologies available at the time. The most reliable and easiest solution to doing work beyond SCUBA depths was to put a human inside a deep diving vehicle.

In the 1950s, the French and US navies developed and operated "bathyscaphs" (from the Greek for "deep boats"). They were invented by the Swiss scientist August Piccard, assisted by his son Jacques (Piccard, 1956). Depending on the mission, they could carry 2–3 persons to depths as great as 20,000 feet, which would permit access to 98% of the world ocean seafloor. And each could carry a significant scientific payload of sensors and physical samplers.

By the end of the 1950s, the two bathyscaphs, *FNRS 3* (France) and *Trieste* (United States), had dived to 13,287 feet (a bit greater than the average depth of the ocean) and 18,150 feet, respectively. Then in January 1960, the Navy's *Bathyscaph Trieste* dove into the deepest place in the world ocean.

This was to a depth of 35,800 feet at Challenger Deep in the Western Pacific's Mariana Trench. Oceanographers now had the capability to be in situ at any place in the world ocean (Piccard and Dietz, 1960).

While the scientific "product" of those early diving operations was somewhat meager, they did demonstrate the feasibility of this ultimate platform for microscale research. Concurrent with developing the operational knowhow, there was an aggressive research and development program for the necessary tools and instrumentation to do work effectively in the oceans' greatest depths. And there was an associated issue with getting oceanographers to use these "innerspaceships." It took a few years to overcome those cultural problems.

This was demonstrated that Challenger Deep was not visited again until 52 years after *Trieste* was there. In 2012, explorer and filmmaker James Cameron dove his one-person *Deepsea Challenger* submersible there. During his hours at the oceans' deepest place, he made important scientific observations while collecting a significant amount of imagery (National Geographic Society, 2012).

In 2019, Texas-based explorer Victor Vescovo developed a two-person, full ocean depth manned submersible, *Limiting Factor.* The vehicle was part of a deep diving system that consisted of a dedicated mother ship and four unmanned "landers," which were also full ocean depth capable.

That year he launched his Five Deeps Expedition (Oceanographic Magazine, 2020). The goal was to make a global circumnavigation to dive into the deepest place in each of the five oceans. Not only were his submersibles equipped with the state-of-the-art samplers, sensors, and imagery systems, the mother ship had a newly developed seafloor mapping system. It was a system that produced hydrographic quality bottom topography imagery. The 42,000-mile, 32 dives, Five Deeps Expedition was significant as it opened up access to the 2% of the seafloor that is the hadal depths, which were largely unexplored. In addition, it mapped 116,000 square miles of seafloor, most of it for the first time.

While *Trieste* and *Deepsea Challenger* only made single dives at Challenger Deep during the Five Deeps Expedition, *Limiting Factor* made four in a period of 8 days with Vescovo piloting the first two dives. This expedition demonstrated a remarkable capability for operational repeatability, system reliability, and safety.

In 2020, Vescovo launched the "Ring of Fire Expedition" that is continuing his exploration of the oceans' hadal depths (Caladan Oceanic, 2020).

During the six decades since *Trieste*'s dive to Challenger Deep, there has been a concurrent development of unmanned submersibles. Initially, this work was done by the US Navy. However, it was not long before there was increasing private sector involvement with these deep ocean work systems (Ocean Engineering Division, 1989). In the 1950s and early 1960s, there was little commercial interest in this area. The operators of deep diving manned submersibles had few, if any, industry support choices. If some new equipment was needed, they had to design and build it in-house. But by the end of the 1960s, dozens of manned submersibles were being designed, with many built by major corporate entities. Several were aerospace companies attracted to the possibilities of "inner space" and a "Wet NASA" to support it. This advancement also benefited through significant technology transfer from work done by government agencies such as the US Department of Defense and NASA.

The most promising deep ocean vehicle development has been the rapid evolution of unmanned submersibles from the early 1960s onward. There are two basic types: the tethered remotely operated vehicle (ROV) and the autonomous untethered vehicle (AUV). Eventually they would replace almost all manned submersible work in the sea.

The US Navy began development of the first ROVs in the early 1960s. Here the researcher was on a surface ship and controlled the submersible through an umbilical cable. Evolution from simple "swimming cameras" to multimission-capable vehicles was rapid. This caught the attention of the private sector. Through technology transfer from the Navy, ROVs began to prove their worth supporting the offshore petroleum industry. Jobs formerly done by divers could now be done more safely and at lower cost with specially configured submersibles. As exploration for and the production from depths evolved beyond diver capabilities, there were few work tasks that ROVs could not do. Eventually, thousands of these vehicles were put into service for offshore industrial applications of all kinds (Research and Markets Report, 2018).

AUVs are especially promising for doing many deep ocean projects in place of manned vehicles. Fast-moving new technologies such as artificial intelligence (AI) will permit these underwater drones to do in situ tasks much better than manned submersibles (Wikipedia, 2020). In the past three decades, the evolution of AUVs has moved from experimental proof of concept to globally capable tools for the oceanographer.

In the future, it is clear that most underwater work will be done by unmanned systems. They are cheaper, more capable, and inherently safer than manned vehicles. While "being there" in manned submersibles always makes good human-interest journalism, it is the much more productive unmanned systems that will do the "heavy lifting" in future deep ocean research operations.

In a relatively short time periods of seven decades, submersibles have opened the ocean depths to the in situ direct or indirect presence of the trained mind and eye. The scientific discoveries made and the work tasks done will all contribute to the full and careful uses of the oceans that is needed for an effective new blue economy.

Another important platform development direction is for the remote sensing of the oceans. In the mid-1960s, NASA began to invest in a program to determine what could be learned about our planet through the use of earth-orbiting platforms. Their goal was to develop the technologies and operational techniques that could be helpful for their programs to study other planetary bodies in our solar system. Using the entire world ocean as a "test site," the oceanographic community got access to new and unique capabilities for large-scale ocean observations.

The founding concept was that the long view from space, through repeated passes over target areas could provide data not obtainable from any other platform. When this work began, there was only a handful of scientists working to find out what was possible. The program was not driven by military requirements but by the larger need to know more about our planet in terms of future commercial opportunities and environmental knowledge.

Today, remote ocean sensing is a powerful tool for oceanographers (Lee-Lueng Fu et al., 2018). It produces unique data and insights that cannot be obtained in any other way. Commercial applications have come from this work. For example, enhanced fisheries, coastal land use planning, and early warning knowledge about maritime generated weather systems.

Essential knowledge that's needed for the new blue economy

Effective commercial development and exploitation of any domain require the best possible knowledge of the place. The better the knowledge, the more attractive, efficient, and profitable its exploitation becomes.

In the late 1980s, NASA proposed its "The Mission to Planet Earth" program (NASA, 1997; Goldstein and Thompson, 2008). Its purpose was to employ diverse satellite platforms in Earth's orbit to undertake a variety of research missions. Missions that would create new knowledge about our planet and its oceans. For many earth scientists, the idea of a "mission" to our own planet was a very compelling concept. This truly would be a "macroscale" platform approach to oceanography permitting repeated and frequent overflying of extensive areas of the world ocean. Measurements could now be made that were not possible with any other scientific platform such as research ships or submersibles. In 1998, the program was renamed "The NASA Earth Science Enterprise" and was given an expanded charter by that agency.

In theory, expanding this NASA construct facilitates a global systematic approach involving space platforms, ships as well as manned and unmanned submersibles. Also, it would include shoreside marine stations and, to a limited degree, aircraft. This is a massive international undertaking as no one nation has the physical and human assets needed for success. To some degree, this is now happening, but it's largely a "work in progress."

Such a program requires massive new investments by the major maritime powers. Regrettably, its investment is not seen to date in the global effort to fully explore the world ocean. Up to 2019, not much has happened. In fact, net global progress seems to have diminished as governments invest their funds in other directions. There are international committees, studies, and work groups, but much of it is simply "fair words and few deeds."

A primary problem for development of the blue economy is that to date, very little of the world ocean has been explored. It's difficult to develop a plan for using ocean resources if one does not know "where" and "what" of what's there.

According to the National Oceanic and Atmospheric Administration (NOAA), only 20% of the 140 million square miles of our planet's seafloor has been "adequately mapped." Other online sites give numbers ranging between 5 and 20% (NOAA, 2018; Wolfl et al., 2019; Nippon Foundation- GEBCO Seabed 2030 Project, 2020a, b). Some of the measurement differences relate to the dates of the estimates and how they were calculated.

"Adequately" is an important word. It refers to the ability to resolve sea bottom features. For example, if resolution is 3 miles, it means that any features smaller than that will not be seen. This is not very helpful for planning resources development.

Since the beginning of the 20th century, maps of the seafloor have been made by interpolating between sparse "soundings" (either from lead lines or single-beam echo sounders) with very limited resolution and spatial coverage. Toward the end of the 20th century, technologies and techniques were developed to estimate seafloor depth from satellite-based measurements, producing the first seafloor maps with global coverage (Nippon Foundation – GEBCO Seabed 2030 Project, 2020a, b; Bauman, 2019). But the resolution is only 12.4 miles. By 2016, this resolution had increased to 3.1 miles, but this is hardly a detailed map (Copley, 2014).

By comparison, NASA has been able to map the surface of Venus to 328-foot resolution and Mars to 66 feet. Closer to Earth, the best moon map resolution is 7 feet. The overused saying, "We know more about the Moon's behind than the oceans' bottom," has a ring of truth.

Some care is needed to differentiate between exploration of the world ocean and mapping of its seafloor. They are not the same thing though seafloor topography is the key element in overall exploration of the oceans.

Highest possible resolution mapping of seafloor topography is vital for support of commercial, scientific, and military uses of the oceans. So, how can this vast somewhat unknown area be mapped with sufficient detail? The classical and prevailing means of ocean mapping is through use of oceanographic research vessels and specialized hydrographic survey ships. However, it is estimated that one ship would need 350 years to accomplish 93% coverage (the remaining 7% is inshore) with current state-of-the-art mapping systems (Copley, 2014).

The estimated cost of such an effort is large, somewhere between 3 and 5 billion dollars (Nippon Foundation-GEBCO, 2020; Wikipedia, 2020). But this is not all that different from the cost of one or two of the unmanned Mars missions. This seems to be a relatively small price to pay to finally have detailed maps of our own planet. They are the key to effective resources uses of the oceans. Rapidly developing technologies may be able to significantly reduce the costs.

An attractive partial solution is to use fleets of long-duration underwater drones programmed to do the multibeam acoustic mapping. They would be large AUV unmanned submersibles. Large numbers of them ("swarms") can be controlled and coordinated by ships. Their data collections would be sent ashore through satellite links. Some AUV users have claimed they could do surveys with the required precision 60% faster than with shipboard systems (Sonardyne Inc., 2018).

An experimental AUV made test voyages across the Atlantic in 2009 (Rutgers University, 2009). But the idea of "autonomy" is not exactly correct. All survey AUVs will require motherships for voyage monitoring, control of navigation, recharging on board power, and downloading survey data. A properly configured mother ship could control and service several AUV hydrographic survey submersibles. This will multiply the amount of seafloor mapping per unit of time compared to what can be done by a single hydrographic survey ship. At some future time, the mother ship might even be another type of drone designed to control its own fleet of AUVs. In 2019, a prototype uncrewed surface vessel (USV) was tested as a mother ship for a single AUV and the operational results were very promising (Ocius Technologies, 2019). Also in 2019, an XPRIZE was awarded for autonomous mapping of the ocean.

Developing new data collection platforms such as spacecraft and AUVs and USVs is but one path to greater productivity. Another is more aggressive global collection and management of existing data through international cooperation. Such data harvesting can be anecdotal from private and commercial sources or professional from hydrographic survey ship operators.

There's now a major international effort to produce a publicly available global seafloor topographic map by 2030. This is the joint Nippon Foundation and GEBCO (General Bathymetric Chart of the Oceans) program begun in 2019 (Nippon Foundation-GEBCO Seabed 2030 Project, 2020a, b). By use of "crowdsourcing" techniques and pooling of all available data, the costs and time for mapping the global ocean seafloor can be significantly reduced.

There are many paths to the future. Continuing and aggressive science and technology programs must be maintained and expanded to effectively use the world ocean. This knowledge will be the essential key to the future new blue economy.

References

Adler, A., 2019. Neptune's Laboratory – Fantasy, Fear and Science at Sea. Harvard University Press, Cambridge Massachusetts.

Bauman, J., 2019. The Race to Make a High-Resolution Map of the World's Oceans. ESRI ArcUser, Spring, Redlands CA, 2020.

Busby, R., 1976. Manned Submersibles. Office of the Oceanographer of the Navy, US GPO.

Caladan Oceanic, 2020. Ring of Fire Expedition. Caladan Inc., Dallas TX.

Copley, J., 2014. Mapping the Deep, the Real Story behind the "95% Unexplored" Oceans. Exploring Our Oceans. University of Southampton, UK.

Forman, W., 2000. History of American Deep Submersible Operations. Best Publishing Company, Flagstaff AZ.

Fu, L.-L., Lee, T., Liu, W., Kwok, R., 2018. 50 years of satellite remote sensing of the ocean. Am. Meteorol. Soc. Monogr. 59, 5.1–5.46. Boston, MA.

Geyer, R., 1977. Submersibles and Their Use in Oceanography and Ocean Engineering. Elsevier Oceanography Series, New York.

Goldstein, E., Thompson, T., 2008. Earth Science: NASA's Mission to Our Home Planet. NASA, Washington DC.

Intergovernmental Oceanographic Commission, 2017. Global Ocean Science Report: The Current Status of Ocean Science Around the World; Executive Summary. UNESCO/IOC, Paris.

Linklater, E., 1972. The Voyage of the Challenger. Doubleday, New York.

National Aeronautics and Space Administration (NASA) Space Science Enterprise, 1997. Mission to Planet Earth. NASA, Washington DC.

National Geographic Society, 2012. James Cameron's Deepsea Challenge. NGS, Washington DC.

National Oceanic and Atmospheric Administration (NOAA) National Centers for Environmental Information, 2018. Mapping Our Planet, One Ocean at a Time. NOAA, Washington DC.

Nippon Foundation-GEBCO, 2020. Frequently Asked Questions. International Hydrographic Organization, Monaco.

Nippon Foundation of Japan-General Bathymetric Chart of the Oceans (GEBCO) Seabed 2030 Project, 2020a. Nearly a Fifth of World's Ocean Floor Now Mapped. Nippon Foundation-GEBCO, International Hydrographic Organization Monaco.

Nippon Foundation of Japan-General Bathymetric Chart of the Oceans (GEBCO) Seabed 2030 Project, 2020b. (Newsletter FAQ's). Which Part of the Ocean Will You Map First and How Much Will it Cost? International Hydrographic Organization, Monaco.

Ocean Engineering Division (Code 94), 1989. Naval Ocean Systems Center Underwater Vehicle History. Naval Ocean Systems Center, San Diego CA. Technical Document 1530.

Oceanographic Magazine (staff), 2020. Five Deeps Expedition Completes Mission to Reach Deepest Ocean Locations. CXD Media Ltd., London.

Ocius Technologies, 2019. OCIUS-DST Underwater Group Achieve Australian First at Summerfest 2019. OCIUS, Randwick, NSW, Australia.

Piccard, J., Dietz, R., 1960. Seven Miles Down. G.P. Putnam, New York.

Piccard, A., 1956. Earth, Sky and Sea. Oxford University Press, New York.

Research and Markets Report, 2018. Underwater Vehicle (UUV) Market to 2025. Research and Markets, Dublin Ireland.

Rutgers University, 2009. The Scarlet Knight's Trans-Atlantic Challenge. Rutgers University New Brunswick, New Jersey.

Scearce, C., 2009. European Fisheries History: Pre-industrial Origins of Overfishing. Proquest, Ann Arbor MI.

Sonardyne Inc., 2018. Let Your AUV Swarm Do the Talking – AUV Swarms. Sonardyne International Ltd., Hampshire UK.

Thomson, C.W., Murray, J., Nares, G., Tourle, F., 1880. Report on the Scientific Results of the Voyage of H.M.S. Challenger during the Years 1873–76. HM Stationary Office, London.

University of Rhode Island, 2002. Discovery of Sound in the Sea. URI, Rhode Island.

Wikipedia, 2020. Autonomous Underwater Vehicles. Wikipedia Foundation Inc.

Wolfl, A.C., et al., 2019. Seafloor Mapping – the Challenge of a Truly Global Ocean Bathymetry. Frontiers in Marine Science, vol. 6. Frontiers Media, Seattle, WA.

CHAPTER 2

Technical advancements to improve ocean understanding

Wendy Watson-Wright[1], Paul Snelgrove[2]
[1]7 Mile Bay Consulting, Albany, PE, Canada; [2]Departments of Ocean Sciences and Biology, Memorial University of Newfoundland, St John's, NL, Canada

Overview

The ocean, covering 71% of the planet, enables human life on this planet. It accounts for 97% of the water, contains at least 96% of Earth's living space, and generates approximately 50% of the atmosphere's oxygen. It has cushioned the blow of climate change by absorbing a third of all anthropogenic carbon emissions and >90% of the heat added to the global system (IPCC, 2014); it influences our weather and provides food for billions of people. The ocean and its resources define a part of our common heritage and an important part of many cultures. The ocean also provides 2/3 of the value of all the natural services offered by the planet, such as climate regulation, carbon sequestration, habitat, and biodiversity, among many others (IOC/UNESCO et al., 2011).

This one vast global ocean, not surprisingly, faces issues that span local spatial scales, such as coastal erosion and loss of local habitat and biodiversity, to global scales such as ocean warming, sea level rise, and ocean acidification. Ocean issues also span timescales from seconds, as in many biological processes, to millennia or longer, such as ocean function under past climates, and use of paleo-oceanographic records to predict the future state of the ocean-atmosphere system.

Yet despite the clear dominance of the ocean on our planet, we echo Elizabeth Mann Borgese's important observation that the ocean differs greatly from the land—its inhabitants know no borders, and its various ecosystems interconnect to form a larger whole (Mann Borgese, 1998). The harsh ocean environment also creates many operational challenges. As noted by Rayner et al. (2019), "much of the economic activity around, on and under the ocean would not be possible without data, information and knowledge derived from sustained ocean observations, measurements and

Preparing a Workforce for the New Blue Economy
ISBN 978-0-12-821431-2
https://doi.org/10.1016/B978-0-12-821431-2.00004-4

forecasts which underpin safe, cost-effective and environmentally acceptable marine and maritime activity." Having said that, we cannot take advantage of the many opportunities the ocean now offers in terms of the new blue economy or respond to the challenges or mitigate coming changes, either catastrophic or otherwise, while achieving sustainable ocean use, without sound management of all sectors of human activities affecting the ocean.

Technology—yesterday and today

The suite of technologies available to sample the ocean today greatly expands the toolbox that was available even 50 years ago, when oceanographers relied almost exclusively on a relatively small range of crude instruments lowered blindly from ships or deployed from shoreline facilities such as wharves of marine laboratories (see chapter by Don Walsh, this volume). Major changes include: (1) greater recognition of the scientific challenges for ocean sciences, such as understanding scale and dynamics of critical processes; (2) increased recognition of the need for measurements in space and time; (3) greater ability to obtain indirect (remote) measurements; and (4) recognition of the value of greatly improved time series and spatial coverage. All of these changes greatly advance our capacity to understand and predict change in the global ocean. Today's marine scientists utilize a wide diversity of platforms, optical and acoustic sensors, genetic tools, digital imaging and automated analyses, improved communication and data transmission capacity, and advanced computational tools to address a much more comprehensive and integrated range of oceanographic problems (Fig. 2.1).

Modern technologies help in addressing the major spatial and temporal gaps left by shipboard sampling, biased in time by research cruises that typically occur in short windows of weeks to months at best, and spatial bias favoring locations near ports, in particular ports in developed countries. Moorings deployed from ships and left in situ helped in covering the temporal bias but not spatial questions and, until recently, depended upon successful mooring recovery before data became available. The development of satellite technologies in the 1970s revolutionized ocean sciences by providing the first truly synoptic measurements of ocean temperature, surface height, and then eventually ocean primary production. However, with a few exceptions (altimetry, ocean currents) these measurements are largely limited to the upper surface layer of the ocean and thus cannot quantify the vast majority of the marine environment.

Figure 2.1 (A) An ocean glider, an autonomous underwater vehicle that achieves propulsion by varying its buoyancy as it slowly glides up and down through the ocean. Their low power consumption allows basin-scale missions measuring variables such as temperature and salinity. (B) An Axys Technologies 1.7 m meteorological/oceanographic buoy at the mouth of Holyrood Bay, Newfoundland, as part of SmartBay Holyrood testbed. The sensors provide near-real-time data on wind speed and direction, air temperature, humidity, dew point, barometric pressure, water temperature, ocean currents, wave height, direction, and period as well as wave spectral information. Despite the remote capabilities of the buoy, the harsh conditions of the North Atlantic require occasional servicing to remove ice buildup. (C) The remotely operate vehicle ROPOS (Remotely Operated Platform for Ocean Sciences), equipped with manipulators and high-resolution cameras gives researchers direct access to habitats at thousands of meters depth. (D) The control room where pilots and scientists work with ROPOS to sample the

Within the last 20 years, cabled observatories, as well as communication relays from surface buoys, Argo floats, gliders and autonomous vehicles, remotely operated vehicles, electronic tags on animals, etc., have reduced reliance on recovered moorings, and these platforms can now deliver real-time or near-real-time data from remote locations for a variety of ocean variables (Barnes et al., 2012). Ocean sensors have advanced at different rates within oceanographic subdisciplines, with stable sensors for physical measurements (temperature, salinity, depth) available for many decades, rapid improvements in chemical sensors (e.g., nutrients, pH, CO_2) within the last 20 years, and biological sensors (other than fluorescence, which has been available for many decades) lagging (Miloslavich et al., 2018). From a biological perspective, advances in two major areas have improved observational capacity. Genetic tools, including those used for population biology and metagenomics, have revolutionized biological oceanography (Scholin et al., 2009). Imaging technologies, including both digital optics and image analysis, have simultaneously leapfrogged the utility of these tools for identifying and quantifying biological diversity (e.g., Davis et al., 2005). Finally, computational capacity has enabled much more powerful models and analysis of much larger data sets (deYoung et al., 2004).

Further, we provide more detail on these advances and how they have improved oceanography. Moreover, we examine improvements in ocean observing technologies in support of the sustainable blue economy by briefly looking at some of the breakthroughs that have occurred in the major technologies as well as key remaining gaps. Specifically, we touch on sampling platforms, sensors, genetics, imaging and automated analysis, and data and modeling.

ocean without leaving the safety of the ship. Satellite linkages also enable remote participation from scientists from their labs or homes. (E) A bottom node for the Ocean Networks Canada (ONC) ocean observatory, which acts as both a "power bar" for plugging in instruments and a data relay for those instruments. (F) "Wally" the ONC bottom crawler moves across the bottom, providing spatial information through its chemical sensors and onboard camera. *(A) Photo courtesy of Marine Technology Society. (B) Photo Credit: Eric Davis, Center for Applied Ocean Technology)/Marine Institute/Memorial University. (C and D) Photo courtesy of P. Snelgrove. (E and F) Copyright Ocean Networks Canada, a University of Victoria initiative.)*

Breakthroughs

Without question, the Argo global array of almost 4000 autonomous profilers that measures temperature and salinity in the upper 2000 m has revolutionized physical oceanography and climate science (Roemmich et al., 2009). Over the past two decades, Argo floats have collected over 2 million profiles of temperature and salinity, four times more than every other ocean observing tool combined. Argo exemplifies international collaboration on a scale rarely seen in the scientific community, with 26 countries deploying some 800 Argo floats each year, and the Argo program will soon expand its sensors to include other disciplines (JCOMM, 2020). They now hold more than 50% of all the oceanographic data EVER collected! The biogeochemical (BGC) Argo project will significantly expand existing Argo capabilities by adding floats equipped with sensors that measure chemical and bio-optical properties (Fennel et al., 2017). They also offer a novel approach for international cooperation on a whole new suite of important oceanographic questions.

Autonomous underwater vehicles (AUVs) offer diverse and versatile configuration possibilities, potentially accommodating different kinds of payloads such as sidescan sonar, multibeam echo sounders, bottom profilers, and cameras. This flexibility, combined with their independent movement, makes AUVs powerful instruments for numerous underwater applications. Many new concepts have emerged in the past few years, from small "swarms" of AUVs to extra-large systems operating in isolation for extended time periods. The military, offshore oil and gas industry, and oceanographic institutions currently provide the major markets for AUVs (Tena, 2016), with growing markets in offshore wind and aquaculture.

Rapid advances in the use of various types of acoustic and optical tools have yielded major breakthroughs in all oceanographic disciplines, from the use of multibeam acoustics to map the geology (and limited biology) of vast swaths of seabed, expanding use of satellites for synoptic ocean region imaging, to the use of radar waves to map local surface waves and currents (Coastal Ocean Dynamics Applications Radar, CODAR), to ultraclear, high-resolution images of living organisms in situ that advance understanding of which organisms live where and how they utilize their habitat. Even digital (or analog) video that produced blurred frame grabs only a decade ago now yields high-definition images that exceed what film still cameras could produce decades ago. Acoustics offer relatively broad-scale coverage and penetrate well through even turbid water, whereas optics penetrate air far better than water and therefore provide invaluable synoptic satellite imagery of the ocean surface but limited subsurface information. Broadly speaking, sensors utilize optics, acoustics, or wet chemistry of different types, to sense ocean variables. Optical and acoustics imagery themselves provide critical oceanographic data, but optics and acoustics also support a variety of other sensors, either by providing raw data that form the basis of the sensor (e.g., sediment particle enumeration) or by transmitting data (e.g., acoustic relays). These tools can collect a large volume of data in short periods of time.

Oceanographers within each subdiscipline depend on an ever-expanding suite of sensors, with increasing stability, miniaturization, sensitivity, and types of variables they detect. Reduced size and power requirements have expanded the utility of conductivity—temperature—depth (CTD) profilers, the "bread and butter" instrument for physical oceanography, so that they can now be used on a broader range of sampling platforms.

Broadly speaking, ocean sensors typically utilize sound (acoustics) or light (optical tools) to characterize ocean variables. Physical oceanographers have used acoustics (e.g., Acoustic-Doppler Current Profilers) to measure ocean currents, thermoclines, and scattering layers for decades, whereas they now also use acoustics to quantify waves breaking, bubble formation, rainfall, wind speed, ice thickness and shape, and sound-related physical phenomena (e.g., Zedel et al., 1999). Turbulence microprofilers add to the physical oceanography toolbox and to understanding turbulence in a wide range of ocean environments (e.g., Ross, 2014). Geologists have used acoustics to quantify seabed composition and morphology for many decades, including habitat mapping (Brown et al., 2011), as well as profiling suspended particles above the bottom, whereas zooplankton biologists today sometimes use acoustics (e.g., zooplankton acoustic profiler, or ZAPs) to profile zooplankton biomass (e.g., Kunze et al., 2006).

Optical sensors also provide a variety of oceanographic sensing solutions. For example, optical sensors that detect low concentrations of inert tracers provide physical oceanographers with novel insights into ocean circulation and geochemists with new understanding of air—sea gas exchange (Stanley and Jenkins, 2013). Geologists have expanded use of optical sensors from bulk measurement of light backscatter to newer instruments that use lasers to measure particle sizes and concentrations (Agrawal, 2017). Geochemists now utilize a much wider range of optical sensors linked to wet chemistry to quantify in situ concentrations and movements of a wide diversity of molecules, including macro- and micronutrients, as well as contaminants (e.g., Brown and Fingas, 2003). These sensors also include microprofilers that can detect a wide range of molecules in environments such as in sediments and at hydrothermal vents (Luther et al., 2008).

Biologists have used optical sensors and fluorescence to quantify abundance of photosynthetic pigments (and thus primary producers) for many decades, but the use of hyperspectral fluorometers (Houskeeper et al., 2020) and, more recently, in situ flow cytometers (Olsen and Sosik, 2007) increases the capacity to differentiate among major groups of primary producers based on pigment composition and cell size. Zooplankton biologists now also use optics (video plankton recorders, or VPRs) to quantify different groups of zooplankton (Davis et al., 2005), whereas benthic ecologists utilize improved optics (see further) in towed cameras and Sediment Profile Imagery (or SPI) to quantify seafloor biota above and below (Solan and Kennedy, 2002) the sediment surface respectively.

Genetic tools represent perhaps the single greatest breakthrough for biological oceanography, combining wet chemistry and optics (Scholin et al., 2008). The development of polymerase chain reaction (PCR) tools revolutionized the use of genetics to enable identification of marine organisms, a major game changer for microbial ecology. This breakthrough resolved our inability to identify different microbes, most of which we cannot currently culture in the lab and lack the necessary range of morphological characteristics to differentiate toxic from nontoxic forms (Scholin et al., 2009). Some microbes play a major role in critical processes such as nutrient cycling and organic matter breakdown, and these tools enable differentiation of major and minor players. Researchers now use DNA barcoding to address a wide range of taxonomic challenges, including differentiating sibling (morphologically similar) species, monitoring composition of fish catches, and even mass sequencing of all organisms in a sample (Keeling et al., 2014). Genomics, proteomics, and metabolomics nowadays enable identification of specific gene expression and presence and form of proteins and metabolites in individual cells and organisms; these tools can quantify when an organism expresses specific genes (e.g., toxicity, nitrogenase) that can significantly alter natural processes and even human health (Zehr and Capone, 2020). Population genetics, which uses microsatellites to differentiate populations, can provide relatively detailed information on geographic linkages among populations of a given species, a critical tool for understanding dispersal and population dynamics of marine species, including those fished commercially (Bradbury et al., 2010). Genetics also presently permits calculation of effective population size in current and past populations of a given species (e.g., Layton et al., 2020). This approach hinges upon some key assumptions but offers a fisheries-independent mechanism to estimate abundances of target species, as well as at-risk species.

With increasing numbers of sensors, platforms, and techniques to acquire data, combined with ever-increasing computer processing power, the diversity and volume of data available to scientists and the general public have increased exponentially in a very short time. Although analyzing diverse data sets and generating projections based on this data may seem overwhelming, it nonetheless creates transformative opportunities. Researchers use Big Data tools to address a diverse range of challenges, and we reiterate the hope opined by Smit (2019) that in this world of plenty, we should see a move from data hoarding to data sharing, collaboration, and relationship building in order to create long-term value. A new emphasis on metadata standards and protocols promises improvement in this area, including increasing adherence to the FAIR (Findable, Accessible, Interoperable, Reusable) principle (Wilkinson et al., 2016), but much work remains.

Artificial intelligence (AI), including machine and deep learning, has caused an explosion of potential, with exciting promise of machine learning solutions for problems that exceed the capacity of traditional methods. Deep learning techniques can help address a great variety of ocean processes and issues, regardless of subdiscipline. Progress in deep learning architecture models primarily drives these recent developments, including flexible convolutional neural networks (CNNs) that help identify features in multidimensional data (Vajradhar, 2019).

DeepSense, a Canadian initiative, illustrates a new strategy to facilitate AI for ocean science. DeepSense was funded explicitly to foster the ocean economy through AI and machine learning. From tracking illegal fishing via Automatic Identification System (AIS) data to large-scale population genomic analyses of economically important marine species to adaptive metric learning for acoustic data, DeepSense supports collaborative projects among businesses in the ocean economy and academic researchers to drive growth in the ocean sector (Deepsense, 2020). The REV Ocean-Established Ocean Data Foundation (ODF) and its Ocean Data Platform (ODP) likewise appear to offer excellent potential to generate solutions for uploading, aggregating, and navigating vast amounts of ocean data from the maritime and offshore energy tech sector, academia, conservation, and government (Ocean Data Foundation, 2020).

Gaps

Intriguingly, but not surprisingly "gap" was mentioned 25 times more often than "breakthrough" in the 3 volumes and 134 chapters of the Ocean Obs' 19 white papers (Lee et al., 2019). It is clear that promoting a new blue economy will require a decided focus on improving that ratio. These papers present the latest thinking on ocean observation, including prioritizing different types of measurements for different subdisciplines and ocean environments (e.g., essential ocean variables [EOVs], see further), recognizing the challenges of regional differences in capacity and opportunity in developing a "well-sensed ocean," and that efforts should prioritize some relatively simple measurements that could be made with minimal resources, as well as more complex measurements that require advanced and (typically) expensive technology.

So what are these key gaps? In terms of platforms, future challenges are to develop AUVs that operate autonomously at distances outside the reach of the operator, and that do not require trained operators for each AUV. Despite decreasing size and increasing autonomy in sampling platforms, they still require research vessels to deploy and recover for many applications and lack the payload to collect some types of samples. Some countries (e.g., Norway, China, Korea) have launched or will soon launch (Ireland, Spain, Sweden) more and increasingly capable vessels, in contrast to countries such as Canada and many developing nations, where a dearth of modern oceanographic research vessels hampers progress (Withers, 2019).

All of the sensors described previously dramatically increase the capacity of oceanographers to characterize the physical and geochemical environment, but major gaps remain in biological sensors. Biodiversity research generally depends on collection of physical specimens because imagery—whether acoustic or optical—cannot resolve all organisms to species. Water clarity and light absorption continue to limit the conditions in which optical sensors can operate, and acoustics cannot produce the detailed images needed to identify individual animals unambiguously. The need for a physical specimen for many applications means continued reliance on century-old technology such as nets, bottles, and grabs. New genetic developments (see further) offer a potential path forward, but many hurdles remain to be overcome.

In situ genetics shows promise. For example, the Environmental Sample Processor can detect the presence of a given type of organism or strain of organism. This technology utilizes a water sample, and thus works effectively for small organisms, and particularly microbes, and has been successfully deployed on ocean observatories to provide real-time monitoring for the presence of toxic dinoflagellates (Scholin et al., 2009). Metagenomics may represent a next step for this tool within the foreseeable future, although primarily for microbes. Recent developments in environmental DNA or eDNA show promise in using metagenomics sequencing of water or sediment samples for the presence of any DNA present, which could include individual cells from fishes or invertebrates swimming by or burrowing through the seafloor. Current challenges include development of a DNA barcode catalog of biota, determining how long DNA can persist in the environment for a given species (and thus how recently the organism was present), and moving past presence/absence information to provide quantification of abundances of different species (Closek et al., 2019).

The aforementioned abundance of data creates considerable challenges in organizing and utilizing data effectively to identify what to look for in vast data sets. Effectively leveraging abundant data to support discovery and innovation requires cross-pollination of ocean scientists with experts in complementary disciplines (such as computer scientists and information managers), other ocean data stakeholders, and disparate disciplines such as medicine where technologies invented there have been successfully applied to marine science (e.g., flow cytometry). Investment in training around interdisciplinarity should accelerate progress (Smit, 2019), and wise investment in initiatives that focus on data pooling and transdisciplinary research would likely lead to breakthroughs (Hussey et al., 2015).

Recent global initiatives on ocean observations, ocean technology, and the blue economy

In recent years, numerous international initiatives have been launched to support the blue economy including those related to ocean observations and ocean data management.

In 2018, the G7 put forward the Charlevoix Blueprint for Healthy Oceans, Seas and Resilient Coastal Communities, wherein one commitment included launching a joint G7 initiative to deploy Earth observation technologies and applications to scale up capacities for the integrated management of coastal zones and to increase both the availability and sharing of science and data (G7, 2018). Canada and other G7 nations committed to the development of the Biogeochemical Argo program and growth of Deep Argo floats that can sample at greater depths than the <2000 m focus of the existing program.

Also during 2018, the signing of the British Commonwealth Blue Charter committed Commonwealth countries to work together to better protect the ocean and manage its use. The eight action groups include a Blue Charter Ocean Observation Action Group with a mandate to explore and harness recent advances in ocean observation technology in order to provide the environmental intelligence needed to support decision-making. The Charter explicitly prioritizes the need to understand ocean processes in developing blue economies (Commonwealth Blue Charter, 2018).

Later in 2018, the first global Sustainable Blue Economy conference, which was convened in Nairobi, focused discussions largely on two pillars: (1) accelerated economic growth, job creation, and poverty alleviation; and (2) sustainability, climate change, and controlling pollution (SBEC Final Report, 2018). Commitments included funding to promote the UN Decade of Ocean Science for Sustainable Development to enhance scientific capacity and help build cooperation among stakeholders in marine affairs, in particular in ocean research and observations. Since then, the number of conferences and meetings focused on the blue economy has continued to increase.

What would it take for a well-sensed ocean?

A well-sensed ocean would provide real-time or near-real-time data on a wide range of ocean variables spanning physics, chemistry, geology, and biology in a format accessible to a wide range of users (Barnes et al., 2012). In the last few years, experts within these disciplines have developed lists of EOVs that encompass those sets of measurements most critical for monitoring and understanding ocean change (e.g., Miloslavich et al., 2018). Despite some key gaps described earlier, the technologies exist to enable a well-sensed ocean, but high cost and limited access mean that few ocean locations could be labeled as well-sensed. For example, the Ocean Networks Canada cabled observatory off the west coast of Canada provides continuous data on a wide range of ocean variables and thus represents one of the best-sensed ocean locations in the world. But the supporting infrastructure for this observatory required investments of many millions of dollars, including hiring a cable ship and the Remotely Operated Platform for Ocean Science, ROPOS (CSSF, 2020)—to deploy the equipment. Sensors also require frequent servicing to correct for drift (a particular problem in biogeochemical sensors), malfunction, damage from fishing gear, or sometimes even changing of batteries. Finally, key gaps in sensor technology noted earlier create a need to supplement those continuous autonomous observations with other types of samples. Quality control and postsample processing and archiving all require dedicated staff, but even then the forms of data outputs often require a high level of expertise to utilize the data, even though it may be openly accessible.

A well-sensed ocean must consider space and time; the ONC observatory off the coast of British Columbia, for example, offers little help in understanding fluctuations in North Atlantic (or even South Atlantic) fisheries. Thus, a well-sensed ocean requires parallel efforts around the world. Although the specific platforms may vary (e.g., cabled observatories vs. AUVs or moorings and surface data relays), the need for basic measurements across disciplines transcends the global ocean. Researchers are now trying to develop strategies to address this need, such as recent analyses on a potential deep-sea observation strategy (Danovaro et al., 2017, 2020) or on basin-scale coordination and integration (deYoung et al., 2019).

The blue economy and ocean observation

Many nations and experts have recently trumpeted untapped opportunities in the ocean sector, framed around the concept of a "Blue Economy" that expands economic opportunity from living and nonliving marine resources, but through sustainable development that maintains ocean health and ecosystem services. This added activity will without question add pressures, and past experience with collapsed marine fisheries and ecosystems clearly links long-term prosperity to sustainability challenges. Sustainability strategies require knowledge, and knowledge requires data. This link punctuates the need for improved ocean observation. However, observation alone will not achieve sustainability—such efforts must also embrace cooperation in data collection and use that feeds transdisciplinary research that brings together natural and social scientists, humanities, society, the private sector, and resource managers to apply understanding of ecosystem change to policy applications. Successful implementation of ocean policy will require outreach and trans-sector engagement to support legitimacy of and compliance with new policy. The revolution in ocean sensors provides the opportunity to move the "blue economy" to the "new blue economy," which emphasizes sustainability by informing wise decisions based on the best available information (Spalding, 2016). Historically, lack of information has thwarted many sustainability efforts, but these ocean technologies could represent a game changer. Finally, ocean observations offer an excellent opportunity for capacity enhancement in nations with weaker infrastructure and scientific capacity (Harden-Davies and Snelgrove, 2020). Such enhancement should include remote training of scientists from developing nations, shared research projects within jurisdictional waters, and transfer of equipment along with support to establish and maintain that equipment.

Acknowledgments

The authors are grateful to Dr. Brad DeYoung (Memorial University of Newfoundland, AtlantOS) for reviewing and providing helpful comments on this manuscript.

References

Agrawal, Y.C., 2017. Measuring sediments: backscatter optics, laser diffraction, acoustics, and combined optics and acoustics. J. Acoust. Soc. Am. 141 (5), 3670.

Barnes, C.R., Best, M.M., Johnson, F.R., Pautet, L., Pirenne, B., 2012. Challenges, benefits, and opportunities in installing and operating cabled ocean observatories: perspectives from NEPTUNE Canada. IEEE J. Ocean. Eng. 38 (1), 144–157.

Borgese, E.M., 1998. The Oceanic Circle: Governing the Seas as a Global Resource. United Nations University Press, Tokyo.

Bradbury, I.R., Hubert, S., Higgins, B., Borza, T., Bowman, S., Paterson, I.G., Snelgrove, P.V., Morris, C.J., Gregory, R.S., Hardie, D.C., Hutchings, J.A., 2010. Parallel adaptive evolution of Atlantic cod on both sides of the Atlantic Ocean in response to temperature. Proc. Biol. Sci. 277, 3725–3734.

Brown, C.E., Fingas, M.F., 2003. Review of the development of laser fluorosensors for oil spill application. Mar. Pollut. Bull. 47 (9–12), 477–484.

Brown, C.J., Smith, S.J., Lawton, P., Anderson, J.T., 2011. Benthic habitat mapping: a review of progress towards improved understanding of the spatial ecology of the seafloor using acoustic techniques. Estuar. Coast Shelf Sci. 92, 502–520.

Closek, C.J., Santora, J.A., Starks, H.A., Schroeder, I.D., Andruszkiewicz, E.A., Sakuma, K.M., Bograd, S.J., Hazen, E.L., Field, J.C., Boehm, A.B., 2019. Marine vertebrate biodiversity and distribution within the central California Current using environmental DNA (eDNA) metabarcoding and ecosystem surveys. Front Mar. Sci. 6, 732.

Commonwealth Blue Charter, 2018. Shared Values, Shared Ocean. A Commonwealth Commitment to Work Together to Protect and Manage Our Ocean. https://bluecharter.thecommonwealth.org/wp-content/uploads/2019/08/Commonwealth_Blue_Charter.pdf.

Danovaro, R., Aguzzi, J., Fanelli, E., Billett, D., Gjerde, K., Jamieson, A., Ramirez-Llodra, E., Smith, C.R., Snelgrove, P.V.R., Thomsen, L., Van Dover, C., 2017. A new international ecosystem-based strategy for the global deep ocean. Science 355, 452–454.

Danovaro, R., Aguzzi, J., Fanelli, E., Billett, D., Carugati, L., Corinaldesi, C., Dell'Anno, A., Gjerde, K., Jamieson, A.J., Kark, S., McClain, C., Levin, L., Levin, N., Ramirez-Llodra, E., Rex, M., Ruhl, H., Smith, C.R., Snelgrove, P.V.R., Thomsen, L., Van Dover, C., Yasuhara, M., 2020. Ecological indicators for an integrated global deep-ocean strategy. Nat. Ecol. Evol. 4, 181–192.

Davis, C.S., Thwaites, F.T., Gallager, S.M., Hu, Q., 2005. A three-axis fast-tow digital video plankton recorder for rapid surveys of plankton taxa and hydrography. Limnol Oceanogr. Methods 3, 59–74.

Deepsense, 2020. https://deepsense.ca.

DeYoung, B., Heath, M., Werner, F., Chai, F., Megrey, B., Monfray, P., 2004. Challenges of modeling ocean basin ecosystems. Science 304, 1463–1466.

DeYoung, B., Visbeck, M., Cunha de Araujo Filho, M., Baringer, M.O., Black, C.A., Buch, E., Edwards, M., Fischer, A., Fritz, J.-S., Ketelhake, S., Muelbert, J.-H., Moteiro, P., Nolan, G., O'Rourke, E., Ott, M., Le Traon, P.Y., Pouliquen, S., Sousa-Pinto, I., Tanhua, T., Velho, F.V., Willis, Z., 2019. An integrated All-Atlantic ocean observing system in 2030. Front. Mar. Sci. 6 https://doi.org/10.3389/fmars.2019.00428. Art 428.

Fennel, K., Greenan, B., participants of the Canadian BGC Argo workshop, 2017. Taking the Ocean's Pulse: A Vision for the Canadian Biogeochemical Argo Program. https://doi.org/10.13155/52451.

G7, 2018. Charlevoix Blueprint for Healthy Oceans, Seas and Resilient Coastal Communities. https://www.international.gc.ca/world-monde/assets/pdfs/international_relations-relations_internationales/g7/2018-06-09-healthy_oceans-sante_oceans-en.pdf.

Harden-Davies, H., Snelgrove, P., 2020. Science collaboration for capacity building: advancing technology transfer through an agreement for biodiversity beyond national jurisdiction. Front. Mar. Sci. 7, 40.

Houskeeper, H.F., Draper, D., Kudela, R.M., Boss, E., 2020. Chlorophyll absorption and phytoplankton size information inferred from hyperspectral particulate beam attenuation. Appl. Optic. 59, 6765–6773.

Hussey, N.E., Kessel, S.T., Aarestrup, K., Cooke, S.J., Cowley, P.D., Fisk, A.T., et al., 2015. Aquatic animal telemetry: a panoramic window into the underwater world. Science 348, 1221–1231. https://doi.org/10.1126/science.1255642.

IOC/UNESCO, IMO, FAO, UNDP, 2011. A Blueprint for Ocean and Coastal Sustainability. IOC/UNESCO, Paris. https://unesdoc.unesco.org/ark:/48223/pf0000215002.

IPCC, 2014. In: Core Writing Team, Pachauri, R.K., Meyer, L.A. (Eds.), Climate Change 2014: Synthesis Report. Contribution of Working Groups I, II and III to the Fifth Assessment Report of the Intergovernmental Panel on Climate Change. IPCC, Geneva, Switzerland, p. 151.

JCOMM, 2020. Argo: A Window into the Ocean. Https://jcomm.maps.arcgis.com/apps/Cascade/index.html?appid=a170a0d522bb42f1a019e4e473cf1bdd.

Keeling, P.J., Burki, F., Wilcox, H.M., Allam, B., Allen, E.E., Amaral-Zettler, L.A., Armbrust, E.V., Archibald, J.M., Bharti, A.K., Bell, C.J., Beszteri, B., 2014. The marine microbial eukaryote transcriptome sequencing project (MMETSP): illuminating the functional diversity of eukaryotic life in the oceans through transcriptome sequencing. PLoS Biol. 12, e1001889.

Kunze, E., Dower, J.F.D., Beveridge, I., Dewey, R., Bartlett, K.B., 2006. Observations of biologically generated turbulence in a coastal inlet. Science 313, 1768–1770.

Layton, K.K., Dempson, B., Snelgrove, P.V., Duffy, S.J., Messmer, A.M., Paterson, I.G., Jeffery, N.W., Kess, T., Horne, J.B., Salisbury, S.J., Ruzzante, D.E., 2020. Resolving fine-scale population structure and fishery exploitation using sequenced microsatellites in a northern fish. Evol. Appl. 13, 1055–1068.

Lee, T., Speich, S., Lorenzoni, L., Chiba, S., Muller-Karger, F.E., Dai, M., Kabo-Bah, A.T., Siddorn, J., Manley, J., Snoussi, M., Chai, F. (Eds.), 2019. Oceanobs'19: An Ocean of Opportunity. Volumes I - III. Frontiers Media, Lausanne. https://www.frontiersin.org/research-topics/8224/oceanobs19-an-ocean-of-opportunity.

Luther III, G.W., Glazer, B.T., Ma, S., Trouwborst, R.E., Moore, T.S., Metzger, E., Kraiya, C., Waite, T.J., Druschel, G., Sundby, B., Taillefert, M., 2008. Use of voltammetric solid-state (micro) electrodes for studying biogeochemical processes: laboratory measurements to real time measurements with an in situ electrochemical analyzer (ISEA). Mar. Chem. 108, 221–235.

Miloslavich, P., Bax, N.J., Simmons, S.E., Klein, E., Appeltans, W., Aburto-Oropeza, O., Andersen Garcia, M., Batten, S.D., Benedetti-Cecchi, L., Checkley Jr., D.M., Chiba, S., 2018. Essential ocean variables for global sustained observations of biodiversity and ecosystem changes. Global Change Biol. 24, 2416–2433.

Ocean Data Foundation, 2020. https://www.oceandata.earth.

Olson, R.J., Sosik, H.M., 2007. A submersible imaging-in-flow instrument to analyze nano-and microplankton: imaging FlowCytobot. Limnol Oceanogr. Methods 5 (6), 195–203.

Rayner, R., Gouldman, C., Willis, Z., 2019. The ocean enterprise – understanding and quantifying business activity in support of observing, measuring and forecasting the ocean. J. Oper. Oceanogr. 12 (Suppl. 2), S97–S110. https://doi.org/10.1080/1755876X.2018.1543982.

Report on the Global Sustainable Blue Economy Conference, November 26–28, 2018. Nairobi, Kenya. http://www.blueeconomyconference.go.ke/wp-content/uploads/2018/12/SBEC-FINAL-REPORT-8-DECEMBER-2018-rev-2-1-2-PDF2-3-compressed.pdf.

Roemmich, D., Johnson, G.C., Riser, S., Davis, R., Gilson, J., Owens, W.B., Garzoli, S.L., Schmid, C., Ignaszewski, M., 2009. The Argo Program: observing the global ocean with profiling floats. Oceanography 22, 34–43.

ROPOS, 2020. Canadian Scientific Submersible Facility. https://www.ropos.com.

Scholin, C., Preston, C., Harris, A., Birch, J., Marin, R., Jensen, S., Roman, B., Everlove, C., Makarewicz, A., Riot, V., Hadley, D., 2008. Development of Ecogenomic Sensors for Remote Detection of Marine Microbes, Their Genes and Gene Products. AGUFM, pp. B31F–B0357.

Ross, T., 2014. A video-plankton and microstructure profiler for the exploration of in situ connections between zooplankton and turbulence. Deep Sea Res. I 89, 1–10.

Scholin, C., Doucette, G., Jensen, S., Roman, B., Pargett, D., Marin III, R., Preston, C., Jones, W., Feldman, J.,., Everlove, C., Harris, A., Alvarado, N., Massion, E., Birch, J., Greenfield, D., Vrijenhoek, R., Mikulski, C., Jones, K., 2009. Remote detection of marine microbes, small invertebrates, harmful algae, and biotoxins using the environmental sample processor (ESP). Oceanography 22, 158–167. https://doi:10.5670/oceanog.2009.46.

Smit, M., 2019. Guest editor's note. J. Ocean Technol. 14 (4), v–vi. https://www.thejot.net/article-preview/?show_article_preview=1105.

Solan, M., Kennedy, R., 2002. Observation and quantification of in situ animal-sediment relations using time-lapse sediment profile imagery (t-SPI). Mar. Ecol. Prog. Ser. 228, 179–191.

Spalding, M.J., 2016. The new blue economy: the future of sustainability. J. Ocean Coastal Econ. 2 https://doi.org/10.15351/2373-8456.1052. Article 8.

Stanley, R.H., Jenkins, W.J., 2013. Noble gases in seawater as tracers for physical and biogeochemical ocean processes. In: The Noble Gases as Geochemical Tracers. Springer, Berlin, Heidelberg, pp. 55–79.

Tena, I., 2016. Autonomous Underwater Vehicles: A View of the Autonomous Underwater Vehicle Market in 2016. Hydro International. https://www.hydro-international.com/content/article/autonomous-underwater-vehicles-2.

Vajradhar, V., 2019. Artificial Intelligence Based Oceanography, Medium. https://medium.com/datadriveninvestor/ai-oceanography-the-fourth-industrial-revolution-comes-in-the-form-of-oceanographic-science-8b14f072a882.

Wilkinson, M.D., Dumontier, M., Aalbersberg, I.J.J., Appleton, G., Axton, M., Baak, A., et al., 2016. The FAIR guiding principles for scientific data management and stewardship. Sci. Data 3, 160018. https://doi.org/10.1038/sdata.2016.18.

Withers, P., 2019. Vessel Shortage Forces Canada to Scrap Key Climate Change Research Mission. https://www.cbc.ca/news/canada/nova-scotia/canada-forced-to-scrap-core-climate-change-mission-1.5357015.

Zedel, L., Gordon, L., Osterhus, S., 1999. Ocean ambient sound instrument system: acoustic estimation of wind speed and direction from a subsurface package. J. Atmos. Ocean. Technol. 16, 1118–1126.

Zehr, J.P., Capone, D.G., 2020. Changing perspectives in marine nitrogen fixation. Science 368 (6492).

CHAPTER 3

Ocean observing requires a new blue economy workforce

Carl Gouldman
United States Integrated Ocean Observing System, NOAA Integrated Ocean Observing System Office, Silver Spring, MD, United States

> *The New Blue Economy is a knowledge-based economy, looking to the sea not just for extraction of material goods, but for data and information to address societal challenges and inspire their solutions.*
>
> ***(Spinrad, 2016)***

Network of networks—the US integrated ocean observing system story

The US Integrated Ocean Observing System (IOOS) is one of the programs "now providing important frameworks for feeding data and information into the New Blue Economy." (see Rick Spinrad, this volume). US IOOS uses ever-improving technologies to provide data, predictions, and information that provide the insight needed for solutions to societal challenges. Decisions based on these insights and solutions have economic impacts that are the foundation of the new blue economy. These insights and solutions are already being delivered in many ways and will continue to grow and evolve as societal needs change and grow over time. We have a diverse and multiskilled workforce today, and that workforce will also need to change and grow over time to deliver ocean information with context, and provided by our networks of technologies, models, people, policies, program frameworks, and partnerships. IOOS is now, and will continue to be, a requirements driver for workforce needs, and this chapter will outline some of the experiences from the past 20 years as to what our workforce specialties are now, why we are expanding our workforce needs, and thoughts on the trends for the next 20 years with respect to the IOOS.

Ocean information applied to mission needs

Ocean, coastal, and Great Lakes data, modeling and analysis, and decision support services are fundamental building blocks for sustainable

Preparing a Workforce for the New Blue Economy
ISBN 978-0-12-821431-2
https://doi.org/10.1016/B978-0-12-821431-2.00032-9

development of our global new blue economy. Ocean ("Ocean" intended to include coasts and Great Lakes) observations (data) underpin business decisions on near-real-time, daily, weekly, monthly, yearly, and decadal timescales as well. Mission areas include weather, commerce, energy, safety, food, fisheries, planning, recreation, and ultimately tie to the health of our populations and the planet. All are critically dependent upon information derived from technologies that provide fundamental observations of physical, biological, and biogeochemical parameters. The development of decision support tools and delivering those tools to decision-makers as a service is foundational to IOOS and foundational to the new blue economy. The "map once, use many times," or "observe once, use many times" philosophy applies here, because ongoing efforts are designed to build an integrated whole system, bigger than the sum of its parts. Yet building such a system requires many skills in our workforce.

In the last 20 years, IOOS has demonstrated value-added contributions to many service delivery successes, where IOOS may be one of many participants without owning the whole production suite. Collaborative solutions built from multiple programs and partnerships are often highly valued by the participants. One feature of the most successful examples is usually a clear customer who has articulated information needs that are clearly being met by bringing together multiple bits of science, observing, data management, modeling, and outreach. When that recipe is successful, it is also incumbent upon the player to honor partnerships and find ways to ensure proper attribution for contributors. Paradoxically, when open data policies and data integration are working very well, that attribution can be more challenging.

Policy and program framework

Within the United States, we have been building an integrated ocean observing system for more than 20 years, beginning with initial planning in the 1990s; to targeted earmarked projects in the early 2000s; to implementation of the Integrated Coastal Ocean Observing Systems Act of 2009 (ICOOS Act of 2009, 2018). The US IOOS coordinates national, international, regional, and local ocean and coastal observation networks, modeling, and data management and communications services to provide the United States with information to safeguard life and property, protect valuable ecosystems, and sustain this country's economic vitality. IOOS combines the resources and reach of the federal government with the efficiency, local know-how, and capability for innovation that our international and regional partners provide. The ICOOS Act requires a federated system to coordinate federal and national program capabilities and fund and lead 11 IOOS Regional Associations designed to meet the specific regional and local needs for ocean, coastal, and Great Lakes information and to ensure coordination of the System with the Global Ocean Observing System (see Ryabinin, this volume). Together the IOOS system components combine to provide new levels of observation, data integration from disparate IOOS and non-IOOS sources, and new decision support tools for federal, state, local, tribal, and private-sector decision-makers to protect lives and property. All IOOS functions underpin the new blue economy.

"Policy neutral, science based, and stakeholder driven" are the tenets of IOOS since its inception within the United States and with the extensive network of nonfederal partners in the 11 IOOS Regional Associations. The IOOS program has employed these key features while building partnerships across multiple agencies and within the National Oceanic and Atmospheric Administration. The ICOOS Act of 2009 formally established IOOS within the National Oceanic and Atmospheric Administration (NOAA) and both IOOS and NOAA are fundamentally aligned with the new blue economy. NOAA's mission is Science, Service and Stewardship: 1. To understand and predict changes in climate, weather, oceans, and coasts; 2. to share that knowledge and information with others; and 3. to conserve and manage coastal and marine ecosystems and resources. The IOOS remit includes integrating capabilities from many federal agencies and nonfederal partners and also to serve the mission of NOAA within the US Department of Commerce, and NOAA's mission areas are perfectly aligned with empowering the new blue economy. IOOS works to develop within this framework and requires extensive and multidisciplinary workforce skills to succeed in applying ocean data to societal needs. Additional benefits extend to other mission areas by enabling science and utilizing science and information from other agencies as well.

Network of networks

The effort to build IOOS relies very much on a networking of networks approach that adapts and adjusts to include more capabilities as advancements and developments in technologies progress. Within this networking of networks approach, which also includes a regional to national approach, we need a strong, multidisciplinary, diverse, and innovative workforce to deliver on the promise of IOOS and subsequently to grow and deliver on the promise of the new blue economy. The IOOS vision is "Ocean, coastal, and Great Lakes information that improves lives and livelihoods" and the mission is "to produce, integrate, and communicate high quality ocean, coastal and Great Lakes information that meets the safety, economic, and stewardship needs of the Nation." (IOOS, 2018; U.S. Integrated Ocean, 2018). IOOS is a collaborative and integrated network of people and technology gathering observing data and developing tracking and predictive tools to benefit the economy, the environment, and public safety at home, across the nation, and around the globe. The functional components of this "system" include observing, data management, modeling and analysis, and decision support tools and services in an end-to-end framework.

From technology to program integration skill

As the world has experienced extensive growth in the volume, quality, and diversity of data, we have seen growth in models, predictive capability, and capacity. And similarly, dramatic technology changes and improved policy and program frameworks have developed during this same period. These technological changes require growth in the depth, diversity, complexity, and integration of knowledge within our workforce that is critical for empowering the new blue economy. With many specialists critical to building capabilities, some are targeted to very specific science areas, technologies, or engineering challenges. However, the integration and collaboration functions require different skills and perspectives, and these skills are equally important to maximize utility of our observing, data, modeling, and decision support investments.

High-frequency (HF) radar use case

This applies to many components of the system such as surface current mapping using the high-frequency (HF) radar network. The HF radar network now includes dozens of operators across a network of over 150 stations in the United States across dozens of institutions. All are working to deliver surface current maps that support oil spill trajectory modeling, search and rescue operations, ecosystem modeling for harmful algal blooms, and other application areas. The HF radar network requires a blended work-force at many levels including technicians, scientists, program managers, modelers, data scientists, communications experts, and political science experts, just to name a few. Building a coalition of these experts and developing common approaches to operate this technology and openly share data and information to maximize the ability to create benefits to society has been a key to success to date within the United States. We also are using international frameworks such as the Global Ocean Observing System Regional Alliances to share and transfer these technologies and lessons to other developing and developed nations. At the same time, the benefits from this component of IOOS are still emerging as the system matures and we "wire in" more connections to deliver these data into other applicable programs, and there is an ongoing need to balance investments among other observing system needs.

Given the definition of the new blue economy in this volume, especially the idea that the NBE is "a knowledge-based economy, looking to the sea ... for data and information to address societal challenges and inspire their solutions," our US IOOS workforce has already emerged as a key enabler that is multidisciplinary, multitalented, and connected locally, regionally, nationally, and globally. Facilitating and growing these talents to evolve over the next 20 years will be critical across the globe while the needs will be varied depending upon maturity of economies and geographic differences.

Research and development and innovation

The past 20–30-year timeframe also included development of new science and technology across a broad spectrum of disciplines. The knowledge and skills to apply these advances to better inform decisions range from network and systems engineering, to mathematics and modeling, to data sciences and software development, to communications, and outreach and education.

Proven technologies to observe have included moorings, satellites, shore stations, bottom mounted sensors, underwater gliders, and others (see, for example, Watson-Wright and Snelgrove, this volume). Meanwhile, the relative "explosion" of new technologies has been rapidly advancing in the past decade to enable more autonomous observing capabilities such as new underwater gliders, autonomous surface vessels, unmanned aerial systems, animal tagging, cube satellites, imaging flow cytometry, artificial intelligence, and machine learning. Within the United States, the people and workforce who have grown and learned over the past 20 years include a wide array of job types and expertise. The range of expertise varies from nonfederal to federal and includes academia, nonprofit, for-profit, and government positions across a spectrum of skills and abilities.

Academic and scientific experts have been actively engaged from many disciplines including, but not limited to, oceanographers, physicists, biologists, chemists, mathematicians, ocean engineers, writers, educators, economists, artists, political scientists, and program managers. Within these disciplines, there are software and hardware developers, modelers, inventors, and entrepreneurial innovators building new and improved capacity to observe, model, predict, and inform decision-makers about changing conditions, risks, and alternatives for taking action.

Integrating and partnering to meet stakeholder needs

The US experience has included a bottom-up approach combined with national and international systems "guidance" and collaboration to connect and incorporate other network components as resources and partnerships allow. The mission areas of IOOS are comprehensive in their reach to all blue economy areas, and the building of a "system" has been about building a coalition of the willing. Within the IOOS enterprise, there has been a major focus on data management and communications (DMAC) as the "glue" that brings disparate systems together. DMAC and IOOS overall have developed with a "do no harm" philosophy as the approach to integrating many existing observations, models and decision support capabilities while allowing innovation and development to improve our collective ability to deliver information and solutions that support the new blue economy. The bottom-up approach includes funding, empowering, and enabling the IOOS Regional Associations, and their principal investigators to conduct stakeholder outreach and engagement to gather, refine, and confirm user needs and requirements at the local, regional, and national and international scales. This work fundamentally requires leveraging and partnering in many ways—vertically, horizontally, and often in a repeatable way to codevelop solutions with stakeholders and customers.

Given this foundational partnering need within the ocean science and technology community, to deliver actionable information that provides economic benefit and value, there are many skills and specialties needed to solve problems and provide innovative solutions. While specialists and experts are needed, it is becoming apparent that some of the most impactful performers are multidisciplinary project management, relationship management, and partnership management experts. These experts need to be conversant in technical and scientific domains and must be able to communicate with experts, policymakers, and stakeholders in order to refine products and services that meet end-user needs. The social sciences are increasingly being recognized as critical subject areas to be included in the decision support and service delivery arena.

Additionally, collaborating and convening are often the way we organize our community to build IOOS and to prioritize requirements and future development. The skills required to convene experts efficiently and create action plans to advance the system are more critical than ever. Therefore, when staffing resources are limited, good, multipurpose players are one of the most critical elements of the new blue economy workforce. This perspective is certainly biased by the experience of trying to implement a national and global system within a government program context. There is a seemingly endless growth path for subject experts to fill roles in this ecosystem of players with high communication and collaboration skills. Some of the most effective participants in this ocean enterprise ecosystem are skilled and accomplished scientists and technologists who also have superb communication and partnership skills. In this context, "superb communication and partnership skills" are necessary because of the complexity of partnerships needed to develop and deliver applied solutions—from observations, to data, to predictions, to information and decision support.

Pipeline for talent expertise and leadership

If you accept the notion that US IOOS is an end point of employment for a vast workforce that powers the blue economy, then we should consider the education, training, and development models that can prepare that workforce. The system within US IOOS already includes a promising structure of academic institutions providing a pipeline of talent through PhD, masters, and increasingly, undergraduate programs. See "RUCOOL Operational Oceanography Masters - workforce development case study" (see Beaird et al., this volume) As universities increase the numbers and types of ocean related majors and elective courses, an emphasis should be made to push cross-disciplinary training, communications, and project management as complementary components for any ocean related degree. The Mid-Atlantic Regional Association Coastal Ocean Observing System example at Rutgers shows the importance of experiential learning, which creates that critical connection for students between their learning and real-world issues and creates the drive to seek a long-term career of applied solutions within the new blue economy.

IOOS Regional Associations and their partner academic institutions are leading the way to fuel a pipeline for our workforce needs. Critical aspects of that pipeline are regionally specific partnerships and diverse approaches that are applicable to geographic differences and cultures. What works in the southern Gulf of Mexico and what works in the Pacific Northwest may be very different and may not work in the Northeast or Pacific Islands or Alaska.

Within our network of networks approach, through the IOOS Regional Associations, we are now hiring "data wranglers" to help coordinate implementation of the Animal Telemetry Network (https://ioos.noaa.gov/project/atn/) to fill a need for coordination between biologists, oceanographers, data scientists, and data management expertise. Similarly, IOOS has partnered with the Ocean Acidification Program (https://oceanacidification.noaa.gov/) to support the Coastal Acidification Network within each IOOS region to increase collaboration, coordination, and communication across communities of shellfish growers, ocean acidification scientists, and local, regional, state, and federal decision-makers. Establishing these relatively new positions demonstrates the recognition of the critical value of collaboration function (https://ioos.noaa.gov/project/atn/) and the need for dedicated staff and collaborative skills.

Pipeline examples

One can find examples of leaders who have emerged from this pipeline across the ocean observing community today and need look no further than the cadre of experts currently on the IOOS Advisory Committee, which is made up of a diverse group of members who are leaders in their fields. Just picking any one or two examples from that group of experts would yield stories of success that include education, job experiences, training, and varied leadership positions in the ocean observing community. These experts have become leaders and multiskilled professionals with career paths applying ocean data, science, and technology to meet society's needs within the new blue economy.

These types of linkage, between education and workforce skills development pipelines, are illustrated by the trajectories of two IOOS Advisory members, Dr. Ruth Perry and Dr. Jyotika Virmani. Both are relatively early in their careers, and both have histories of starting out earning oceanographic degrees that included work within an IOOS Regional Association or its affiliated universities, before moving on to successful careers in the private sector and nonprofit ocean foundations respectively.

Dr. Perry received her PhD in physical oceanography at Texas A&M University and started her ocean observing and technology career with Gulf of Mexico Coastal Ocean Observing System (GCOOS) before forging a career path in the energy sector and has ever since been building partnerships between private industry and other members of the ocean observing community. She has been pivotal in expanding the offshore energy industry's role as ocean partners in today's new blue economy.

Dr. Perry represents someone who grew up within the system of IOOS and moved on in her career, but then returned to "give back" through contributions in various ways. Dr. Perry has pushed innovative partnerships from her private sector position that combine private sector (energy and technology), government (defense and civilian), and IOOS regions. These types of partnerships demonstrate that the power of the network of network approach IOOS relies upon to increase the richness of expertise and technology being integrated to deliver benefits for society. Dr. Perry now continues to give back by serving on the GCOOS Board of Directors and on the IOOS Federal Advisory Committee, which provides recommendations to both NOAA and the Interagency Ocean Observation Committee.

Dr. Jyotika Virmani, similarly, received her PhD in marine science and physical oceanography at University of South Florida, before an early career position as Executive Director of the Florida Coastal Ocean Observing System—an entity that operated within Florida and worked closely with both the Southeast Coastal Ocean Observing Regional Association (SECOORA) and GCOOS. Following a position at the UK MetOffice, she then served as an associate director in academia at the Florida Institute of Oceanography and as a Board Member for SECOORA. From there, Dr. Virmani joined the XPRIZE Foundation where she held various director-level positions to develop and operate incentive competitions for marine and global ecosystem technology innovations, including running the Shell Ocean Discovery XPRIZE as the Executive Director of Planet & Environment and Prize Operations. She left XPRIZE after 6 years to become the first Executive Director of the Schmidt Ocean Institute.

Both Dr. Perry and Dr. Virmani are known leaders in their fields and are called upon as experts frequently to serve as spokespersons for ocean innovation and for discovering new ways to apply knowledge to support societal needs in an ever-changing world, with an ever-changing ocean and new blue economy. Dr. Perry has been a leader and a voice for the value of private–public partnerships and for the energy industry to lead by example in terms of sustained ocean observations. Dr. Virmani has been a leader and a voice for transformative technology breakthroughs and cross-disciplinary approaches to problem-solving through her work at XPRIZE.

When considering private, nongovernmental, academic, and government sector staff, management, and leadership needs, the new blue economy is going to need cross-disciplinary experts. Those experts are going to need to include highly skilled project management and partnership management experts with additional science or technical domain expertise to truly get at the end goal of inspired solutions to the problems of today and tomorrow.

The ocean enterprise as employer

The companies that enable sustained ocean observation, measurement and forecasting, and deliver its benefits as commercial services, combine to create a unique and growing industry cluster called the Ocean Enterprise (Rayner et al., 2018). As described in some detail in other parts of this volume (see Rayner, this volume), the US IOOS Office partnered with The Maritime Alliance (see Jones and Fischel, this volume) and ERISS Corporation to conduct and publish the first national-level assessment of the scale and scope of the "ocean enterprise," defined as for-profit and not-for-profit businesses, which support ocean measurement, observation, and forecasting. The study surveyed 410 US companies and showed that the ocean enterprise, the for-profit and not-for-profit firms that support ocean measurement, observation, and forecasting, accounts for $7 billion of the US economy annually and provides up to 30,000 jobs specific to the ocean enterprise focus of the study. The total number of jobs within those companies included in the survey was closer to 225,000–250,000 when counting all of their market sectors. This estimate was based on analyzing the provider companies and intermediary companies, but did not attempt to capture the third major component of the enterprise—the end users. The end users are primary beneficiaries of the ocean enterprise, but were not included because of the complexity and scale of measuring the end uses. Prior to the 2016 report, the ocean enterprise had been rarely recognized on its own because the work done by these companies is often only a portion of what they do and it is widely distributed across a range of disciplines not typically classified as maritime operations. US IOOS and ERISS Corporation conducted another study in 2020 and plan to issue a report in 2021 comparing trends since the first study.

Workforce needs and pipeline of tomorrow

The variety and types of companies, markets, and employment opportunities are still in the early stages of development for the maritime and ocean sector compared to other business sectors, including the weather enterprise. As the information demands and services grow in the ocean enterprise, the workforce needs will grow accordingly as well. An ongoing and pressing challenge for IOOS is to better define the operational service delivery of ocean, coastal, and Great Lakes products and services at varied timescales. As we evolve IOOS to enable more mature ocean, coastal, and marine service delivery, we should continue to include social scientists and increase their roles to design the most impactful decision support systems including metrics of evaluation and success.

Additionally, as the pace of technological advancement continues to accelerate, we need experts and analysts who can discern ways to incorporate new hardware and software systems without breaking what has already been built. We need them to learn how to take advantage of the new swarming technologies and small form observing assets when older technologies still provide critical long time series data sets that are being used operationally by a multitude of customers on a daily basis. We need them to determine how to meet a regional or local need built upon both federal and nonfederal data and information. We need them to apply design new ways to create machine learning and artificial intelligence to analyze and solve questions from large data holdings and to take critical quality assurance and initial computing closer to the sensor to enable faster delivery of actionable data and information for end users.

Now and into the future with this successful recipe (technology, programs, and resources)

Recent events designed to bring ocean observing, science, and technology communities together to plan for the next 10 years and beyond include Ocean Obs '19 (https://www.oceanobs19.net/) as well as planning workshops for the U.N. Decade of Ocean Science for Sustainable Development (see Ryabinin, this volume). Both have revealed strong signals calling for greater diversity in our community and workforce. This demand signal is documented in community white papers (https://www.oceanobs19.net/community-white-papers/) for Ocean Obs '19 and the follow-on living action plan (https://www.oceanobs19.net/living-action-plan/), and also in the U.N. Decade (https://www.oceandecade.org/) workshop reports. The demand for more skills diversity in our workforce stems from the ongoing and growing need for ocean information for societal benefit. The demand includes recognition that the pace of change in our ocean and the global environment is going to require blending of skills in ways never seen before. It is also based upon recognition that the new blue economy touches all shores, and there is not yet shared access to ocean information services from networks such as IOOS in all countries.

The broad ocean science community has identified the need for greater listening skill—to take feedback and listening skill to determine how partnering could be beneficial for both parties and what project elements need to be agreed upon for progress to be made. The community has begun to discuss new terminology such as "capacity sharing" instead of "capacity building" because of the need to work together with tribal, indigenous, and traditional knowledge networks. "Building" risks implying those communities don't have valuable ocean insights to share and worse, that those who visit remote communities and populations are only interested in building something new that doesn't exist, rather than sharing insights and solutions built from traditional knowledge and experience.

As we look forward, it is exciting and daunting to imagine the ways our ocean observing community will meet the societal needs of the next 20 years. We must build upon the successes of the past and evolve with focused effort to meet the needs of decision-makers at many timescales. We will need to continue to observe, learn, evolve, adapt, and make choices that will empower society to look to the sea for the solutions it needs. We will need a new blue workforce for 2040 that grows from the seeds of today.

References

Integrated Coastal and Ocean Observation System (ICOOS) Act of 2009, 33 U.S.C. §§ 3601-3610, 2018. Retrieved from: https://www.congress.gov/111/plaws/publ11/PLAW-111publ11.pdf.

Rayner, R., Gouldman, C., Willis, Z., 2018. The ocean enterprise – understanding and quantifying business activity in support of observing, measuring and forecasting the ocean. J. Oper. Oceanogr. 12 (Suppl. 2), S97–S110. https://doi.org/10.1080/1755876X.2018.1543982.

Spinrad, R., 2016. The new blue economy: A vast oceanic frontier. EOS 97. https://doi.org/10.1029/2016EO053793.

U.S. Integrated Ocean Observing System, 2018. U.S. IOOS Enterprise Strategic Plan 2018–2022". U.S. Integrated Ocean Observing System Program, NOAA, Silver Spring, MD. Retrieved from: https://cdn.ioos.noaa.gov/media/2018/02/US-IOOS-Enterprise-Strategic-Plan_v101_secure.pdf.

CHAPTER 4

Ocean modeling

Eric P. Chassignet
Center for Ocean-Atmospheric Prediction Studies (COAPS), Florida State University, Tallahassee, FL, United States

Introduction

Ocean modeling is a relatively new course of study within the field of oceanography that emerged in the 1960s and has experienced rapid growth due to the exponential increase in computer processing capabilities (Chassignet et al., 2019). Over the past decade, improved understanding, numerics, grid configurations, spatial discretization, parameterizations, data assimilation, environmental monitoring, and process-level observations/modeling have led to significant advances in ocean circulation modeling (see Fox-Kemper et al. (2019) for a review). Enhanced computational capabilities and greater physical consistency in numerical formulations (Griffies et al., 2000) have made routine resolution of oceanic flows possible at the mesoscale on global scales, at the submesoscale on regional scales, and to detailed harbor structures in the coastal ocean (e.g., Álvarez-Fanjul et al., 2018; De Mey-Frémaux et al., 2019). Ocean circulation models have expanded our ability to identify and characterize complex and diverse physical mechanisms in the ocean. They can also be incorporated into data assimilative frameworks to produce forecasts on a variety of timescales. They can be used to produce seasonal to decadal forecasts when coupled to atmospheric circulation models, and when fully integrated into earth system models (ESMs), ocean circulation models are an essential component of climate modeling. ESMs seek to simulate all relevant aspects of the Earth system, and by integrating the interactions of atmosphere, ocean, land, ice, and biosphere, they can be used to estimate the state of the ocean under a wide variety of conditions. In practice, ocean numerical models consist of a discretized solution to a set of partial differential equations describing the ocean dynamics that are based on an approximated version of Navier–Stokes equations (Griffies, 2004; Griffies and Treguier, 2013). Ocean numerical models are typically written in FORTRAN, consisting of 20,000–200,000 lines of code, and they can take up to a decade in community development before they are fully functional (Fox-Kemper et al., 2019).

Preparing a Workforce for the New Blue Economy
ISBN 978-0-12-821431-2
https://doi.org/10.1016/B978-0-12-821431-2.00055-X

Ocean models can be used to generate idealized or realistic configurations that allow us to explore fundamental ocean mechanisms. However, in order to build a physically consistent estimate of the ocean state, its evolution, and ultimately reanalysis products that can describe past evolution, ocean models are combined with data assimilation techniques to extrapolate in space and time the sparse oceanic observations (Ferry et al., 2012; Carton et al., 2018; Wunsch, 2018). The same data assimilation techniques are used to generate the initial state that is used to perform short-term forecasts (Chassignet and Verron, 2006; Dombrowsky et al., 2009; Schiller and Brassington, 2011; Bell et al., 2015; Chassignet et al., 2018). Observational data via data assimilation pave the way for model state estimates and forecasts (Chassignet et al., 2009), the quality of which is strongly dependent upon an ocean numerical model's ability to accurately represent not only the ocean dynamics that is resolved by the model grid, but also the unresolved physics via subgrid-scale parameterizations.

Operational oceanography

Operational oceanography is defined as the systematic and long-term routine measurement of the oceans and their rapid interpretation and dissemination. Ocean forecasting, as a component of operational oceanography, is based on the near-real-time collection of ocean observations that are assimilated into ocean numerical models to provide forecasts (5–10 days) (Schiller et al., 2018; Davidson et al., 2019). Ocean numerical models are a key component of global, regional, and coastal operational oceanography systems. A comprehensive operational system contains models that solve governing equations for ocean currents (u, v), sea levels, temperature (T), salinity (S), sea ice, surface waves, and concentrations of tracers relevant to environmental or biogeochemical processes. Biogeochemical forecasting systems are a combination of existing physical forecasting systems and biogeochemical models developed either for climate research or for ecological modeling (Ford et al., 2018).

In order for operational oceanography to be of maximum use, an accurate representation of large-scale upper ocean structure and mesoscale features, such as eddies and meandering fronts, is required (Fig. 4.1) (Hecht and Hasumi, 2008). Most of the improvements in the representation of these features during the past 10 years is the direct result of more powerful computing platforms that allow for increased horizontal resolution. We can now perform basin-scale simulations with grid spacing on the order of 1 km

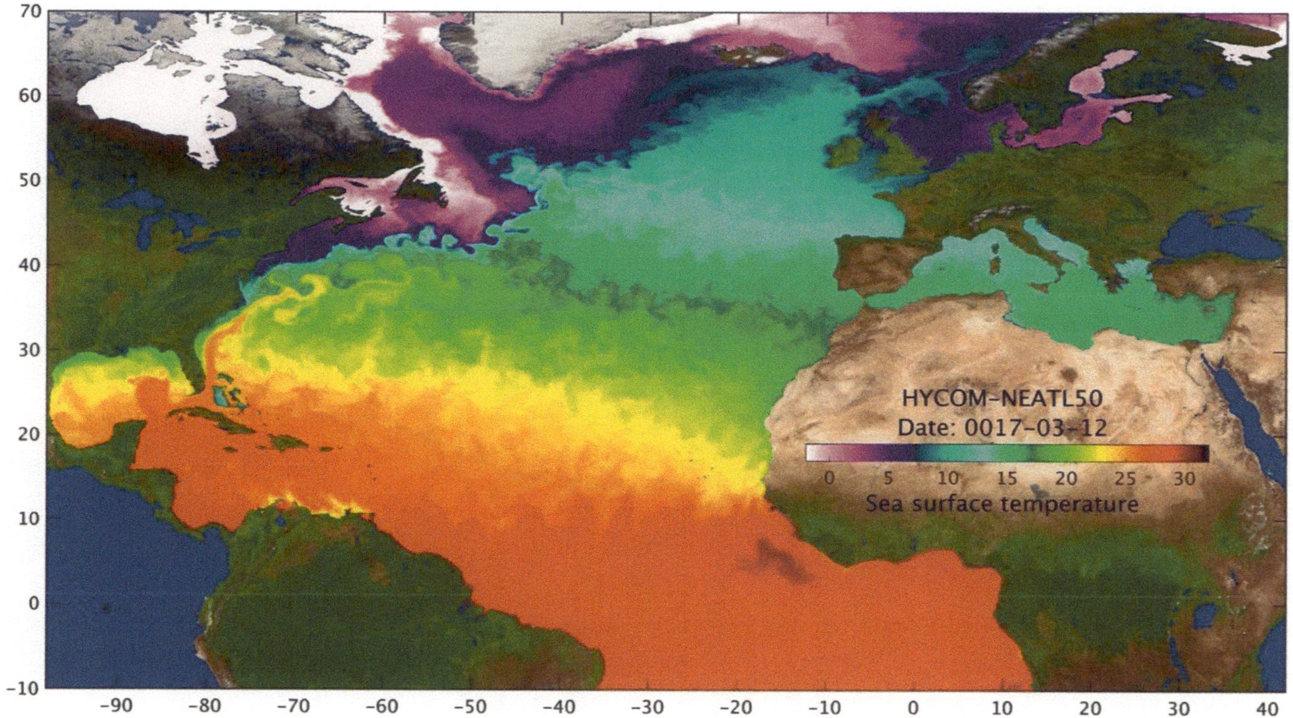

Figure 4.1 Sea surface temperature snapshot for the North Atlantic and Equatorial Ocean from a 1/50° (∼1.5 km at mid-latitudes) HYCOM numerical simulation (Chassignet and Xu, 2017). Noteworthy are the Loop Current in the Gulf of Mexico, the Gulf Stream separating at Cape Hatteras, the Gulf Stream penetrating, meandering, and shedding eddies into the ocean interior, and tropical instability waves in the equatorial region.

(submesoscale resolving) (Chassignet and Xu, 2017; Ajayi et al., 2020, 2021) and regional simulations with grid spacing on the order of 100 m (Capet et al., 2008). However, these models are not able to account explicitly for the full range of scale interactions that regulate ocean circulation. Because oceanic flows are turbulent in nature, ocean circulation at a given scale is fundamentally dependent on oceanic motions at scales ranging from global (of order 10,000 km) to dissipative (of order 1 cm). Further, the scales of motions explicitly represented in a particular model's solution will be constrained by the predetermined grid resolution of the model configuration. Subgrid-scale parameterizations are needed to account for the important physical processes that are not resolved. Parameterization choices have a profound impact on the physical representation of boundary currents (Chassignet and Xu, 2017) and on biogeochemical processes since they are quite sensitive to small-scale processes (Ford et al., 2018).

In addition to capturing and predicting large-scale ocean features, operational systems are needed to help predict the array of spatial and temporal scales that are found near coastlines (De Mey-Frémaux et al., 2019). Coastal oceans are geometrically constrained by jagged coastlines and irregular bathymetry; they are driven (internally, laterally, and at the surface) on a large range of space/timescales by buoyancy, wind, and tidal forces (Haidvogel et al., 2000). The subsequent coastal circulation patterns include both persistent and time-variable fronts, intense currents with strong spatial (offshore and/or vertical) dependence, coastal trapped waves, internally generated mesoscale variability, large horizontal water mass contrasts, strong vertical stratification, and regions of intense turbulent mixing in both surface and bottom boundary layers. This results in an observational as well as a modeling challenge and requires the development of coastal ocean forecasting systems capable of supporting societal and management decisions and policy (Kourafalou et al., 2015a,b). Coastal models, therefore, need to be able to resolve interactions between nearshore, estuarine, and shelf processes (required resolution of 10–100 m), as well as open ocean processes (required resolution of 1–10 km). Consequently, coastal numerical models are typically nested within a regional or global model configuration (Katavouta and Thomson, 2016; Holt et al., 2017; De Mey-Frémaux et al., 2019). Coastal regions present an advantage in that permanent, multivariate instrumented sites are easier to set up than global ocean observing systems and coastal observatories can employ many data sources that are not available for the open ocean. For example, technologies such as telemetering moorings and fixed platforms, profiling floats, autonomous underwater vehicles (AUVs), Lagrangian drifters, and surface current measuring radar can be deployed regionally. These observational technologies are complemented by global satellite observing networks and direct in situ observations of the subsurface ocean. By combining coastal modeling systems and comprehensive observational networks, coastal ocean forecasting systems facilitate the constant monitoring of variations in the coastal ocean and support forecasting activities. This, in turn, allows for the delivery of useful and reliable ocean services, such as forecasting, sustainable management of ecosystems, shipping efficiency, mitigation of storm damage, etc.

As stated by Schiller et al. (2018), further advances in operational oceanography will require increasingly multidisciplinary efforts in physics, chemistry, biology, geomorphology (especially in the littoral zone), and information technology/visualization, as well as the exploitation of "big data" expertise given the petabytes of model outputs generated. Apart from the scientific challenges, continued expansion of operational oceanography will require the recognition of the collective benefit of all its constituents (observations, data management, prediction system, production/service delivery, and clients). In particular, the satellite and in situ elements of physical observing systems, as well as the sustainability and expansion of biological and biogeochemical observing systems, are vital components of ocean forecasting systems. The introduction of "intelligent" in situ sensors, sensor networks/webs, and new and/or improved remote sensing technologies will also create new opportunities. Finally, the development of open-access web data services aligned with community conventions for metadata descriptions has been shown to foster systems' data exchange and model usage, accelerate testing, validate the acceptance cycle for modeling system enhancements, simplify the addition of new data streams, enable operational monitoring, and allow novice users to view and download model outputs to become the basis for higher-level ocean information products (Wilkin et al., 2018).

Earth system modeling (ESM)

Variations of ocean circulation are dynamically coupled with that of sea ice, waves, and atmospheric physics. Ocean circulation models are only one component of an ESM. The coupled systems that form an ESM allow for improved representation of physical processes across oceanic boundaries, which is necessary for short-range forecasts as well as for seasonal to climate timescales (Brassington et al., 2015). As stated by Harris et al. (2018), the value of genuinely coupled models is that adjustments in one model can immediately and proximately influence the other model. For example, the lack of any feedback in ocean-only models on atmospheric forcing variables, such as temperature, winds, or humidity, can result in erroneousness (see e.g., Griffies et al., 2009). This is especially true if the atmospheric forcing used is obtained from an atmosphere model integrated with a significantly different ocean surface boundary condition. Differences in areas of ice cover are especially problematic in this regard because heat and moisture fluxes over the open ocean are very different from those over sea ice. Also, a high-resolution coupled system of the coastal ocean that takes into account the nonlinear feedback between the atmosphere and the upper ocean via a wave interface will reduce ocean prediction errors, particularly under extreme conditions (Staneva et al., 2017; De Mey-Frémaux et al., 2019). Deterministic modeling systems are fairly accurate when applied to short-term weather time scales. However, longer subseasonal-to-seasonal scales require ensemble-based modeling systems that can represent the inherent uncertainty and predictability. Several competing priorities exist as such modeling systems continue to evolve. First, is the need for refined resolution to improve the fidelity of resolved flow features. Second, the representation of otherwise missing processes and feedbacks will require increased component complexity/capability. And third, extraction of a signal from the noise will require more ensemble members (Hewitt et al., 2017). Further, in order to fully incorporate biogeochemical models in the forecasting systems, additional work will be needed (see next section for an example).

Initially, the scales at which ocean circulation models were able to represent oceanic properties and physical processes were significantly larger than the mesoscale (horizontal scales in the neighborhood of 100 km and approximately 3-month timescales). However, in order to be of maximum use, the ocean circulation model component of an ESM must be able to represent the wide range of oceanic physical processes that today's ocean models are capable of as a result of increases in computing power and of the improved physical consistency of their formulations. For example, today's global ocean circulation models are able to resolve oceanic flows down to the submesoscale (horizontal scales on the order of 10 km or less; Chassignet and Xu, 2017) and to account for internal wave and internal tides (Shriver et al., 2012). Fig. 4.2 shows the impact of increasing horizontal grid spacing on the representation of small-scale features in the Gulf of Mexico.

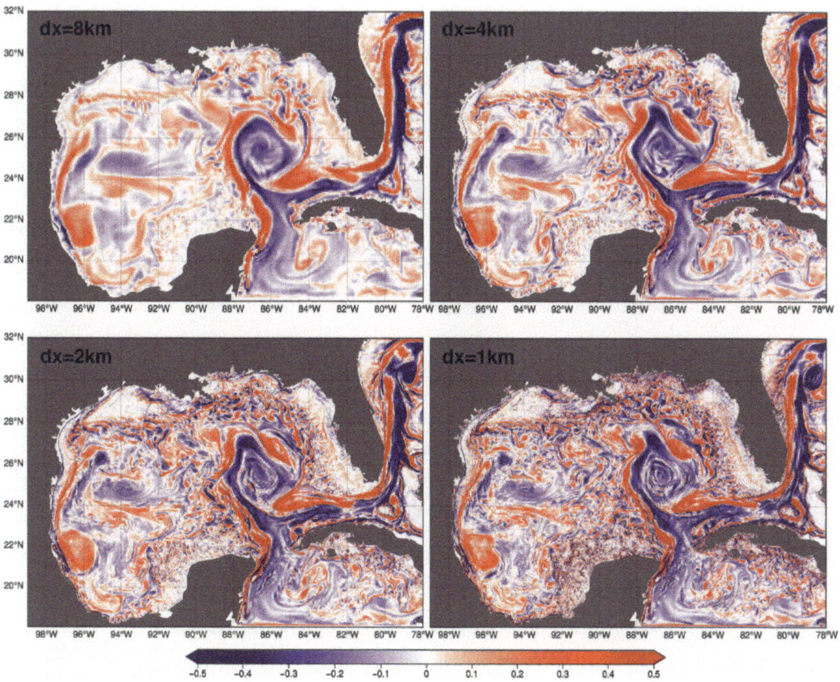

Figure 4.2 Modeled Gulf of Mexico surface relative vorticity (ζ/f with $f = 10^{-4}\ s^{-1}$) depicting the flow rotation and shear as a function of horizontal grid spacing (8, 4, 2, and 1 km, respectively). In addition to the loop current penetrating in the Gulf interior and large separated loop current eddies, one can observe an increased number of small-scale features that arise when the model horizontal resolution is increased.

Example: Gulf of Mexico oil spill modeling

The 2010 Deepwater Horizon (DwH) oil spill occurred in an area that encompasses a broad variety of oceanographic environments and ecosystems. The DwH spill originated over the deep outer shelf, where it could potentially have been transported seaward over the abyssal region of the Gulf of Mexico or onto shelves of varying width and toward coasts with sandy shores or wetlands. One of the most important components of this area's circulation, hydrography, and biogeochemistry is the Mississippi River. Among other things, the river supplies nutrients fueling productivity, freshwater that contributes to upper ocean stratification, and sediments that can interact with oil in the water column. Over the Texas–Louisiana shelf to the west, seawater properties (temperature, salinity, turbidity, and currents) are very different than what is found in the deeper location of DwH, and low oxygen (hypoxic) conditions occur each summer. While analysis of the DwH surface oil transport shows that a relatively small fraction of oil from the spill drifted over the Texas–Louisiana shelf, under different forcing regimes, a substantial quantity of the oil could have advected over this shelf and been deposited on the seafloor through sedimentation/flocculation. Further, it is presumed that microbial activities, which are quite significant and diverse in these areas, have a substantial impact on biodegradation and the accumulation of petroleum in the water column and marine sediments of the deep ocean. Given that toxic oil constitutes unknown threats to benthic organisms, many of which are harvested for human consumption, the ability to predict the eventual fate of oil and its impact on ecosystems is critical. Additionally, ongoing oil and gas extraction activities taking place in the Gulf of Mexico increase the likelihood of large spills in the future.

Ocean circulation models increase our ability to understand this complex system and associated diverse physical mechanisms. And now that critical novel interactions, such as hydrocarbon biodegradation and flocculation, can be represented in the coupled system (Morey et al., 2020), a modeling framework can be used to simulate oil in the marine environment, including its interaction with ecosystems and sediments and predict where ecosystems may be affected. So, whereas biodegradation might have been included in oil models for removal of oil from the system as a simple decay parameterization, in reality it is a complex process mediated by diverse microbial taxa that convert hydrocarbons into biomass, carbon dioxide, and more refractory forms of organic carbon. Similarly, flocculation is commonly included via parameterizations in sediment transport models, but the formulations used have rarely accounted explicitly for either the biological constituents of the water column or the presence of oil. Yet, flocculation is likely to be an important link facilitating transport between oil in the water column and the sediments at the seafloor, both in deep water and over the shelf. The role of flocculation is especially important within turbid river plumes and during storms when significant resuspension occurs. Fig. 4.3 shows a snapshot from the DwH oil model developed by the Consortium for Simulation of Oil-Microbial Interactions in the Ocean (CSOMIO, http://CSOMIO.org). CSOMIO has developed a modeling system that dynamically couples components for simulating ocean hydrodynamics, oil transport, dispersion and weathering, oil-mineral aggregate (OMA) formation, flocculation and settling, and the lower trophic-level marine ecosystem (Dukhovskoy et al., 2021). It is an adaptation and extension of the Coupled Ocean-Atmosphere-Wave-Sediment

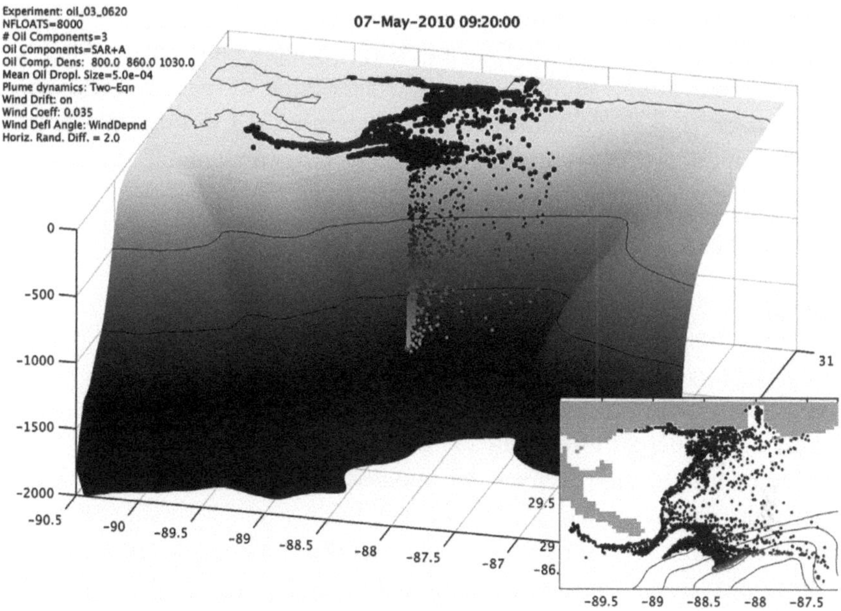

Figure 4.3 Modeled vertical and horizontal distribution of oil on May 7, 2010. The CSOMIO oil plume model is based on a Lagrangian approach that describes the oil plume dynamics by advecting and diffusing individual "floats" representing a cluster of oil droplets. The chemical composition of oil is described in terms of three components (saturates, aromatics, and heavy oil including resins and asphaltenes). Oil droplets are characterized by size and chemical structure (mass fraction of compounds). The oil plume model simulates rise velocity of oil droplets based on ambient ocean flow and density fields and density and size of the oil droplets. The oil model also includes weathering processes and surface wind drift. A key component of the CSOMIO model is Lagrangian–Eulerian and Eulerian–Lagrangian mapping of the oil characteristics. This mapping is necessary for the interaction between the ocean-oil module with the sediment and biology models.

Transport (COAWST) modeling system (Warner et al., 2010). A biogeochemical modeling component incorporating a microbial model (Genome-based EmergeNt Ocean Microbial Ecosystem (GENOME); Coles et al., 2017) is implemented in the system and adapted for the presence of hydrocarbons. The sediment transport component of COAWST (the Community Sediment Transport Modeling System, CSTMS) is modified to include computationally efficient flocculation parameterizations for OMAs developed from laboratory experiments. The ocean modeling component of COAWST (the Regional Ocean Modeling System, ROMS) is modified to simulate three-dimensional oil transport and compositional changes (weathering). These modeling components are linked together using a two-way Lagrangian–Eulerian mapping technique enabling interaction between all of the modeling components for tracking of hydrocarbons from a source blowout to deposition in sediment, microbial degradation, and evaporation while being transported through the ocean. This component can be run offline to increase computational speed.

Perspective

Ocean circulation model design is an active field of study that has matured, in large part, because of strong cross-disciplinary collaborations and the consolidation of efforts among different groups involved in the various applications and dimensions of ocean circulation and ESMs. Ocean circulation model design often calls for entire scientific teams focused on the development or improvement of models. Continued improvement to ocean circulation models depends heavily upon ocean observing networks and the critical role that sustained satellite observations and in situ observations play in the routine assessment of model skills and limitations over different timescales ranging from days to decades. A great deal can be learned from targeted field observations geared toward documenting specific oceanic processes so that they can be better represented in ocean models.

In order to meet the evolving needs of operational oceanography and ESM end users, ocean circulation models will need to be able to account explicitly for a comprehensive range of physical scales due to the multiscale nature of oceanic flows. Improving the performance of ocean circulation models on modern high-performance computing platforms will allow for broadening the spectrum of resolved scales in the models by increasing the horizontal and vertical resolution. The current trend is toward massively parallel machines with heterogeneous multicore architectures (Giles and Reguly, 2014). But while modern platforms can deliver a peak performance in the Petaflops range, existing ocean circulation models are unable to fully exploit this potential. The computational intensity (i.e., floating point operations per memory access) of ocean models is very low because they are dominated by stencil operations (typical of discretized partial differential equations) and typically run at ~5% of the peak speed of the system. Also, ocean models have a very small vertical dimension, typically O(10), so they scale more like 2-D domains than 3-D domains. The ability of an ocean model to take advantage of massively parallel computers is limited by the communication overhead, load imbalance, latency, and inputs/outputs overhead inherent to the spatial 2-D domain decomposition cores one has to make to distribute the tasks evenly to all the processors. Sustained collaborations between ocean modelers and computer scientists will be needed to overcome single processor and scalability issues (Chassignet et al., 2019). Practical approaches will likely involve more hybrid parallel programming in order to more efficiently exploit memory hierarchy and innovative algorithms for solving the set of partial differential equations that govern ocean dynamics, for instance, parallelization in time in addition to spatial domain decomposition (Schreiber et al., 2017).

A quick Internet search for "ocean modeling jobs" shows that there is great demand for ocean modeling computing skills. This is driven by several factors. In academia, the emphasis on climate research not only implies strong cross-disciplinary collaborations, but a need for students and postdocs with good programming skills to further develop and refine ESMs. The main challenge facing academia in attracting young and talented programmers is an incompatibility between model developments that may require many years to reach fruition and high rates of publication now becoming the norm for advancement of early career scientists.

There is also a strong need for qualified applicants in government laboratories because there is a push to develop the next generation of ESMs to address national needs. For example, in a bid to regain global leadership in weather forecasting, the National Oceanic and Atmospheric Administration (NOAA) is establishing an Earth Prediction Innovation Center (EPIC) to address longstanding challenges in translating research advances into operational forecasts (https://owaq.noaa.gov/Programs/EPIC). This center will accelerate community-developed scientific and technological advancements into the operational applications for numerical weather prediction by supporting a Unified Forecast System (UFS) community model. NOAA is working closely with entities in the weather enterprise (public, private, and academic) to accelerate the transition of research to operations, share the current status and future of community-based ESM, and identify emerging technologies. The Energy Exascale Earth System Model (E3SM) project at the US Department of Energy (DOE) is another example of an ongoing, state-of-the-science earth system modeling, simulation, and prediction project that optimizes the use of DOE laboratory resources to meet the science needs of the nation and the mission needs of DOE. It focuses on (1) the hydrological cycle and how more realistic portrayals of features important to the water cycle (clouds, aerosols, snowpack, river routing, land use) will affect simulations of river flow and associated freshwater supplies at the watershed scale; (2) uptake of CO_2 by land and ocean ecosystems; and (3) cryosphere—ocean interactions and the impacts of ocean—ice shelf interactions on the melting of the Antarctic ice sheet, the global climate, and sea level rise (https://climatemodeling.science.energy.gov/program/earth-system-model-development).

Ocean modeling employment opportunities in the private sector fall into two categories, one that works closely with government agencies as contractors in order to supplement the federal workforce to address national needs as described earlier and one that provides operational oceanographic products and ocean services to the public at large. Many of these products and services cover a wide range of applications; address real-world issues related to public safety, the maritime industry, and environmental conservation efforts; and are provided and/or disseminated in a systematic and routine manner to end users. Examples include warnings (coastal floods, ice and storm damage, harmful algal blooms and contaminants, oil spill, etc.), optimum routes for ships, and prediction of seasonal or annual primary productivity, ocean currents, ocean climate variability, etc. The field's evolution and fast-paced advances mean that the growth potential for this sector is high, with many future training and job opportunities (see Op-Ed in this volume on the Rutgers Operational Oceanography program). It is fair to say that the multidisciplinary nature of operational oceanography makes it an exciting scientific career path, particularly for those interested in research or the assimilation of new technological advances and methods.

Acknowledgments

This chapter is heavily influenced by review articles and by books written or edited by the author. The author would like to thank T. Ippolito for editing the chapter, X. Xu for providing Fig. 4.1, A. Bozec for providing Fig. 4.2, and D. Dukhovskoy for providing Fig. 4.3.

References

Ajayi, A.O., Le Sommer, J., Chassignet, E.P., Molines, J.-M., Xu, X., Albert, A., Cosme, E., 2020. Spatial and temporal variability of the North Atlantic eddy field at scale less than 100 km. J. Geophys. Res. 125. https://doi.org/10.1029/2019JC015827. e2019JC015827.

Ajayi, A.O., Le Sommer, J., Chassignet, E.P., Molines, J.-M., Xu, X., Albert, A., Dewar, W., 2021. Diagnosing cross-scale kinetic energy exchanges from two submesoscale permitting ocean models. J. Adv. Model. Earth Syst. (in press).

Alvarez Fanjul, E., et al., 2018. Operational oceanography at the service of the ports. In: Chassignet, E., Pascual, A., Tintoré, J., Verron, J. (Eds.), New Frontiers in Operational Oceanography. GODAE OceanView, pp. 729–736. https://doi.org/10.17125/gov2018.ch27.

Bell, M.J., Schiller, A., Le Traon, P.-Y., Smith, N.R., Dombrowsky, E., Wilmer-Becker, K., 2015. An introduction to GODAE OceanView. J. Oper. Oceanogr. 8 (Suppl. 1), s2–s11. https://doi.org/10.1080/1755876X.2015.1022041.

Brassington, G.B., Martin, M.J., Tolman, H.L., Akella, S., Balmaseda, M., Chambers, C.R.S., Chassignet, E.P., Cummings, J.A., Drillet, Y., Janssen, P.A.E.M., Laloyaux, P., Lea, D., Mehra, A., Mirouze, I., Ritchie, H., Samson, G., Sandery, P.A., Smith, G.C., Suarez, M., Todling, R., 2015. Progress and challenges in short- to medium-range coupled prediction. J. Oper. Oceanogr. 8 (Suppl. 2), s239–s258. https://doi.org/10.1080/1755876X.2015.104987.

Capet, X., McWilliams, J.C., Molemaker, M.J., Shchepetkin, A.F., 2008. Mesoscale to submesoscale transition in the California current system. Part I: flow structure, eddy flux, and observational tests. J. Phys. Oceanogr. 38, 29–43.

Carton, J.A., Chepurin, G.A., Chen, L., 2018. SODA3: a new ocean climate reanalysis. J. Clim. 31, 6967–6983. https://doi.org/10.1175/JCLI-D-18-0149.1.

Chassignet, E.P., Verron, J. (Eds.), 2006. Ocean Weather Forecasting: An Integrated View of Oceanography. Springer, p. 577.

Chassignet, E.P., Hurlburt, H.E., Metzger, E.J., Smedstad, O.M., Cummings, J., Halliwell, G.R., Bleck, R., Baraille, R., Wallcraft, A.J., Lozano, C., Tolman, H.L., Srinivasan, A., Hankin, S., Cornillon, P., Weisberg, R., Barth, A., He, R., Werner, F., Wilkin, J., 2009. U.S. GODAE: global ocean prediction with the hybrid coordinate ocean model (HYCOM). Oceanography 22 (2), 64–75.

Chassignet, E.P., Xu, X., 2017. Impact of horizontal resolution (1/12° to 1/50°) on Gulf stream separation, penetration, and variability. J. Phys. Oceanogr. 47, 1999–2021. https://doi.org/10.1175/JPO-D-17-0031.1.

Chassignet, E.P., Pascual, A., Tintoré, J., Verron, J. (Eds.), 2018. New Frontiers in Operational Oceanography. GODAE OceanView, p. 811. https://doi.org/10.17125/gov2018.

Chassignet, E.P., Le Sommer, J., Wallcraft, A.J., 2019. General circulation models. In: Cochran, K.J., Bokuniewicz, H.J., Yager, P.L. (Eds.), Encyclopedia of Ocean Sciences, third ed., vol. 5, pp. 486–490. https://doi.org/10.1016/B978-0-12-409548-9.11410-1.

Coles, V.J., Stukel, M.R., Brooks, M.T., Burd, A., Crump, B.C., Moran, M.A., et al., 2017. Ocean biogeochemistry modeled with emergent trait-based genomics. Science 358, 1149–1154.

Davidson, F., Azcarate, A.A., Barth, A., Brassington, G.B., Chassignet, E.P., Clementi, E., De Mey-Fremaux, P., Divakaran, P., Harris, C., Hernandez, F., Hogan, P., Hole, L.R., Holt, J., Liu, G., Lu, Y., Lorente, P., Maksymczuk, J., Martin, M., Mehra, A., Mo, H., Moore, A., Oddo, P., Pascual, A., Pequignet, A.-C., Kourafalou, V., Ryan, A., Siddorn, J., Smith, G., Spindler, D., Spindler, T., Stanev, E., Staneva, J., Storto, A., Tanajura, C., Vinayachandran, P.N., Wan, L., Wang, H., Zhang, Y., Zhu, X., Zu, Z., 2019. Synergies in operational oceanography: the intrinsic need for sustained ocean observations. Front. Mar. Sci. 6, 450. https://doi.org/10.3389/fmars.2019.00450.

De Mey-Frémaux, P., Ayoub, N., Barth, A., Brewin, R., Charria, G., Campuzano, F., Ciavatta, S., Cirano, M., Edwards, C.A., Federico, I., Gao, S., Garcia Hermosa, I., Garcia Sotillo, M., Hewitt, H., Hole, L.R., Holt, J., King, R., Kourafalou, V., Lu, Y., Mourre, B., Pascual, A., Staneva, J., Stanev, E.V., Wang, H., Zhu, X., 2019. Model-observations synergy in the coastal ocean. Front. Mar. Sci. 6, 436. https://doi.org/10.3389/fmars.2019.00436.

Dombrowsky, E., Bertino, L., Brassington, G.B., Chassignet, E.P., Davidson, F., Hurlburt, H.E., Kamachi, M., Lee, T., Martin, M.J., Mei, S., Tonani, M., 2009. GODAE systems in operation. Oceanography 22, 80–95.

Dukhovskoy, D.S., Morey, S.L, Chassignet, E.P, Chen, X., Coles, V.J., Cui, L., Harris, C.K., Hetland, R., Hsu, T.-J., Manning, A.J., Stukel, M., Thyng, K., Wang, J., 2021. Development of the CSOMIO coupled ocean-oil-sediment-biology model. Front. Mar. Sci. Submitted.

Ferry, B.N., Parent, L., Garric, G., Bricaud, C., Testut, C.E., Galloudec, O.L., et al., 2012. GLORYS2V1 global ocean reanalysis of the altimetric era (1992–2009) at mesoscale. Mercator Q. Newslett. 44, 29–39. https://doi.org/10.1016/j.ocemod.2018.07.009.

Ford, D., et al., 2018. Marine biogeochemical modelling and data assimilation for operational forecasting, reanalysis and climate research. In: Chassignet, E., Pascual, A., Tintoré, J., Verron, J. (Eds.), New Frontiers in Operational Oceanography. GODAE OceanView, pp. 625–652. https://doi.org/10.17125/gov2018.ch22.

Fox-Kemper, B., Adcroft, A., Böning, C.W., Chassignet, E.P., Curchitser, E., Danabasoglu, G., Eden, C., England, M.H., Gerdes, R., Greatbatch, R.J., Griffies, S.M., Hallberg, R., Hanert, E., Heimbach, P., Hewitt, H.T., Hill, C.N., Komuro, Y., Legg, S., Le Sommer, J., Masina, S., Marsland, S.J., Penny, S.G., Qiao, F., Ringler, T.D., Treguier, A.M., Tsujino, H., Uotila, P., Yeager, S.G., 2019. Challenges and prospects in ocean circulation models. Front. Mar. Sci. 6, 65. https://doi.org/10.3389/fmars.2019.00065.

Giles, M.B., Reguly, I., 2014. Trends in high-performance computing for engineering calculations. Phil. Trans. Math. Phys. Eng. Sci. 372 (2022) https://doi.org/10.1098/rsta.2013.0319, 20130 20319–20130 319.

Griffies, S.M., Böning, C., Bryan, F.O., Chassignet, E.P., Gerdes, R., Hasumi, H., Hirst, A., Treguier, A.-M., Webb, D., 2000. Developments in ocean climate modelling. Ocean Model. 2, 123–192.

Griffies, S., 2004. Fundamentals of Ocean Climate Models. Princeton University Press.

Griffies, S., et al., 2009. Coordinated ocean-ice reference experiments (COREs). Ocean Model. 26, 1–46. https://doi.org/10.1016/j.ocemod.2008.08.007.

Griffies, S.M., Treguier, A.M., 2013. Ocean Circulation Models and Modeling. International Geophysics, vol. 103. Elsevier, pp. 521–551.

Haidvogel, D.B., Blanton, J., Kindle, J.C., Lynch, D.R., 2000. Coastal ocean modeling: processes and real-time systems. Oceanography 13 (1), 35–46. https://doi.org/10.5670/oceanog.2000.51.

Harris, C., 2018. Coupled atmosphere-ocean modelling. In: Chassignet, E., Pascual, A., Tintoré, J., Verron, J. (Eds.), New Frontiers in Operational Oceanography. GODAE OceanView, pp. 445–464. https://doi.org/10.17125/gov2018.ch16.

Hecht, M.W., Hasumi, H., 2008. Ocean modeling in an eddying regime. AGU Geophys. Monogr. Ser. 177, 409.

Hewitt, H.T., Bell, M.J., Chassignet, E.P., Czaja, A., Ferreira, D., Griffies, S.M., Hyder, P., McClean, J., New, A.L., Roberts, M.J., 2017. Will high-resolution global ocean models benefit coupled predictions on short-range to climate timescales? Ocean Model. 120, 120–136. https://doi.org/10.1016/j.ocemod.2017.11.002.

Holt, J., Hyder, P., Ashworth, M., Harle, J., Hewitt, H.T., Liu, H., et al., 2017. Prospects for improving the representation of coastal and shelf seas in global ocean models. Geosci. Model Dev. 10, 499–523. https://doi.org/10.5194/gmd-10-499-2017.

Katavouta, A., Thompson, K.R., 2016. Downscaling ocean conditions with application to the gulf of Maine, scotian shelf and adjacent deep ocean. Ocean Model. 104, 54–72. https://doi.org/10.1016/j.ocemod.2016.05.007.

Kourafalou, V.H., De Mey, P., Le Henaff, M., Charria, G., Edwards, C.A., He, R., et al., 2015a. Coastal Ocean Forecasting: system integration and evaluation. J. Oper. Oceanogr. 8, S127–S146. https://doi.org/10.1080/1755876X.2015.1022336.

Kourafalou, V.H., De Mey, P., Staneva, J., Ayoub, N., Barth, A., Chao, Y., et al., 2015b. Coastal ocean forecasting: science foundation and user benefits. J. Oper. Oceanogr. 8, 147–167.

Morey, S.L., Chassignet, E.P., Dukhovskoy, D., Harris, C., Coles, V., Stukel, M., Hetland, R., Thyng, K., Hsu, T., Manning, A., Mason, O., Ye, L., Cui, L., 2020. A coupled modeling system for simulating oil-biological-sediment interactions in the ocean. In: GOMOSES 2020 Conference, OS-001-004.

Schiller, A., Brassington, G.B. (Eds.), 2011. Operational Oceanography in the 21st Century. ISBN 9789400703315.

Schiller, A., et al., 2018. An overview of operational oceanography. In: Chassignet, E., Pascual, A., Tintoré, J., Verron, J. (Eds.), New Frontiers in Operational Oceanography. GODAE OceanView, pp. 1–26. https://doi.org/10.17125/gov2018.ch01.

Schreiber, M., Peixoto, S.P., Haut, T., Wingate, B., 2017. Beyond spatial scalability limitations with a massively parallel method for linear oscillatory problems. Int. J. High Perform. Comput. Appl. https://doi.org/10.1177/1094342016687625.

Shriver, J.F., Arbic, B.K., Richman, J.G., Ray, R.D., Metzger, E.J., Wallcraft, A.J., Timko, P.G., 2012. An evaluation of the barotropic and internal tides in a high-resolution global ocean circulation model. J. Geophys. Res. 117, C10024. https://doi.org/10.1029/2012JC008170.

Staneva, J., Alari, V., Breivik, O., Bidlot, J.-R., Mogensen, K., 2017. Effects of wave-induced forcing on a circulation model of the North Sea. Ocean Dynam. 67, 81–101. https://doi.org/10.1007/s10236-016-1009-0.

Warner, J.C., Armstrong, B., He, R., Zambon, J.B., 2010. Development of a coupled ocean-atmosphere-wave-sediment transport (COAWST) modeling system. Ocean Model. 35 (3), 230–244.

Wilkin, J., et al., 2018. A coastal ocean forecast system for the U.S. Mid-Atlantic Bight and Gulf of Maine. In: Chassignet, E., Pascual, A., Tintoré, J., Verron, J. (Eds.), New Frontiers in Operational Oceanography. GODAE OceanView, pp. 593–624. https://doi.org/10.17125/gov2018.ch21.

Wunsch, C., 2018. Towards determining uncertainties in global oceanic mean values of heat, salt, and surface elevation. Tellus A Dyn. Meteorol. Oceanogr. 70, 1–14. https://doi.org/10.1080/16000870.2018.1471911.

CHAPTER 5

Propelling the new blue economy with safer marine transportation and infrastructure

A.J. Reiss, Darin Figurskey, Christine Bassett, Thomas Cuff
United States National Oceanic and Atmospheric Administration (NOAA), Washington, DC, United States

Introduction

American industries that power our nation's economy are increasingly looking to the sea to meet fast-rising consumer needs. In 2015, over 80,000 vessels made port calls in US ports alone. According to a 2019 report to Congress, the Bureau of Transportation Statistics at the Department of Transportation (DoT) reported that the leading mode of transportation for freight arriving and departing the United States is via water (U.S. DoT, 2020). As of 2018, sea-going vessels moved 41.9% of the value and 70.7% of the weight of US international trade in 2018. According to the International Maritime Organization (IMO), there is no other more efficient mode of "en masse" transportation of the world's goods and raw materials. Consequently, over 90% of the world's trade is carried by sea. Satellite data reveal how increasingly busy global shipping lanes have become, as vessel traffic has increased fourfold in the past 20 years (Fig. 5.1; Tournadre, 2014). And it's forecasted to grow. A February 2020 Global Industry Analysts, Inc. report projects the cargo shipping market worldwide to grow from today's values by 2.9 billion metric tons and to reach 6.8 billion metric tons by 2025 (Global Industry Analysts, 2019).

In 2020, data released by NOAA and the Bureau of Economic Analysis shows a total marine economy that is growing faster than the US economy as a whole (Nicolls et al., 2020). It contributes nearly $373 billion to the gross domestic product (GDP) and supports 2.3 million jobs. Globally, today's ocean economy is valued at approximately $2 trillion annually and is projected to grow to $3 trillion by 2030 (Nicolls et al., 2020). Given such projected growth, protecting this vital engine of the marine economy becomes imperative. Investments in marine technology that modernize and enable life and cargo saving decision-making at sea will be needed and can

Preparing a Workforce for the New Blue Economy
ISBN 978-0-12-821431-2
https://doi.org/10.1016/B978-0-12-821431-2.00003-2

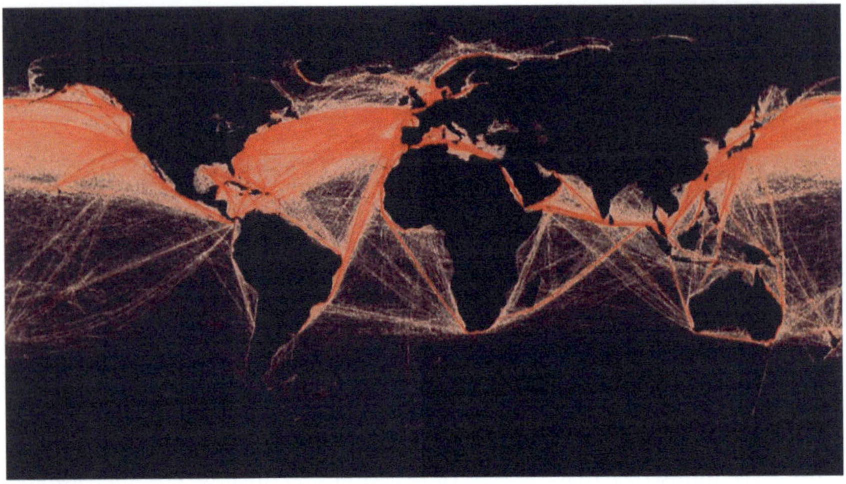

Figure 5.1 Satellite imagery of vessel traffic (marked in red) reveals a 400% increase in shipping in the past 20 years; Image attribution: (B) S. Halpern (T. Hengl; (D) Groll)/ Wikimedia Commons/CC BY-SA 3.0.

enable entrepreneurial opportunity in what this book defines as the new blue economy (see Spinrad, this volume). Surprisingly, however, despite major advances in marine weather prediction, high-speed global communications, and improvements in vessel design and engineering, the technology used to disseminate and receive hazardous weather and ice warnings on the high seas is still stuck in the last millennium.

Help over the horizon—national marine weather service providers

Did you know that most vessels on the high seas only receive warnings of hazardous conditions in their general vicinity by text bulletins (Fig. 5.2) and only at an approximate interval of every 6 h? The land-based equivalent would be to drive on a cross-country trip and, once every 6 h, pause at a rest stop, pick up a newspaper, and read the national weather forecast! Additionally, in international waters there are no state or province boundaries for reference, so weather communications rely on geographic coordinates that must be plotted by hand on paper charts. With today's constant stream of map-based weather information and warnings transmitted through various applications on smartphones and watches, triggering automated alerts and notifications, it is hard to imagine the cross-country driving scenario described earlier that faces most of today's mariners—however, it's a reality for many.

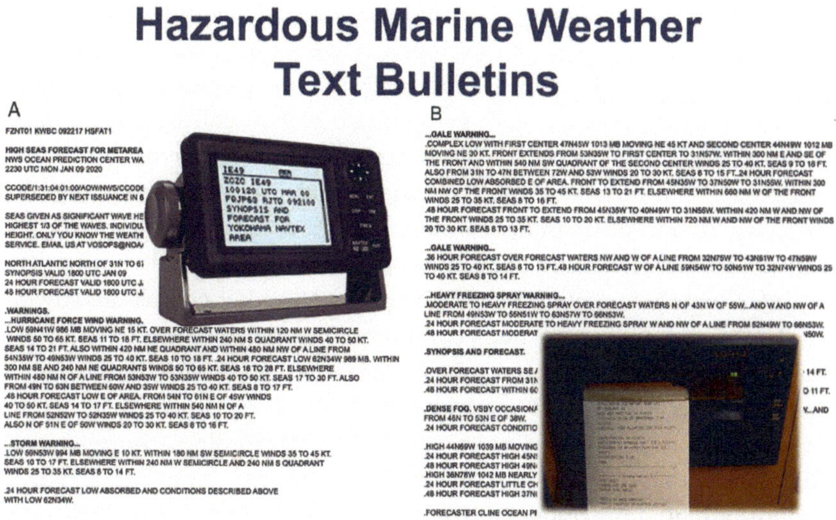

Figure 5.2 Images (A and B) depict a NAVTEX receiver, which is the current means of receiving marine weather warnings on many ships. Warnings are disseminated via text (such as aforementioned) with the coordinates of an approaching hazard, which can take time to plot versus the ship's position. *(Image from NWS-OPC.)*

Every vessel in the maritime arena, small or large, is exposed and vulnerable to the elements. Unlike our land-based scenario, vessels have nowhere to shelte from hazardous weather on the high seas. Vessels are relatively slow, usually far slower than the storms closing in on them. Additionally, these storms can span hundreds and even a thousand miles over the open ocean. Consequently, vessels must receive warnings and take action early—often days in advance—to avoid undesirable conditions. Although improved forecast accuracy and timely delivery have resulted in a decrease in the number of lives and property lost at sea, accidents at sea still primarily result from severe and extreme maritime weather. When these events occur, the consequences are catastrophic to the crew, cargo, and vessel and can also cause costly environmental disasters.

World meteorological organization

The World Meteorological Organization (WMO) is a specialized agency of the United Nations (UN) that serves as the UN's authoritative source of information on weather, water, and climate. Its 193 Member States and Territories, divided into six regional associations, are responsible for coordinating meteorological, hydrological, and related activities within their region. WMO assists National Meteorological and Hydrological Services (NMHSs), especially those of developing and least developed countries, in their efforts to contribute in the most effective manner to national development and to become full partners in global collaborative efforts. Consequently, contracting governments provide weather data, analyses, warnings, and forecasts of weather, waves, and ice, which benefit global efforts to, among other things, protect life and property and promote the long-term observation and collection of weather, water, and climate data.

Extending these services to include marine hazards is complicated by the fact that many such hazards transcend national boundaries into international waters, where previously no single entity was tasked with providing forecasts and warnings. To address this issue, the Joint WMO/IOC Technical Commission for Oceanography and Meteorology coordinates 21 METAREAS (Fig. 5.3), or geographically designated marine areas, where maritime safety information (MSI), forecasts, and warnings are provided for mariners through the IMO's Global Maritime Distress Safety System (GMDSS). These requirements are codified by amendments to the International Convention for the Safety of Life at Sea (SOLAS), the origin of which dates back to the 1912 Titanic disaster.

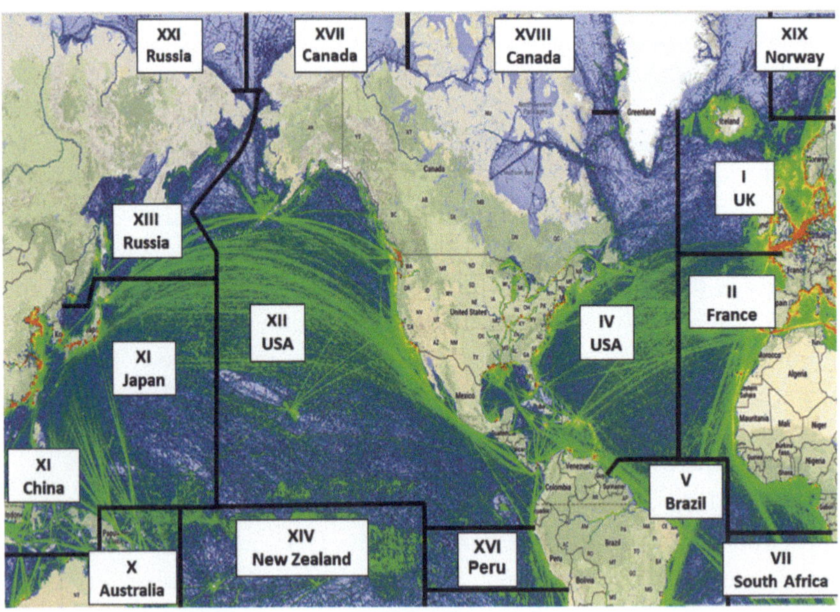

Figure 5.3 Geographic designations of Meteorological Areas (METAREAs) for the Pacific and Atlantic Oceans overlaid on vessel tracks. *(Image from NWS-OPC; background courtesy of MarineTraffic.com.)*

The national weather service—the US' NMHS provider

For the United States, under the National Oceanic and Atmospheric Administration (NOAA), the National Weather Service's (NWS) National Hurricane Center (NHC) and Ocean Prediction Center (OPC) are part of a global network of NMHSs providing support to the IMO's GMDSS by transmitting hazardous weather warnings for METAREAS IV and XII. The NHC also serves as one of 12 Regional Specialized Meteorology Centers (RSMC) and Tropical Cyclone Warning Centers (TCWC) participating in the WMO Tropical Cyclone Program, which will not be addressed specifically here.

While most people are aware of the threat tropical cyclones pose to coastal communities and mariners, few are aware of the threats "extra"-tropical cyclones pose. Each year in the north Atlantic and Pacific oceans alone, over 80 hurricane-force-wind-generating extratropical storms typically occur outside of the tropics and generally during the winter months (NOAA press release, 2020a). High winds, mountainous waves (steepened if there are opposing strong currents), dense fog, severe thunderstorms, sea ice, and freezing spray all conspire to make marine transportation in extratropical latitudes a very high-risk enterprise. Nonetheless, the ocean and seas provide a sustainable transportation route for the global economy—not only in terms of goods and natural resources, but also for the approximately 1.2 million seafarers and 26 million cruise ship passengers each year.

Using a suite of satellite and in situ observations, along with weather and ocean models, teams of meteorologists (Fig. 5.4) collectively apply

Figure 5.4 Professional marine meteorologists such as the Ocean Prediction Center's Frank Musonda examine a marine weather forecast. *(Image from NWS-OPC.)*

hundreds of years of cumulative specialized experience in the maritime environment to provide the best marine forecasts and critical decision-support services for mariners. These products include current and forecast conditions related to wind, sea state, temperature, and navigational hazards such as sea ice, freezing spray, fog, thunderstorms, large waves, and high winds. These services ensure the safety of the extensive commercial and recreational fishing, boating, and shipping activities in offshore waters. The current product suite offers surface analyses and graphical forecasts of upper air, wind, and wave conditions out to 4 days. The surface forecasts with hazards can be offered out to 5 days in advance and work is underway to extend this lead time out to 10 days to give mariners a wider time window to track the trending intensity of conditions and make routing decisions to avoid or mitigate the maritime hazards well in advance.

Blue data—observations

"Red sky at night, sailor's delight. Red sky in morning, sailor's warning." At the heart of this ancient weather proverb is the fundamental principle that marine weather observations underpin the protection of life and property at sea. But whereas ancient mariners could only apply safe navigation principles using observations within their line of sight, now more than ever observations from many sources can be collected to inform not only mariners' decisions in the moment.

Observations play a major role in forecaster situational awareness of the marine environment, accuracy of Numerical Weather Prediction (NWP) models that forecasters rely upon to make forecasts and issue warnings, along with understanding of the role of our oceans in controlling climate. The resulting weather models that drive marine forecasts not only give mariners insights well beyond the horizon in both space and time, they also allow them to consider oncoming weather in developing voyage plans. Just as a robust marine transportation system is crucial to a vibrant marine economy, a robust marine observation network is an essential component of safe marine transportation and, therefore, to a burgeoning new blue economy.

Ship observations

The oldest recorded marine meteorological observations are from ships at sea that date back as far as 1853, where such observations were utilized to inform decisions regarding the vessel safety as well as climatological monitoring purposes (e.g., Maury, 1854; Mallory, 1855; IMO, 2018). Today, that legacy not only continues to enable life and cargo saving meteorological services for mariners, but has also preconditioned a whole network of connectivity and data that will continue to propel the new blue economy. The importance of ship observations also serves as an impetus for international cooperation in collecting and sharing ship observations.

The Safety of Life at Sea Convention, Regulation 5, "Meteorological Forecasts and Warnings," specifies provisions whereby contracting governments are encouraged to arrange for a selection of ships to be equipped with tested marine meteorological instruments and to take, record, and transmit meteorological observations at the main standard times for surface synoptic observations (IMO, 2002). Regulation 5 also requests that participating governments encourage other ships to record and transmit modified observations in areas where data can be difficult to collect (Smith et al., 2019). This also serves to encourage governments to utilize these observations in issuing warnings of gales, storms, hurricane-force wind events, and tropical cyclones.

The Voluntary Observing Ships' (VOS) scheme (VOS, 2020), a panel of the Ship Observations Team jointly sponsored by the WMO and the Intergovernmental Oceanographic Commission (IOC, see Ryabinin, this volume) through the Global Ocean Observing System (GOOS), provides the governance by which ships are recruited by NMHSs for making and transmitting meteorological observations (Kent et al., 2010). The contribution that VOS reports make to operational meteorology, to marine meteorological services, and to global climate studies is unique and irreplaceable (VOS). The increasing recognition of the role of the oceans in the global economy and climate system has placed even greater emphasis on the importance of marine meteorological and oceanographic observing systems.

One of the major contributions of the VOS scheme is that it provides data in vast areas of the world's oceans that are also data sparse. VOS meteorological reports are a unique and invaluable contribution to operational meteorology and marine meteorological services (Fletcher, 2008), filling critical data gaps and providing support for basic weather forecasting, the provision of marine and meteorological and oceanographic services, and climate analysis and research. While the new generations of meteorological satellites also help to overcome these problems, ship observations provide ground truth for the satellite observations, important information that the satellites cannot observe. Moreover, these observations are also essential contributions to the data input for the NWP and provide real-time reports that can be used immediately in services specifically for the mariner in the form of forecasts and warnings that are transmitted through the GMDSS.

Several modern examples exist of ship observations having assisted in providing timely meteorological data for improved forecasting and, consequently, immediate decision-making at sea. This is especially true for high-impact events (e.g., Stern and Davidson, 2015). As noted by forecasters at the NWS's OPC, and published by Smith et al. (2019), a critical ship observation was instrumental in enabling the upgrade of a high seas forecast to a hurricane-force wind warning on July 29, 2015 (Fig. 5.5). Many other examples of the value of observations occur on a daily basis in weather that is not nearly as extreme. For example, on December 30, 2019, a timely ship report with winds of 37 knots and seas to 9.8 feet validated satellite observations received around the same time. On July 8, 2020, ship observations that included fog in the Pacific Ocean helped forecasters add confidence to forecasts and to determine the type of model guidance to use for later forecast periods. Later, on August 28, 2020, a pair of ship observations helped to validate expected conditions over the western Atlantic, although both observations were below the thresholds for any wind hazards (Figurskey, personal correspondence). Also of importance is that ship observations are essential components of the WMO's Global Climate

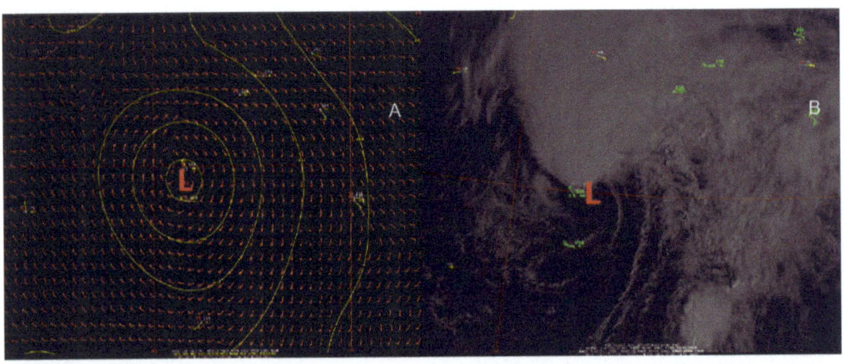

Figure 5.5 (A) GFS model July 29, 2015 0600 UTC 6-hour forecast, valid 1200 UTC July 29, 2015, with available ship observations using standard synoptic station notation. Isobars are drawn with yellow lines and the model wind field is indicated by wind barbs in orange. The low pressure center is located near 39°N, 58.5°W. (B) Visible satellite image with available ship observations (overlain in green and yellow using standard synoptic station notation) in the central Atlantic Ocean (1200 UTC, July 29, 2015). Based on this image, the low pressure is centered near 40°N, 58°W. *(Reprinted with permission, Frontiers in Science.)*

Observing System (GCOS), used by regional and global climate analyses, as well as the growing field of operational oceanography with regard to predictions and a better understanding of ocean parameters including temperature, salinity, and currents. While NWP skills are certainly improving (Bauer et al., 2015), human guidance still adds value to daily forecasts, watches, and warnings.

Other important observations from ships include those from the Ship Of Opportunity Program (SOOP) and the Automated Shipboard Aerological Program, ASAP (WMO, 2020). There are more than 1200 vessels globally participating in SOOP that provide real-time data and millions of measurements every year of things such as ocean temperature, salinity, pressure, and carbon dioxide (SOOP). These observations help to estimate the heat contained in the upper ocean and the surface currents, which drive the sea surface temperatures, a critical ocean variable determining the locations of high and low atmospheric pressure systems. They also initialize weather and climate forecast model results, validate satellite observations, and provide critical information to determine ocean acidification assessments. The ASAP involves the generation of upper-air profile data from data sparse ocean areas using automated sounding systems carried on board merchant ships. Most of the soundings are presently from the north Atlantic and northwest Pacific oceans. The soundings are important in understanding more about the upper levels of the atmosphere for forecasting and research purposes (WMO, Automated Shipboard Aerological Program).

Finally, ship observations have set an early example for open-access data: WMO (1995) Resolution 40 Annex 1 lists marine data among "data and products to be exchanged without charge and with no conditions on use." For near-real-time applications, the data are delivered to users via the WMO's Global Telecommunication System (GTS). The GTS is a global data management network that facilitates and coordinates the collection, processing, and dissemination of observations from weather stations, satellites, and numerical weather prediction centers (WMO). Through this, the GTS allows the World Weather Watch (WWW) to collect and distribute critical information in a timely, reliable, and cost-effective way, ensuring full access and the real-time exchange of information critical for forecasting and warnings of hydrometeorological hazards in accordance with approved procedures.

Data buoys, satellites, and uncrewed vessels

Data buoys, such as ship observations, are another significant component of the WMO's GOOS Data Buoy Cooperation Panel (WMO-IOC, 2020).

Buoys are reliable sources of near and open ocean meteorological data. Among other parameters, depending on the instrumentation available, buoys can provide data on winds, waves, air pressure, air and water temperature, dew point, swell, and salinity. Buoys can also detect changes in sea level out in the deep ocean, which is helpful for determining tsunamis. Data buoys measure and transmit automatically in a predictable and controlled way, communicating in real time via satellite telecommunication systems. Data buoy observations make significant contributions to our ability to model, understand, and describe global weather and climate on all spatial and temporal scales. The data collected complements and validates data from other platforms, such as from VOS and weather satellites.

There are two primary types of buoys, moored and drifting. Moored buoys are anchored at fixed locations and regularly collect observations from many different atmospheric and oceanographic sensors attached to the buoy. These buoys can vary from a few meters in height and breadth to over 12 m and are usually deployed to serve national forecasting needs, maritime safety needs, or to observe regional climate patterns. Typically, they are relatively large and expensive platforms, usually upgraded and/or serviced annually. Drifting buoys, also known as "drifters," are generally attached to some form of drogue or sea anchor. They are easy to deploy, relatively inexpensive to operate, and reliably measure the atmosphere and ocean surface conditions for an average of 18 months. Drifting buoys have a long history of use in oceanography, principally for the measurement of currents; however, they can also provide some meteorological data, particularly air pressure and, in some cases, wind speed and direction.

Satellite-based sensors are another valuable method of measuring and recording ocean surface wind speed and wind direction, along with wave height and sea-surface height anomaly data. These data are typically taken by polar-orbiting satellites and gather information in "strips" as they pass continuously over the Earth (Fig. 5.6). One series of polar-orbiting satellites, "Jason," flies in a 9.9-day repeat track orbit, meaning the satellite will make observations over the same ocean point once every 9.9 days. These data are important for forecast and warning purposes, as well as to understand more about climate change. Satellites to some degree also help fill in the immense gaps in the ocean observations where there are no ships or buoy data. The Jason series of satellites is planned to continue flying through the next decade with newer sensors and even more precise measurements of sea level.

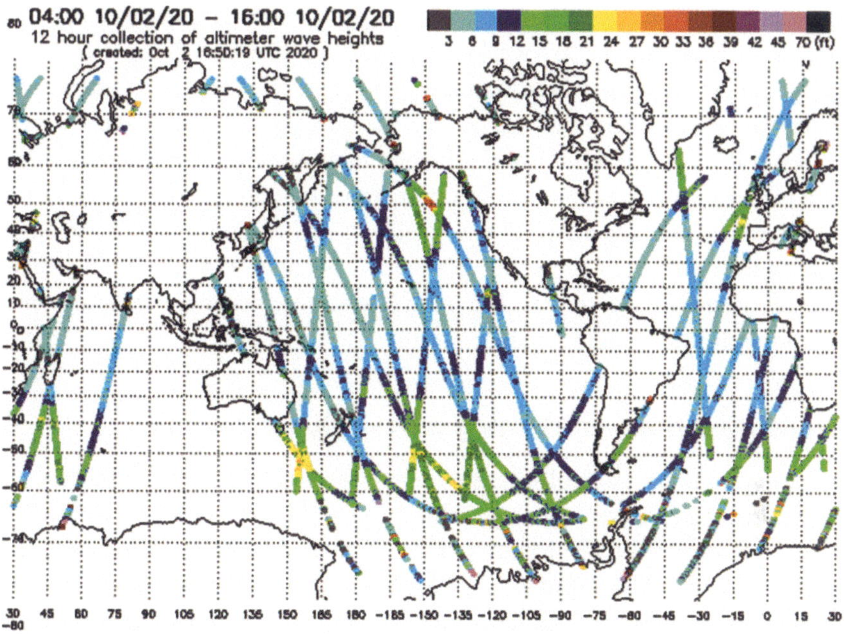

Figure 5.6 Wave heights as measured by the Sentinel-6/Jason Continuity of Service (CS) Mission. *(Image from NOAA.)*

Finally, newer technologies such as uncrewed marine vehicles (Fig. 5.7) can prove helpful in providing key data for models and marine forecasters in areas where ships and buoys do not usually exist and where satellite data are intermittent. In the case of ocean gliders, measurements of the heat content of the upper layers of the ocean can help intensity forecasting for tropical cyclones. With measurements of ocean temperature and salinity, gliders in

Figure 5.7 A NOAA employee from the Office and Marine and Aviation Operations pilots an uncrewed ocean surface vehicle. *(Image from NOAA.)*

the right location can help forecasters better identify areas where tropical systems can strengthen or weaken. In 2020, NOAA and the US Navy signed a new agreement to jointly expand the development and operations of uncrewed maritime systems in the nation's coastal and world's ocean waters (NOAA, 2020b). By strengthening this ongoing partnership, NOAA will be better positioned to transition uncrewed maritime technologies into autonomous operational platforms that will gather critical environmental data, which will help grow the new blue economy and specifically sense and report real-time critical data at the ocean—atmosphere interface that drives hazardous marine weather.

Transformation—sustaining a new blue economy

As the volume of ocean observations grows, especially at sufficient temporal and spatial scales, so does the opportunity to apply these data to a myriad of new blue economy applications. This kind of robust ocean observation network would immediately improve the skill of advanced earth system prediction capabilities and lead to a highly accurate four-dimensional database (i.e., in three-dimensional space plus time) for the first time over the ocean at the same robustness that it exists over land today. While NMHS centers will certainly improve their basic safety of life and property at sea services with such an advanced four-dimensional database, they will not have the capacity to develop and deliver services that meet every nuanced need in an expanding global ocean economy projected to grow to $3 trillion by 2030 (Nicolls et al., 2020). Instead, an unprecedented opportunity will exist for a new and booming maritime support industry that turns these data into numerous and nuanced on-demand decision-making applications for the maritime environment to ensure safety and enhance commerce, as well as to provide a better understanding and monitoring of our changing climate—highlighting new potential in the era of the new blue economy.

Ship avoidance practices

As an "ocean of things" concept grows, a corresponding "maritime WiFi" will be needed to move data quickly back to marine forecasters and prediction systems to capitalize on the data in real time for subsequent watches and warnings. For marine surface or near surface observations, one such robust and reliable communications path already operating today is the Automated Identification System (AIS). Primarily designed for vessel collision avoidance, AIS may prove useful as a proxy "Marine Internet" in being able to collect and transmit weather and limited oceanographic information from more ships around the world. AIS may also be a two-way means by which to send marine hazard watches and warnings, especially as tropical cyclone track forecasts continue to improve and marine winds and seas forecasts associated with storm or hurricane force extratropical low-pressure systems improve as well. With adequate bandwidth and lower cost communication means such as AIS, the potential exists for industry to advance more sophisticated automated vessel routing services for not only hazard avoidance but also for maximum fuel efficiency as discussed further in a following section.

The NHC and OPC sometimes use graphics such as these to demonstrate the effect of its work in protecting life and property at sea. In September of 2018, Florence, a Category 4 hurricane, bore down on the Carolina coast. Its path toward landfall—an area of 165,000 square nautical miles, shaded in red—was completely void of vessel traffic (Fig. 5.8A). Similarly, in August of 2020, Hurricane Isaias, a Category 1 hurricane, was paralleling the Florida coast as it headed toward the Carolinas. Its path toward landfall is also void of vessel traffic (Fig. 5.8B). In both cases, the ability of mariners to clear the area and travel to safer waters was due in part to the OPC's and the NHC's ability to communicate up-to-date and trending weather and sea-state information to mariners at sea.

Marine and ice hazards integration

Increased maritime traffic is expected in the Arctic in response to climate change and decreased summertime sea ice coverage (NOAA's Arctic Action Plan, 2014; see Gallaudet et al., this volume). Unlike relatively short-lived extreme weather events, sea ice poses a constant and oftentimes hidden threat. In anticipation of this change, the NOAA component of the US National Ice Center (USNIC) was realigned from the National Environmental Satellite, Data, and Information Service (NESDIS) to be collectively managed through the NWS's OPC, the US Coast Guard, and US Navy in May of 2020. This realignment better positions the USNIC to achieve its future goals, which include a greater focus on ice prediction capabilities. It also enables OPC, in partnership with NWS's Alaska Region and Arctic Sea Ice Program, as well as with the Canadian Ice Services and other international partners, to lead to an integrated full-spectrum approach for marine weather and ice prediction and impact-based decision-support services across the Arctic and Antarctic, the Great Lakes, and the

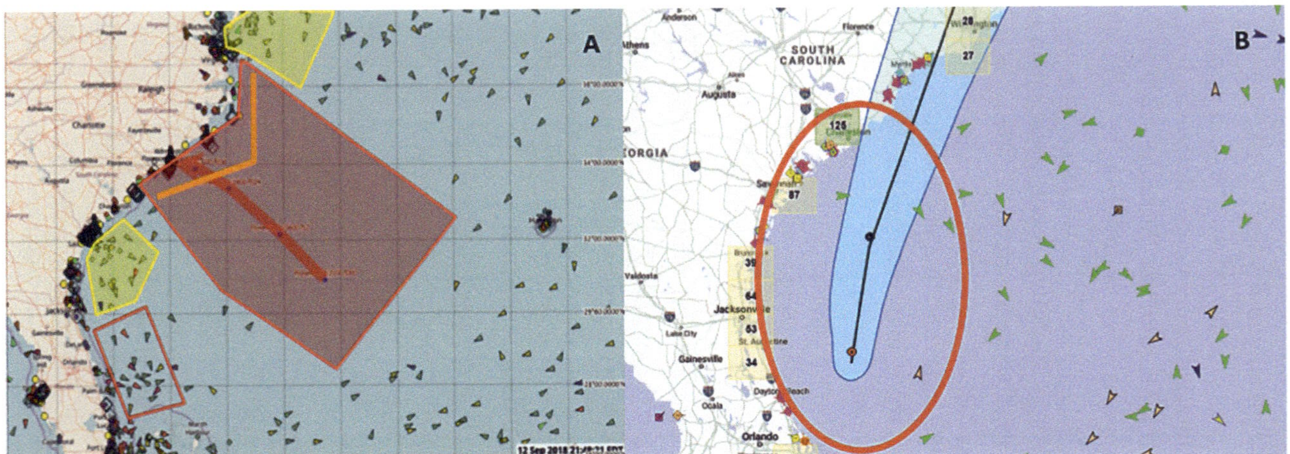

Figure 5.8 (A) Vessel heavy weather avoidance during Hurricane Florence (September 2018). Yellow polygons indicate vessels anchored or underway at very slow speed, and red polygons indicate sortied ships. (B) Vessel heavy weather avoidance during Hurricane Isaias (August 2020). *(Image from NWS-OPC.)*

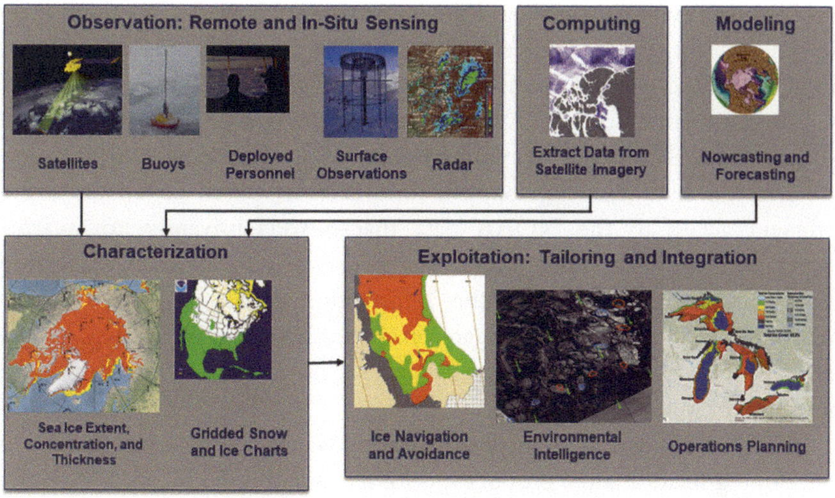

Figure 5.9 Ice analysis and forecast process at USNIC. *(Courtesy of the USNIC.)*

Chesapeake and Delaware Bays. In the end, the mariner transiting through ice or ice-infested waters would have a wholly consistent set of marine weather and sea ice guidance. Further, the integration will lead to more efficient use of sea ice information in developmental and operational earth system prediction models. Fig. 5.9 illustrates USNIC's current suite of ice products and services. All in all, if successful, this strategy will effectively mitigate the increased risks to shipping and the highly sensitive operating environment of the polar oceans and enhance the protection of life and property at sea, expanding and strengthening the nation's new blue economy in Polar Regions and the Great Lakes.

X-files

Perhaps the most revolutionary technological advancement that would immediately modernize the international maritime community is the development of marine hazards polygon and gridded data sets for display on shipboard electronic navigation (e-Nav). Nearly all vessels use e-NAV to plan and execute their navigation routes. These displays are usually front and center on the vessel's bridge and frequently consulted several times throughout a crew's watch for situational awareness. Automated notifications are already built into some e-Nav applications for navigation hazards (e.g., shoal waters and closing vessels) and could be easily modified to alert for extreme maritime weather hazards. The integration of polygons and gridded data with e-Nav would prevent mariners from finding themselves in the same situation as our earlier example of a land-based cross-country road trip without timely weather information and warnings.

This initiative at OPC falls within the larger scope of the National Ocean Services' (NOS) Precision Marine Navigation project, which aims to create a new one-stop shop online services portal to enable more efficient access to the NOAA data that powers private-sector marine navigation products. The goal of this initiative is to improve navigation safety by readily presenting the information mariners need to make informed decisions as they navigate our nation's waterways, enabling more efficient coastal route planning by leveraging the latest innovations in integrated shipboard systems.

The components of the Precision Marine Navigation project that OPC is developing include three product specifications, collectively called "S-41X" (Fig. 5.10), that are coordinated through the WMO's Standing Committee on Marine Meteorological and Oceanographic Services (SC-MMO). S-41X is being developed within the framework specified in the International Hydrographic Organization (IHO) S-100 Universal Hydrographic Data Model and the International Standards Organization (ISO) 19100 series of standards. These product specifications will outline the content, structure, and metadata needed for encoding maritime weather and sea condition analyses and forecasts for situational awareness, route planning, and hazard

S-412	S-413	S-414	S-411
Wave and Weather Hazards	**Wave and Weather Conditions (features e.g. fronts)**	**Wave and Weather Observations**	**Ice and Iceberg Hazards**
Polygons	**Graphics & Gridded Data**	**Point Based Data**	**Polygons, Graphics & Gridded DataPoint, Based Data**

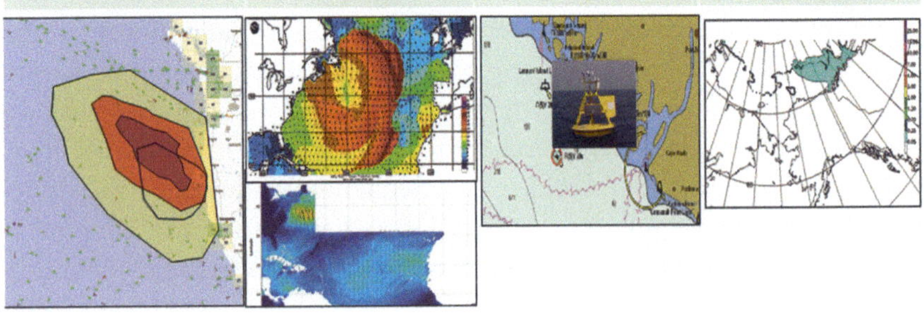

Figure 5.10 S-100 formatted data layers and grids under development. *(Image from NWS-OPC.)*

mitigation decisions. Additionally, these products will be fully compatible as overlays on future electronic navigation systems including Electronic Chart Display and Information Systems (ECDIS). S-41X contains the following: S-412: Weather and Wave Hazards; S-413: Weather and Wave Conditions; and S-414: Weather and Wave Observations.

Separately, the International Ice Charting Working Group (IICWG), formed in October 1999, is a working group of the world's national ice services who collaboratively promote coordination of operational sea ice and iceberg information services to better meet the needs of their national and international maritime clients. Through this group, S-411 ice and iceberg hazards are being examined for operational implementation among members along with other marine hazards in polar waters. The OPC, with its formal ties to the US National Ice Center (USNIC), is uniquely positioned to bring the full spectrum of integrated marine and ice warning services right onto the mariner's navigation displays, the equivalent of a "Heads Up Display." Like big data observations and model data, these baseline products also present new potential for tailored navigational and marine/ice data services in the era of the new blue economy.

Global rapid refresh: supercomputing (supersized and scaled)

Marine meteorology and oceanography models continue to advance in skill through novel transitions of research into operations. For example, early in 2020, the NWS's Environmental Modeling Center (EMC) implemented an improved data assimilation system developed by NOAA's Geophysical Fluid Dynamics Laboratory (GFDL) called the Finite-Volume Cubed-Sphere Dynamical Core (FV3) (Fig. 5.11), which is a scalable and flexible dynamical core capable of both hydrostatic and nonhydrostatic atmospheric simulations. By 2021, EMC will have given marine forecasters at OPC and NHC a global atmospheric model that is fully coupled to a global wave model at 21 km resolution or better, including ensembles, the Global Ensemble Forecast System version 12.0. Meanwhile, NWS is forging ahead with a Unified Forecast System (UFS) that sets the stage to ultimately deliver its vision of even higher-resolution earth system prediction capabilities that generate fully coupled air-land-sea-ice forecasts from hours out into the seasonal-to-subseasonal (S2S) timescales.

While computing resources are constrained today when faced with the aforementioned computing challenge, it is only a matter of time before innovation makes it feasible. For example, many innovative initiatives are

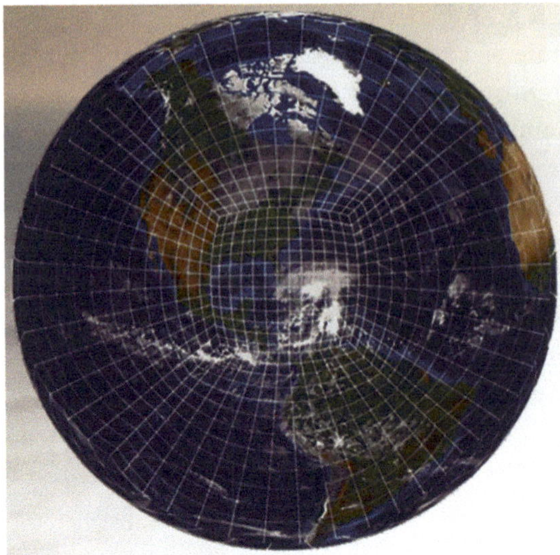

Figure 5.11 The finite-volume cubed-sphere dynamical core (FV3). *(Courtesy of Geophysical Fluid Dynamics Laboratory (GFDL).)*

underway now to improve supercomputer performance through Graphical Processing Unit (GPU) accelerators, as well as to creatively partner with artificial intelligence and cloud storage and computing. In a not-too-distant future, it is probable that a marine forecaster might have at his or her disposal data from a global, high-precision (~3 km), rapidly updating (hourly), and fully coupled land-air-sea-ice prediction system.

From a marine forecasting perspective, the possibilities of application are numerous and varied. For example, global rapid refreshes of larger domains where deep convection is simulated could be used in conjunction with real-time satellite data to generate warning polygons for large complexes of thunderstorms 3–6 h into the future. In other words, the equivalent of today's Severe Thunderstorm Watches and Warnings on land could be extended out into offshore waters and on the high seas! Now imagine delivering that warning polygon into a vessel's navigation system and the crew either adjusting course and speed to avoid the hazard all together or taking early action to secure cargo and crew and maneuver for best ride through the hazard. This kind of capability would also have utility for oil rigs and static energy platform providers.

Safe and efficient vessel routing for a safe and sustainable future

As the global economy continues to grow and marine traffic increases in response, so does concern surrounding the potential for adverse effects on global climate and marine environments. Of these potential environmental impacts, carbon emissions and contamination in the event of an accident are of immediate concern. As previously noted, nearly all ships develop voyage plans prior to leaving port and adapt them as needed in response to marine weather and navigational hazards. Consequently, concerns over the environmental impact of the shipping industry open an opportunity for both impact-based forecasts and value-added meteorological services (e.g., onboard and shore-based route optimization and decision-support services).

Many immediately understand the need for marine weather forecasts to increase the safety of life and property at sea. However, few consider that marine forecasts can result in greater operational efficiency that leads to reduced emissions. For example, route optimization and decision-support services can assist captains in avoiding hazardous weather conditions that require more fuel to navigate through. This also allows masters and captains to better balance safe weather routing and maximum efficiency. To date operational efficiency is the best means from emissions reduction until shipping companies are able to develop and implement low-carbon fleets.

Finally, in avoiding hazardous marine weather conditions, vessels not only increase the safety of life and property and fuel efficiency, they can avoid costly and devastating environmental disasters. Having access to both real-time conditions in ports and harbors and impact-based marine weather forecasts aids in the prevention of vessel loss. Moreover, in the event of an environmental disaster, accurate weather and ocean forecasting are requisite for modeling the temporal and spatial diffusion, evaporation, or evolution of surface, subarea, and atmospheric contaminants. These services are particularly useful in remote locations where aerial surveillance and satellite coverage of contaminant transport are limited.

Conclusion

As the marine economy expands, more and more vessels will traverse the oceans and a myriad of new industry platforms will go to sea, some permanently offshore. Consequently, exposure to hazardous marine conditions will continue to increase. However, increased exposure does not have to equate to increased risk of loss of life and property at sea. Instead, improvements in observation systems, earth prediction systems, forecasting tools, dissemination paths, and decision-support tools can greatly drive that risk down and support a growing and sustainable new blue economy. A 2019 international symposium hosted by the IMO and WMO on "Extreme Maritime Weather: Toward Safety of Life at Sea and a

Sustainable Blue Economy" (WMO-IMO, 2020) demonstrated great enthusiasm and support from governments and international agencies toward this effort. The report also demonstrated that service providers and stakeholders alike clearly recognize the economic value of early, accurate, and actionable warnings of extreme maritime weather when they lead to decisions that prevent catastrophic loss of life and property at sea and the costly environmental disasters that usually accompany them.

The global nature of extreme weather hazards, especially as our climate continues to change, will require enormous change and innovation. The global marine weather enterprise is well positioned to revolutionize the way stakeholders receive and integrate marine weather information into onboard and shore-based decision-making and the US' NWS is at the forefront of this effort. However, to do so will require investments into initiatives that (1) expand ocean observing programs; (2) advance the accuracy, resolution, and timescales of marine forecast systems; and (3) improve the dissemination and decision-support tools used in the maritime industry. The most successful and prosperous nations will be those who seize the opportunities of a new blue economy and invest in these initiatives and then create policies and incentives for industry and or public—private partnerships to capitalize on them. Further, they will build societies that are ready, responsive, and resilient, or "Weather-Ready," against the threat of marine weather hazards and their risk to life and property at sea, while also enhancing sustainable economic and environmental security through safe and efficient vessel routing and response to environment emergencies and coastal disasters.

References

Bauer, P., Thorpe, A., Brunet, G., 2015. The quiet revolution of numerical weather prediction. Nature 525, 47—55. https://doi.org/10.1038/nature14956.

Business Wire, 2019. Global Cargo Shipping Market Analysis, Trends, and Forecasts Report 2019—2025. Global Industry Analysts, Inc. Available from: https://www.businesswire.com/news/home/20191105005678/en/Global-Cargo-Shipping-Market-Analysis-Trends-Forecasts. (Accessed 1 October 2020).

Fletcher, J., April 2008. Meteorological Observations from Ships. Seaways, pp. 7—10. Available from: http://sot.jcommops.org/vos/documents/seaways-vos-200804.pdf.

International Maritime Organization (IMO), 2018. Participation in the WMO Voluntary Observing Ship Scheme, MSC.1 Circular 1293. Available from: http://sot.jcommops.org/vos/documents/msc_1293.pdf. (Accessed 1 October 2020).

International Maritime Organization (IMO), 2002. Safety of Life at Sea, Chapter V. Available from: http://www.imo.org/en/OurWork/facilitation/documents/solas%20v%20on%20safety%20of%20navigation.pdf. (Accessed 1 October 2020).

Kent, E.C., Ball, G., Berry, D.I., Fletcher, J., Hall, A., North, S., et al., 2010. The voluntary observing ship (VOS) scheme. In: Hall, J., Harrison, D.E., Stammer, D. (Eds.), Proceedings of the "OceanObs'09: Sustained Ocean Observations and Information for Society" Conference (Vol. 2), Venice, Italy, 21–25 September 2009. ESA Publication WPP-306. https://doi.org/10.5270/OceanObs09.cwp.48.

Mallory, 1855. Origin of the Wind and Current Chart and Sailing Directions. Report, 32 Congress, The Senate of the United States, Rep. Com. No 443, 29 January 1855. Available from: https://books.google.co.uk/books?id=WacFAAAAQAAJ&pg=PP7#v=onepage&q=Maury&f=false.

Maury, M.F., 1854. Maritime Conference held at Brussels for devising a uniform system of meteorological observations at sea, August and September, 1853. In: C., E., Biddle, J. (Eds.), Explanations and Sailing Directions to Accom- Pany the Wind and Current Charts, sixth ed., pp. 54–96 Philadelphia, PA.

Nicolls, W., Franks, C., Gilmore, T., Goulder, R., Mendelsohn, L., Morgan, E., Adkins, J., Grasso, M., Quigley, K., Zhuang, J., Colgan, C., 2020. Defining and Measuring the U.S. Ocean Economy. https://www.bea.gov/system/files/2020-06/defining-and-measuring-the-united-states-ocean-economy.pdf.

NOAA, 2014. NOAA's Arctic Action Plan – Supporting the National Strategy for the Arctic Region. U.S. Department of Commerce, National Oceanic Atmospheric Administration, Silver Spring, Md.

NOAA, 2020a. 2019/20 North Atlantic Hurricane Force Wind Events: Highlighting the North Atlantic Hurricane Force (HF) Wind Event Season from June 1, 2019 through May 31, 2020. Department of Commerce, National Oceanic and Atmospheric Administration, Silver Spring, MD. Available from: https://storymaps.arcgis.com/stories/83038d799d7249b68e81be1aa7ff84fd. (Accessed 1 October 2020).

NOAA, 2020b. NOAA, US Navy Will Increase Nation's Unmanned Maritime Systems Operations. Department of Commerce, National Oceanic and Atmospheric Administration, Silver Spring, MD. Available from: https://www.noaa.gov/media-release/noaa-us-navy-will-increase-nation-s-unmanned-maritime-systems-operations. (Accessed 1 October 2020).

Smith, S., Alory, G., Andersson, A., Asher, W., Baker, A., Berry, D., Drushka, K., Figurskey, D., Freeman, E., Holthus, P., Jickells, T., Kleta, H., Kent, E., Kolodziejczyk, N., Kramp, M., Loh, Z., Poli, P., Schuster, U., Steventon, E., Swart, S., Tarasova, O., de la Villéon, L., Vinogradova-Shiffer, N., 2019. Ship-based contributions to global ocean, weather, and climate observing systems. Front. Mar. Sci. 6 https://doi.org/10.3389/fmars.2019.00434. https://www.frontiersin.org/article/10.3389/fmars.2019.00434.

SOOP. The Ship of Opportunity Program, National Oceanic and Atmospheric Administration Atlantic Oceanographic and Meteorological Laboratory Miami, Florida, USA. Available from: https://www.aoml.noaa.gov/phod/goos/docs/SOOP_brochure_recent.pdf. (Accessed 3 October 2020).

Stern, H., Davidson, N.E., 2015. Trends in the skill of weather prediction at lead times of 1–14 days. Q. J. R. Meteorol. Soc. 141 https://doi.org/10.1002/qj.2559.

Tournadre, J., 2014. Anthropogenic pressure on the open ocean: the growth of ship traffic revealed by altimeter data analysis. Geophys. Res. Lett. 41, 7924–7932. https://doi.org/10.1002/2014GL061786.

U.S. Department of Transportation, 2020. Bureau of Transportation Statistics, Port Performance Freight, Statistics in 2018, Annual Report to Congress 2019. https://doi.org/10.21949/150 4598. Washington, DC.

VOS, 2020. United States Voluntary Observing Ship Program. Available from: https://www.vos.noaa.gov/. (Accessed 3 October 2020).

WMO, 1995. Resolution 40 (Cg-XII): WMO policy and practice for the exchange of meteorological and related data and products including guidelines on relationships in commercial activities. In: Twelfth World Meteorological Congress, 30 May—21 June, Abridged Final Report with Resolutions, pp. 125—130. https://library.wmo.int/pmb_ged/wmo_827_en.pdf.

WMO, 2020. Automated Shipboard Aerological Program. Available from: https://www.wmo.int/pages/prog/amp/mmop/JCOMM/OPA/SOT/asap.html. (Accessed 1 October 2020).

WMO-IMO, 2020. Report of the WMO/IMO International Symposium 'Extreme Maritime Weather: Towards Safety of Life at Sea and a Sustainable Blue Economy'. World Meteorological Organization, Geneva, Switzerland. Available from: https://library.wmo.int/doc_num.php?explnum_id=10305. (Accessed 1 October 2020).

WMO-IOC Data Buoy Cooperation Panel (DBCP), 2020. Data Buoy Types. Available from: http://www.jcommops.org/dbcp/platforms/types.html. (Accessed 1 October 2020).

SECTION II

Defining the new blue economy

The premise of the New Blue Economy is defined and described by **Richard Spinrad**, including case studies of emerging markets. The concept is built around data, information, and knowledge of the oceans being applied to targeted products and services for many of those components of the long-standing blue economy (e.g., oil and gas, shipping, etc.) and other applications. **Ralph Rayner** presents an assessment of the ocean enterprise including the market analyses that are essential to future economic development. **Judith Kildow's** op-ed is an important perspective on definitions and interpretations of what exactly are the ocean economy, the blue economy, the coastal economy, and the New Blue Economy.

CHAPTER 6

The new blue economy

Richard W. Spinrad
Oregon State University, Corvallis, Oregon, United States

Evolution of the blue economy to date (how has the economic exploitation of the sea changed over time?)

Summary descriptions of several of the major components of the blue economy are included here. More detailed discussions are included in the following chapters.

Material resource extraction—protein

For millennia, humankind has recognized the value of food resources coming from the sea. From age-old traditional or artisanal fishing to modern commercial efforts, extraction of protein (mostly fish at the highest trophic levels) has been a well-recognized example of economic exploitation of the sea. In 2016, the global total capture fisheries production was 81.2 million metric tons, and this extraction has grown from approximately 20 million metric tons to the current levels in 65 years (http://www.fao.org/state-of-fisheries-aquaculture). While significant efforts are underway to ensure sustainable extraction of healthy fish stocks (e.g., https://www.fisheries.noaa.gov/topic/sustainable-fisheries), technology can provide increased efficiencies resulting in overall increases in catch per unit effort in some fisheries (Maunder et al., 2006). Global fish exports were valued at $152B in 2017 (http://www.fao.org/3/i9540en/i9540en.pdf).

Notably, the demand for ocean-based protein has stimulated increased investment and general activity around aquaculture—onshore, nearshore, and offshore. In the United States alone, most (~90%) of the seafood consumed is imported, and over half of that portion comes from aquaculture, with leading providers being shellfish aquaculture producers from Asia (https://www.fisheries.noaa.gov/national/aquaculture/us-aquaculture).

Clearly, protein extraction from traditional fisheries and next-generation aquaculture activities will continue to be a major contribution to the blue economy. The UN Food and Agriculture Organization (FAO) expects aquaculture production to grow by nearly 40% between 2016 and 2030 (http://www.fao.org/3/i9540en/i9540en.pdf).

Preparing a Workforce for the New Blue Economy
ISBN 978-0-12-821431-2
https://doi.org/10.1016/B978-0-12-821431-2.00042-1

Material resource extraction—energy

Another notable extractive industry is associated with the production of energy. Oil and gas exploration in the sea started in 1896 off California, with the discovery of relatively shallow water reserves and the engineering developments that allowed placement of extracting platforms on site. Currently, more than 50 nations engage in offshore oil and gas extraction from their Exclusive Economic Zones, EEZs (https://www.eia.gov/todayinenergy/detail.php?id = 28492), and annual global production from these energy sources is estimated at 10 billion barrels (https://www.americangeosciences.org/geoscience-currents/offshore-oil-and-gas). The estimated global workforce, employed within the offshore oil and gas industry, is approximately one million workers in 2019, with an annual labor market estimated at over $60B (https://www.beroeinc.com/category-intelligence/offshore-labor-market/). In recent years, new technologies for operations in hostile environments (esp. the Arctic), deep-water and directional drilling, robotic platform support (autonomous underwater vehicles for pipeline inspection and maintenance), and exploration (more sophisticated acoustic technologies) have expanded the potential workforce (in balance, obviously with global oil demand), and as a result, this component of the blue economy has continued to be an important and evolving sector.

Just as aquaculture has diversified the component of the blue economy originally defined by fisheries, the expansion of the market for marine renewable energy sources has broadened consideration of offshore energy. As commercial offshore wind energy farms expand, as well as early developments for tidal, wave, and thermal, the energy sector will undoubtedly continue to grow as an economic sector of the blue economy (see, for example, https://www.energy.gov/eere/water/marine-and-hydro kinetic-energy-research-development).

Material resource extraction—other minerals and compounds

Sand and gravel are key components for a number of industries—especially construction and coastal protection/restoration—and are major marine extractable resources. The commercial mining of gravel, sand, and related minerals (Baker et al., 2014) is generally restricted to shallow waters within national EEZs. In 2016, the revenues from marine-derived limestone, sand, and gravel were estimated at $1.6B in the United States (https://coast.noaa.gov/data/digitalcoast/pdf/econ-report.pdf).

In the 1960 and 1970s, a significant amount of attention was focused on the potential for extracting minerals from deep ocean deposits (Glasby, 2000). This interest was largely spurred by discoveries of vast fields of manganese nodules sitting proudly on the deep abyssal plains of the Clarion Clipperton Zone in the North Pacific Ocean (International Seabed Authority, 2010). For a variety of reasons—especially technological challenges and the high operational costs—the efforts to build an economy around deep ocean mining were diminished and remained dormant for roughly the last half century.

However, recent increased demands for rare earth materials (critical in the manufacture of many modern electronic devices), coupled with geopolitical challenges of limited land-based sources (80% of global supplies of rare earths were extracted from mines in China in 2017, https://prd-wret.s3-us-west-2.amazonaws.com/assets/palladium/production/atoms/files/mcs-2019-raree.pdf), have stimulated renewed interest in extracting minerals from the deep ocean. With the technological developments of systems that can operate at "full ocean depth" (i.e., typically about 6500 m, or pressures of about 10,000 psi) there is now a meaningful economic sector emerging around the market for deep ocean minerals (see, for example Miller et al., 2018).

Material resource extraction—bioprospecting

There is an established, albeit not well-known economic sector defined by the unique compounds and materials that derive from marine organisms. These materials may serve a variety of important applications: medical therapies, cosmetics, food industries, household products, etc. (Wynberg, 2015). In 2016, the global market just for marine-derived drugs was estimated to be $8.6B (http://www.bccresearch.com/pressroom/phm/global-market-marine-derived-drugs-reach-nearly-$8.6-billion-2016). The field of marine bioprospecting—that is, the extraction of materials from marine organisms for direct applications or chemical duplication/synthesis can be expected to continue growing as new missions of exploration are undertaken, new technologies (e.g., deep-water autonomous underwater vehicles) are deployed, and new species are discovered; it is estimated that 91% of species in the ocean still await description (Mora et al., 2011).

Shipping and tourism

While not a sector involved in the extraction of material resources, transportation in the sea (to include commercial shipment of products and the commercial cruise line industry) has grown to a nearly half-trillion-dollar industry ($380B in freight shipping and $46B in tourism: http://www.ics-shipping.org/shipping-facts/shipping-and-world-trade/world-seaborne-trade and https://cruisemarketwatch.com/market-share/) in 2019. What started the millennia in the past in moving goods over relatively short distances on small wind- or human-powered vessels has now evolved into an industry in which G-class container ships of 400 m length can carry in excess of 20,000 TEU (20 foot equivalent units), transported on transoceanic voyages at maximum speeds in excess of 20 knots, with a crew of about 20–30. In total, roughly 130M TEU of goods are shipped per year through oceanic shipping lanes (http://www.worldshipping.org/about-the-industry/global-trade/trade-statistics). Over the last two centuries, major engineering advances (the Panama and Suez Canals, for example), improvements in power plant technology, and increased consumer demand for global trade of goods across national borders have all resulted in dramatically expanded levels of activity for ocean-based shipping.

Advances in robotics, communications, and automated control are now introducing extraordinary levels of autonomy into the commercial oceanic shipping industry. At present, there is considerable potential for operational autonomous capabilities for short-sea shipping (defined as commercial waterborne transportation that does not transit an ocean), as discussed in Ghaderi (2018). One can expect that this technology may soon yield a capability for fully robotic transoceanic shipping with no personnel on board—dramatically reducing many of the operational costs for this sector.

Similarly, there has been considerable growth of the cruise line industry (from 17.8M passengers in 2009 to 30M in 2019, for example; https://cruising.org/-/media/research-updates/research/clia-2019-state-of-the-industry-presentation-(1).pdf). Larger vessels, more ports of call, and increased on-board services have all contributed to the expansion of this sector of the blue economy.

What characterizes the new blue economy?

The blue economy, as characterized earlier, has developed around tangible resources (e.g., food and energy) and critical services (e.g., shipping). Data, per se, serve a supporting function at best. For example, to ensure a sustainable yield for a particular fishery, one needs a measurement of the concentration and distribution of those species. Similarly, economic benefit from the extraction of minerals requires geophysical characterization of the size, precise location, and quality of the relevant reserves. Thus, data have been necessary, but not central to the build-out of the traditional blue economy. In the future, the role of data—specifically data characterizing the status and trends of the environment of the ocean—will become the economic driver for the new blue economy. So, what then is the definition of the new blue economy? We offer the following:

The new blue economy is a knowledge-based economy, looking to the sea not just for extraction of material goods, but also for data and information to address societal challenges and inspire their solutions (Spinrad, 2016).

Before discussing the specifics of how the new blue economy will evolve, it is important to define the elements of data and information that will be most relevant to this emerging paradigm.

Data

Volume

The utility of data is highly dependent on the amount of data available. Historically, in the ocean community, data paucity has been a limiting factor in the ability to draw scientific conclusions or develop meaningful predictive capabilities. However, over the last two decades or so, with the emergence of new platforms and sensors (described further), the volume—and equally important, the spatial distribution—of the data describing the ocean has increased dramatically. For example, the holdings of the World Ocean Database (WOD18) now include roughly 16M oceanographic casts. Volume of data, alone is not sufficient to provide improved capabilities, but its increase is a necessary component of process improvements around predictive capabilities.

Quality

Quality assurance and quality control (respectively, QA and QC) provide the guarantee that the data used to initialize predictive models, for example, are consistent, accurate, and reliable. Historically, oceanographic data were often highly inconsistent across the research community (Mantyla, 1994) or across different observational instruments, so development of seamless characterizations of ocean properties across large areas of the ocean, or spanning long periods of time was—at best—challenging. Through aggressive efforts by a number of global organizations, notably the Intergovernmental Oceanographic Commission (IOC), standards and protocols were established to ensure adequate intercomparability of data from different sensors, sources, and methods. The need for ensuring the QA and

QC of oceanographic data was made even more apparent with the advent of satellite-based remote sensing since the 1960s. The capability to observe and measure (with high resolution) the temperature of the surface of the ocean using radiometric techniques from sensors placed hundreds of miles above the sea surface demanded a means for merging those data with historic records of sea surface temperature taken from buckets of water pulled from the sea or intake temperatures of sea water used for cooling ship's engines. Ensuring QA and QC of oceanographic data has become a critical component in considering the economic value of ocean information.

Diversity

Most early oceanographic research cruises focused on a limited set of phenomena, processes, or parameters. Historically, for example, the first true oceanographic research explorations (e.g., the Challenger expedition of 1872–76 and the 1910 North Atlantic expedition of John Murray and Johan Hjort) were highly descriptive in nature, acquiring data and information in the most opportunistic fashion. From these efforts, the first sense of biomass concentrations and gross differences in ocean water temperature and salinity were acquired. In the mid-19th century, an international conference was held in Belgium, led by US Navy LT Matthew Fontaine Maury, in which seagoing nations of the time agreed to share oceanographic data (Williams, 1969); Maury, in his 1855 classic volume, entitled "The Physical Geography of the Sea," states that "… an appeal was addressed to mariners, there was a flight up into the garrets, and a ransacking of time-honored sea chests in all the maritime communities … for old logbooks and sea journals." And even with all that ransacking of data, the holdings of that volume are limited to about a dozen parameters, including "Galvanic Properties of Gulf Stream Waters" (Maury, 1855).

Over time, a combination of technological advances (e.g., conducting wire) and increased demands for information (e.g., for national security) allowed for acquisition of different types of data. More than just counts of fish in nets, or measurements of temperature from a thermometer in a bucket, seagoing scientists could now make detailed measurements of the chemistry of seawater, the bathymetry of the seafloor, and the dynamic conditions of the ocean. Data could also be taken from the full water column and at locations heretofore impossible to access. Over time, the diversity of observations grew to the point where, today, various groups associated with the organization and management of ocean observing have identified significantly larger numbers of critical ocean variables (for example, the US Integrated Ocean Observing System, IOOS, has identified 34 "core variables": https://ioos.noaa.gov/about/ioos-by-the-numbers/).

Availability/access

But the value of those data depends on their availability and access. On the one hand, given the abundant social benefits of the application of data, and the fact that their collection is already "paid for," oftentimes with public resources through taxes and government appropriations, an argument can be made that the easiest solution is to make all data acquired with public resources openly available in a timely manner. At the same time, the commercial value of these data (and, of course, in a completely different vein, the value of the data for national security perspectives) argues for limiting access to those who choose to purchase or license their use. This issue is central to the concept of the new blue economy and must be carefully considered through policy, regulations, and business practices as described elsewhere in this volume (see, for example, the contributions from Whitehouse, Kildow, and Rayner).

Models/predictive capability/capacity

The heart of the new blue economy is the application of data to predictive models (see Chassignet, this volume for a more detailed discussion). The modeling effort is where innovation and entrepreneurship come to bear. A predictive model with consistent skill is a monetizable commercial product, the value of which is defined by its utility within a particular economic sector, as well as its adaptability to changing natural conditions or emerging user needs.

Complexity

Models can be quite simple or complex. For example, a tidal prediction system that only feeds in phases of the moon and sun might serve the needs of a particular community quite adequately, even though it might embody a significant error in sea height or time of tide. By incorporating detailed physics of fluid mechanics, hydrography, and local meteorology, however, that simple model can become quite complex, with—presumably—a significant improvement in skill (albeit with an incumbent limitation to its geographic applicability). One might argue that it is in the accommodation of specific user needs, and the resulting requirement for modeling complexity, that the market potential lies for the new blue economy.

Accuracy

Not to be confused with resolution (e.g., to how many decimal points does the model provide output), accuracy is often used synonymously with "skill" in discussing predictive models. Model accuracy needs to be attended to in close consonance with the user requirements. Does the US Coast Guard need a wave height model that is accurate to 3 cm (~1 inch) at 12 h? Probably not. But that same predictive capability with a >90% skill at 30 cm (~1 foot) is probably of great utility. One might argue, however, that if you could get the 3 cm (accuracy) forecast as easily as the 30 cm forecast—with no loss in skill—why not do so. And, indeed, that's a fair question, but for the fact that doing so almost always depends on dramatically increased computational resources (cycles and time). One rule of thumb is that for each doubling in resolution of a model one needs a factor of 10 times more computing power (https://scied.ucar.edu/longcontent/climate-modeling). In this regard, recent developments in cloud computing may provide a dramatic improvement in modeling capacity. Moving to an exascale (1 exaflop = a billion billion, or one quintillion calculations per second) computing environment is an important focal point for the new blue economy.

Diversity

As with the issue of data diversity, we are seeing a dramatic growth in the diversity of predictive models. These models address a full spectrum of oceanographic, meteorological, climatological, ecological, hydrological, and cryospheric phenomena. At the Environmental Modeling Center of the US National Oceanic and Atmospheric Administration (NOAA), for example, more than 20 prediction systems are currently listed as operational. Given that the associated models are to be applied to issues of safety, economics, and environmental stewardship, we are also seeing a growing set of social, economic, and behavioral modeling capabilities.

Interoperability

The true utility of a model lies in its incorporability into a larger system of models, wherein the critically interdependent inputs and outputs of each seamlessly feed the comprehensive predictive capability. For example, to fully understand the dynamics of a natural ecosystem—such as a marine reserve—one needs to understand how changes in the physical regime affect and are affected by changes in the chemistry and biology of that system. How does the growth of algae affect the thermal structure of the surface waters? Is there feedback between the algal growth and the thermal structure of the surface waters? Does that feedback respond to or affect the vertical circulation of local waters? The ability to accommodate what is often a nonlinear relationship between physics, biology, chemistry, and geology in a model system is often the factor that best defines the predictive skill of that system.

Product generation/dissemination

The ultimate and most critical element of any modeling effort is defined by how the outputs of the model are interpreted and disseminated to the user community. Recent technological advances are making this translation and transfer function better. Web-enabled services and cloud services, used so commonly in commercial applications (e.g., Amazon Web Services), are being used increasingly more by the environmental modeling community (https://gcn.com/articles/2019/12/23/noaa-cloud-data.aspx). Apps, available for portable (or wearable) devices, are also proving to be a useful, tunable (to particular user needs), and economically viable market tool. In 2018, XPRIZE conducted a competition around the development of ocean-related apps. While the total prize value was only $100,000, there were 80 entrants into the competition, with app development ranging from shipping and trade to conservation. In 2016, the global leader in geographic information systems (GIS), ESRI, issued a conceptually unique product called Ecological Marine Units (EMU), defined as "baseline 3D mapped ecosystems of the ocean that have been classified through statistical clustering" (Sayre et al., 2017). While not tied directly to modeling and predictive capabilities, EMUs serve as a fundamentally new way to present complex ocean information to a range of applications. Model outputs could be easily incorporated into an EMU framework, thus presumably making predictive products even more accessible and interpretable to a broader community.

What enables the new blue economy?

The argument then holds that we have a rich economic portfolio in the blue economy of traditional products and services and that over the last several decades, we have seen growth in the potential for new products and services. The question, then, is whether we are now enabled to construct a new blue economy. Several elements are critical to the realization of such an effort.

Technology

One factor enabling our move into a new paradigm is the dramatic improvement in technology for measuring, modeling, and predicting the ocean.

Platforms

Historically, we measured the ocean using ships or buoys at sea. In the 1960s and ensuing decades, we developed satellite platforms and a small set of drifting floats and buoys. We also saw the emergence of purpose-built oceanographic research vessels. In the United States, this evolved into the fleet of US Navy oceanographic ships and the University National Oceanographic Laboratory System (https://www.unols.org/what-unols/history-unols) wherein mostly publicly funded and constructed ships are operated by academic institutions. Since 2000, we have seen several interesting trends. First, private philanthropy has come to the fore and is contributing to the fleet of seagoing research vessels (e.g., https://schmidtocean.org/rv-falkor/, https://www.paulallen.com/rv-petrel, and http://www.oceanx.org). More significantly, however, we have seen the growth in availability, size, capability, capacity, and applicability of autonomous platforms. Having emerged from developments with tethered systems (i.e., remotely operated vehicles), the portfolio of platforms from which to make measurements, collect, and transmit data, or conduct subsea operations now includes autonomous underwater vehicles (AUVs, sometimes called underwater robots), autonomous surface vehicles (ASVs), unmanned aerial systems (UAS) (Verfuss et al., 2019), and even hybrid air—sea designs (Kirkwood, 2019). These new platforms will serve as key components in feeding the large data needs of the new blue economy.

Sensors

As indicated earlier, for centuries, the types of data to be collected from the sea were limited to the most fundamental physical, biological, chemical, and geological parameters (temperature, depth, conductivity, fish counts, etc.). This limitation was mostly defined by the types of sensors available. Nowadays, as discussed in detail by Watson-Wright and Snelgrove (this volume), we are able to measure scores of variables, largely due to development of new sensors. Solid-state electronics, microminiaturization, as well as whole new technological breakthroughs have been at the heart of these new capabilities. A good example of the most recent manifestation of this kind of capability is the measurement of "environmental DNA" (eDNA). Effectively a fingerprint of the genetic makeup of a sample of seawater (i.e., "who's been here"), eDNA measurement was originally demonstrated with an oil-drum-sized package lowered from a ship into the sea, just a few years ago. Now, that same capability is being incorporated into AUVs for application just about anywhere in the ocean (Pargett et al., 2015). See Ausubel and Curry (this volume). With these dramatic improvements in sensing capabilities come abundant opportunities for economic applications.

Models

Perhaps the most significant development in environmental modeling comes under the header of "earth system modeling." Traditional approaches of building a predictive system around the physics, or biology, or chemistry, are being replaced with fully coupled physical and biogeochemical models, much more realistically reflective of the natural earth system. Operational earth system models are still in their infancy, but this is clearly the major technological development that will drive modeling capabilities that feed the new blue economy in the future. Additional technological developments in data assimilation, high-performance computing (to include cloud computing and quantum computing, perhaps) will, undoubtedly, provide improved modeling capabilities in the near future as well.

Policy

The emergence of a new blue economy is also, of course, facilitated (and can be accelerated) with national, regional, and global policy developments. In the chapter of this book provided by US Senator Sheldon Whitehouse, there is good evidence that such activities have and will continue to grow with legislative and regulatory support. Some examples of policies that have stimulated recent economic growth include the National Oceanographic Partnership Act (including its legislated program, the National Oceanographic Partnership Program, NOPP: https://www.nopp.org) and the set of activities around data standards of the International Oceanographic Data and Information Exchange, IODE (https://www.iode.org), and the World Meteorological Organization, WMO (https://public.wmo.int/en).

Programs and resources

Similar to policy initiatives and guidance, there have been several important program developments in the last two decades that are now providing important frameworks for feeding data and information into the new blue economy. Several examples—some of which are discussed in greater detail elsewhere in this volume—are the Global Ocean Observing System (GOOS, http://goosocean.org), the US Integrated Ocean Observing System (IOOS, https://ioos.noaa.gov), the European Union's Horizon 2020 marine programs (https://ec.europa.eu/programmes/horizon2020/en/area/oceans-and-seas), and the development in Europe of Mercator/MyOcean (http://marine.copernicus.eu/about-us/about-mercator-ocean/) and Copernicus Marine Environment Monitoring Service (CMEMS, http://marine.copernicus.eu/the-copernicus-marine-environment-monitoring-service-delegated-to-mercator-ocean/). Again, the emergence and growth of the new blue economy would simply not be possible without these programmatic developments of recent years.

What does the new blue economy look like now and what is its potential?

Examples of extant new blue economy activities

This new blue economy is not an imaginary conceit; we already have potent examples of how it can be built, used, managed, and proven economically, from a variety of sectors:

Wave forecasting

A number of user communities draw benefit from having timely, accurate, and reliable wave predictions: commercial and recreational fisheries, recreation (e.g., surfers and boaters). To this end, several companies have emerged to provide tailored forecast products tuned to specific needs. Some examples are Surfline (https://www.surfline.com/category/forecast), Surf-forecast.com (https://www.surf-forecast.com), and Magicseaweed (https://magicseaweed.com/Oregon-Surf-Forecast/15/). These companies' products pricing is highly variable: some provide free forecasts (effectively repacking publicly available forecast information) and others are available as subscriptions or purchased apps for mobile devices (with typical monthly subscriptions for premium services running less than $10 per month).

Recreational fishing

The sale of a limited set of proprietary predictive products is another established commercial enterprise. Recreational sportfishing, including the tournament fishing for trophy species, can be highly competitive. Any information assisting the competitors to locate and catch the target species can provide a meaningful competitive edge. In 1986, Mitch Roffer recognized this target client base, as well as an opportunity to develop predictive products, which might provide a "force multiplier" to the competitors' fishing efforts (see interview with Roffer, this volume). The resulting ROFFS Recreational Fishing Analysis is now used by many competitive fishers, and according to ROFFS website, their products are contributing factors to more than 30 fishing records (https://www.roffs.com). The cost for a typical 7-day analysis is $60 each, and a discounted unlimited preseason package costs $2400 (all in 2019 dollars).

Commercial shipping

Naval operations and commercial shipping are critically dependent on accurate forecasts of surface ocean conditions. In the US Navy, the capability to predict these conditions serves as the basis for a sustained operational capacity called Optimum Track Ship Routing, or OTSR. OTSR serves the function of optimizing time to waypoint for the fleet, by exploiting winds, surface currents, waves and tides, as well as reducing losses from excessive fuel consumption. The same benefits are seen in commercial shipping, where gaining a small advantage from avoiding headwinds or utilizing surface currents can take time (and expenses) off transoceanic transits. By some estimates, ship track optimization through the utilization of ocean forecasts can save the commercial shipping industry an average of 4%–8% of costs per year (http://oceanweatherservices.com/benefits_of_optimum_ship_routing), which would translate to approximately $0.4M–$0.8M savings on just one transit by a 10,000 TEU vessel from Shanghai to Northern Europe (UNCTAD, 2019).

Other domains

There are other examples, in different domains, of the application of ocean data to create economic benefit. Aquaculture and mariculture firms are using ocean observations to locate/relocate their stocks and to preemptively treat intake water (in the case, for example, of shellfish hatcheries) to mitigate against thermal and chemical shock to their larvae. Search and rescue operations, as well as insurance investigations, regularly depend on highly accurate forecasts of drift and calculations of back trajectory (i.e., working backward to the site of a capsizing, allision, or collision) to recover/rescue accident victims or to attribute cause or fault.

Case studies for the future

The examples of the new blue economy as cited earlier, in conjunction with the scientific and technological developments of the recent past, yield enormous opportunity for building a robust set of commercial ventures for which the "currency of the realm" is the output products of the most skillful predictions and projections. To illustrate these opportunities, several hypothetical case studies are presented further. In each example, the particular opportunity is briefly described in terms of the need/requirement and the market opportunity. A recap is then provided, summarizing the diverse market opportunities and—to ground the case studies in reality—some very specific definitions of the particular products that would be needed in each case.

Natural resource damage assessment

In the aftermath of the Macondo Prospect oil spill in the Gulf of Mexico (more commonly known as Deepwater Horizon) in April of 2010, legal/financial settlements of $20.8B were assessed, BP was fined $4B in the largest criminal penalty in US history (https://www.noaa.gov/explainers/deepwater-horizon-oil-spill-settlements-where-money-went), and more than 80,000 square miles of commercial and recreational fishing grounds were closed (https://www.noaa.gov/education/resource-collections/ocean-coasts-education-resources/gulf-oil-spill). This man-made disaster, while particularly large (by any metric), was likely not the last of its kind. In the course of conducting the natural resource damage assessment, as required by law and regulation, one important fact was made clear: there was insufficient characterization of the baseline biota, both in the water column and in the seabed. Consequently, determining the specific impacts of the spill was difficult and often ambiguous. Had there been detailed surveys of the bathymetry, floral and faunal assemblages, as well as sustained longitudinal characterizations of their trends (in biodiversity and concentrations, for example), a much clearer assessment would have been easier to conduct. Such an assessment would also have been more efficient, in terms of identifying well before any accident where the most sensitive and vulnerable environments were to be found. The mapping, charting, and analysis of ecosystem distributions represent a valuable, information-based market, which is not within scope for purely federal (e.g., NOAA or the US Bureau of Ocean Energy Management, BOEM) activities. Every in-water installation subject to regulatory controls associated with environmental protection would benefit from regularly updated, highly resolved, accurate geospatial characterizations of the ecological construct of the marine environment potentially impacted by activities of those installations.

Optimized commerce

Consider, for example, the commercial shipping activity associated with a port like Portland, Oregon. Portland's main commodity is grain; the US Department of Agriculture values wheat exports from Oregon, Washington, and Idaho at $1.3B, or 40% of US wheat exports annually (https://www.pnwa.net/wp-content/uploads/PNWA_Handout.pdf). In 2010, the Columbia River saw over 4000 ship transits (NOAA, 2010) many of which transited from Portland, down the Columbia River to the Pacific Ocean (a distance of 105 miles) and then to receiving ports around the world. Given the tidal mode of that part of the world, the transit down the Columbia River from Portland/Vancouver typically takes about one tidal cycle (6–8 h). There are five bridges under which the ships must pass, and a shipping channel of typically 42 feet depth is maintained. The US Pacific Northwest is a region of considerable precipitation, imposing significant hydrological impact on the Columbia River (with the consequent changes in relative buoyancy of the ships depending on the periodic injection of freshwater into the system). Given that the goal of the grain exporters is to optimize the loadout of their products, there must be careful consideration given to factors such under-hull clearance, air gap (under bridges), as well as the important factors of surface currents and winds, especially in the broad lower Columbia River. In short, the challenges with determining how much grain to load, when to debark from Portland/Vancouver, and what speeds to use in the transit down the river are significant. A reliable predictive model of all the environmental factors (tides, currents, winds, waves, salinity, etc.), coupled with a vessel-specific load-out model, would be of enormous value to the shipping community. And, of course, such a system of models will be unique to each port environment; Portland is different from New Orleans, Baltimore, Boston, etc. Again, such an economic opportunity critically depends on the availability of publicly funded data sources, but can be developed in a tailored, client-specific, fully commercializable manner.

Coastal community resilience

Superstorm Sandy hit the east coast of the United States in late October of 2012 and was one of the deadliest, most destructive storms to hit the United States with more than 200 deaths and nearly $62B of damages assessed (https://www.dosomething.org/us/facts/11-facts-about-hurricane-sandy). A large percentage of the damage done within the New York/New Jersey coastal areas was a consequence of storm surge (in some areas almost 9 feet; https://www.climatecentral.org/news/sandys-surge-climate-change-20776). Research (Lin et al., 2016) indicates that with consequences of climate change, a storm like Sandy could occur every 20–25 years toward the end of this century. Adapting to the impacts of the storm surge as seen in Sandy might be addressed with a number of engineering solutions or natural infrastructure architecture. As an example, the application of "living breakwaters" (also known as oyster-tecture, in one example, see www.scapestudio.com) has been proposed as a naturally enhanced coastal improvement that is environmentally sustainable, economically self-supporting, and offers important buffering protection from Sandy-magnitude storm surges. The challenge, of course, is in building the models and simulations to support such investments. The ability to predict reliably how to build out such natural infrastructure, predict its benefits and weaknesses, and "manipulate" the design of such to assess different scenarios represents an important economic opportunity. One could imagine a full catalog of such modeling and simulation products, which address the site-specific ecosystem services valuations to address a spectrum of natural hazards (hurricanes, superstorms, Nor'easters, tsunamis, as well as general sea level rise). As in the previous example, the breadth of the economic opportunity lies in the diversity of predictive products that would be required for the full range of global coastal geographies.

Natural hazard protection

In addition to the primarily physical nature of naturally occurring storms, there are primarily biological/chemical natural hazards as well, the most notable of which is harmful algal blooms, HABs (popularly known as Red Tides). Another important example, of course, is the emergence of dead zones and local impacts of deoxygenation. In the United States, NOAA has begun a program of issuing forecasts for HABs in a very small number of specific, vulnerable coastal regions (https://tidesandcurrents.noaa.gov/hab_info.html). These HAB forecasts provide the most fundamental information about when these blooms might occur and what their temporal and spatial extent might be. While extremely useful, these forecasts are highly limited in their number and do not provide the decision support that is needed for each particular user (tourism community, public health sector, food security community, etc.). As a result, two problems emerge: (1) many areas of the coast go without the benefit of any forecast product ... this may become particularly insidious given the current trend of increasing magnitude and frequency of these blooms (Anderson, 2012); (2), the current set of products are not tuned to the particular needs of specific users (e.g., "what seafood is safe?", "Will my beach be hit as hard as one that is 1 km away?"). Consequently, there is an opportunity to develop, commercially, a set of forecast products to adapt to predicted HABs. Not unlike the current spectrum of weather products, for which one can purchase tailored forecasts (ski areas, for example, do this all the time), it's not difficult to imagine a market development around client-specific needs for decision support regarding naturally occurring biological hazards.

Infrastructure resilience

While the portfolio of impacts consequent to global climate change is enormous (consider, for example, the details of Topic 2 of the Fifth Assessment Report of the Intergovernmental Panel on Climate Change, of 2014), perhaps one of the most impactful is sea level rise, and its consequences for local and regional coastal flooding. There are many examples of where inadequate information or understanding regarding the impending (less than decades) consequences of sea level rise resulted in costly and sometimes unusable public infrastructure investments. One classic example is the installation of a light rail system in Norfolk, Virginia, where sea level rise is among the most accelerated in the world (at over 4.5 mm per year Ezer, 2013), the increased frequency of local flooding has resulted in that system being subject to increasing disruption of service (https://www.pilotonline.com/news/transportation/article_d1e705b2-3cf4-11e8-896f-5f19b50fa044.html). Sea level rise, while occurring globally, is a very localized phenomenon, due to geographic/bathymetric influences, local meteorological conditions, and other factors. As a consequence, the kinds of general projections of sea level rise produced through national or international climate change assessments may be of very limited utility in major decisions about what kind of infrastructure to build in a particular area. The author John Englander, in his 2012 volume entitled "High Tide on Main Street," however, cogently points out that, "Rising sea level will present us with boundless opportunities to demonstrate our intellect ..." (Englander, 2012).

Again, climate models tuned to the particularities of the oceanographic/environmental conditions of a specific locale—and including a meaningful risk assessment—could serve an important function in making municipal decisions about how to deal with very regionally specific issues associated with sea level rise.

These five case studies provide a very brief, but diverse, set of examples of where new markets and associated products can be developed. A summary of how those might play out more specifically in terms of targeted markets and products is captured here, in Table 6.1:

The short description of these five opportunities, their incumbent products and markets, is intended to stimulate thinking about the full set of examples where the new blue economy built around data and information might develop. Clearly there are many other examples tied to the markets of fisheries, aquaculture, public health, tourism, energy, water security, national/homeland security, and other areas. But there are some important elements that need attention before the new blue economy can be fully realized. Later in this volume, for example, there are important treatments of the market assessments and economic analyses of the potential for the new blue economy (see Kildow and Rayner, this volume).

Table 6.1 Examples of New Blue Economy opportunities and products.

Case study	Market opportunity	Product
Deep-water horizon	Regulatory compliance Response/restoration service and support	High-resolution, archived, and real-time observational data in KML formats
Portland grain shipping	Commodity-specific load-out and transit intelligence	Short-term, coupled models (ocean, river, hydrological)
Superstorm Sandy	Natural infrastructure architecture	Site-specific ecosystem services valuation
Harmful algal blooms	Derivatives for public health, tourism, seafood safety, etc.	Coupled ecological and behavioral forecasts
Sea level rise	Resilience planning and guidance	Down-scaled sea level rise projections Risk/translation

What is needed to make the new blue economy a reality?

In order to realize the potential of this new blue economy, there are some critical ingredients: (1) scientific and technological development; (2) market assessments and build-out; (3) market stimulation and investment; (4) education and workforce development; (5) policy; and (6) measures of effectiveness.

Science and technology (S&T)

The current capabilities and needs for S&T have been discussed earlier, but there is an important additional element to incorporate in this required component. In addition to exploiting the latest research findings and inventions in sensors, platforms, models, and computing our community of ocean scientists and technologists must "get out of our comfort zone" and look to seemingly disparate scientific arenas. Several decades back, for example, great strides were made in understanding phytoplankton ecology by looking to the field of medical diagnostic technologies for new approaches to measure single cells. The outcome in that case was the application of flow cytometry to marine biology, a dramatic change in analytical capabilities for ocean scientists. What developments can we exploit and adapt from such communities as advanced materials, additive manufacturing, behavioral genetics, chemoinformatics, or cognitive neuroscience? What venues do we have or do we need to facilitate the kind of cross-fertilization that we need?

Market assessments and build-out

No investor would consider a potential market without the benefit of a robust assessment. Fortunately, this issue is not new in regard to the new blue economy, and any number of global groups are working actively on such assessments. These include the Organization for Economic Cooperation and Development (OECD, 2016) and the US IOOS Program (IOOS, 2016). In addition, over the last decade, the emergence of marine clusters (e.g., The Maritime Alliance—Blue Tech Cluster Alliance—https://www.bluetechclusters.org/the-maritime-alliance, Marine Industries Science & Technology—https://www.mistcluster.org, Marine and Oceanographic Technology Network—https://www.motn.org, and Canada's Ocean Supercluster—https://oceansupercluster.ca) has proven to be an effective platform for capturing all of the potential "players" in the blue economy and providing important venues for defining the way forward in building out the new blue economy. More details on the specific activities associated with market assessments are captured in the contributions to this volume by Kildow and Rayner.

Market stimulation/investment

The profile of investment to make the new blue economy a reality will need to be fundamentally different from the way the more traditional blue economy has been resourced historically. With minimal commercial presence (other than in specific sectors, such as oil and gas or national security), the bulk of research investment in ocean products and services has been by the public sector: mostly federal funding agencies. In the United States that has been accommodated by roughly a dozen agencies, with the majority of investments coming from NOAA, the National Science Foundation (NSF), the US Navy (e.g., through the Office of Naval Research), and the National Aeronautics and Space Administration (NASA). The total US federal funding of academic research and development in geosciences in 2016 was approximately $2B and has only increased by about 0.1% annually between 2007 and 2016 (Table 5.5, NSB, 2018) By comparison, that growth rate was 3.8% between 1997 and 2006 (Table 5.6).

In recent years, the investment profile for ocean science and technology has had some interesting and notable changes. For example, philanthropy now represents a considerable slice of the pie. Private benefactors such as the Schmidt Ocean Institute, Vulcan, and the Paul Allen Foundation are making significant investments ($100 Ms) in infrastructure (e.g., ships), people, and research. How these investments integrate with public resources is an important issue that needs clarity of definition.

We've also seen the emergence of venture capital, angel investment, and new venues, such as the Ocean Exchange (https://www.oceanexchange.org) where investors and entrepreneurs can provide seed funding to accelerate the movement of ideas into new capabilities with commercial potential.

New approaches to building out capability are also being introduced to the oceanographic community. XPRIZE, for example, came onto the scene in the 2000s with a set of competitions designed to push the limits of technology. These competitions focused on specific issues, including: oil cleanup (with sponsorship from Wendy Schmidt); ocean health (Wendy Schmidt); and ocean mapping (Shell and NOAA). Given that the mission of XPRIZE is to stimulate "industry-changing technology that brings us closer to a better, safer, more sustainable world," these investments could prove critical to stimulating the new blue economy.

As important as the "How" of economic stimulation is the "Who." In this regard, and in addition to the sustained effort of the well-established community of ocean research community, we have seen the emergence of a class of "disruptors" playing an important role, especially in terms of technology development. People like Joe Rizzi and Roger Hine (Liquid Robotics), Eric Stackpole and David Lang (OpenROV) and Boyan Slat (The Ocean Cleanup) all came from outside the traditional pipeline of marine technologists, and each has provided dramatic improvements on our capability to measure, monitor, or clean the ocean. Recognizing this important role of "disruptors" will be a factor in building the infrastructure for workforce development for the new blue economy.

Education and workforce development

This whole volume is intended to focus on workforce development, for which education is a critical component. Numerous other contributions in this volume address the details of education (see section on education and training). The one important point to call out here is tied to the concept introduced earlier regarding "seemingly disparate" technologies/communities. It is important for us, when looking to the educational needs for the new blue economy to be sensitive to the full set of skills required, including social sciences, business acumen, as well as technical training.

Policy changes

To the extent that laws and regulations can both guide and stimulate development of the new blue economy, our primarily technical community must embrace an early and active engagement with policymakers and legislators. To this end, the contribution in this volume from US Senator Sheldon Whitehouse provides excellent context of how this relationship has benefited society in the past and—perhaps more importantly—how policy development will play a critical role in the future. But this issue goes beyond just the actions of individual national legislative bodies. By virtue of the global connectedness of the world's oceans, the role of international bodies cannot be understated. The Intergovernmental Oceanographic Commission, the International Maritime Organization, the International Hydrographic Organization, and the International Seabed Authority are just a few examples of the institutions who must be actively engaged in the development of the new blue economy. And private sector institutions that work in ocean market development can play an important role. The Economist World Ocean Summit, as just one example, provides a neutral venue for global leaders to debate and conduct dialog that will help frame policies critical to the global new blue economy.

Measures of effectiveness

As with any economic sector, the new blue economy must be accountable to the clients it serves as well as the investors who support it. While the particular metrics may not be altogether different from other economies, the "subelements" of those measures may be specific to an ocean economy.

Levels of investment

The status and trends of public funding are critical indicators. If demand for products and services grows, one would expect total federal funding, numbers of research and educational projects funded, and diversity of programs (e.g., basic or applied research, product development, new business development. etc.) to grow. Similarly, the proportion of private investment (philanthropic and corporate) should have well-established target levels and schedules of performance.

Workforce

While growth in the workforce should be considered carefully (an increase in total employees is not always a reliable indicator of health of a sector), the diversity—by every measure, including sector, disciplinary focus, geography, demographics, etc.—must be included.

Market size (sales)

Obviously, if the new blue economy is to succeed in the most traditional sense of a market, the size, breadth, sustainability, and trends in sales are critical indicators.

New business development

This metric is at the very heart of the new blue economy. If investment, revenues, workforce size, and market size increase, but the sector is defined by the same types of companies as have existed historically, then there really is not a new blue economy. It is in this area that business clusters may be a particularly useful tool, as those activities are directly connected to economic development and should be a good conduit for defining the volume and nature of new business development.

Risk tolerance

What is the community profile regarding risk tolerance in both research and business practices? While there is most likely not a pat answer to such a question, what should not be in doubt is that this question should continually be asked at both the level of the community and the individual entity. Risk tolerance is a very "personal" element, but one, nonetheless, which is—as Norm Augustine has captured so well in this volume (pp. XX–YY)—essential.

TRL distribution curve

The concept of Technology Readiness Levels (TRL) was introduced in the 1970s by NASA (https://www.nasa.gov/pdf/458490main_TRL_Definitions.pdf) and represents a nine-level characterization of the maturity of technology moving from basic research to full operations (or commercialization). The TRL framework (and its various modified versions) provides a good shorthand method for assessing the schedule, pace, and viability of moving a product through research, development, testing, and evaluation. As such, an assessment of the TRL status and trends of a particular technology provides a meaningful metric for how fast and how well a concept is moving into the new blue economy.

Conclusion

In summary, the new blue economy as envisioned here is one that is now possible and which holds great promise for the future. Estimates may vary regarding the full potential of the new blue economy, but even the most conservative treatments suggest that the potential for this sector far exceeds that of its closest comparator, the commercial weather services industry, currently valued at roughly $7B per year (https://www.weather.gov/media/about/Final_NWS%20Enterprise%20Analysis%20Report_June%202017.pdf). Enabled by science, technology, policy, and workforce development, the new blue economy can be a significant and impactful development if appropriately managed and built out. We stand at a critical time to build this new blue economy.

References

Anderson, D., 2012. HABs in a changing world: a perspective on harmful algal blooms, their impacts, and research and management in a dynamic era of climactic (sic) and environmental change. In: Harmful Algae 2012: Proceedings of the 15th International Conference on Harmful Algae, pp. 3–17.

Baker, E., Gaill, F., Karageorgis, A., Lamarche, G., Narayanaswamy, B., Parr, J., Raharimananirina, C., Santos, R., Sharma, R., 2014. Chapter 23. Offshore Mining Industries.

Englander, J., 2012. High Tide on Main Street. The Science Bookshelf, Boca Raton.

Ezer, T., 2013. Sea level rise, spatially uneven and temporally unsteady: why the U.S. East Coast, the global tide gauge record, and the global altimeter data show different trends. Geophys. Res. Lett. 40 (20), 5439–5444.

Ghaderi, H., 2018. Autonomous technologies in short sea shipping: trends, feasibility and implications. J. Transp. Rev. 39 (1), 152–173.

Glasby, G.P., 2000. Economic geology: lessons learned from deep-sea mining. Science 289 (5479), 551–553.

International Seabed Authority, 2010. A Geological Model of Polymetallic Nodule Deposits in the Clarion-Clipperton Fracture Zone and Prospector's Guide for Polymetallic Nodule Deposits in the Clarion Clipperton Fracture Zone. Technical Study.

IOOS, 2016. The Ocean Enterprise: A Study of US Business Activity in Ocean Measurement, Observation and Forecasting. NOAA, Washington.

IPCC, 2014. Climate Change, 2014: Synthesis Report. Contribution of Working Groups I, II and III to the Fifth Assessment Report of the Intergovernmental Panel on Climate Change. IPCC, Geneva.

Kirkwood, W.J., 2019. Shearwater: A Homogeneous Platform for Ocean Applications. Australian Society for Underwater Technology, Perth.

Lin, N., Kopp, R.E., Horton, B.P., Donnelly, J.P., 2016. Hurricane Sandy's flood frequency increasing from year 1800 to 2100. In: Proceedings of the National Academy of Sciences.

Mantyla, A., 1994. The treatment of inconsistencies in Atlantic deep water salinity data. Deep Sea Res. Oceanogr. Res. Pap. 41 (9), 1387–1405.

Maunder, M.N., Sibert, J.R., Fonteneau, A., Hampton, J., Kleiber, P., Harley, S.J., 2006. Interpreting catch per unit effort data to assess the status of individual stocks and communities. ICES (Int. Counc. Explor. Sea) J. Mar. Sci. 63 (8), 1373–1385.

Maury, M.F., 1855. Physical Geography of the Sea. Harper and Brothers.

Miller, K.A., Thompson, K.F., Johnston, P., Santillo, D., 2018. An overview of seabed mining including the current state of development, environmental impacts, and knowledge gaps. Front. Mar. Sci. 4, 418.

Mora, C., Tittensor, D.P., Adl, S., Simpson, A.G.B., Worm, B., 2011. How many species are there on Earth and in the ocean. PLoS Biol. 9 (8).

NOAA, 2010. Estimating Economic Benefits from NOAA PORTS Information: A Case Study of the Columbia River. NOAA Technical Report. https://tidesandcurrents.noaa.gov/publications/Columbia_River_PORTS_Economic_Study_pdf.

NSB, 2018. Science and Engineering Indicators 2018. National Science Foundation, Alexandria.

OECD, 2016. The Ocean Economy in 2030. OECD Publishing, Paris.

Pargett, D.M., Birch, J.M., Preston, C.M., Ryan, J.P., Zhang, Y., Scholin, C.A., 2015. Development of a Mobile Ecogenomic sensor. OCEANS 2015. MTS/IEEE, Washington.

Sayre, R.G., Wright, D.J., Breyer, S.P., Butler, K.A., Graafeiland, K.Van, Costello, M.J., Harris, P.T., Goodin, K.L., Guinotte, J.M., Basher, Z., Kavanaugh, M.T., Halpin, P.N., Monaco, M.E., Noel, C., 2017. A three-dimensional mapping of the ocean based on environmental data. Oceanography 30 (1), 90–103.

Spinrad, R.W., 2016. The new blue economy: A vast oceanic frontier. In: EOS Proceedings of the American Geophysical Union, 97. American Geophysical Union. https://doi.org/10.1029/2016EO053793.

UNCTAD, 2019. Review of Maritime Transport - 2019. United Nations Conference on Trade and Development, New York.

Verfuss, U.K., Aniceto, A.S., Harris, D.V., Douglas, G., Fielding, S., Jiménez, G., Johnston, P., Sinclair, R.R., Sivertsen, A., Solbø, S.A., Storvold, R., Biuw, M., Wyatt, R., 2019. A review of unmanned vehicles for the detection and monitoring of marine fauna. Mar. Pollut. Bull. 140, 17–29 (Elsevier).

Williams, F.L., 1969. Matthew Fontaine Maury, Scientist of the Sea. Rutgers University Press.

Wynberg, R., 2015. Marine Genetic Resources And Bioprospecting in the Western Indian Ocean. The Regional State of the Coast Report: Western Indian Ocean, Nairobi, pp. 407–418. UNEP.

CHAPTER 7

Market analysis

Ralph Rayner[1,2]
[1]London School of Economics, London, United Kingdom; [2]National Oceanic and Atmospheric Administration, Washington, DC, United States

Introduction

This chapter considers the markets that benefit from new blue economy products, services, data, and information. It does so recognizing that nonmarket benefits are also an important dimension of the value delivered by new blue economy activities. Nonmarket benefits delivered by the new blue economy are discussed in greater detail in the following op-ed from Judith Kildow.

The ocean is a key and growing source of food, energy, and minerals and is the primary medium upon which global trade takes place. Coastal waters and regions are the location of a large proportion of the global tourism and recreational industries. Approximately 40% of the world's population lives in coastal regions and three-quarters of the world's large cities are located on the coast. A large proportion of economic activity occurs in coastal regions. For example, US coastal regions accounted for 83% of total US GDP in 2018 (NOEP, 2020).

The ocean also provides vital natural assets and ecosystem services, which play a pivotal role in human society (Spalding et al., 2016). Hundreds of millions of people depend directly on the oceans for their food and livelihoods. We all depend on the oceans for provision of much of the oxygen that we breathe and for their controlling and moderating role in weather and climate.

In contrast to the terrestrial environment, the ocean represents a difficult and harsh environment in which to operate. Much of the economic activity around, on, and under the ocean would not be possible without data, information, and knowledge derived from sustained ocean observations, measurements, and forecasts.

As the needs for ocean data and information have grown in scale, the provision of the technological means to conduct ocean observations and measurements has become an important economic activity in its own right. The growth in use of the ocean has also driven the development of an important and growing service industry comprising intermediary businesses,

Preparing a Workforce for the New Blue Economy
ISBN 978-0-12-821431-2
https://doi.org/10.1016/B978-0-12-821431-2.00038-X

which add value to public and private ocean observations, measurements, and forecasts, tailoring them for specific end uses.

The organizations, both public and private, that enable sustained ocean observation, measurement, and forecasting and deliver their benefits comprise the new blue economy. The new blue economy critically underpins ocean research and supports the safe, cost-effective, and sustainable use of the ocean and ocean resources. In addition, it provides critical information to help manage and protect coastlines and coastal infrastructure and is a key input to weather forecasts and climate projections, delivering socioeconomic benefits far inland.

The ocean environment is subject to an increasingly complex range of pressures. Foremost are those related to ocean health, overexploitation of marine resources, pollution, rising ocean temperatures and sea level, ocean acidification, and loss of biodiversity. Unsustainable use of the ocean and its resources threatens the basis on which much of the world's welfare and prosperity depend. New blue economy activities play a key role in underpinning the scientific basis for national and international legislation regulating sustainable use of the ocean and protection of the ocean environment. They also provide the basis for monitoring of compliance and effectiveness of ocean and coastal protection measures.

The ultimate beneficiaries of the new blue economy are organizations whose activities or businesses benefit from ocean data and information in terms of better scientific understanding of the ocean, improved safety, economic efficiency gains, effective regulation of ocean use, and the protection of the ocean and wider environment. These end users of new blue economy products and services constitute the markets for new blue economy activity and are the recipients of new blue economy benefits, whether as public goods or through commercial transactions.

In summary, the end-use beneficiaries of the new blue economy comprise:

- **Science end users** who use or purchase new blue economy technologies, data, or information to support research;
- **Policy end users** who use or purchase ocean observation technologies, data, or information to support policy formulation, monitoring of policy compliance, and assessment of policy effectiveness;
- **Operational end users** who use or purchase new blue economy technologies, data, or information to support safety, economic efficiency, and environmental protection and regulatory compliance needs associated with their activities.

- **Public end users** who use or purchase new blue economy technologies, ocean data, and information in support of their leisure activities or recreational pursuits.

Science end users employ new blue economy products and services to support many different areas or research. However, the role of the scientific community in the new blue economy is actually twofold. Science is not only an end user of new blue economy outputs; it is also a major producer of the ocean observations that underpin it. Scientific interest initiates ocean observation and often defines what, where, and how data are collected. The needs of scientific research also motivate the development of measurement devices, from individual sensors to complex observing systems. Examples of science end use are the use of ocean data and information in support of research into understanding physical, chemical, and biological processes in the ocean or the scientific study of the role of the oceans in weather and climate. As end users of ocean observations, scientists develop methods to edit and analyze data and derive insights into the ocean and its dynamics. These methods and insights contribute to society's knowledge pool and are used to develop, for example, analytical tools, forecasts, and assessments. Through these activities, science lays critical groundwork for delivery of the wider benefits of the new blue economy.

Policy end users depend on new blue economy products and services to help inform the drafting of effective legislation to support safety of life or property, protection of the environment, or regulation of the use of ocean space and ocean resources. New blue economy products and services are employed to deliver and monitor compliance with the resulting legislation, for example, monitoring to determine beach closures under the US Environmental Protection Agency (EPA) Clean Water Act. New blue economy products and services also deliver benefits in supporting measurement of policy effectiveness. For example, determining the effectiveness of a policy to reduce concentrations of a harmful pollutant requires long-term monitoring technology, data, and information to determine whether the policy is delivering on this goal.

Operational end users employ new blue economy products and services to support strategic decision-making and day-to-day operational planning. For example, an offshore oil and gas company might use information products derived in part from ocean data and information to support the optimum and safe design of a facility or to help plan safe drilling activities.

Public end users are important beneficiaries of the new blue economy. Recreational users of the ocean and coast increasingly make use of public and commercial information products to ensure their safety and enhance their enjoyment. Examples are the use of specialist surf forecasts or provision of tailored information to support sport fishing.

Supporting all of these end uses is the technological means to make ocean observations and measurements, the production of the observations and measurements themselves, and the capacity to turn the resulting data into useful actionable information. These are the key elements of the new blue economy:

- **Providers** of observing infrastructure;
- **Producers** of ocean observations;
- **Intermediate users** who tailor ocean data or information for a specific end use through data integration, data analysis, and modeling.

Providers of observing system infrastructure include manufacturers of sensors, instruments, and platforms; those building, launching, and operating ocean observation satellites; providers of the cyber infrastructure that interconnects observing system components; and organizations that develop and maintain the data management systems, software tools, and models that are used to help turn ocean data into useful information.

Producers of ocean observations are primarily public organizations delivering ocean observations, measurements, and modeled data and information on a free of charge or cost of distribution basis as a public good. They encompass a wide range of national, regional, and international institutions and initiatives operating on different spatial scales. Many countries have installed marine research infrastructure and ocean observatories. Regions have combined their efforts to observe different parts of the ocean on a collective basis, sharing and combining the collected data and information. Globally, frameworks have been developed to foster the growth of existing ocean observing systems and implement observing capability in regions that are underobserved.

Efforts for establishing national ocean observing systems have been continuously increasing. Almost every coastal country is involved in marine research and activities related to ocean observation, although their systems are not equally developed. Whereas some countries demonstrate stronger expertise in observing activities, others are in the process of gaining experience. Examples of developed ocean observing systems are the Australian Integrated Marine Observing System (IMOS), Canadian Ocean Networks and the Canadian Integrated Ocean Observing System (CIOOS), the Japan Oceanographic Data Center, the European Global Ocean Observing System (EuroGOOS), the European Union Copernicus

Marine Environment Monitoring Service (CMEMS), the European Marine Observation and Data Network (EMODnet), and the United States' Integrated Ocean Observing System (IOOS) and the National Oceanic and Atmospheric Administration (NOAA) Global Ocean Observing and Monitoring Program.

In order to link different national ocean observing capabilities, multinational ocean observing frameworks and systems have been set up.

Most national and regional ocean observing systems contribute to the Global Ocean Observing System (GOOS). GOOS operates under the auspices of the Intergovernmental Oceanographic Commission of the United Nations Educational, Scientific, and Cultural Organization (UNESCO-IOC) in partnership with the World Meteorological Organization (WMO). It provides coordination for global ocean observation and links with the Global Earth Observing System of Systems (GEOSS).

Today, GOOS comprises a steering committee, three discipline-based GOOS Expert Panels, which provide scientific oversight on physics and climate, biogeochemistry, and biology and ecosystems, and 13 GOOS regional alliances (GRAs). The GRAs are coalitions of national or regional ocean observation initiatives that follow GOOS principles. The GRAs support marine research and operational oceanographic services to different extents, depending on the degree of development of each of the systems. They also help to promote collaboration and capacity building at local, national, and regional scales.

In addition to public sector activity, many business end users of ocean observation, measurement, and forecast information products commission their own data collection, analysis, and modeling to support operational needs, which cannot otherwise be supported through use of public good operational ocean data and information. In these instances, end users generally place contracts with specialist commercial producers of ocean data, who undertake such work on their behalf. In some instances, a single organization will undertake all of the measurement, analysis, and modeling work required, as well as generating the information products needed to meet the specific end-user need. In others, contracts for measurement, data analysis, modeling, and the generation of specific tailored information products are let to multiple organizations, each delivering a part of the value chain leading to meeting the specific end-user need. In some cases, public organizations also contract private companies in similar ways. A more recent development has been towards public-private partnerships, under which private companies undertake extensive ocean data collection in partnership with public bodies and agencies.

Intermediate users are organizations that add value to ocean data and information tailoring it for specific end uses. In this context it is important to recognize that delivery of end-user benefit is frequently not a simple linear end-to-end chain. Benefits are often delivered by multiple organizations merging and mashing different sources of data and information, often not exclusively marine or exclusively derived from ocean measurements, observations, and models, to derive an information product used by a further intermediary in the value chain and, at the end of the chain, a service tailored to a specific end-use.

The markets for new blue economy products and services

In describing the markets for the new blue economy, it is important to distinguish between the broader socioeconomic benefits of new blue economy activities and market-based revenue generated through sale of new blue economy products and services. By definition, markets are areas or arenas in which commercial dealings are conducted.

New blue economy activities also underpin vital public goods, which deliver considerable societal value. An example is hurricane forecasts derived in part from new blue economy activities and delivered free of charge by public bodies. Commercial new blue economy businesses may build on these public good services to generate revenue from specific target markets. Coastal infrastructure risk assessments, built on public good services, and sold to the insurance and reinsurance markets by value-added commercial intermediaries, are a good example.

The developing data and information needs of the ocean and coastal economy are the major driver of growth of the commercial markets for new blue economy products and services. The OECD Ocean Economy 2030 report (OECD, 2016) describes the global ocean economy and its likely future growth trajectory through description of the components of the ocean economy and the methodologies employed to predict its future development. This report contains an extensive bibliography of published work associated with understanding and quantifying the ocean economy, providing a valuable resource for those seeking a better understanding of the opportunities and challenges associated with its future development.

The OECD report describes the ocean as the new economic frontier, holding the promise of immense resource wealth and with great potential for boosting economic growth, employment, and innovation. It also recognizes the role of the ocean in many of the global challenges facing the planet in the decades to come, from food security and climate change to the provision of energy, natural resources, and improved medical care. While the potential of the ocean to help meet these challenges is considerable, the report recognizes that the ocean is already under stress from overexploitation, pollution, declining biodiversity and climate change and that realizing the full potential of the ocean will therefore demand responsible, sustainable approaches to its economic development.

Accordingly, the OECD report addresses the activity of ocean-based industries, for example; shipping, fishing, offshore wind and marine biotechnology; and the natural assets and ecosystem services that the ocean provides (for example, fish, shipping lanes, CO_2 absorption). These two aspects of the ocean economy are inextricably interlinked in what is often referred to as the "blue economy" (see, for example, Kildow, this volume). As Kildow describes, there is considerable debate concerning the exact definition of this term. The World Bank defines it as sustainable use of ocean resources for economic growth, improved livelihoods and jobs, and ocean ecosystem health. Other definitions are more expansive, with some extending to the inclusion of all water-related economic activity, including freshwater supply and wastewater treatment and disposal.

The OECD defines the overall ocean economy as the sum of the economic activities of ocean-based industries, together with the assets, goods, and services provided by marine ecosystems, a definition broadly equivalent to the World Bank definition earlier.

It is important to put the OECD estimates of ocean market-based economic activity into context with the natural capital and ecosystems services the ocean provides. These have been estimated at US$ 24 trillion as a part of the estimated US$ 125 trillion in natural capital and ecosystems services provided by the Earth as a whole (De Groot et al., 2012).

While economic activity on, in, under, and around the ocean is expanding rapidly, driven primarily by growth in global population, economic growth, trade, rising income levels, and technology, this growth is constrained by any significant deterioration of ocean health and the US$ 24 trillion of nonmarket natural capital and ecosystem services the ocean provides (Hoegh-Guldberg et al., 2015).

The global ocean economy, measured in terms of the ocean-based industries' contribution to economic output and employment, is significant. Preliminary calculations on the basis of the OECD's Ocean Economy

Table 7.1 Established and emerging ocean-based industries.

Established	Emerging
Capture fisheries and aquaculture	Open water aquaculture
Seafood processing	Deep- and ultradeep-water oil and gas
Shipping	Offshore wind energy
Ports	Ocean renewable energy
Shipbuilding and repair	Marine and seabed mining
Offshore oil and gas (shallow water)	Maritime safety and surveillance
Marine manufacturing and construction	Marine biotechnology
Maritime and coastal tourism	High-tech marine products and services
Marine business services	Others
Marine R&D and education	

Adapted from OECD, 2016. The Ocean Economy in 2030. OECD Publishing, Paris.

Database value the ocean economy's contribution in 2010 at US$ 1.5 trillion, or approximately 2.5% of world gross value added (GVA). Offshore oil and gas accounted for one-third of the total value-added of the ocean-based industries, followed by maritime and coastal tourism, maritime equipment and ports. Direct full-time employment in the ocean economy amounted to around 31 million jobs in 2010. The largest employers were industrial capture fisheries, with over one-third of the total, and maritime and coastal tourism with almost one-quarter (Table 7.1).

Looking to 2030, many ocean-based industries have the potential to outperform the growth of the global economy as a whole, in terms of both value-added and employment. The OECD projections suggest that between 2010 and 2030 on a "business-as-usual" scenario basis, the ocean economy could more than double its contribution to global value-added, reaching over US$ 3 trillion. Particularly strong growth is expected in marine aquaculture, offshore wind, fish processing, and shipbuilding and repair. Ocean industries also have the potential to make an important contribution to employment growth. In 2030, they are anticipated to employ approximately 40 million full-time equivalent jobs in the business-as-usual scenario. The fastest growth in jobs is expected to occur in offshore wind energy, marine aquaculture, fish processing, and port activities (Table 7.2).

A number of countries are engaged in the production of satellite accounts seeking to describe and quantify ocean economy-related activity at a national level. The US National Ocean Economics Program has been tracking those components of the US national ocean economy contained in

Table 7.2 Estimates of industry-specific growth rates in value-added and employment between 2010 and 2030 (OECD, 2016).

Industry	Compound annual growth rate for GVA between 2010 and 2030	Total change in GVA between 2010 and 2030	Total change in employment between 2010 and 2030
Industrial marine aquaculture	5.69%	303%	152%
Industrial capture fisheries	4.10%	223%	94%
Fish processing	6.26%	337%	206%
Maritime and coastal tourism	3.51%	199%	122%
Offshore oil and gas	1.17%	126%	126%
Offshore wind	24.52%	8037%	1257%
Port activities	4.58%	245%	245%
Shipbuilding and repair	2.93%	178%	124%
Maritime equipment	2.93%	178%	124%
Shipping	1.80%	143%	130%
Average of the total ocean-based industries	3.45%	197%	130%
Global economy between 2010 and 2030	3.64%	204%	120%[a]

[a]Based on projections of the global workforce, extrapolated with the UN medium fertility rate.

the databases of the Bureau of Labor Statistics since 1990 (US National Ocean Economics Program, 2020). Data for all years beyond 2004 have been generated by the NOAA Coastal Services Center, leading to production of the NOAA Report on the US Ocean and Great Lakes Economy for 2016 (NOAA, 2019) and corresponding results for 2018 (BEA, 2020). In Europe, data is compiled annually by the Maritime Affairs and Fisheries Department of the European Union (European Union, 2019). Over the coming years, the majority of maritime nations are expected to publish similar national and regional ocean economy accounts.

There are several useful regional studies such as the World Bank and United Nations report on the potential for increasing long-term benefits of the sustainable use of marine resources for small island developing states and coastal least developed countries (World Bank, 2017), the blue economy handbook of the Indian Ocean region (Atri and Bohler-Muller, 2018), and the policy briefs of Partnerships in Environmental Management for the Seas of East Asia (PEMSEA) summarizing different segments of the ocean economy in East Asian Seas (PEMSEA, 2020).

A number of important meetings have been held to seek to harmonize the basis for studies of the ocean economy. These include the Center for the Blue Economy Ocean Summit on reaching consensus on national ocean accounts hosted by the Middlebury Institute of International Studies in 2015, a follow-up meeting sponsored by the Chinese National Marine Data Information Center in 2016, and the new approaches to evaluating the ocean economy symposium held at the OECD in November 2017, with lessons learned and outcomes summarized in the OECD report Rethinking Innovation for a Sustainable Ocean Economy (OECD, 2019). A further important meeting of the Global Ocean Accounts Partnership was held in 2019 (ESCAP, 2019).

Quantifying new blue economy business activity

Recent work by the OECD has sought to collate and summarize the existing literature concerning the benefits of sustained ocean observations (OECD, 2019).

Science remains a crucial driver for most ocean observations. Observations and measurements derived from diverse platforms (e.g., in situ, research vessels, satellite remote sensing) contribute to advancing fundamental knowledge on the ocean, weather, and the climate, directly and via their use in driving, calibrating, and verifying ocean, atmospheric, and climate models. In the Intergovernmental Oceanographic Commission's (IOC) Global Ocean Science Report (UNESCO, 2017), around 80% of data centers that provide ocean observation data, products, and services named scientific communities as their most important end users. Many of the societal benefits associated with improved science are not readily associated with economic value, partly because they do not flow through markets and do not generate economic benefits in and of themselves. For this reason, the literature has often considered ocean observations data to be a public good, the benefits of which are difficult to identify and value. Despite the relative complexity of valuing societal benefits, a number of recent studies have used a range of methodologies to do so. Further valuation of societal benefits is of particular importance to undertaking a thorough assessment of the overall value of the new blue economy and is of crucial importance to any future overall socioeconomic assessment.

There are a wide diversity of operational products and services based on sustained ocean observations. Based on the OECD literature review, weather forecasts (36%), sea state forecasts (21%), and climate forecasts (7%) are the products and services most taken up for operational use. Some of the traditional operational end-user groups include navies and coastguards, the offshore oil and gas industry, commercial shipping, fisheries, and aquaculture. User domains benefiting from ocean observations and covered the most by the literature do not mirror the distribution of these traditional user groups. This is because much of the work on quantifying these areas exists only in the "gray" literature rather than as peer-reviewed material.

The most comprehensive study of national new blue economy commercial market activity is the US Ocean Enterprise Study, conducted by the National Oceanic and Atmospheric Administration (Rayner et al., 2019). This study sought to quantify the 2015 scope, scale, and value of provider and intermediate business activity in the United States. The study built on methodologies used in a national assessment of new blue economy business activity in the United Kingdom (Society of Maritime Industries, 2018). This less detailed annual national survey to determine the extent and scope of UK new blue economy activity has been undertaken annually by the Society of Maritime Industries since 2009 (SMI, 2018). This survey is based on interviews conducted at Oceanology International and Ocean Business, the two principal UK trade shows for new blue economy businesses.

The US Ocean Enterprise Study set out to evaluate the scale, scope, and characteristics of new blue economy activity in the United States, through a comprehensive survey of US businesses engaged in provision of new blue economy products and services. It also explored the ways in which these provider and intermediary businesses interact with the US IOOS.

The first stage of the study was devoted to obtaining and compiling lists of companies who were potentially engaged in new blue economy activities. Lists of companies were obtained from a variety of sources including; professional organizations, industry-specific email and contact lists, professional and industry conferences, Federal agencies, and US IOOS stakeholders. The final database of target businesses comprised 410 companies.

Initial qualitative interviews were conducted with nine of these, representing both providers and intermediaries, as well as varied company sizes

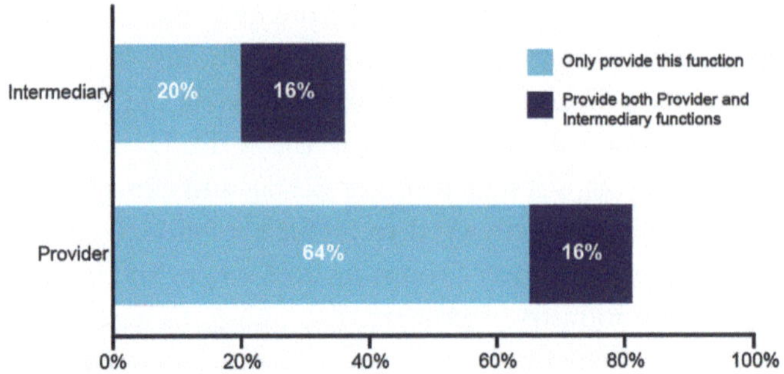

Figure 7.1 Business functions of respondent companies.

and geographical locations. The results of these interviews informed the content of a quantitative survey instrument and provided context for the quantitative survey results. These interviews took place during the fall of 2014. The survey of the 410 identified companies was conducted in 2015 by means of a web questionnaire.

The survey sought to categorize respondents' activity between providers of new blue economy infrastructure, such as sensors, instruments and platforms, and intermediaries that make use of ocean, coastal, and Great Lakes measurements, observations, and models as an input to the creation or enhancement of value-added information products in support of specific ocean and coastal end-use markets. Fig. 7.1 provides a breakdown of the survey results across these two types of business activity.

The majority of respondent firms (64%) were providers, involved solely in developing infrastructure and equipment related to the new blue economy. 20% of respondents identified themselves as solely performing activities as an intermediary; using ocean data and information to create or enhance value-added data products in support of end-user requirements. 16% indicated they performed both functions, yielding a combined 80% of respondents performing activities as a provider and a combined 36% providing services as an intermediary.

Respondents who indicated they filled the role of a provider of infrastructure for ocean observation, measurement, and forecasting were asked a question regarding the types of new blue economy infrastructure they sold. Nearly half (47%) provide physical oceanographic products of some type (Fig. 7.2). Approximately one-third offer hydrographic survey products. These results are consistent with what might be expected from the more mature mission areas associated with mapping, charting, tide prediction, and general oceanography.

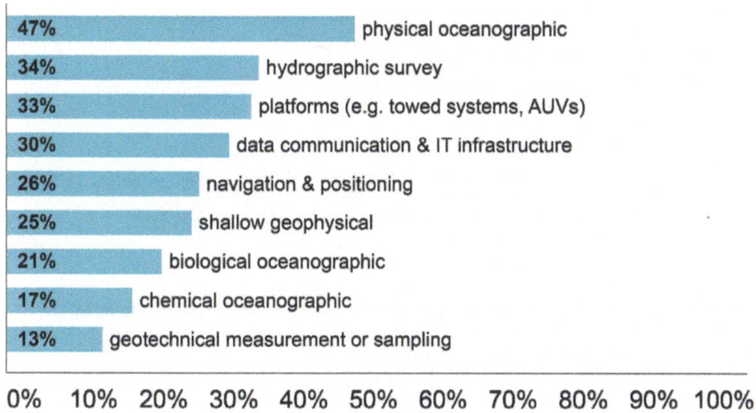

Figure 7.2 Types of new blue economy infrastructure sold.

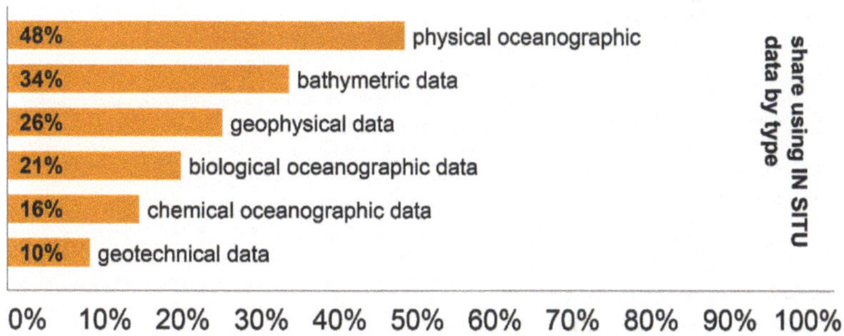

Figure 7.3 Use of *in situ* data by data type.

Organizations that indicated they were intermediaries were asked a series of questions regarding their use of in situ and remotely sensed data types. As shown in Fig. 7.3, nearly half use physical oceanographic data, with one-third using bathymetric data and one-quarter reliant on geophysical data streams. The widespread use of physical oceanographic data is not surprising given that most intermediaries would need this information to combine with other data type(s) to serve their service or product purposes.

Use of remotely sensed data is less common than in situ data. Remotely sensed data is often not as easy to access or utilize and is applied to more specialized versus general purposes as compared to in situ data. While approximately two-thirds of intermediaries rely on in situ data, 41% employ data that is remotely sensed. As shown in Fig. 7.4, shore and satellite observations are both used by one-third of respondents, while airborne observations are used by just less than one-fifth.

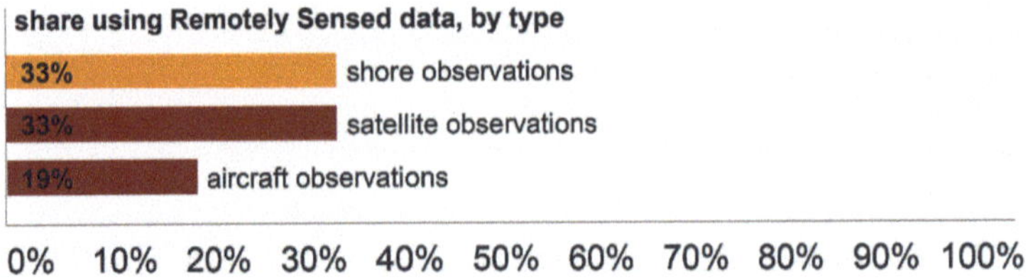

Figure 7.4 Use of remote-sensed data by data type.

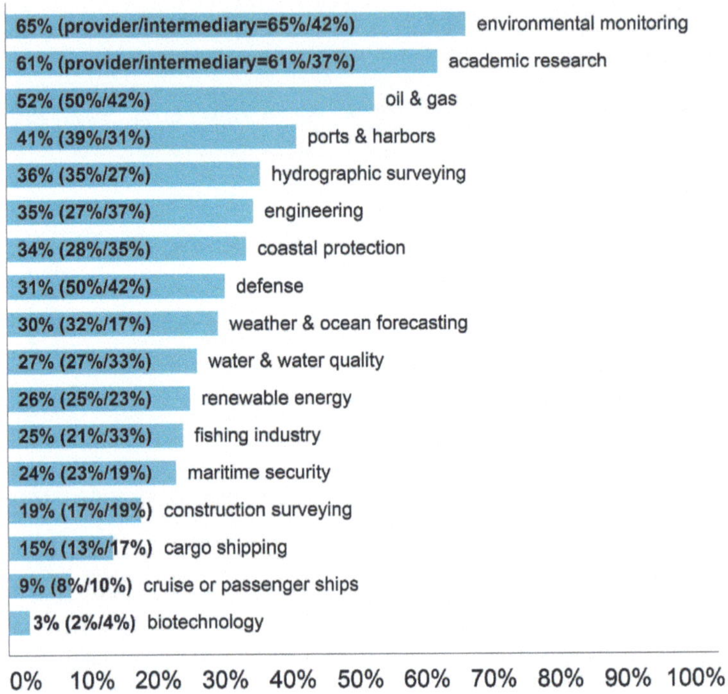

Figure 7.5 Market sectors served by respondent businesses.

Respondents were asked to report the market sectors with which their company is engaged. As shown in Fig. 7.5, the Ocean Enterprise space is incredibly diverse in the end-use markets that it serves. Three end uses are a target for at least half of the respondents; environmental monitoring, academic research, and offshore oil and gas, and a further seven are foci for at least one-third of respondents. The 25% threshold was associated with 12 different market sectors.

The number of market sectors chosen by respondents provides an indicator of diversity of function. As shown in Fig. 7.6, 14% of respondents

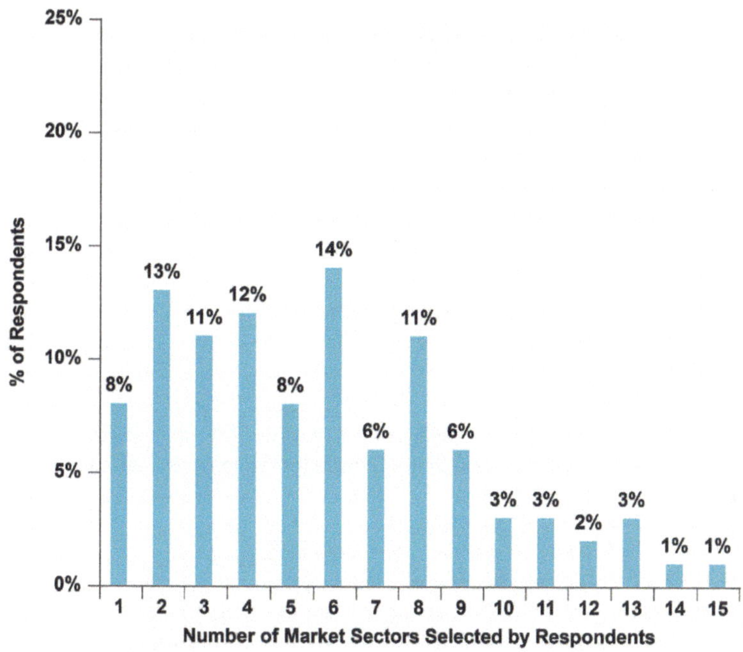

Figure 7.6 Number of market sectors served.

reported their company was engaged in six of the listed market sectors. A combined 48% of respondents reported their company was engaged in more than five of the listed market sectors. Only 8% of respondents reported their company was engaged in only a single listed sector.

Respondents with Ocean Enterprise export activity were asked to identify the geographical markets that are a source of at least 5% of revenues for their businesses and were allowed to select as many markets as applied. Unsurprisingly, for a survey of US businesses, the most mentioned market was North America. Nearly all firms (96%) report customers in North America, with 93% of those reporting customers in the United States. Of particular interest in the responses about foreign markets is the equal importance of Europe and Asia and the Pacific Rim, with 44% reporting sales in both regions. Summary results for the analysis of the export markets for survey respondents are provided in Fig. 7.7.

One of the key goals of the study was to provide estimates of revenues based on the self-reported data from the survey respondents and other public sources of information.

The proportions of revenue attributed to general maritime and ocean enterprise or new blue economy activity are shown in Fig. 7.8. Over half of

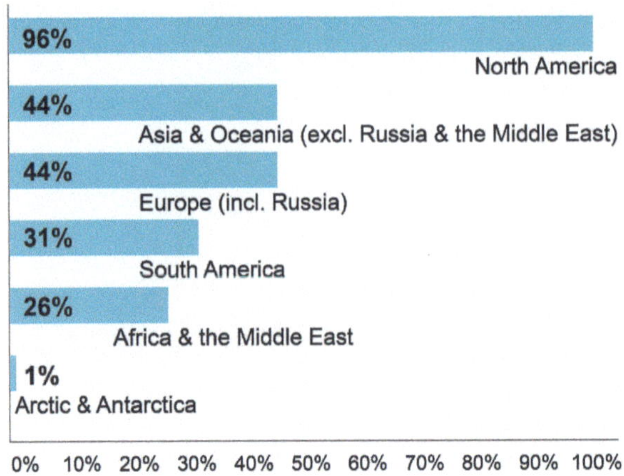

Figure 7.7 Geographical markets that account for at least 5% of total new blue economy revenue.

respondents (58%) reported that the majority (75%–100%) of their revenues were from marine and maritime business (top pie chart in Fig. 7.8). When asked what proportion of their maritime revenue was due to Ocean Enterprise or new blue economy activities, just under half (46%) reported that Ocean Enterprise revenues accounted for the majority (75%–100%) of their overall maritime activity (middle pie chart in Fig. 7.8). Respondents were further asked to report what proportion of their Ocean-Enterprise-related revenue was from export sales. 27% of respondents reported that foreign sources accounted for at least half of their Ocean-Enterprise-related revenues, while 49% reported that export sales contributed to less than one-quarter of their total Ocean-Enterprise-related business (bottom pie chart in Fig. 7.8). One-fifth reported that overseas customers accounted for none of their Ocean-Enterprise-related revenue.

Fig. 7.9 provides the overall estimates of revenues for businesses that have Ocean Enterprise activities. Total revenue for these businesses is estimated at $58 billion, with $14 billion of these overall revenues being attributable to maritime-related activities, and $7 billion of these maritime revenues being attributable to Ocean Enterprise business activities. Finally, it is estimated that $1.4 billion of Ocean Enterprise revenue is associated with export activity. The survey results indicate that companies that functioned as both Provider and Intermediary accounted for the majority of export revenue ($1 billion), which accounted for almost 35% of their Ocean Enterprise revenue ($3 billion). Not surprisingly, these dual-role

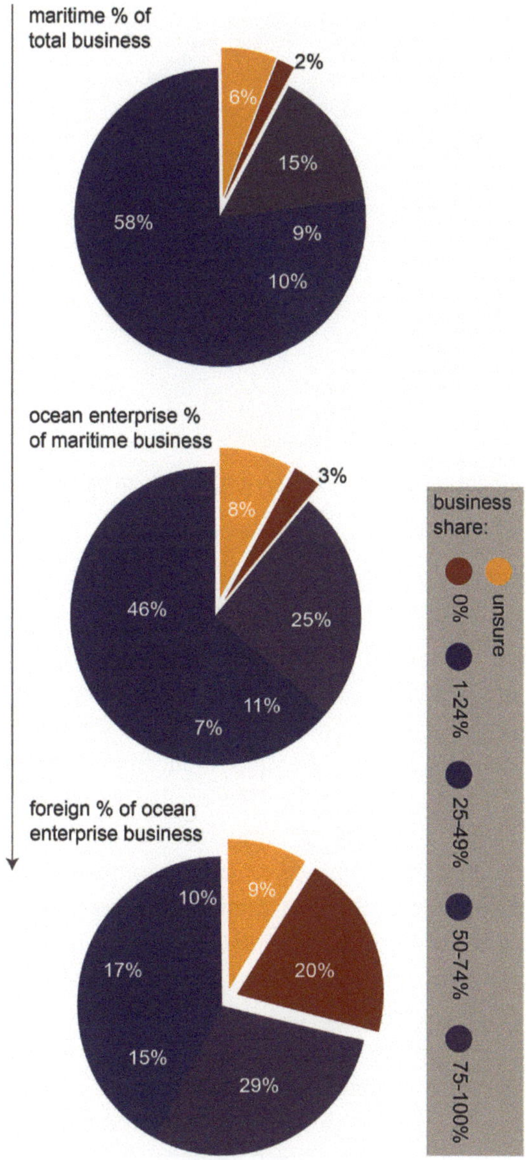

Figure 7.8 Proportions of revenue attributable to general maritime, domestic ocean enterprise (new blue economy), and ocean enterprise (new blue economy) export activity.

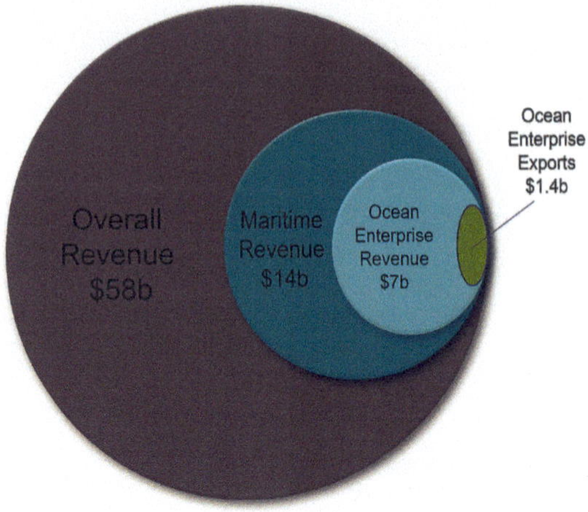

Figure 7.9 Components of overall revenues attributable to ocean enterprise (new blue economy) activities.

businesses tended to be larger and more diversified than those that served only one function.

Respondents were asked to report the employment levels of their organization. Using a similar methodology to that used for revenues, the employment data from the survey respondents was supplemented with data from secondary sources to project employment for all businesses thought to be involved in the Ocean Enterprise. The total employment is estimated to be between 223,000 and 268,000 employees in the United States. This employment estimate is for all business activity of companies who identified that they are active in provision of Ocean-Enterprise-related products and services.

The survey asked a series of questions that related to expectations about the future. Respondents were generally optimistic with regard to growth of their Ocean-Enterprise-related business activities in the forthcoming 12 months. As shown in Fig. 7.10, half of respondents expected growth for the Ocean-Enterprise-related market sectors of their business. Intermediaries were less likely to project growth as compared to providers and more likely to predict their Ocean-Enterprise-related business would decrease or stay relatively the same. This optimistic outlook for Ocean-Enterprise-related revenue mirrors the expectations for growth of the overall ocean economy.

For the same survey year as the US Ocean Enterprise study, the number of companies identified in the UK survey (SMI, 2018) was approximately one-quarter of the number of companies in the United States. UK new

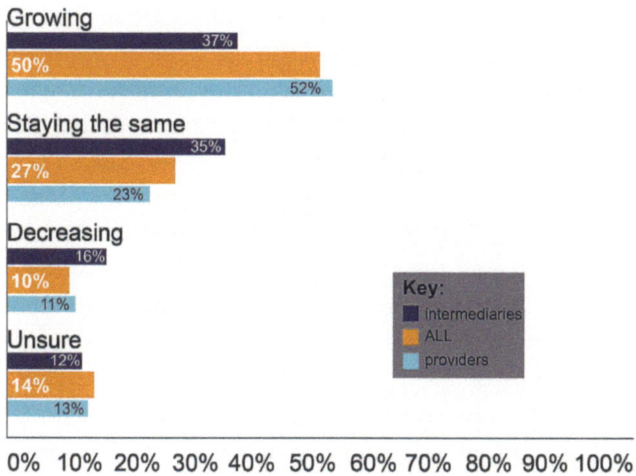

Figure 7.10 Business expectations.

blue economy revenue for 2015 was estimated at £1.8 billion ($2.75 billion at the average exchange rate in 2015), compared with the $7.0 billion for the United States; consistent with the smaller number of companies. Export activity represented a far larger proportion of the revenue for UK companies than for the US companies at £778 million ($1.2 billion at the average exchange rate for 2015) or 43% of total revenue, compared with $1.4 billion or 20% of total revenue.

Overall the characteristics of the two countries' Ocean Enterprise business activity are similar. Both have significantly more producers than intermediaries and provide services to similar markets with academic research, defense, offshore oil and gas, and ports being the top four for each. Both studies point to the global nature of the new blue economy with North America, Europe, and Asia—Pacific identified as the most important geographical markets.

The principal difference between the results of the two studies is the relative level of export activity, with UK exports being a far larger proportion of total revenue and having an absolute value close to that for the United States. This reflects the large differences in scale of the domestic markets in each country.

The US Ocean Enterprise Study has been repeated in 2020 (IOOS, 2021) allowing changes over the five year period since the first study to be determined. Other nations are also planning to undertake surveys of their new blue economy activities including a Canadian study for 2020, which is using a survey methodology similar to that of the US work.

Based on the results of the 2015 surveys conducted in the United States and the United Kingdom, it is possible to calculate an approximate 2015 global revenue estimate for new blue economy business activity based on relative 2015 GDP estimates for developed nations. Using International Monetary Fund 2015 GDP estimates (IMF, 2020) based on Purchasing Power Parity across developed economies, global new blue economy revenue is estimated at between US$20 billion and US$ 28 billion on the assumption that revenues for each developed national economy lie somewhere in the range of the proportion of total national GDP represented by US and UK activity.

If the projections of 2030 value-added derived in the OECD Ocean Economy 2030 hold true, and the mix of ocean economy markets demands similar levels of new blue economy commercial activity as in 2015, then new blue economy business revenue might be expected to approximately double by 2030, to of order US$ 50 billion.

Apart from the obvious limitations of the simplistic basis used to derive this possible global growth estimate, there are important constraints on delivering this level of growth in new blue economy business activity.

At the time of writing, the world is in the grips of the Coronavirus pandemic. The global impacts of this crisis have the capacity to radically reshape markets, rendering "business as usual" scenarios of future change grossly inaccurate.

Potentially even more disruptive are the impacts of climate change. As anthropogenic carbon emissions have risen over time, the ocean has absorbed a significant proportion of the excess carbon dioxide emitted into the atmosphere, leading to ocean acidification. Ocean temperatures and sea levels are rising and ocean currents shifting, resulting in biodiversity and habitat loss, changes in fish stock composition and migration patterns, and higher frequency of severe ocean weather events. The prospects for future ocean development are further aggravated by land-based pollution, in particular agricultural run-off, chemicals, and macro- and micro-plastic pollutants that feed into the ocean from rivers, as well as by overfishing and depleted fish stocks in many parts of the world.

Global economic and environmental disruptions such as the Coronavirus pandemic and anthropogenic climate change will result in difficult-to-predict changes to markets. There will be winners and losers. For example, in the case of climate change, it must be hoped that exploitation of offshore oil and gas will be a declining activity, while other markets such as the exploitation of offshore wind will increase in importance. Major disruptions will also accelerate the growth of new markets for ocean data and information products. For example, the market for data and information needed to incorporate climate change into the design of coastal assets is likely to grow in response to increased risks associated with sea level rise and greater storm frequency (Rayner, 2019).

Most importantly, delivery of the economic potential of the ocean, and in turn the potential of the new blue economy, fundamentally depends on ensuring that the use of the ocean and ocean resources is sustainably balanced with protecting the natural capital and ecosystems services that the ocean and our planet provide.

References

Atri, V.N., Bohler-Muller, N., 2018. The Blue Economy Handbook of the Indian Ocean Region. Africa Institute of South Africa, Pretoria, South Africa.

BEA, 2020. Defining and Measuring the U.S. Ocean Economy. Bureau of Economic Analysis. https://www.bea.gov/system/files/2020-06/defining-and-measuring-the-united-states-ocean-economy.pdf.

De Groot, R., Brander, L., Sandervander, P., Costanza, R., Bernard, F., Braat, L., Christie, M., Crossman, N., Ghermandi, A., Hein, L., Hussain, S., Kumar, P., McVittie, A., Portela, R., Rodriguez, L.C., Brinkm, P., Beukering, P., 2012. Global estimates of ecosystems and their services in monetary units. J. Ecosys. Serv. 1, 50–61.

ESCAP, 2019. Global Dialogue on Ocean Accounting and First Annual Meeting of the Global Ocean Accounts Partnership. https://www.unescap.org/sites/default/files/Global_Dialogue_on_Ocean_Accounting_Meeting_Report_12-15Nov2019.pdf.

European Commission, 2019. The EU Blue Economy Report 2019. Publications Office of the European Union, Luxembourg.

Hoegh-Guldberg, O., et al., 2015. Reviving the Ocean Economy: The Case for Action - 2015. WWF International, Gland, Switzerland., Geneva, 60pp.

IMF, 2020. International Monetary Fund World Economic Outlook Database. https://www.imf.org/external/pubs/ft/weo/2015/02/weodata/index.aspx.

IOOS, 2021. https://ioos.noaa.gov/project/ocean-enterprise-study/.

NOAA, 2019. National Oceanic and Atmospheric Administration (NOAA), Office for Coastal Management. 2019. NOAA Report on the U.S. Ocean and Great Lakes Economy. NOAA Office for Coastal Management, Charleston, SC. http://coast.noaa.gov/digitalcoast/training/econreport.html.

NOEP, 2020. National Ocean Economics Program. https://www.oceaneconomics.org/Market/.

OECD, 2016. The Ocean Economy in 2030. OECD Publishing, Paris.

OECD, 2019. Rethinking Innovation for a Sustainable Ocean Economy. OECD Publishing, Paris.

PEMSEA, 2020. Policy Briefs for the Blue Economy. http://pemsea.org/publications/policy-briefs.

Rayner, R.F., 2019. Incorporating climate change within asset management. In: Lloyd, C., Corcoran, M. (Eds.), Asset Management: Transforming Asset Dependent Businesses. Thomas Telford Press, pp. 143–161.

Rayner, R.F., Gouldman, C., Willis, Z., 2019. The ocean enterprise -understanding and quantifying business activity in support of observing, measuring and forecasting the ocean. J. Oper. Oceanogr. 12 (2), S97–S110.

Society of Maritime Industries, 2018. Annual Review of UK Marine Scientific Industries. http://www.maritimeindustries.org/CoreCode/Admin/ContentManagement/MediaHub/Assets/FileDownload.ashx?fid=179875&pid=10160&loc=en-GB&fd=False.

Spalding, M.D., Brumbaugh, R.D., Landis, E., 2016. Atlas of Ocean Wealth. The Nature Conservancy Arlington, VA.

UNESCO, 2017. Global Ocean Science Report - the Current Status of Ocean Science Around the World. UNESCO Publishing, Paris.

US National Ocean Economics Program, 2020. https://www.oceaneconomics.org/.

World Bank, 2017. The Potential of the Blue Economy: Increasing Long-Term Benefits of the Sustainable Use of Marine Resources for Small Island Developing States and Coastal Least Developed Countries. World Bank, Washington DC.

OP/ED: The importance of the new blue economy to a sustainable blue economy: an opinion

Judith T. Kildow

Introduction

The purpose of this op-ed is to clarify and expand our understanding of the importance of the economy that depends on the ocean. While the other chapters of this book focus mainly on a narrow set of fundamentals that contribute to the ocean's value, the new blue economy (NBE), this opinion piece takes a broader look at the ocean-dependent economy and puts the NBE in context, so that readers can understand its importance within the larger framework of commercial and industrial market activities, public goods, and the often-neglected nonmarket benefits associated with environmental and ecological assets and services.

The ocean-dependent economy has become an important part of the conversation in international circles, because the ocean floor, water column, and coastal waters have become essential alternatives to declining land options that have traditionally met most societal needs. Oceans and shorelines now draw increasing commercial, industrial, and environmental attention, not only for traditional uses but for new uses and enhanced, more efficient, and cleaner traditional uses.

Why has the blue economy only recently risen to the surface of international discussion as a panacea for economic growth? Rationally and ideally nations should have discovered the potential for offshore economies in the early 1980s when they were awarded 200 mile exclusive economic zones (EEZs)—an extension of their national jurisdictional boundaries—under the UN Convention on the Law of the Sea (UNCLOS, 1982). With some vision, this bounty should have prompted nations to focus on the offshore potential and develop methods to track and reflect the impressive revenues generated by coasts and oceans at the time, as well as the economic potential of this newly acquired expanse of territory. However, few had that vision. Although the first efforts at identifying the role of oceans in national economies took place in 1974 (Nathan Associates, 1974), it would take another 25 years for nations to estimate the ocean's value through national income accounts, recognizing their importance as a new frontier for meeting societal needs when compared to the limited potential of the land. Now, turning their attention offshore, nations are apparently surprised at the size of the additional space they acquired from the UNCLOS agreement and are now eager to maximize revenues by engaging in the blue economy.

Definition of terms

The field of ocean economics is still evolving after several decades of effort, and the use of terms remains uneven and confusing. However, there are international efforts to reach a consensus-based, more uniform set of definitions and terms, industrial categories, and geographies at particular points in time. The following definitions reflect the author's interpretation of terms from the literature and current community usage at meetings.

- *The Ocean Economy* describes economic activity that indirectly or directly uses the ocean as an input or from the production side (The National Ocean Economics Program, NOEP 2020); it can also describe economic activity from the consumption side, often applied to the commercial and recreational fishing industry (NOAA, 2016).

- *Marine Economy* is used as a synonym for Ocean Economy.
- *Maritime Economy* includes marine transportation and port activities, although some nations use "maritime economy" synonymously with Ocean Economy and Marine Economy (Park and Kildow, 2015).
- *The Blue Economy* is the large economic engine that drives the value of the oceans. Its exact definition has provoked considerable debate, but for this discussion, it will be defined to include three types of account values—economic, environmental, and ecological, which together describe a sustainable ocean economy in which economic activity and environmental impacts are balanced with the long-term capacity of ocean ecosystems to support economic activity while remaining resilient and healthy (OECD, 2016). Elaborating further, the Blue Economy is:
    (1) economic activity of ocean-based industries, e.g., shipping, fishing, offshore energy, construction, and marine biotech, together with the innovative technologies that are going to be needed: e.g., nonfossil fuel ship power, ways to make renewable energy generation cheaper, discovering new materials for fishing gear, and other uses that are degradable and lightweight to get plastics out of the ocean, as well as developing new coral reef restoration and regeneration strategies (Colgan, 2016);
    (2) natural environmental assets, e.g., renewable resources—fish and plant protein and nonrenewable—mineral resources, water; and
    (3) ecosystem services, e.g., oxygen production, climate regulation, shipping lanes, and CO_2 absorption.

Much rarer definitions for the blue economy include all water-related economic activity, including freshwater supply and wastewater treatment and disposal, although the latter use is rare.

- *The Global Ocean Economy*, as defined by the OECD at its 2016 meeting and its accompanying report (OECD, 2016), is the sum of the economic activities of ocean-based industries, together with the assets, goods, and services provided by marine ecosystems, a definition equivalent to the World Bank definition of the Blue Economy. There is some question whether this also includes all ocean industries including some traditional ones that continue to impose unacceptable environmental damages, such as offshore oil and gas production and deep sea mineral production.

- *The New Blue Economy (NBE)* is defined and described in great detail in this book, in the chapters by both Richard Spinrad and Ralph Rayner, respectively. Briefly, it is the data derived from sustained ocean observations, monitoring, communications, and measurements, informing much of the economic activity around, on, and under the ocean. It is also known as the knowledge-based, digital economy, the innovative generation of scientific information delivered by communications technology products (ICT). It generates a substantial economic activity on its own, feeding key information to other industries, such as weather forecasting, insurance pricing, agricultural scheduling, and fish stock predictions (Colgan, 2020).
- *The Ocean Enterprise* is "the business activity [linked to] ocean observation, measurement and forecasting," according to a 2016 study (IOOS/NOAA, 2016).

As the operational component of the NBE, it consists of a unique and growing cluster of industries 410 firms in 36 states that generated more than $7 billion toward the US GDP (IOOS/NOAA, 2016) that delivers the means to collect, communicate and use ocean data and information on a sustained basis through models, assessments, and forecasting. All of these provide a wide range of societal benefits underpinning the safety, operational, and compliance needs of beneficiaries that operate around, on, and under the ocean. That cluster also is an essential input to marine scientific research that provides effective protection of the ocean environment, while also delivering significant economic and employment benefits (Rayner et al., 2019).

- *The Coastal Economy* describes all economic activity that takes place in geographically defined coastal areas: states, regions, watersheds, counties, zip codes, etc (National OceanEconomics Program, 2020). While these data are not generally collected by many nations, coastal economies have become increasingly important because of their large and disproportionate contribution to national economies. They are also exposed and vulnerable to some of the largest threats from climate-induced impacts as well as the results of societal overuse and misuse of natural resources. Two of the most important ocean-related industries, coastal ocean tourism and recreation and fishing, are most at risk from these threats, so tracking coastal economies helps nations determine the knowledge, technologies, and strategies they need to protect their shoreline economies for the future.

Measuring the economy of the oceans

All of these definitions assume that nations have metrics by which they can measure economic progress or lack thereof. Selecting which metrics to use as indicators of growth and decline is yet another definition issue. Since each nation has its unique economic strengths and weaknesses, each has its own preferences about what to measure. Before selecting indicators to measure performance, they need to know what they are measuring and, if comparing one nation or industry to another, there should be some uniformity among nations for categories of economic activities: what industrial sectors and industries are included, the geographical attributes, and the time frames that are being estimated. Once some precision is reached regarding important indicators, metrics, geographies, and time frames, measurement of these activities is possible. To ensure that a true picture of the ocean and coastal-dependent economy is revealed, nations are taking the additional steps of estimating the values of both environment and ecological costs and benefits derived from impacts of the economic activities on the natural resources and systems upon which most of them rely for their future.

During the 1990s, the United Kingdom, China, and a US-funded private-university-based group, the National Ocean Economics Program (NOEP), began to address this important ocean slice of the national economy. Canada, France, Australia, Ireland, and a few others followed the example of these nations. The NOEP went further than the other nations. Rather than producing only one time or periodic reports, the NOEP created an online public presence, publishing annual government data that were categorically, spatially, and temporally consistent across multiple years for both ocean and coastal economic sectors and individual industry data (NOEP, 2020). Over the years this platform has received global attention from more than 190 nations. By 2014, more than 30 nations were compiling ocean sector data. Today, not only North American but also European, Asian, and African nations, some Island nations, and a few Latin American nations either have created or are seeking assistance to create ocean satellite accounts to better understand their current and future prospects for supplementing their land-based economy with offshore growth.

As this global awakening to the importance of the coastal and ocean-dependent economy has occurred, there has been a growing desire for greater uniformity among nations. Several international meetings in the past 7 years, initiated in the United States, followed by China and Europe, have facilitated a growing consensus among the participating nations about primary or "core" accounts, which include the following categories: marine transportation and ports, offshore mineral resources, living resources, marine construction, marine research and education, ocean-related professional and technical services, utilities, marine tourism and recreation, ship and boat building and maintenance, and ocean-related government programs. Beyond these categories, the OECD has identified "emerging industries," which may not be included separately in most national ocean accounts but which they believe will be significant in the near future and need tracking (OECD, 2016). More nations are also using "extended accounts," which represent natural resource values that are not reflected in the market system as well as damage costs to the economy from environmental misuse and overuse of natural resources—or "natural capital" as these values are often referred to. Some nations are attempting to value the social benefits from ecosystem values and incorporate those into their national ocean accounts. Through these added accounts, net values are generated for the contribution of the oceans to the economy. The European Union and the Partnerships in Environmental Management for South East Asia (PEMSEA) member states have been leaders in extended accounts. These accounts are made possible with the knowledge and products of the NBE and Ocean Enterprise, facilitating the distribution of important environmental data that can be modeled to estimate environmental values and human benefits. The NBE is an important sustainability link for these efforts.

Ocean accounts globally are works in progress. No two are alike in content, category, geography, scale, or detail. Because so many nations now see the need to have ocean accounts and lack the capacity to create them on their own, complicated by the diversity of attempts, the UN recently stepped in with efforts to standardize a basic framework for data collection and interpretation of the ocean economy with the goal of achieving a Global Ocean Product (GOP) in the future.

We know that the oceans and coasts are important generators of revenue for global, national, and regional economies, as evidenced by both expanded interest and successful investments over the past decades. For example, according to National Ocean Economics Program (NOEP) statistics, coastal states represented ~83% of the US gross regional product, employment, and earned income in 2018, although claiming ~57% of land mass. Counties that are considered part of the US coastal zone represented ~49% of the US economy, yet represent ~20% of the land. In many coastal nations, these numbers are even higher, indicating coastal economies are critical to national economies and warrant tracking and attention.

Using a different set of metrics, ocean-dependent industries located in coastal states accounted for >3% of the US GDP in 2017, a seemingly small but actually large amount of revenue and employment. A UN report issued a few years earlier indicated the GOP was more than $3 T, or >5% of the Global Product. These are impressive numbers that had not been published previously (UN, 2015). Adding to its importance, the economy of the ocean has demonstrated greater resilience, stability, and strength than the national economy as a whole during economic declines. During the large recession in 2008, the US ocean economy didn't dip as far and came back faster than the US national economy (NOEP, 2020).

Sustainability and the high risk of loss

Concerns over sustainable practices have accompanied this growth in past decades as understanding of the oceans and its rapid changes has provided important information about threats to ocean health. The interpretation of what is "sustainable" is not always clear and is sometimes disputed. Some nations, newly discovering the ocean's potential for them, often see it as nations that have seen it for centuries: a place for extraction, transport, and development to make lots of money in the short term, neglecting environmental concerns and disregarding future implications. These nations, desperate for revenue, believe they must act expeditiously while deferring efforts to deal with sustainable implications downstream. Others take a more enlightened view, considering the massive environmental changes underway in the ocean and along its shores that demand a changed perspective, balancing revenue generation with conservation and preservation of ocean well-being. However, increasingly nations use the UN Sustainable Development Goals (SDGs) as their framework, which is a hopeful sign (UN, 2015).

Interest in the sustainability of the oceans accelerated in 2015 and 2016 when two reports drew the attention of world leaders. These reports have been integrated into many national planning efforts in recent years:

1. The UN SDGs, especially #14, Life Below Water, known widely as the Global Goals, were adopted by all United Nations Member States in 2015 to "protect the planet and ensure that all people enjoy peace and prosperity by 2030." Goal #14 calls for nations "to sustainably manage and protect marine and coastal ecosystems from pollution, as well as address the threat of ocean acidification." It summarizes the ocean's contribution to our very lives by reminding us that we

are a blue planet, one whose oceans (a) cover three-quarters of our planet's surface, (b) produce >50% of the oxygen we breathe, (c) absorb >30% of the CO_2 we emit, and (d) hold 97% of Earth's water. Additionally, it highlighted the economic importance of the ocean by listing the importance of the ocean's biodiversity, which supports livelihoods for >3 billion people, including 200 million fishermen worldwide. It also noted the harm from subsidies that encourage depletion of fisheries, which the UN estimated results in >50 billion dollar losses annually (UN General Assembly, 2015).

2. The 2016 OECD Report: "The Ocean Economy in 2030," provided a forward-looking assessment of the ocean economy to 2030 and beyond, with particular emphasis on the development potential of established and emerging ocean-based industries, including implications for the ocean environment and its management. The data collection of evidence provided in this report, its analyses, and forecasts have reverberated throughout discussions and reports from national efforts to value the oceans.

At the time these two reports were released, few people had recognized the urgency of cries for sustainability of the world's oceans and coasts. However, these reports ignited interest and enthusiasm among officials of both international and national institutions, major corporations, and the media. It became clear that the flourishing blue economy with all of its promise was at great risk of loss from natural changes such as climate effects exacerbated by human-induced pollutants. Coastal areas are probably most at risk. As sea levels continue to rise and more intensive storms occur, unprecedented flooding and inundation of physical as well as natural capital could deeply affect human capital and populations that depend on economic activity along coasts.

The oceans are also faced with major threats. They have absorbed much of the CO_2 emitted by modern societies, changing the chemistry and affecting marine communities; increasing carbon changes the chemistry of the ocean, dramatically increasing its acidity, which in turn alters the habitat for many ocean resources. Polar ice is rapidly melting, creating water temperature and salinity changes and rising seas. All of these are altering the environment for key environmental and economic resources such as coral reefs and commercial fisheries as well as altering the entire marine food web.

In the United States, a billion-dollar oyster industry along the coast of Washington State crashed several years ago from a combination of acidification and warming waters. Growers lost huge sums of revenue and had to spend even more to work around this problem (Welch, 2012). Another localized example on the US East coast was a bay scallop die-off in Peconic Bay, New York in 2019, attributed to unusually warm summer seas in the area that killed off the adult populations. The local fishing communities around the bay were hard hit economically and a staple food for the area was largely erased (Booty, 2019). Around the world, island nations are experiencing rising and warming waters, watching their fisheries and other coastal industries decline as a result.

In addition to these cases of private industry loss, there are the enormous losses to governments and the public incurred by damage to federal, state, local, and private coastal structures. An example of a potentially huge loss to both the public and private sectors is the withdrawal by many insurers from coverage of coastal property in the face of threats from sea level rise and other flooding because it is no longer profitable. They warn that coastal property values will drop precipitously in the next few years as a result, and property will be unsaleable. The loss of tax revenue from devalued, uninsurable properties could be catastrophic to communities, states, regions, and the federal government, not to speak of the ripple effects throughout society (Kusisko and Campo-Flores, 2018).

Some local communities are already feeling the pinch: in Florida "Miami is investing $400 million to raise streets and install water pumps, and St. Petersburg is investing $360 million to adapt its wastewater treatment system for heavier rainstorms, flooding, and seawater intrusion. In South Carolina, mayors of smaller coastal towns and cities are being asked to assess the infrastructure costs they face in adapting to the coastal impacts of climate change" (Scorse and Helvarg, 2019).

Globally, large migrations away from the coasts will occur, propelled by droughts, inundation, starvation, and homelessness. Some have already begun.

Globally, ocean policies must consider both the rich potential and the devastating possibilities the ocean has to offer. The growth of ocean innovation and knowledge colliding with the twists and turns of climate threats and impacts provides a perfect opportunity to overcome some of the technical problems of the past that have impeded our response to these existential threats. One obstacle remains that has stymied progress toward solutions: communicating widely and clearly the urgency and importance of the decline in ocean health and the increasing number of coastal risks that threaten our survival on this planet. This paradigm shift in our behavior and attitude toward the environment has been underway for 50 years, but the scale and scope of the effects at all levels of global economies are not yet broadly understood. Negative impacts on the blue economy will overwhelm the entirety of the economy unless steps are taken soon to protect the positive contributions of coastal and ocean economic activities.

Encouraging sustainable practice

Impressive efforts in community building toward sustainable coastal management can be found in The Partnership in Environmental Management for the Seas of East Asia (PEMSEA), a regional partnership of 11 ASEAN nation states that for the past 25 years has been working toward incorporating sustainable practices into management of their coasts through partnerships.

PEMSEA promotes integrated coastal management among its partners and encourages public and private investment in sound sustainable practices. PEMSEA maintains a Sea of Knowledge Bank with a library of literature regarding sustainable cases and solutions (http://www.pemsea.org/publications/case-studies). Most recently, on PEMSEA's 25th anniversary, a report was issued describing these cases with the hope they might be models for others to follow. As the report observed:

> *The private sector plays a critical role in building a blue economy. Across the entire economy, the top 100 environmental externalities for business are estimated to cost around $4.7 trillion a year. Companies increasingly recognize the value inherent in managing environmental and social issues, which could be as high as 25%-70% of earnings .... More than 50 studies now demonstrate that companies leading on environmental and social performance financially outperform their less sustainable peers*
>
> ***Whisnant and Reyes (2015).***

PEMSEA is an unusual organization that has been ahead of its time, setting the standard and example for other organizations and nations to follow. Its members are encouraged with effective incentives to promote sustainable practices as they implement sustainable blue economies by incorporating the best information and practices through an integrated approach to considering multiple activities and their impacts—as opposed to the silo fashion of operating that for so long has dominated the philosophies of the industries that comprised the traditional blue economy.

The drive for sustainable oceans is at the forefront of international organizations now. The Indian Ocean Rim nations and their associated regional organizations, in addition to PEMSEA, are planning major initiatives for a thriving and sustainable blue economy. The World Bank, European Union, OECD, and the UN are heavily engaged in helping nations pursue sustainable goals for the ocean through economic accounting methods and strategies that promote the health of the oceans and the economy.

Conclusion

The combination of high-value assets serving societal needs and high-level threats with a prospect of major financial losses demands recognition and commitment to the sustainable uses of the oceans, using the discoveries, innovations, and strategies generated by the NBE and transferred to the larger blue economy. Just as scientists warn to cut carbon emissions now to avoid disasters, policy-makers should urge the development and implementation of economic plans soon to avoid the unaffordable costs that ocean and coastal changes will impose.

This is an area where the work of science and engineering can be powerful. New technologies are essential to achieve sustainability. Measurable metrics are required to compare differing scenarios under different time frames and to track the health of both the oceans and the economy dependent on it. Translating the rapidly growing knowledge base generated by the NBE, along with its ocean enterprise industries that enable it through advanced observation and communication technologies into useable forms is crucial to a sustainable ocean world. Equally important is the urgent transformation of innovative technologies such as new materials, energy sources, renewable fuels, and efficient and safe production strategies into useful products for ocean-dependent industry practices, converting them from unsustainable to sustainable users of the world's oceans as they struggle to address the rapid changes threatening oceans and coasts.

Actions are needed now as crises unfold, not after the crises occur. We need to address the current crisis, envision the future, and plan for it, or the effects of the NBE on the larger blue economy, that global economic engine, will disappoint. The state of the ocean's health and the unpredictable changes underway there that threaten it and the entire planet may represent the most existential threat to society in our lifetimes.

References

Booty, P, November 13, 2019. Peconic Estuary Program Connects Scallop Disaster to High Bay Temperatures and Global Warming. SAGHARBOREXPRESS.com.
Colgan, C.S., May 2020. Conversations.
IOOS, NOAA, 2016. A Study of US Business Activity in Ocean Measurement, Observation and Forecasting. www.ioos.noaa.gov/oceanstudy.oceanstudy@noaa.gov.
Kusisko, L., Campo-Flores, A., April 20, 2018. Rising sea levels reshape Miami's Housing market. Wall St. J. https://www.wsj.com/articles/climate-fears-reshape-miamis-housing-market-1524225600.
Nathan Associates, 1974. Gross Product Originating from Ocean-Related Activities. U.S. Department of Commerce, Washington D.C.
National Ocean Economics Program, 2020. Oceaneconomics.org.
NOAA, 2016. Fisheries Economics of the United States. https://www.fisheries.noaa.gov/content/fisheries-economics-united-states-2016.
OECD, 2016. T. Paris: The Ocean Economy in 2030. OECD Publishing. https://doi.org/10.1787/9789264251724-en.
Park, K.S., Kildow, J., 2015. Rebuilding the Classification System of the Ocean Economy. J. Ocean Coast. Econ. 2014 http://cbe.miis.edu/joce/vol2014/iss1/4/?utm_source=cbe.miis.edu%2Fjoce%2Fvol2014%2Fiss1%2F4&utm_medium=PDF&utm_campaign=PDFCoverPages.
PEMSEA. http://www.pemsea.org/our-work/blue-economy.
Rayner, R., Gouldman, C., Willis, Z., 2019. The Ocean Enterprise – understanding and quantifying business activity in support of observing, measuring and forecasting the ocean. J. Oper. Oceanogr.
Scorse, J., Helvarg, D., March 13, 2019. Putting the Blue in the New Green Deal (Commentary). Mongabay.
United Nations General Assembly, 1982. United Nations Convention on the Law of the Sea. https://www.un.org/depts/los/convention_agreements/texts/unclos/unclos_e.pdf.
United Nations General Assembly, 2015. UN Resolution 70/1, 2030. SDG Agenda, #14.
Welch, C., October 18, 2012. Oysters in Deep Trouble: Is Pacific Ocean's Chemistry Killing Sea Life? Seattle Times. https://www.seattletimes.com/seattle-news/oysters-in-deep-trouble-is-pacific-oceans-chemistry-killing-sea-life/.
Whisnant, R., Reyes, A., 2015. Blue Economy for Business in Asia: Towards an Integrated Understanding of the Blue Economy. PEMSEA. http://www.pemsea.org/sites/default/files/PEMSEA%20Blue%20Economy%20Report%2011.10.15.pdf.

Further reading

Colgan, C.S., 2016. Measurement of The ocean economy from national income accounts to the sustainable blue economy. J. Ocean Coast. Econ. 2 (2) https://doi.org/10.15351/2373-8456.1061. Article 12.

SECTION III

Programs that deliver

The value proposition of the New Blue Economy is realized in the form of traditional sectors as well as emerging development of new markets. Starting with fisheries, **Josh Kohut et al.,** express the value of research in terms of this sector, especially the potential role of the New Blue Economy in the context of traditional commercial fisheries management. The interview with **Mitch Roffer** (whose business was oriented to recreational fishing) provides an example of the development of ocean information for use in commercial operational oceanography. In an interview, **Perry Raso** discusses implications of data and information for aquaculture. The energy sector is another important area of application for the New Blue Economy. **Simon Geerlofs** discusses the application of ocean information to the emerging industry of marine renewable energy. The government perspective is provided by **Bill Brown**, who addresses environmental considerations for siting and permitting of ocean energy extraction activities. Oil and gas industry perspectives are provided in the interview with **Sam Walker** (with attention to oil spill response) and the chapter by **Jan Van Smirren** (focusing on the industry's support activities). The intersection of information about the world's oceans and attention to human health is introduced by **Paul Sandifer et al.,** with special attention to historic work in the context of the New Blue Economy in an interview with **Rita Colwell.** Recent highlighting of concerns about harmful algal blooms (often known colloquially as "red tides") is addressed in an op-ed by **Barb Kirkpatrick.** And the relatively new capacity for measuring the environmental DNA of the ocean is presented by **Alan Curry and Jesse Ausubel**. **Tom Drake** addresses the military application of operational oceanography, specific to the US Navy. Finally, a broader set of environmental applications that will benefit from build-out of the New Blue Economy are addressed from a number of different perspectives. **Tim Gallaudet et al.** focus on implications for the Arctic, an important geography of particular concern/interest in light of global climate change. Concerns about emerging developments with marine litter are expressed in an interview with **Emily Smail.** A case study for water mapping and the use of autonomous technology—especially in the context of oil spills—is presented in an interview with **Oscar Garcia-Pineda.** Natural hazard damage mitigation is addressed in a treatment on tsunami prediction by **Vasily Titov,** while **John Englander** demonstrates the value of data and information in addressing sea level rise. The recreational applications of ocean information are addressed, with a focus on needs for enhanced data science, in an interview with **Ben Freeston.** This discussion of "Programs that Deliver" concludes with the broadly applicable treatment of the importance of seafloor mapping, by **David Millar.**

Fisheries

CHAPTER 8

A new blue economy: connecting ocean information with the needs of the recreational and commercial fishing industry

Josh Kohut[1], Matthew J. Oliver[2], Kevin Wark[3], Matthew Breece[2], Laura Nazzaro[1]

[1]Rutgers, The State University of New Jersey, New Brunswick, NJ, United States; [2]School of Marine Science and Policy, University of Delaware, Lewes, DE, United States; [3]FV Dana Christine II, Barnegat, NJ, United States

Introduction

While commercial and recreational fishing has been an important part of our ocean-based economy for centuries, there is an opportunity for these sectors to be a foundational component of a new blue economy. Many coastal communities around the country were founded on fisheries that directly and indirectly support many components of our economy. On local and regional scales, these economies structure livelihoods that depend on the sustained commercial and recreational fishing. Linked supply chain industries such as vessel construction and service, bait, fuel, and equipment along with downstream restaurant and wholesale markets extend the reach of these local communities around the world. In 2016, commercial and recreational fishing supported over 1.7 million American jobs. They contributed over $244 billion in sales and added an estimated $110 billion into the US economy (National Marine Fisheries Service, 2018). However, a dynamic-ocean climate system also affects these jobs. From 1996 to 2017, it is estimated that variability in ocean temperatures, indexed by the North Atlantic Oscillation, reduced count-level fishing employment in New England by 16% (Oremus, 2019), indicating the relevance of dynamic ocean features in the new blue economy.

Here we consider the new blue economy as the commercialization of ocean-informed products and services delivered to the various stakeholder communities associated with the recreational and commercial fishing enterprise. The new blue economy responds to an increased demand for

Preparing a Workforce for the New Blue Economy
ISBN 978-0-12-821431-2
https://doi.org/10.1016/B978-0-12-821431-2.00056-1

ocean observing, prediction, and research to inform planning and decision-making. The rapid evolution of the Integrated Ocean Observation System (IOOS, see Gouldman, this volume), made possible through inter-disciplinary partnerships and networked data sharing, provides descriptions of coastal ocean hydrography and hydrodynamics at fine time and space scales over large regional spatial extents. This enables hydrography and hydrodynamics to be placed at the foundation of dynamic seascape ecology that can inform fisheries-based decision-making in a timely way. IOOS not only provides the ocean data required to further develop the discipline of seascape ecology but also the infrastructure and expertise to operationalize it for regional ecosystem assessment and adaptive comanagement. Finally, regional IOOS associations are cooperative partnerships of academic, government, and private industry experts from diverse fields and interests. As a result, the IOOS culture can foster the collaborative development of an inter-disciplinary seascape science that is more likely to lead to effective and less adversarial strategies of regional ecosystem comanagement that operate at space-time scales more closely matching those of the ecosystem itself.

One challenge in realizing this new blue economy is the time mismatch between ocean science and technology research and development (R&D) and the market needs of the commercial and recreational fishers, potential investors and developers, and the regulatory fisheries managers. It sometimes takes years for scientific studies or technological R&D to be properly vetted before they can be utilized in products that benefit the broad fishing community. If management decisions are delayed due to the slow peer review process, it can be frustrating for fishers who need to make daily decisions on how to best manage the resource and their businesses. Additionally, the failure of ocean science and technology to keep pace with the demand could lead to regulatory decisions that do not fully incorporate available environmental resources. This missed opportunity will limit the impact of these ocean resources to the decision-making process, leading to more conservative regulation with significant downstream impacts throughout the recreational and commercial fishing enterprise. To help overcome this time mismatch, product development must be founded on frequent and effective communication among all the experts familiar with the oceanic, ecological, and economic aspects of both the regulation and the impacted fisheries markets. Through a collaborative community-based development process, products and services can benefit from the latest science and technology to directly respond to client needs, an important aspect of the new blue economy (Hobday et al., 2019).

Here we present two case studies as examples of how seascape ecology and ocean observing can support fisheries products as part of a new blue economy. These examples establish workgroups with broad participation across the expert communities including the fishing industry, academia, federal fisheries managers, and government fisheries scientists. The industry participation from multiple sectors and ports contributed their knowledge of the specific fishery, the ecosystem, and the economic benefit of the delivered product. The academic expertise consisted of social scientists, physical and biological oceanographers, and fisheries and outreach scientists. Federal participation included expertise in fisheries habitat/ecology, population dynamics, and management from the NOAA/NMFS, the Fisheries Management Councils, and the ocean observing community. Each case study describes the client need and the community-based product development that rapidly worked toward an ocean-based solution that could be commercialized as part of the new blue economy.

Case study #1: Atlantic butterfish stock assessment

In the first case study, we summarize the work conducted by a collaborative workgroup focused on a regulatory assessment of the Atlantic butterfish (*Peprilus triacanthus*) in the Mid-Atlantic Bight. Individual stock assessments are based on many factors including a mathematical stock assessment model that considers long-term trends in fish caught by the commercial fishing industry and by fishery-independent bottom trawl surveys. More than a year before the actual regulatory process was scheduled to begin, a workgroup was assembled to identify the research priorities that should be addressed to ensure that the upcoming assessment was informed by the latest in ocean and fisheries science. To ensure that the most relevant expertise contributed to this process, the workgroup represented relevant stakeholder communities including the fishing industry, fisheries scientists, oceanographers, regulators, and fisheries managers. Together this team identified the need to better represent dynamic Atlantic butterfish habitat in the stock assessment model, which depends partially on a lengthy time series of species abundance derived from a multidecadal federal bottom trawl survey conducted twice a year since the late 1960s. Along with other factors, trends in these survey data inform stock status and in turn regulation of the associated fishing industry. Specifically, the workgroup identified the need to develop a time-varying estimate of suitable Atlantic Butterfish habitat that overlaped with the fishery surveys to be considered in the population assessment model. The multidisciplinary workgroup of experts used data from seven field surveys throughout the Northwest Atlantic Ocean to develop a species-specific realized thermal niche model (Manderson et al., 2011). This niche model of Atlantic butterfish temperature preference was coupled to a modeled ocean hindcast of daily bottom water temperature derived from a regional numerical ocean model (ROMS) in order to project Atlantic butterfish thermal habitat in the Northwest Atlantic on a daily basis over the last 40 years (Fig. 8.1). The hindcast of thermal habitat suitability was used to estimate the proportion of thermal Atlantic butterfish habitat that overlapped with each

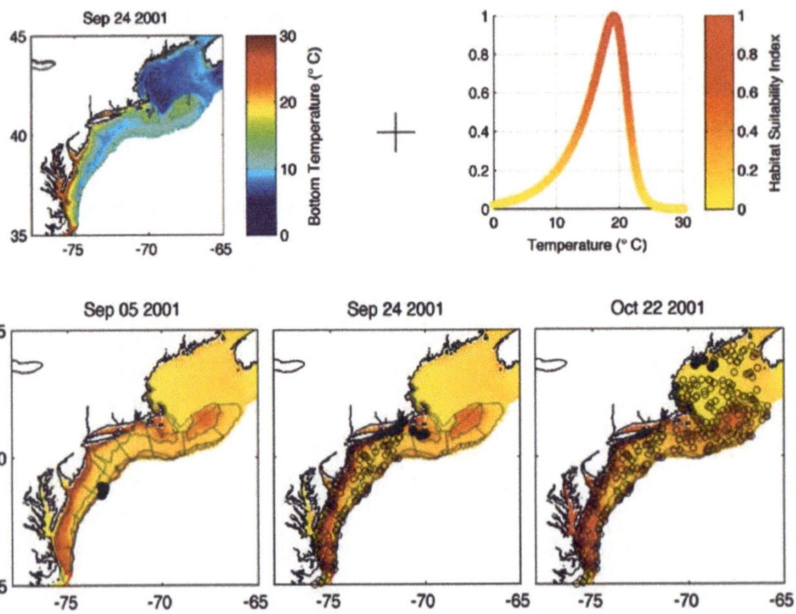

Figure 8.1 Butterfish thermal habitat suitability was projected in space and time by coupling the niche model rescaled from 0 (unsuitable habitat) to 1 (highly suitable habitat) to daily hindcasts of bottom temperature. A bias-corrected Regional Ocean Model System (ROMS) bottom water temperature hindcast daily average for the median date of the 2001 fall NEFSC survey (September 24th; top left) was coupled to a realization of the thermal niche model (top right) to produce an estimate of thermal habitat suitability for butterfish at the beginning (September 5th), middle (September 24th), and end (October 22nd) of the 2001 fall survey (bottom row). The habitat suitability projections were used to calculate the proportion of the total habitat suitability in the regional sea sampled at each station on the day of sampling. These values were summed across survey stations to estimate the availability of the butterfish stock to the survey as a function of the total available thermal habitat that was sampled. *(Reproduced from Adams et al. (2015).)*

fishery-independent survey conducted during analysis period. These dynamic habitat-based estimates of stock availability to the surveys were integrated into the stock assessment model accepted by the reviewers of the 59th NEFSC stock assessment review, as well as the Mid-Atlantic Council's Scientific and Statistical Committee. The contribution of the habitat availability estimate (along with an estimate of detectability) allowed for the development of fishery reference points, a change in stock status from unknown to known, and the establishment of a directed fishery with an allocation of 20,000 metric tons of quota. This dramatic change in fishery status and available quota to the industry was made with the latest available science. The workgroup facilitated the necessary crossover between the fishery, the regulators, and the ocean data and prediction communities. The result was a product that addressed a specific need, a product that is based on an ocean information infrastructure that includes ocean temperature observations and predictions at the scales relevant to the fishery, and, finally, a product that has the potential to be commercialized as part of a new blue economy.

Following the review of the habitat product, "the Panel strongly recommended continued refinement of the butterfish habitat model, consideration of other potential covariates, and exploration of its use for assessments of other species." Following the panel's advice, this methodology has been included in subsequent assessments of bluefish, scup, and Atlantic mackerel. This approach highlights the opportunity to support a new blue economy that bridges the gaps between communities to co-develop environmentally informed products that leverage publicly available dynamic ocean information in support of an environmentally informed regulatory process, an outcome that could have wide-reaching impacts across the recreational and commercial fishing enterprise.

Case study #2: Atlantic sturgeon

Instead of using ocean observing to better assess a specific fishery, the second case study is focused on better predicting the spatial distribution of a protected species that interacts with other legal fisheries. Protected under the Endangered Species Act since 2012, Atlantic sturgeon (*Acipenser oxyrinchus oxyrinchus*) need management strategies to reduce interactions with various industries (United States Federal Registry, 2012). As a result of overfishing and inadequate conditions in spawning rivers, Atlantic sturgeon populations were dramatically reduced in the early 20th century and have not recovered. It is estimated that Atlantic sturgeon have declined >99% from historic abundance (Secor and Waldman, 1999; Smith and Clugston, 1997; Atlantic Sturgeon Status Review Team, 2007) in rivers such as the Delaware. Their low population status combined with late maturation and longevity means that an individual mortality event through bycatch can have significant effects on the population for decades. Future bycatch taking limits are expected to increase regulatory pressure on the Delaware Bay commercial fishing industry. Furthermore, when Atlantic sturgeon interact with gillnets, they can cause significant damage to gear and costly downtime for commercial fishermen. Therefore, the commercial fishing industry has strong regulatory and economic incentives to avoid Atlantic sturgeon as part of their daily, operational decision framework.

In response to this need, a community of experts developed an Atlantic Sturgeon Risk Model (ASRM) that produced daily nowcasts, 1–3 day forecasts, and climatological averages of the risk associated with encountering Atlantic sturgeon in Delaware waters (http://robots.ceoe.udel.edu/shiny/sturgeon/; Breece et al., 2021). Dynamic management or stakeholder decision-making is quickly becoming a useful method to reduce interactions while minimizing burdens (Brodie et al., 2017; Hazen et al., 2018; Welch et al., 2018). This species distribution model, created on dynamic environmental data, was designed to assess encounter risk and deliver an easily interpretable forecast product to stakeholders who are interested in avoiding an endangered species while minimizing economic losses. This model serves as an example of the economic benefit of real-time and future forecast products of species-specific encounter risk. ASRM was built on association between historic Atlantic sturgeon locations (2009–12) linked to sea surface temperature, ocean color, water depth, and day of year (Breece et al., 2018). The ASRM forecast system depends on real-time environmental data from satellites, processed at the University of Delaware in collaboration with the Mid-Atlantic Regional Association of Coastal Ocean Observation Systems (MARACOOS). Besides a web interface (link above), this application also generates SMS messages, emails, and weblinks to images that are fed into the DNREC Blackboard text messaging and email system (Fig. 8.2). The SMS text messages are organized by four regions outlined in the 2016 Atlantic States Marine Fisheries Council (ASMFC) Delaware River Sustainable Fishing Plan for American shad, but are used for commercial reporting in other fisheries in the state. The regions include the Upper Delaware Bay, Mid Delaware Bay, Lower Delaware Bay, and Atlantic Ocean (Fig. 8.2).

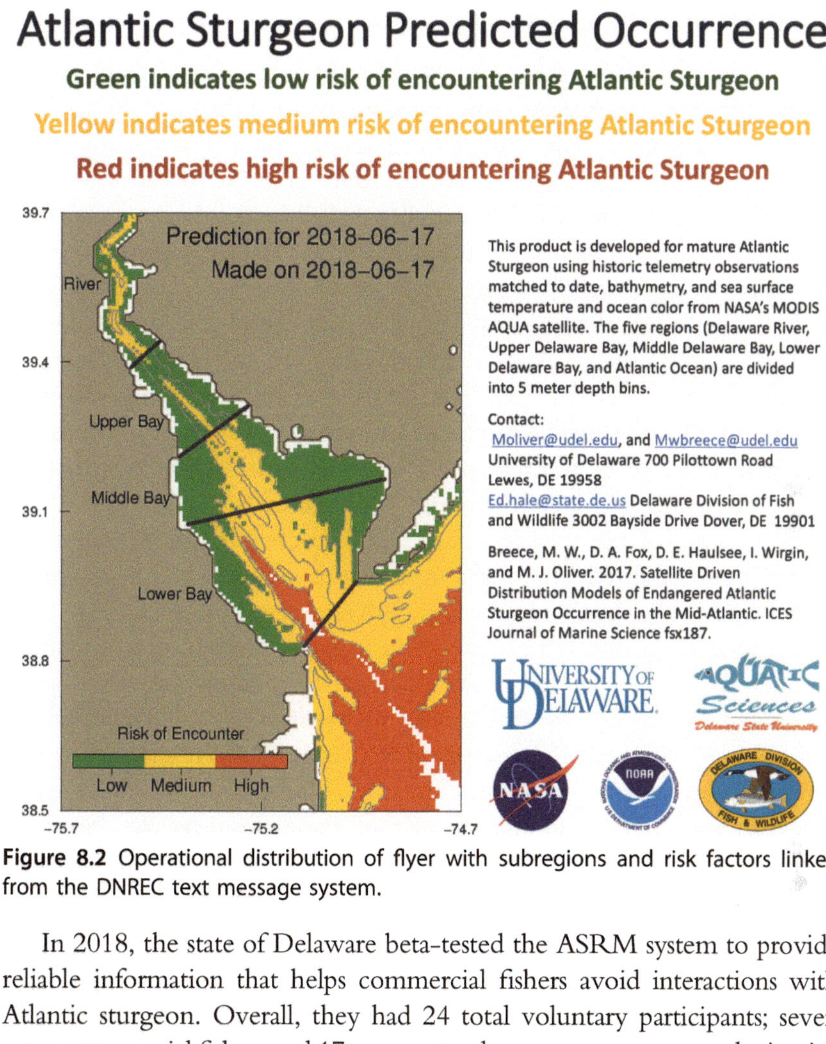

Figure 8.2 Operational distribution of flyer with subregions and risk factors linked from the DNREC text message system.

In 2018, the state of Delaware beta-tested the ASRM system to provide reliable information that helps commercial fishers avoid interactions with Atlantic sturgeon. Overall, they had 24 total voluntary participants; seven were commercial fishers and 17 were natural resource managers and scientists working for state and federal agencies. There are approximately 40 active commercial gill netters in Delaware Bay, many of which hold multiple permits. The seven commercial fishers in our group represented 17.5% of the active fishers in Delaware Bay and 6.3% of the total number of permitted gill net fishers. Our partners at DNREC surveyed our users, and 11 managers and five commercial fishers responded. The managers did not show a clear preference for one of the three delivery methods, but the commercial fishermen overwhelmingly (80%) preferred the text delivery system over the flyer maps and web application. The majority of commercial fishers that responded to the survey following project implementation were satisfied with the frequency and amount of information in the warnings.

Summary

These case studies represent just two recent examples of how fishing communities are working to link the resources needed to build accurate and timely dynamic habitat products. Coupled mechanistic biophysical models supported by rapidly evolving and available ocean observing networks enable products and approaches that allow the dynamic nature of the habitat to be tracked over time (Mills et al., 2017). Ecosystem assessment and management in the sea are based upon interdisciplinary science that considers physical, chemical, and biological ocean processes, including feedbacks with human ecological systems, that structure and regulate marine ecosystems. Therefore, the management of human fishing activities, from decisions made on the bridge of an individual ship to a room full of regulators, needs to be informed by a broad-scale habitat ecology that reflects the dynamic realities of the ocean. Through these case studies we have shared a couple of examples meant to represent categories of products that can be commercialized as part of a new blue economy. The integration of publicly available data and forecasts provides resources to a new blue economy to build directed products that support the economic engine of commercial and recreational fishing. This new economic sector can serve the fishing and regulatory sector with various forms of dynamic habitat classification in a way that serves the client's needs. We have shown just two different forms these products can take. Through community-based codevelopment, the economic benefit of these products can be more rapidly developed and delivered in a way that is impactful and informative to all the economies linked to commercial and recreational fishing. Ocean science and technology are now well placed to provide relevant and well-resolved environmental observations and predictions in support of a new blue economy. The new blue economy can properly commercialize these types of products that leverage available dynamic ocean information in a way that supports decision-makers who must make difficult decisions that have widespread downstream economic impacts.

References

Adams, C.F., Miller, T.J., Manderson, J.P., Richardson, D.E., Smith, B.E., 2015. Butterfish 2014 stock assessment. US Dept Commer. Northeast Fish Sci. Cent. Ref. Doc. 15 (06), 110. https://doi.org/10.7289/V5WM1BCT.

Atlantic Sturgeon Status Review Team, 2007. Status Review of Atlantic sturgeon (Acipenser oxyrinchus oxyrinchus). In: Report to National Marine Fisheries Service, Northeast Regional Office, p. 175.

Breece, M.W., Oliver, M.J., Fox, D.A., Hale, E.A., Haulsee, D.E., Shatley, M., Bograd, S.J., Hazen, E.L., Welch, H., 2021. A satellite-based mobile warning system to reduce interactions with an endangered species. Ecological App. Rev. (in-press).

Breece, M.W., Fox, D.A., Oliver, M.J., 2018. Environmental drivers of adult Atlantic sturgeon movement and residency in the Delaware Bay. Mar. Coast. Fish. 10, 269–280.

Brodie, S., Hobday, A.J., Smith, J.A., Spillman, C.M., Hartog, J.R., Everett, J.D., Taylor, M.D., Gray, C.A., Suthers, I.M., 2017. Seasonal forecasting of dolphinfish distribution in eastern Australia to aid recreational Fishers and managers. Deep Sea Res. Part II Top. Stud. Oceanogr. 140, 222–229.

Hazen, E.L., Scales, K.L., Maxwell, S.M., Briscoe, D.K., Welch, H., Bograd, S.J., Bailey, H., Benson, S.R., Eguchi, T., Dewar, H., Kohin, S., Costa, D.P., Crowder, L.B., Lewison, R., 2018. A dynamic ocean management tool to reduce bycatch and support sustainable fisheries. Sci. Adv. 4, 1–7.

Hobday, A.J., Hartog, J.R., Manderson, J.P., Mills, K.E., Oliver, M.J., Pershing, A.J., Siedlecki, S., 2019. Ethical considerations and unanticipated consequences associated with ecological forecasting for marine resources. ICES J. Mar. Sci. https://doi.org/10.1093/icesjms/fsy210.

Manderson, J.M., Palamara, L., Kohut, J., Oliver, M., 2011. Ocean observatory data are useful for regional habitat modeling of species with different vertical habitat preferences. Mar. Ecol. Prog. Ser. 438, 1–17. https://doi.org/10.3354/meps09308.

Mills, K.E., Pershng, A.J., Hernández, C.M., 2017. Forecastng the seasonal timing of Maine's Lobster fishery. Front. Mar. Sci. https://doi.org/10.3389/fmars.2017.00337.

National Marine Fisheries Service, 2018. Fisheries economics of the United States, 2016. In: U.S. Dept. Of Commerce, NOAA Tech. Memo. NMFS-F/SPO-187, p. 243.

Oremus, K., 2019. Climate variability reduces employment in New England fisheries. Proc. Natl. Acad. Sci. U.S.A. 116 (52), 26444–26449. https://doi.org/10.1073/pnas.1820154116.

Secor, D.H., Waldman, J.R., 1999. Historical abundance of Delaware Bay Atlantic sturgeon and potential rate of recovery. In: Proceedings of the Symposium Conservation of Long-Lived Marine Animals. Monterey, California, USA 24 August 1997, vol. 23. American Fisheries Society Symposium, pp. 203–216.

Smith, T.I.J., Clugston, J.P., 1997. Status and management of Atlantic sturgeon, Acipenser oxyrinchus, in North America. Environ. Biol. Fish. 48, 335–346.

United States Office of the Federal Registry, 2012. Endangered and threatened wildlife and plants; threatened and endangered status for distinct population segments of Atlantic sturgeon in the Northeast region. U.S. Off. Fed. Reg. 77, 5880–5912.

Welch, H., Hazen, H.L., Bograd, S.J., Jacox, M.G., Brodie, S., Robinson, D., Scales, K.L., Dewitt, L., Lewison, R., 2018. Practical consideration for operationalizing dynamic management tools. J. Appl. Ecol. 1, 1–11.

Interview: Recreational fishing

Mitch Roffer

How relevant/important is the "data revolution" (i.e., "big data," "cloud computing," etc.) to the business of commercial operational oceanography?

The proliferation of satellites with numerous channels resulting in large data sets requires larger storage capacity, employing cloud computing, and increasingly faster computers and networks. The huge amount of data in its raw form is too difficult for an individual to use. Commercial operations could provide services to offer access to real-time data in a more consumable manner by more users. They would do the initial data manipulation including size reduction. However, such companies would need to make those data publically available in near-real time, otherwise the data loses its value. Companies will need to establish partnerships with governmental agencies to procure initial funding and access to raw data sets and in exchange share or release the data in a noncompeting way.

What are the biggest challenges (technological, policy, logistical, etc.) to the business of commercial operational oceanography?

With respect to satellite data, the biggest challenges are:

(1) Obtaining data in real time, i.e. within 30–60 min from a satellite pass.

(2) Lack of data sharing between some nations, for example, India and China.

(3) The cost of very high resolution data and synthetic aperture radar.

When it comes to available buoy data, it boils down to operation and maintenance. There are simply not enough funds provided for the NOAA's National Data Buoy Service to adequately maintain and update their network of buoys.

Some of the new arrays, such as the Pioneer Array off the New York Bight funded by NSF, have assembled equipment and scientists, which are producing a really great array from the surface to the bottom, including sea surface mid- and bottom temperature, chlorophyll, and currents. However, this effort is led by the research community so there is no real interest to get those data out in real time. By the time they release the data, usually a month or 2 months later, at best, the data are of little to no use to the operational oceanographic community. They are useful when evaluating ocean model performance.

I believe there should be a policy change requiring the government-funded research community to make the uncalibrated data available immediately. Buyer beware for certain, however the messy, uncalibrated data can reveal sudden or dramatic changes in wind and/or current speed or direction for example, which could be extremely valuable in the moment to the operational oceanographic community. Real-time data on thermocline and pycnocline depth is critically important to the operational community.

Do we have the right sensors?

I think we have the right sensors from the surface to the thermocline in terms of temperature and wave height. But standard ocean color—chlorophyll, measurements should be part of any surface array. As time goes on, people will realize the increasing importance of subsurface measurements in operational oceanography. Subsurface oceanography and thus subsurface sensing are very important to oil and mining companies and are an area of needed development. These lacking subsurface data are important for understanding demersal fisheries dynamics as well. Acoustic, optical, and eDNA sampling is needed as well for the surface to bottom layers of the ocean.

Will nongovernmental sources of data become more important? Do we have the right policies and systems in place to acquire those private data?

Right now there's no formalized format as far as I am aware. I know people in the weather industry, and they have weather buoys up and down the east coast of the United States. It took them 5 years to get NOAA to be interested in those data, even though that data set is a highly calibrated and useful data set. After 5 years of negotiation, NOAA did agree to acquire the data; however, this should not be the path forward. This process should be streamlined and an overarching method to acquire private data should be established. The more professionally collected data that are available in the system, the better.

Are there examples of best practices from other sectors that we might emulate?

One example might be terrestrial applications people using Google Maps, ArcGis, and open source data along with Google satellite data and looking at building elevation mapping along with land use and land change mapping. These people are very good at being able to very quickly vectorize the data. Google is doing a little bit of this in the ocean now, but there has not been as much tool development as I would have thought. I don't think they see the value in it for the oceanographic community. But with plastic and other pollution mapping along with combating illegal fishing, such tools are needed.

There is great value in using an open source or commonly used application or code so even though a final product may be proprietary, the engine or core of the product is interoperable, inexpensive to maintain or could be repurposed for other uses and reach broader communities.

From your experience in the industry, what were the most valuable educational experiences (e.g., courses, hands-on experiences) for your workforce?

When first thinking about employing people with my company, I initially thought that we would really need people with the standard Master's/PhD level backgrounds (biology, chemistry, and physics). However, I then began to realize that as long as you have an intelligent person with an advanced science background, someone who knows how to do research, think logically, and write well, then they were even more useful. Hiring from that point forward, we searched for people who are interdisciplinary in nature and who could speak well interactively. We've hired people with no biology backgrounds who have done quite well because they were basically relatively intelligent, then we could train. Also a working knowledge of graphics and statistics is important.

Field experience is always a plus. I think it is critical for people to understand the culture on a ship, or what a captain or deckhand is trying to convey, and sometimes urgently convey, to be useful as part of the full operation. Understanding the culture of the fishers and fishing industry is important.

Are there other fields we could consider when hiring?

Submarine communications and cables enable most of the communications we are reliant upon in the 21st century. People with terrestrial cable and connectors experience could easily consider work at sea, and vice versa. The same would apply to mining. So I think those kinds of experiences on land are very useful when people move into the ocean. The ability to use unmanned vehicles in the air and in the water is important as is artificial intelligence. Visual storytelling will never go out of style.

Mitch biography

Mitchell Roffer, Ph.D., is President of Fishing Oceanography, Inc., a scientific consulting company that focuses on fisheries oceanography and remote sensing along with environmental education and advocacy, as well as sustainable alternative energy, waste disposal, and bioremediation solutions, that was founded in 2019 in Melbourne Beach, Florida. Prior to this he was founder (1989) and President of Roffer's Ocean Fishing Forecasting Service, Inc. (ROFFS). He received his doctorate in Biological Oceanography—fisheries oceanography at the University of Miami's Rosenstiel School for Marine and Atmospheric Science (RSMAS).

While many in the fisheries world know Dr. Roffer for his pioneering and continued work with operational fishing forecasting analyses and habitat classification, he has received significant national and international acclaim for the innovative oil oceanographic analyses products after the British Petroleum DeepWater Horizon explosion and oil spill, as well as other pollution monitoring products after hurricanes, Katrina, Rita, and Harvey. Roffer has been an active member of NASA's prestigious Biodiversity and Ecological Forecasting Science Team since 2003.

Interview: Ocean observing and the new blue economy impacts on business—aquaculture

Perry Raso

Have recent technological advancements improved or impacted the aquaculture industry?

The reality is that the advancements of marine technology have led to the need for aquaculture. If you think about how many people have fished throughout history, the more people went fishing and technologies to catch fish improved, the more fish they were catching. During World War II, we developed sonar and radar and applied those technologies to capture our fish that certainly changed the efficiency and capacity to catch fish, to the point that it is too fast for the ocean to replenish. So you can really argue that advancements in marine technology have driven the need for aquaculture. Now that we have this need for aquaculture, it's more about I'd say the advancements in agricultural technology that we've applied to the ocean and to modern-day aquaculture.

Does the increasing acidification of ocean water or any other water quality issues impact your industry?

Increasing ocean acidification has created an impact on the Pacific coast more than has the Atlantic coast so far and it's being noticed mostly in the hatcheries, whereas it's more difficult for the larvae to form their shell. It's also being noticed in some hatcheries on East Coast, but it is not noticeable to a farmer such as myself who's not actually studying the shell development.

There are other parameters that are of concern on the East Coast that might not necessarily be on the West Coast for producing the oysters such as water temperature. For example, last summer we experienced a couple of very hot days followed by a huge downpour for a few days and that created issues, and we lost a lot of our oyster seed as the combination of very hot water followed immediately by drastically reduced salinity was too much variability for the juveniles to take.

If someone walked into your office and said I can offer you the following X predictive product to improve your business, what would you want that product to predict?

Chlorophyll-a. A user-friendly way to tell how much food is present in my farm by measuring the chlorophyll-a and figure out how much is being depleted in a certain area and also within containment to figure out how much food is being allowed to go through the system and into these containers and these bags to see how much food and water flowing into and through the farm system.

Looking at your industry from the workforce development angle, is there a need for a certification process for the aquaculture industry?

About 5 years ago in Rhode Island, they started a certification process where anybody applying for aquaculture license has to take this course, which teaches them the basics of shellfish aquaculture. Most importantly focused on the importance of the safety since the product is being consumed by humans, but also the safety of the environment, to prevent invasive species and disease. I think it's really important process and was helpful when I started this business.

As a business person, what polices would you like to see change to ensure sustainable development of fisheries and/or aquaculture?

The first thing that comes to mind is probably not what you're envisioning of an answer, but I think in order for aquaculture to flourish in the blue economy, is acceptance from society. People like the idea of it, they like the product, a surprisingly large percentage of people know that shellfish are good for the environment, and more and more people are consuming oysters. However, there is a not in my backyard mentality associated with aquaculture. For some reason, people can look out at hundreds of lobster pot buoys and think it's a romantic, beautiful thing on the water. But if there are floating oyster cages, they're really peeved by it. The culture needs to shift and be proud to have a shellfish farm on a water body and take ownership of rather than upset by it. I think is something perception-wise that has to change in order for aquaculture to grow as a business and also shellfish aquaculture, which is good for the environment.

In your opinion, is there another culture of community that you cite as an example of where you feel aquaculture is embraced by the local culture?

France. I like to stick to shellfish aquaculture because that's what I know, but France is a place that is really growing the industry. They have towns where most people in the town are employed by a shellfish farm, either directly farming or with the logistics of shipping them all over the country. And the more they grow, the more people seem to consume and the price is still at a premium to drive the growth. So much so that French farmers are expanding into other countries to meet demand.

They've done such a good job of keeping people safe eating their product and growing their farms and creating acceptance and even creating tourism around shellfish farms. I think we were stepping in that direction, but over the last 5 years, I think the growth of the industry has been slow to see, in Rhode Island for sure.

I've been seeing on the West Coast more and more resistance to increase farming because there's user group issues. And so as farmers, we got to be mindful of that people want to water ski, they should be able to water ski, they want to fish, they should be able to fish, they want recreationally harvest shellfish, these things all have to be allowed. Aquaculture has to be a part of that.

So, in the end, I don't know that we need a policy to change in order to do that. But the people that are deciding on whether a farm should be allowed need to make science-based decisions and not decisions based on emotion. You do have to take social impact into consideration because of recreational importance. But you also have to factor in what space should be allotted for what is needed most and especially with food considerations. And using the space for environmentally friendly shellfish aquaculture is a win-win for all.

Perry Raso biography

Perry Raso, owner of Matunuck Oyster Bar and an oyster farmer, holds bachelor's and master's degrees from URI in aquaculture and fisheries technology. He founded Matunuck Oyster Farm in 2002, in East Matunuck (South Kingstown), Rhode Island, and the Pond to Plate vision was formed.

The restaurant and oyster farm have received national acclaim in publications including the New York Times, National Geographic, USA Today, Yankee Magazine, the Boston Globe, and Chicago Tribune. His business was featured on the PBS show "Start Up." In 2016, he gave a TedX talk in Providence sharing his story and outlining the "Importance of Sustainable Aquaculture in Our Future."

Energy

CHAPTER 9

Marine energy and the new blue economy

Simon Geerlofs
Pacific Northwest National Laboratory, Seattle, WA, United States

An ocean full of energy

Marine energy technologies convert waves, tidal currents, ocean currents, and ocean thermal gradients into electricity and other usable forms of energy. Marine energy has two distinct roles within the new blue economy. First, it is an emerging ***sector*** that has the potential to provide renewable power and grid services to fast-growing coastal population centers. Marine energy is also positioned to serve ***other sectors*** within the blue economy, providing power at sea for offshore industries and energy resilience for isolated coastal and island communities.

Abundant energy has always been foundational to economic growth and human prosperity; energy is foundational to a prosperous blue economy as well. The ocean functions as a battery, storing energy from the sun, the rotation of the Earth, and the gravitational pull of the moon, cycled and distributed through waves, currents, tides, and thermal gradients. In one form or another, this energy is always available and can be harnessed in many different ways—day and night, at the surface, at depth, near shore, offshore, in the arctic, at the equator, and everywhere in between. A sustainable blue economy that taps into this ocean battery as a clean, affordable, and self-sustaining source of energy will unlock new economic, ecological, and social benefits—on land and at sea.

The successful development of marine energy faces many technical, social, and market challenges. Meeting the full potential of marine energy will require innovation and codevelopment across energy and ocean technology sectors; an improved understanding of ocean physical, biological, and social systems; and new public and private sector partnerships to coordinate and leverage research and development investments.

Preparing a Workforce for the New Blue Economy
ISBN 978-0-12-821431-2
https://doi.org/10.1016/B978-0-12-821431-2.00037-8

Marine energy

Marine energy resources are diverse, predictable (over the long term), forecastable (over shorter terms), and sizable—the total US technical resource potential is approximately 1250–1850 TW-hours of generation per year (Baldwin et al., 2015) from ocean wave (Jacobson et al., 2011), ocean current (Haas et al., 2013), ocean tidal (Haas et al., 2011), and river current energy (Ravens et al., 2012). With 40% of the population of the United States living in coastal counties (NOAA, 2020), and with increasing growth in energy-hungry blue economy industries, marine energy is an emerging part of our future energy mix.

Over the past decade, marine energy technologies have graduated to testing venues and pilot sites, demonstrating the feasibility of single devices at full scale. For example, as of the end of 2019, Atlantis Energy's MeyGen project in Scotland's Pentland Firth (Simec Atlantis Energy, 2020) has exported nearly 25 GW-hours of tidal-generated electricity to the grid. Scaling beyond single-device demonstrations to grid-scale arrays, however, has proven to be challenging. Marine energy has struggled to achieve commercial success because of high costs and challenges associated with working in the ocean environment. Producing and selling power to the electrical grid is competitive—marine energy must compete with not only traditional low-cost power options such as hydropower and natural gas, but also with solar, wind, and other renewables that are further along their development paths. There are significant benefits to commercializing a new suite of renewable resources, especially given the complementary nature of marine sources relative to wind and solar. However, high costs mean that near-term revenue and private investment may require new commercial pathways and markets beyond the grid—those that are better suited to the unique technology and locational attributes of marine energy.

In April 2019, The DOE Water Power Technologies Office (WPTO) published *Powering the Blue Economy: Exploring Opportunities for Marine Renewable Energy in Maritime Markets* (LiVecchi et al., 2019). The report made a public, high-level case that marine energy has the potential to provide unique value to a set of coastal and offshore markets and that these "maritime markets" could provide near-term commercial opportunities for the marine energy industry. The report argued that the marine energy industry and its associated R&D partners would be well served to engage directly with public and private sector groups organizing around innovation in the blue economy.

Success in blue economy markets requires that energy innovators work directly with ocean and coastal stakeholders to understand and innovate around their unique energy needs and use cases. Doing so requires a shift in thinking for energy innovators accustomed to considering only the value of their technology serving large grid-scale energy commodity markets.

The benefits of marine energy

Grid-scale marine energy could create benefits similar to offshore wind—jobs and supply chain opportunities within traditional maritime industries as well as R&D and technology sectors. Marine energy could also provide value to the energy grid as part of a diverse renewable energy system, particularly as renewables continue to grow as a proportion of energy generation in response to policy and market forces. Locational benefits, timing, and the ability to forecast wave and tidal resources (made possible by data and information generated in the new blue economy) could provide stability and reduce the need for energy storage in areas that have a high penetration of variable resources, such as wind and solar. Better understanding of the magnitude and drivers of these value streams will be critical to planning and operating marine energy systems in the future.

In addition, the emerging focus on serving blue economy markets and end users creates new opportunities to achieve goals and values beyond bulk power. As blue economy sectors move farther offshore, as energy technologies adapt to new opportunities for distributed generation and storage, and as the demand for decarbonization in blue economy sectors such as maritime transportation grows, we are likely to see interest in marine energy deployed in new and novel ways—integrated with solar, wind, batteries, and microgrid technologies.

The development of offshore blue economy technology sectors such as ocean observation, autonomous underwater vehicles, offshore aquaculture, and ocean-derived critical materials and renewable fuels corresponds to a parallel revolution in the energy sector toward smaller-scale, distributed, renewable resources, energy storage, and smart microgrid technology. Marine energy is uniquely well suited to provide off-grid, offshore power as part of an integrated energy system that serves as enabling technology to achieve market goals. Marine energy may be well positioned between solar and wind—providing abundant, reliable power at a scale more suitable to activities that have low-to-medium power requirements.

One example is ocean observing: the foundation of the new blue economy rests on knowledge about the physical, chemical, and biological ocean. According to the National Oceanic and Atmospheric Administration (NOAA), ocean observing and the broader "ocean enterprise" generate $7 billion in revenue per year (Gouldman et al., 2016) in the United States, and demand is growing as new industries seek to understand and use the ocean in new ways. The trend in ocean observing is toward distributed, ubiquitous, and autonomous sensor platforms operating far from shore with little human intervention. Batteries can power these technologies, but batteries are expensive and require replacement—and systems are often sized (and scientific capabilities limited) by the constraints of the battery system. As a result, oceanographers for decades have had to wrestle with a fundamental trade-off between the data they would like and the energy required to collect it. Additional power sources, regenerating and located in situ, have the potential to open completely new strategies for studying the ocean. Researchers have for years harnessed renewable energy to charge batteries at sea, and solar panels are near ubiquitous. But solar panels foul, don't function well at higher latitudes, can only generate power during daylight hours, and require a surface expression. Marine energy could potentially provide additional power to complement power available from solar and wind and enable new capabilities in the new blue economy. Collecting data over longer time periods, processing those data onboard, and having the ability to stay on station to observe ephemeral events (such as subsea volcanic eruptions), to function in harsh arctic environments, and to collect and process new kinds of data and access new communication technologies all require energy above and beyond what is currently available to the ocean observing industry.

Companies such as Liquid Robotics have used wave energy for propulsion, and others such as Wirewalker have used wave energy to drive profiling floats—these companies have proven that marine energy can be used as part of an ocean observing system. There are opportunities to codevelop observing platforms with integrated energy generation, as well as opportunities to use wave and tidal power as part of a multipurpose floating platform that has an offshore microgrid, thereby providing electricity to charge autonomous vehicles and run sensor, data processing, and communications systems. Recognizing this value, marine energy to power ocean observing is a focal area for a collaboration between DOE and NOAA. In 2019, the two agencies initiated a set of "prize" activities to develop this technology and partnerships between the ocean science and ocean energy communities (DOE, 2020a).

Closer to shore, the United Nations' sustainable development goals have highlighted the growing challenges island and remote coastal communities face to obtain an affordable and secure supply of energy, water, and food. Some of these communities pay close to $1 per kilowatt hour with prices fluctuating with the cost of diesel fuel. Remote and island communities are investigating renewable energy as part of the solution—and are likely to achieve close to full renewable energy penetration much sooner than large conventional power systems. This means they need a diverse portfolio of resources that can generate power consistently and reliably throughout the day and the year. Wave and tidal power could offer benefits as part of an energy system, balancing load from wind and solar and reducing fuel use, particularly for communities at higher latitudes. Ocean thermal energy technology (OTEC) and related technologies such as seawater air conditioning (SWAC) could provide similar benefits for tropical island communities.

The WPTO is investigating technologies that use wave power to produce desalinated water to serve remote coastal and island communities through another prize, "Waves to Water" (DOE, 2020b). Desalination is energy-intensive. Wave power has the ability to use the pumping motion of waves to directly pressurize seawater across a reverse osmosis membrane, producing water without the need to convert power to electricity. A number of companies and communities are pursuing this opportunity, and again, the benefits of success include water security as well as smaller-scale market opportunities for wave energy entrepreneurs. New data and information are needed to understand the potential impacts of desalination, particularly at large scale. Data and modeling are also needed to characterize wave resources at local scales to inform site selection and technology design. Data and information needs vary by region, and each region brings a unique set of challenges and opportunities.

Knowledge required to realize these opportunities

Marine energy has the ability to produce new knowledge for the blue economy by enabling technologies such as ocean observing systems. But developing marine energy effectively and sustainably also requires knowledge about the following:

- System performance: We need to continue to improve the technological capability to harness marine energy resources efficiently and reliably. Wave and current energy converters have come a long way, but advances in controls systems, materials, and power take-off systems—and even entirely new designs—offer the potential to further improve performance and reduce costs.

- Resource characterization: We need to understand the metocean resource at a variety of timescales and spatial resolutions, including annual timescales and coast-wide resolutions to understand the opportunity, weekly to hourly at the local scale to forecast energy availability and interoperability, and minutes to seconds ahead at the device scale to inform control strategies and device efficiency/survivability.
- Environmental impacts: Much research has been done to investigate the potential environmental impacts of individual devices to assure that pilot and test projects are carried out responsibly. As deployments move toward arrays, more information will likely be required to develop a baseline characterization of environments in which marine energy is likely to be deployed and to understand the cumulative impacts of multiple devices. Environmental assessment and resource characterization depend on ocean observing and monitoring technology and products developed through partnerships in the new blue economy.
- Social acceptance and local benefits: To be developed responsibly—and successfully—marine energy projects must be integrated into local uses of and interactions with the ocean. It is essential to understand how benefits flow from development of marine energy resources and to assure transparency and trust between other ocean user groups and coastal communities. Spatial planning has been used to understand spatial conflicts and highlight promising areas for marine energy development; further development of data sets and tools to inform the planning process is needed to incorporate a broader range of energy and non-energy values.
- End-user needs: For marine energy to help power a wide range of ocean technologies and missions, systems must be designed with the end use in mind. For example, the most efficient energy harnessing system in the world will not work for an ocean observing platform if the power generation interferes with the observation equipment. The marine energy community must understand the end-user needs for which it is designing and, in some cases, must optimize for a different suite of system characteristics, beyond maximizing energy production.

Conclusion

Marine energy offers the possibility of tremendous value to the blue economy, both as part of a rapidly growing renewable ocean energy sector creating coastal economic development opportunities and as an enabler of a suite of new technologies, scientific applications, and commercial opportunities in the ocean. Affordable, easily accessible energy is a major constraint at sea and in remote communities. The ocean offers a wide range of persistent, clean sources of energy. The challenge and opportunity is to put those sources to work to serve where they are best suited to meet the economic, ecological, and social aspirations of the blue economy. If successful, marine energy could enable entirely new fields of science or commerce, from tracking ocean species by traces of their DNA left in the water column to catalyzing deep-water aquaculture to feed the planet.

Many technical challenges hinder the development of cost-effective, reliable marine energy systems, and to date, most wave and current energy converters have been designed for large-scale grid power. New missions such as powering the blue economy offer much-needed potential for near-term revenue and investment, but marine energy systems will need to be optimized for a new suite of end uses. In some cases, this may mean small adjustments, in others new design approaches.

The new blue economy depends on working across disciplines and industry stovepipes. This is as evidenced by the growth in regional and national blue economy–focused clusters, incubators, and accelerators (see contributions from Jones and Fischel, Jugan and Baygents, and Hanlon, this volume). These organizations offer new connections, partnerships, access to capital, and access to entrepreneurial services that are critical to the success of grid-scale marine energy as well as offshore and remote market opportunities. In all cases, the marine energy community will need to work closely with those it seeks to serve. From researchers developing new metocean buoys to companies at the cutting edge of rechargeable underwater vehicles, to communities looking to wave or tidal power either as a key part of the generation mix or a source of local maritime investment and jobs—everyone who stands to benefit from marine energy development must have a say in how the technologies are designed, tested, and deployed.

References

Baldwin, S., Bindewald, G., Brown, A., Chen, C., Cheung, K., Clark, C., Friley, P., 2015. Quadrennial Technology Review: An Assessment of Energy Technologies and Research Opportunities. US Department of Energy, Washington DC, pp. 1–505. Technical Report.

DOE (U.S. Department of Energy), 2020a. Powering the Blue Economy™: Ocean Observing Prize. Energy Efficiency & Renewable Energy, Water Power Technologies Office, Washington, D.C. https://americanmadechallenges.org/oceanobserving/.

DOE (U.S. Department of Energy), 2020b. Waves to Water Prize. Energy Efficiency & Renewable Energy, Water Power Technologies Office, Washington, D.C. https://americanmadechallenges.org/wavestowater/.

Gouldman, C., Rayner, R., Bosley, K., 2016. In: Tally, S., Jones, M.B. (Eds.), The Ocean Enterprise, a Study of US Business Activity in Ocean Measurement, Observation and Forecasting. ERISS Corporation, San Marcos, TX.

Haas, K., Fritz, H., French, S., Neary, V., 2013. Assessment of Energy Production Potential from Ocean Currents along the United States Coastlines. Georgia Tech Research Corporation, Atlanta, GA. http://energy.gov/sites/prod/files/2013/12/f5/energy_production_ocean_currents_us_0.pdf.

Haas, K., Fritz, H., French, S., Smith, B., Neary, V., 2011. Assessment of Energy Production Potential from Tidal Streams in the United States. Georgia Tech Research Corporation, Atlanta, GA. http://energy.gov/sites/prod/files/2013/12/f5/1023527.pdf.

Jacobson, P., Hagerman, G., Scott, G., 2011. Mapping and Assessment of the United States Ocean Wave Energy Resource. Report Number 1024637. Electric Power Research Institute, Palo Alto, CA. http://www.epri.com/abstracts/pages/productabstract.aspx?ProductID=000000000001024637.

LiVecchi, A., Copping, A., Jenne, D., Gorton, A., Preus, R., Gill, G., Robichaud, R., Green, R., Geerlofs, S., Gore, S., Hume, D., McShane, W., Schmaus, C., H. Spence, H., 2019. Powering The Blue Economy; Exploring Opportunities for Marine Renewable Energy in Maritime Markets. U.S. Department of Energy, Office of Energy Efficiency and Renewable Energy, Washington, D.C.

NOAA (National Oceanic and Atmospheric Administration), 2020. Fast Facts: Economics and Demographics. Office for Coastal Management, Charleston, SC. https://coast.noaa.gov/states/fast-facts/economics-and-demographics.html.

Ravens, T., Cunningham, K., Scott, G., 2012. Assessment and Mapping of the Riverine Hydrokinetic Resource in the Continental United States. Report Number 1026880. Electrical Power Research Institute, Palo Alto, CA. http://www.epri.com/abstracts/Pages/ProductAbstract.aspx?Produc-tId=00000000000102688.

Simec Atlantis Energy, 2020. Tidal Stream Projects – MEYGEN. Edinburgh, Scotland. https://simecatlantis.com/projects/meygen/.

CHAPTER 10

Using science to safely tap energy and minerals from the sea: role of the US Bureau of Ocean Energy Management

William Yancey Brown
Bureau of Ocean Energy Management, Department of the Interior, Washington, DC, United States

Mission of the Bureau of Ocean Energy Management

BOEM's mandate rests in the Outer Continental Shelf Lands Act (OCSLA), 43 U.S.C. §§ 1331 et seq. OCSLA tasks the Secretary of the Interior with managing oil, gas, renewable energy (including wind), and non-energy mineral exploration and production on the offshore federal outer continental shelf (OCS). OCSLA declares that the OCS is "a vital national resource reserve held by the Federal Government for the public, which should be made available for expeditious and orderly development, subject to environmental safeguards..." 43 U.S.C. § 1332(3). The statute defines the geographic jurisdiction for this program as the area between the state seaward boundary and the seaward extent of US jurisdiction, which was defined in judicial rulings and codified in the Submerged Lands Act, 43 U.S.C. §§ 1301 et seq.

The bureau's geographic jurisdiction begins 3 nautical miles (nm) from the coastal baseline (or 9 nm for Texas and the west coast of Florida) and extends to the 200 nm edge of the Exclusive Economic Zone (EEZ) or farther if authorized under Article 76 of United Nations Convention on Law of the Sea (UNCLOS). The confirmed area amounts to nearly 2.5 billion acres. Under the Interior Department's current interpretation of the statute, the OCS does not include the US continental shelf offshore the Nation's territories, although legislation has been advanced in the Congress to add roughly another 1 billion acres to BOEM's jurisdiction (Fig. 10.1 maps this and the OCS).

The Secretary's administration of OCSLA has evolved. OCSLA responsibilities were initially implemented by the US Geological Survey

Preparing a Workforce for the New Blue Economy
ISBN 978-0-12-821431-2
https://doi.org/10.1016/B978-0-12-821431-2.00035-4

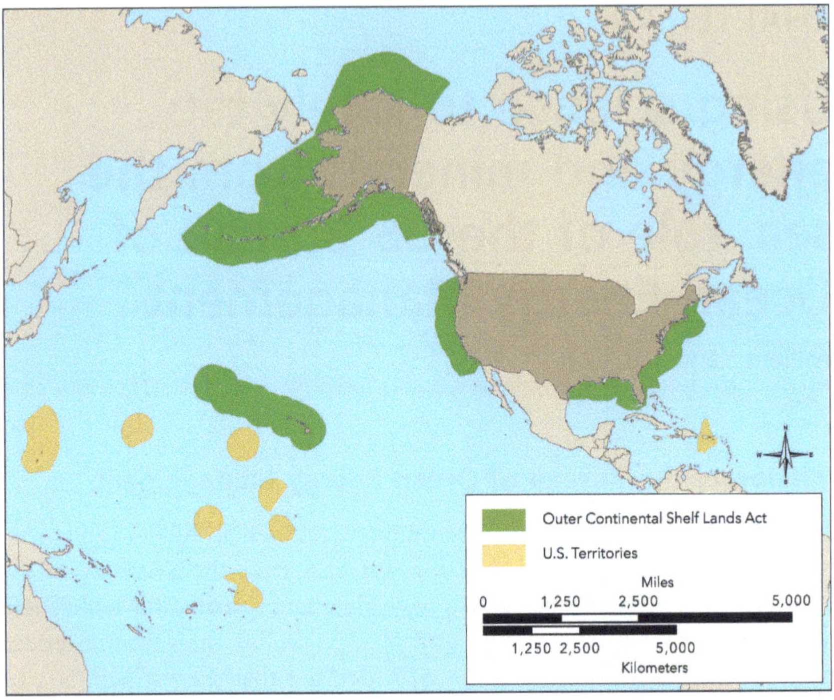

Figure 10.1 Areas in green indicate the bureau's geographic jurisdiction, beginning at 3 nautical miles (nm) from the coastal baseline (or 9 nm for Texas and the west coast of Florida) and extending to the 200 nm edge of the Exclusive Economic Zone (EEZ) or farther if authorized under Article 76 of United Nations Convention on Law of the Sea (UNCLOS). The confirmed area amounts to nearly 2.5 billion acres. The OCS does not include US territories, shown in yellow.

and Bureau of Land Management, bureaus within Interior, but in 1982 Secretary James Watt established a new bureau at Interior, the Minerals Management Service (MMS) (Secretary Order 3071, 1982) to carry out the task. MMS administered OCSLA for nearly 30 years, until the 2010 explosion and fire at the *Deepwater Horizon*, a deep-water drilling rig operating over BP's Macondo Prospect in the Gulf of Mexico, brought criticism and reorganization. In May 2010, Secretary Ken Salazar issued Secretarial Order 3299 that separated and reassigned MMS' responsibilities into three new management structures: BOEM, the Bureau of Safety and Environmental Enforcement (BSEE), and the Office of Natural Resources Revenue (ONRR), whose role is limited to financial matters. BOEM handles the planning and approvals that lead up to operations, and BSEE inspects operations, enforces environmental and safety requirements, and oversees decommissioning when platforms reach the end of productive years. BOEM's work fundamentally divides into three programs: oil and gas, offshore renewable energy, and non-energy marine minerals (e.g., sand, aggregate, critical minerals).

Oil and gas

BOEM has long managed oil and gas exploration and production through a multistep process: a five-year national lease sale schedule, lease sales themselves, lessee exploration plans, and lessee development plans if discovered oil or gas warrants production. In conjunction with the National Oceanic and Atmospheric Administration (NOAA) National Marine Fisheries Service, BOEM reviews and may approve exploration activities such as permits to conduct seismic airgun surveys. This type of survey is conducted to help determine sub-sea oil or gas resources but has been controversial because of potential impacts on marine mammals and other wildlife.

To date, industrial energy production from the OCS has been limited to oil and gas, with nearly all production from the Gulf of Mexico (over 99% in 2019), and the remainder from Southern California and Alaska (BSEE, 2020). In Fiscal Year 2019, the OCS provided about 16% of total domestic US oil production and about 3% of domestic natural gas production (Bureau of Ocean Energy Management, 2020b). That year, companies paid the US Treasury more than $5.5 billion in bonuses, rent, and royalties to extract oil and natural gas in federal waters. Portions of that revenue fund conservation initiatives such as the Land and Water Conservation Fund and the Historic Preservation Fund. In Fiscal Year 2019, the offshore oil and gas industry supported more than 277,000 jobs (DOI, 2019).

However, greenhouse gas emissions from combustion of offshore oil and gas are also a significant concern in domestic and international consideration of climate change. On January 27, 2021, President Biden ordered that

> *"[t]o the extent consistent with applicable law, the Secretary of the Interior shall pause new oil and natural gas leases on public lands or in offshore waters pending completion of a comprehensive review and reconsideration of Federal oil and gas permitting and leasing practices in light of the Secretary of the Interior's broad stewardship responsibilities over the public lands and in offshore waters, including potential climate and other impacts associated with oil and gas activities on public lands or in offshore waters."*
>
> ***(E.O. 14008).***

The temperature-raising effects of greenhouse gas emissions are undeniable, and many technical reports show that climate change threatens the well-being of our Nation and global society generally. E.O. 14008 also provides that "[i]n conducting this analysis, and to the extent consistent with applicable law, the Secretary of the Interior shall consider whether to adjust royalties associated with coal, oil, and gas resources extracted from public lands and offshore waters, or take other appropriate action, to account for corresponding climate costs." This reference to climate costs is particularly significant because in another executive order issued on the day of his inauguration, President Biden established a working group on the social cost of greenhouse gases with the direction that "agencies capture the full costs of greenhouse gas emissions as accurately as possible, including by taking *global* damages into account" (E.O. 13990).

Marine minerals

BOEM manages non-energy minerals, currently mostly sand, on the OCS. Hurricanes and other storms scrub sand from the coast, working in combination with rising seas and, in the Gulf of Mexico, subsidence, that increase erosion. Often haphazard human settlement including coastal structures such as jetties magnifies coastal erosion. To mitigate these unintentional consequences, OCS sand is utilized to restore vital habitats for birds and sea turtle nesting, beach mice, crabs, and other invertebrates, as well as replenishing beaches where people live, play and work.

According to the Middlebury Institute Center for the Blue Economy, in 2019, about 80% of the nation's population lived in coastal states and nearly 44% of US GDP was in shore-adjacent counties, even though only 17% of the nation's landmass is in shore-adjacent counties.

Millions of cubic yards of sand are needed for beach, dune, and wetland renourishment, and using sand from within the 3-nm zone of state-owned seabed ringing our coast is not always feasible. Using near-shore shoals may have a negative ecological impact, may increase storm wave impacts and flooding potential, or may not be the best match for the area to be restored. BOEM is increasingly collecting data on resources further offshore. BOEM is developing a National Offshore Sand Inventory to identify sand resources on the OCS. BOEM authorizes removal of OCS sand for beach and coastal restoration projects through leasing, often partnering with the US Army Corps of Engineers to oversee the dredging of sand from the OCS and to deliver it to where it is needed. Major Gulf of Mexico restoration projects using OCS sand have leaped forward in recent years. In 2019, for example, BOEM and the State of Louisiana's Coastal Protection and Restoration Authority signed two agreements to provide about 10 million cubic yards of sand from federal waters to restore beach, dune, and marsh habitat along

Trinity-East Island, Timbalier Island, and West Belle Barrier Headland (BOEM, 2019).

Critical minerals occurring offshore such as manganese, cobalt, and rare earths complement sand in BOEM's marine minerals program. The reason for a hard look is clear: some minerals are essential for sophisticated technologies—such as cobalt for batteries in electric vehicles—are almost entirely imported by the United States. The US relies on imports from countries like the Democratic Republic of the Congo and China, in the case of cobalt, whose continuing supply chain cannot be guaranteed. However, several hurdles stand on the road before development: Current law does not allow for exploration licenses with development rights, limiting commercial interest. The US Geological Survey, another Interior bureau, is researching whether critical minerals are available on land in the United States where access may be easier. One may also ask whether critical minerals are available from reliable US allies at less expense or less environmental harm than seabed mining.

Offshore wind

The Energy Policy Act of 2005, Pub. L. 109—58, amended OCSLA to authorize renewable energy development on the OCS. With that, BOEM developed regulations (30 CFR Part 585) and launched a program for offshore wind. BOEM's authority also includes regulating ocean current, ocean wave, and tidal (all of which are examples of hydrokinetic energy sources) (see Geerlofs and Moreno, this volume) and offshore solar development. The wind process has no national leasing schedule as for oil and gas but includes similarities. It progresses from state expressions of interest, state task force collaboration, identification of wind energy areas, calls for information and nominations, lease sales, lessee site assessment plans (SAPs), and construction and operation plans (COPs) covering projects through decommissioning. All of this involves extensive environmental review and federal, state, and tribal coordination and consultation. As with the oil and gas program, there is extensive opportunity for public comment. As of February 1, 2021, BOEM had 16 active leases on the Atlantic from Massachusetts to North Carolina (leased area shown in Fig. 10.2) with 11 construction and operation plans (COPs) under review and up to five more expected in the next year. Although just two 6 megawatt turbines are currently operating on the OCS (offshore Virginia), BOEM has concluded that installation of some 22 gigawatts (GW) is foreseeable by 2030 (BOEM, 2020a). In economic terms, the National Offshore Wind Strategy has

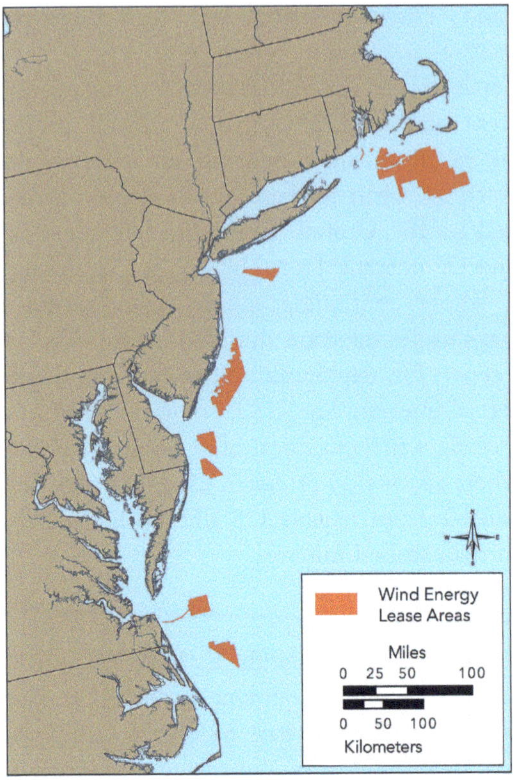

Figure 10.2 As of February 1, 2021, BOEM had 16 leases on the Atlantic from Massachusetts to North Carolina, shown here in orange. This includes 11 construction and operations plans (COPs) under review and up to 5 more expected in the next year. Although just two 6 megawatt turbines are currently operating on the OCS (offshore Virginia), BOEM has concluded that installation of some 22 gigawatts (GW) is foreseeable by 2030.

identified significant potential public benefits associated with offshore wind development, including an estimated $440 million in annual lease payments into the US Treasury, approximately $680 million in annual property tax revenues, and 160,000 gross jobs in coastal regions.

Offshore wind sits at an earlier stage on the West Coast. Deeper waters in the Pacific will require floating turbines rather than those that will be mounted in the seabed on the East Coast; that technology is less proven and less deployed. Furthermore, conflicts with Department of Defense activities are more of an issue, which BOEM actively works to resolve. But, California, Oregon, and Washington have dramatic goals for increasing the share of renewable electrical energy to serve highly dense population areas while safeguarding key natural protected areas on the coast. The prognosis for energy from West Coast wind turbines is good; meanwhile, scientists and engineers are testing marine hydrokinetic technologies to capture energy from the ocean's movements. BOEM issued a lease for the first wave energy research project in federal waters offshore the US West Coast in February 2021. The overall prognosis for offshore wind has also been advanced by the Biden Administration's priority for addressing climage change and increasing reliance on renewable energy. In particular, Executive Order (E.O.) 14008 (Executive Order 14008, January 27, 2021 86 FR 7619, February 1, 2021) establishes a goal of doubling offshore wind by 2030 (section 207). BOEM's current plate of wind farm applications suggests that this goal is likely to be exceeded significantly.

Using science and observational data in the new blue economy

How do science and observational data contribute to the new blue economy? Establishing baseline data and conducting long-term monitoring are key to understanding the ecosystem and employing mitigation measures to protect the environment. Following are a few examples of BOEM information needs and how the data can help BOEM responsibly fulfill its offshore energy, economic, and environmental mission:

- Data to create a 4-D map of resources on the OCS encompassing the dimensions of space and time is useful for developing energy and mineral resources while protecting resources such as marine mammals, fish, and seafood.
- Data to understand the oil and gas reservoirs beneath the sea floor and the sand and other minerals on the seabed surface.
- Data to understand the wind that blows above the sea surface and whether it will spin turbine blades strong and steady.
- Data to understand the ecosystems encompassing the sea floor, sea, and air above that may be harmed by the activities BOEM allows.
- Data on sensitive areas—banks and canyons rich in corals and fish; places most important for whales; subsistence hunting grounds for Indigenous people—that may require exclusion from any development.
- Data that may indicate a need for protective measures, mitigation, such as placing limits on timing and loudness of sounds from platform and turbine construction and decommissioning, caps on air pollutants from engines on facilities, and constraints on the kind and place of discharges from drilling and production of oil and gas.

- BOEM is at the forefront of research on traditional ecological knowledge of indigenous people and local knowledge, and has effectively collaborated with Native peoples over the years. In the last 5 years, the Bureau has sponsored a dozen projects (e.g., marine mammal observations, social networks, harvest patterns) that have engaged indigenous peoples. The bureau plans for even more robust engagement with Native peoples in the future through collaborative and co-productive research efforts, which will likely require a greater investment of agency resources. Understanding how bureau activities may affect traditional ways, subsistence, and indigenous cultural resources is a key element to the bureaus decision-making processes.

BOEM addresses research needs for environmental science and observation through its Environmental Studies Program (ESP). Section 20 of OCSLA mandates studies to assess and manage impacts on the human, marine, and coastal environments from offshore energy and marine mineral development. The section specifically calls for studies to monitor human, marine, and coastal environments and to provide time series and trend information for identifying significant changes in the quality and productivity of those environments and identifying the causes of these changes. Since its inception in 1973, the ESP has provided over $1 billion for research to meet its science needs and disperses about $30 million annually. BOEM's environmental program integrates the ESP with environmental assessment and policy to ensure that environmental protection is a foremost concern and an indispensable requirement in BOEM's decision-making. Other entities such as BSEE also use BOEM's environmental science in fulfilling their research and analysis missions.

BOEM's studies cover diverse subjects and those doing the research are primarily federal agencies, academic institutions, and private experts that may work with nonprofit or for-profit entities. The funding reinvests federal dollars into economic activity and is channeled through three different vehicles: interagency agreements for federal agencies; cooperative agreements with coastal state-affiliated educational institutions and potentially with nonprofits; and competitive contracts with any sources that meet all requirements and stand out against others seeking to conduct the research. Graduate students often gain a stipend by participating in research, which helps pave the way to future careers and maintain the nation's science workforce.

In 2015, BOEM contracted with the National Academies of Science, Engineering, and Medicine (NASEM) to establish a standing Committee on Offshore Science and Assessment (COSA) to provide independent, scientifically credible, and objective information on issues relevant to the bureau's environmental program. For programs that fund and use science in decision-making, excellence requires independent advice and scrutiny. That serves two purposes. One is securing information and perspectives from experts who do not have a stake in internal agency dynamics and incentives, which may conflict with objectivity. The other purpose is credibility. Presenting scientific thinking to independent experts and hearing their reaction, in public, gives credibility.

COSA (operating under the Academy's distinct Federal Advisory Committee Act legislation) typically meets three times annually with BOEM support. The meetings are public, accessible live to anyone with Web access, and recorded. COSA advises on the study profiles that the bureau develops. At a higher level, the committee provides suggestions on the ESP strategic framework that guides ESP funding decisions (Bureau of Ocean Energy Management, 2020b). The strategic framework identifies categories of impacting activities and resources that BOEM needs to understand and evaluate, as well as information needed to monitor and assess status, trends, and resilience of resources and confirm compliance with environmental laws.

BOEM has identified strategic questions in the framework to guide development of the ESP research portfolio over the next 5–10 years, at all levels (national, regional, and program), and for both short- and long-term needs:

- How can BOEM best assess cumulative effects within the framework of environmental assessments?
- What are the acute and chronic effects of sound from BOEM-regulated activities on marine species and their environment?
- What are the acute and chronic effects of exposure to hydrocarbons or other chemicals on coastal and marine species and ecosystems?
- What is the effect of habitat or landscape alteration from BOEM-regulated activities on ecological and cultural resources?
- What are the air emissions impacts of BOEM-regulated activities to the human, coastal, and marine environment and compliance with the National Ambient Air Quality Standards (NAAQS) and Prevention of Significant Deterioration (PSD) increments?
- How will future ocean conditions and dynamics amplify or mask effects of BOEM-regulated OCS activities?
- How does BOEM ensure the adequate study and integrated use of social sciences in assessing the impacts of OCS activities on the human environment?

- How can BOEM better use existing or emerging technology to achieve more effective or efficient scientific results?
- What are the best resources, measures, and systems for long-term monitoring?

The framework then establishes seven criteria for BOEM to follow in prioritizing study funding decisions, summarily listed here:

1. Need for Information in BOEM Decision-Making
2. Contribution to Existing Knowledge
3. Research Concept, Design, and Methodology
4. Cost-Effectiveness
5. Leveraging Funds
6. Partnerships
7. Multiregional and Strategic Utility

Although the private sector receives less than a quarter of BOEM's environmental research funding directly, that sector also supports the federal agencies and academic institutions funded by the bureau. More significantly, private sector investment in offshore energy depends on BOEM's authorizations. The stakes are high. Consider offshore wind, for example: a March 2020 study by the American Wind Energy Association estimates that the offshore wind industry will invest \$28–57 billion in the US economy and will support 45,000–83,000 jobs in the next 10 years, based on a range of development varying from 20 to 30 GW along the East Coast of the United States (AWEA, 2020). As noted earlier, BOEM estimates that 22 GW of development is reasonably foreseeable on the Atlantic coast by 2030.

Given this context, consider BOEM's ESP project: "Real-time Opportunity for Development Environmental Observations" (RODEO). RODEO began with the Block Island wind farm in state waters offshore Rhode Island—BOEM-authorized cable passage through federal waters—and is expanding to address other wind projects to the north and south. RODEO's purpose is to collect observations during the installation and early operation of offshore wind farms to improve predictions of impacts from development. This includes recording activities as they occur, such as sound (airborne and under water), seafloor disturbance and recovery, benthic monitoring, scour monitor testing, and submarine cable laying. Block Island was the first installation of marine offshore turbines in the United States, in State waters, but many more farther offshore are planned for the near future, and RODEO is providing important actual rather than conjectural information about spatial and temporal impacts for application to the potential several thousand turbines of the future. It helps us to protect the environment but not unnecessarily burden this burgeoning blue industry.

When the first wind turbines were installed in federal waters, offshore Virginia in June 2020, a bubble curtain was placed around the planned site for one of the two turbines but not around the other. The turbines were installed—a noisy affair—and hydrophone monitoring determined that the bubble curtain significantly reduced sound passing into the waters beyond. Who benefits, besides the marine mammals hearing less? BOEM learned that the bubble curtain helps meet the bureau's mandate to avoid harm and comply with environmental laws. The project team received funds directly for project work, but they also gained experience and qualifications for a much larger market serving the environmental information needs of the companies that plan to build and operate offshore wind farms. Last, but hardly least, the information gained from RODEO helps BOEM establish conditions of approval that will mitigate environmental impacts and, in turn, provide companies and their investors confidence that the projects and the money spent will not turn to dust. Tens of billions of dollars are at stake.

BOEM's long-term environmental goals

In addition to the ESP strategic framework, BOEM is pursuing a number of long-term environmental goals addressing both research and assessment, four of them more than relevant to this volume's theme.

The first and overarching goal is to be first in class. That means that the bureau will focus on, invest in, and champion BOEM's environmental program until its science, assessment, and regulation are recognized as "first in class" among peer federal agencies. BOEM is engaging the NASEM for advice on this, including but not limited to advice from COSA, and anticipates developing benchmarks for comparing its environmental program to other federal agencies in its class; making that comparison; communicating success and results; searching for opportunities to improve, and taking steps to close gaps where it falls behind or needs to advance; and remeasuring progress and communicating outcomes at appropriate points.

A second goal is to establish one or more centers of expertise that apply highly specialized knowledge and skills in key technical areas to serve and benefit all BOEM programs and regions. These centers will optimize capabilities and management, eliminate redundancy, enhance BOEM-wide consistency, facilitate emerging technology, maintain institutional knowledge in key areas, and expertly translate analysis into products for decision-making. The first is the Center for Marine Acoustics, established by BOEM in 2020.

Among the other goals is to advance emerging technologies to answer key scientific questions concerning our activities. The primary means is by asking potential funding recipients to present opportunities for using the technologies in their proposals, rather than BOEM making substantial investment in the technologies *per se*. Research and development of new technologies stimulates the economy and has long-term return-on-investment potential in ways hard to fully imagine. BOEM's goal is to find and use high-quality information more efficiently, cheaply, safely, and with less adverse effect on the marine environment or the people who use it. In particular, we will seek to leverage and promote promising emerging environmental sensors that use the electromagnetic spectrum (optical, infrared, RADAR, LIDAR), sound, and eDNA (environmental deoxyribonucleic acid, see Ausubel and Curry, this volume); platforms for those sensors including autonomous unmanned vehicles in the atmosphere and sea; and a network of small satellites accessed through both commercial vendors and National Technical Means. BOEM is also increasingly using artificial intelligence methodologies to identify species.

A final goal is advancing the sophistication of our communication with key audiences, both internal and external. We are developing ways to use illustration graphics, imagery, and videos to visually communicate complex scientific information through simple and engaging story-driven products. We want to do more than a great job at communicating the results of our research and decisions once made—we want to embed in that research and decision-making the best that communications art and science can give us.

Implications for the "New Blue Economy"

What do these goals mean for the "New Blue Economy" explored by this book? These goals—in search of excellence for BOEM's research and regulation—can help define and refine the methods and results needed for investment security and return in the controversial and competitive realm of ocean energy. By requesting eDNA methodology in proposals to monitor marine ecosystems, BOEM indirectly helps reduce costs for companies that must monitor and bring clients to the technical firms that can help those companies do the work. By exploring satellite imagery to assess air pollution concentrations, BOEM obviates the need for expensive equipment on research facilities at sea, with similar beneficial results for investors and good care for the environment. By developing and deploying a well-vetted and widely accepted acoustic model, BOEM may bring some peace to stakeholders on all sides of the equation seeking certainty about impacts to marine mammals. And don't mistake this: uncertainty significantly raises costs.

References

American Wind Energy Association, 2020. U.S. Offshore Wind Power Economic Impact Assessment. https://supportoffshorewind.org/wp-content/uploads/sites/6/2020/03/AWEA_Offshore-Wind-Economic-ImpactsV3.pdf.

Bureau of Ocean Energy Management, 2020a. Vineyard Wind 1 Offshore Wind Energy Project Supplement to the Draft Environmental Impact Statement. https://www.boem.gov/sites/default/files/documents/renewable-energy/Vineyard-Wind-1-Supplement-to-EIS.pdf.

Bureau of Ocean Energy Management, 2020b. Oil and Gas Energy Fact Sheet. BOEM. https://www.boem.gov/sites/default/files/documents/oil-gas-energy/BOEM_FactSheet-Oil%26amp%3BGas-2-26-2020.pdf.

Bureau of Ocean Energy Management, 2020c. Center for Marine Acoustics. https://www.boem.gov/center-marine-acoustics.

Bureau of Safety and Environmental Enforcement, 2020. Data Center: Outer Continental Shelf Oil and Gas Production. https://www.data.bsee.gov/Production/OCSProduction/Default.aspx.

Bureau of Ocean Energy Management, 2019. Press Release 06/04/19. BOEM Announces Restoration Project for Louisiana's Gulf Coast. https://www.boem.gov/newsroom/press-releases/boem-announces-restoration-project-louisianas-gulf-coast.

Department of the Interior (2019). U.S. Department of the Interior Economic Contributions Report, Fiscal Year 2019.

Environmental Studies Program Strategic Framework 2020. (https://www.boem.gov/sites/default/files/documents/about-boem/ESP-Strategic-Framework-Final-FY20.pdf).

Executive Order 13990, January 20, 2021 (86 FR 7037, January 25, 2021).

Executive Order 14008, January 27, 2021 (86 FR 7619, February 1, 2021).

Secretary's Order 3071, January 19, 1982 (47 FR 4751, February 2, 1982).

Secretary's Order 3299, Issued on May 19, 2010, and Amended on June 18, 2010, and August 29, 2011.

https://www.boem.gov/center-marine-acoustics, 2020.

Interview: Ocean data use in the oil and gas industry

Sam Walker

Energy

How has the availability of ocean observing data and enhanced operations and decision-making in your area?

It's lumped into a couple of categories one being obvious—the safety and efficiency of operations. That's always at the forefront of what we think about offshore, from evacuating platforms to dispatching vessels with product around the world. And there's a tieback to risk management. For example, where in the world you might locate the particular asset, or where you might look to conduct business, or where you might look to divest some of your portfolio because the risk level is too high or just doesn't make sense from an economic standpoint, it is essentially situational awareness.

How are these data used in spill response?

I would say it's not altogether different than normal operations in terms of how data would be used, just a bit more focus and with more emphasis on the modeling. I would tend to focus more on the repeatable kinds of things that are taking place offshore as opposed to stochastic events. There are just so many unknowns in a spill response, the size and location of the spill, the weather conditions, etc. It's more about getting the right kinds of data in a timely fashion. And do you have the right kinds of people who can interpret and help the decision-making process in the broader context? But it is not really that much different than normal operations. You would have the same kinds of tools that you would be using to ingest the data. A primary difference is the pace of decision-making.

Can you discuss natural resource damage assessment? There were a lot of frustrations about the lack of adequate baseline information in dealing with the aftermath of the Deepwater Horizon to conduct an adequate natural resource damage assessment

So that's a data problem—different kind of data—but it's a data problem. Let me try to answer it more broadly as opposed to any one specific incident. The baseline piece is definitely something that we pay a fair bit of attention to here and all around the world.

As a company, we've made some significant biodiversity commitments here this year and have a newly appointed executive VP for sustainability. So that baseline piece is critical. Not just for damage assessment, but for where to site a new operation or modify an existing one, laying pipeline, survey hazards to preemptively mitigate damages. We use camera-driven ROVs (remotely operated vehicles) and detailed autonomous systems and ship-based technology to make sure that infrastructure installation operations are not going to damage something like a coral system, but also to make sure that you're not going to lay something down on top of say unexploded ordinance or some other kind of geological hazards.

So, what do you need to know about your environmental context in the baseline in the course of your day-to-day operations? That's very likely going to be the same kind of information that you would need in a damage assessment or active spill response, where you're trying to prioritize the deployment of certain kinds of mitigations.

And so, back to your point about data, I would say it's as much about managing the data and providing accessibility to it than it is about whether or not we have it. Now that's not to say that we couldn't have more data or more timely data or more derivative information.

For example, there would be a large cost to conducting a high-resolution bathymetric survey of the entire Gulf of Mexico. However, that cost would not come anywhere near the cost that was ultimately spent on the Gulf of Mexico after the Deepwater Horizon spill. Orders of magnitude lower. So why aren't we conducting collective surveys to obtain data like this? Everyone would benefit.

Data available from acoustic Doppler current profilers (ADCPs) mounted on offshore structures and operated by oil companies in the Gulf of Mexico were shared during the Deepwater Horizon blow out, an example of a private–public partnership. Using this as a model, can you think of other industry-operated/generated data sets that could help fill data gaps to help build a more robust picture of the Gulf of Mexico?

I think Europe has a relatively easier time of it as the relationships between government, academia, and the private sector flow a little bit more naturally there.

During Deepwater Horizon, some academics were reticent to share data because they were worried that data would be embargoed and they wouldn't be allowed to publish. That was not the case, but the rumor spread rapidly. As a result, a lot of additional data were collected that maybe didn't have to be, or a lot of extra analytical work was conducted when the effort could have been streamlined if better relationships were in place ahead of time.

I guess the point is there are real challenges to forming mutualistic partnerships. But without question there's value and efficiency that can be derived from shared collection and use of data. We do this frequently in our global projects—partnering with local institutions to help guide science-based efforts and to share the results.

Partnerships can take many forms and we are increasingly looking at building long-term relationships with cities—where much of the demand in energy growth is projected to occur. We've long held partnerships with research universities around the world, and then there are the big strategic alliances with data-centric and information-management companies where we can mutually leverage expertise. We also have important relationships within the nonprofit sector including Flora and Fauna International, EDF, and others. For a company like ours where operations are at the heart of our activities—the need to bring data into the decision-making space is critical so if that need can be met while also addressing, say research needs for a university that can make a really powerful impact and built trust at the same time.

How important is the Big Data concept in your work?

We have a couple of the largest high-performance computing centers in the world within our portfolio. One of them is just down the road here in Houston. We've got another one in the United Kingdom. We've been crunching big data in various forms for a long time, especially in the geoscience space. We've got some extremely skilled people in this space not just on the acquisition side but also on the interpretation side. It's changed pretty dramatically from the sort of early stages of digitization, where we had a massive library in our buildings where people could go and look at that data. Now it's almost exclusively digital using massive computing power. And that's just on the geosciences side.

On customer-facing businesses—gasoline stations and other kinds of services—we provide huge amounts of data about the transactions. BP has one of the largest trading floors in the industry and it's world-class. All they do is crunch data, it looks like a Wall Street investment house.

And then on the offshore side, we're constantly looking at circulation models and have a customized weather capability as well. And our annual Energy Outlook that BP publishes on behalf of the industry requires huge amounts of data being crunched. So we're constantly trying to upgrade that capability and stay current.

When you talk about the human dimension, what is your view on the value of certification and credentialing?

I think it's a component. If you have, for example, someone coming out of the Coast Guard or the Navy or some other branch of the military and they may only have a high school diploma but a lot of life experience, a lot of professional experience, or a military occupational specialty (MOS code) of some type. But then their resume is alongside someone who's got a Master's degree. If they have a certification in marine technology now they stack up. So I think it needs to be a component and maybe the trick is to make those certifications available through nontraditional means. The reality is that higher education is getting a wake-up call here in 2020. If they didn't have it already from a financial standpoint, they're certainly going to get it now due to the global pandemic. So, what could and should be done? I think without question there need to be certifications. I like this idea of microcredentialing and maybe there are different pathways to it. There's a test-based one, but also maybe a knowledge-based one where someone is, for example, presented with a problem and they have to solve it to demonstrate capability. I do think it's important to have those opportunities for people that may just need to augment their resume.

I think it's time for a little bit more progressive view. A college degree or graduate degree should still be worth plenty and it is an earned credential and a hard-earned one in most cases. But what about professional experience? Where does that stack up, and this microcredentialing is somewhere in the middle that would allow people to move up or into a different career without having to go back to school.

How unique is your own career track?

I've come to appreciate it over the years. I have a diversity of experience in different sectors. Academia, government—at the state and federal levels—, nonprofit, when I was with IOOS (the Integrated Ocean Observing System; see Gouldman, this volume), and in the commercial sector. And so how unique am I in that regard? I would say more unique than most. The path was conscious on my part because I anticipated there would be opportunities across different sectors and that's proven to be true.

But I do think a more interdisciplinary background will be more common in the future because there's going to be a requirement to work across sectors more and more. I think you don't pick a single career path. Frankly maybe right up to my parents' generation, certainly my grandparents' generation, you picked a profession and stayed in the same job and maybe even with the same company all your career. I won't say it's a thing of the past but that's becoming the more unique situation. It is more likely you're going to have multiple careers not just multiple jobs.

Would you recommend that sort of career diversity to young folks coming up in the community?

I certainly would. I think the world is evolving too quickly to put all your eggs in one basket. There are careers or jobs that you could bet on for 30 or 40 years. I personally wouldn't make those bets, but it also depends on where you are in the world. If you're in a deep poverty and you don't have any job, you might be more than happy to take a steady reliable job. But that's different than saying would you consider having a kind of multitrack career over time? And yes, absolutely I'd recommend it.

Are you looking toward other sectors, disparate kinds of sectors, as breeding grounds for your workforce?

Without question. We (BP) have made some significant changes as a company in the last year and a number of them were already in place before COVID hit. The whole company has been reorganized and we're not quite done with that yet. But in the course of doing that it's become quite evident that the executive leadership of the company is very committed to bringing in other kinds of expertise. For example, we just brought in somebody who headed up the data side in the early days for Uber and then moved over to Google and was running logistics on the data side. So that's maybe not quite as big as step is say something like medical diagnostics, but in the robotic space we have a very competent innovation and engineering team.

We don't have the traditional upstream and downstream parts of the business anymore. It's just production and operations under one group. So, we were looking to bring people in that to understand how to operate in a flatter organization that's better integrated as opposed to a more siloed kind of organization.

Even in the skill set arena, we're looking for different types, trying to recruit a more diverse work staff and we have been explicit in defining diversity for the company that is not strictly based on balancing gender or race, it includes geographies and includes nationalities and includes skill sets, different age groups, and so on. I think this is actually one of the big strengths for BP.

Do you foresee a shift in skill sets required for technicians in your industry given the increasing inclusion of real-time data in decision-making (data literacy skills)?

This is right at the heart of it in terms of the capability that needs to be developed. The capacity piece is a little bit harder to chase down right now because there's so much going on in digitization and robotics, etc. so it's hard to say. Twenty years ago, it was much easier to say how many people you're going to need for a particular activity for the next 10 years. It's harder to do so now when focusing more on the capability piece.

There is a skillset shift taking place right now and I think it's going to continue to evolve quite rapidly. There are some obvious things such as digital skills, visualization skills, and data synthesis skills, but also how to interpret and discuss data. I believe it will take a balance of skills, whether that comes in the form of individuals that are working together or people that have a more Renaissance-type background. I imagine it'll be some combination.

But one thing I'm quite certain is that we need to keep the human cognition piece in as an important element. The danger of Big Data and automation is that you start to rely on computing almost exclusively. You could imagine scenarios unfolding particularly in a big data environment where you've got really high-quality data, that's very timely and it spits out a graph or an annotated map or whatever the graphic format might be, and you're like, "wow, look at that area!" All red on the map and then when you look at the scale, you have to ask, does that really matter? So that's why I think we're going to have to consciously maintain that human component.

How does the predicted decline of oil in the world's energy portfolio impact BP operations? Is the company transitioning to include more renewable energy sources? Do you expect the data needs for the growing marine renewable energy sector will be different?

Where do I start with this one? I could have spent the entire time on this question. We already have a pretty significant renewables (low-carbon) portfolio. We have a big biofuels business in Brazil, we have quite a significant onshore wind business in the United States, we are moving into offshore wind along the US Atlantic seaboard with a commitment to 50 GW of new low-carbon energy production by 2030. That's less than a decade away. In Europe we have huge solar presence, we're moving into the electric vehicle charging arena and looking to scale that up significantly in the coming years. We are making commitments of a billion dollars a year capital investment in alternative energies. We've got our new net-zero ambition that is really guiding the company now, so the executive team has made an "all in" play on transitioning from being an international oil company to an integrated energy company.

So rather than belabor that, I'll switch over to the second part of your question, which is more around the data needs. Short answer, yes, data needs will be very different.

From offshore wind, biological, acoustical, reservoir management, seismic surveys to dealing with intermittency of energy load in the power grid, these applications all demand increased data needs and capabilities. Greener energy, cleaner energy, and more consistent energy is the goal and will be data-driven.

What opportunities do you see?

Just like we're seeing in the terrestrial environments with advancements in battery technology, for example, there will be a need for significant power in supporting autonomous ocean systems. Much of the ocean-borne commerce, even more traditional marine commerce areas could be done robotically in the years to come. It is more efficient and safer for people, avoiding hopefully a lot more accidents and collisions at sea. So I think that entrepreneurial sort of space is potentially a target audience.

I think of the weather service analogy and how they've improved things over time, thinking about the early days when the big focus was on predicting and tracking storms. And now, those predictive capabilities are quite good. Then there was more of a shift toward the ability to predict the strength of those storms and the specific levels and types of impact they may or may not have.

It seems to me that you can make a step change in an existing marine commercial space such as aquaculture, which hasn't really been taken to scale at a global level yet despite plenty of history. There are plenty of operations that are working and profitable but how do you really take it up to the next level? Is that part of the New Blue Economy? Probably, because if you take things to scale, then you can really see the efficiencies and then you're going to need that larger workforce and the application of more technological tools, etc. It's a fascinating subject.

While we're on the subject of new opportunities, consider the reallocation of high-value kinds of resources, those specialized metals and other kinds of materials, like all the subsea cabling that's around the world. There's a lot of high-value metals down there. How can that be reutilized? There almost certainly is an opportunity there for offshore salvage that's going to grow dramatically over time. How do you do that in a cost-effective manner, so you're not taking the same path as decommissioned vessels and marine equipment that have often been simply scrapped and not always in the safest manner. That's a very fascinating space—applying the principles of a circular economy to reuse, recycle, and redeploy resources.

CHAPTER 11

Influences and impact of the energy industry on the new blue economy and its workforce development

Jan van Smirren
Ocean Sierra, LLC, Sugar Land, TX, United States

The movement of the energy industry offshore

Prior to 1890, oil and gas exploration was an entirely onshore industry; however, in 1896, as the first offshore wells were drilled from 100 m long piers constructed on the Santa Barbara, CA coast, this marked the beginning of the relentless march of the oil and gas industry into deeper water in search of hydrocarbons. By the 1930s, offshore platforms were constructed in the shallow waters of the Gulf of Mexico. These platforms had piles and elevated decks to prevent hurricane waves from overtopping the structure. In 1947, Kerr McGee contracted the engineering company Brown and Root to build a platform 10 miles out to sea, this was the first occasion a platform had been built out of sight of land (Morton, 2016). In 1962, Shell drilled the first well from a floating drilling rig (semisubmersible), the Blue Water 1, which was held in position using an eight-anchor mooring system. Semisubmersible drilling rigs made it possible to drill in increasing water depths, but as drilling moved into ultra-deepwater (greater than 1500 m), the mooring system became restrictive, as the depth a rig could drill in was limited, and installation and recovery of the anchors were a time-consuming and costly process. To overcome this, the SEDCO 445 drillship was built in 1971. This was the first dynamically positioned rig, held on location using marine thrusters rather than a mooring system, and therefore capable of drilling in unlimited water depths. Today the majority of the ultra-deepwater drilling fleet is made up of dynamically positioned drillships.

As drilling depths increased, the designs of production platforms had to change. Traditional gravity-based structures became unfeasible and as with

Preparing a Workforce for the New Blue Economy
ISBN 978-0-12-821431-2
https://doi.org/10.1016/B978-0-12-821431-2.00023-8

drilling rigs, were replaced by a variety of different moored designs, including floating production systems, tension leg platforms, semisubmersibles, and spars. These technological advances have allowed the industry to move progressively into more remote and hostile areas and deeper water. Today deepwater wells have been drilled in water depths exceeding 3400 m and the deepest permanent platform is Shell's Perdido, in the Gulf of Mexico, which is moored in a water depth of 2450 m.

A demanding environment for scientists

This migration of the oil and gas industry offshore introduced new challenges and uncertainties. To allow a well to be safely drilled and for a platform to be efficiently designed, installed, and operated, it is necessary to understand the nature of the seafloor sediments, local bathymetry, seabed relief, and the physical meteorological and oceanographic (metocean) conditions. This information—the core element of the new blue economy—is also needed for the design and installation of the subsea infrastructure, pipelines, and coastal facilities.

Oil and gas companies recognizing the requirement to understand the conditions needed the skills of geotechnical engineers, surveyors, geophysicists, meteorologists, and oceanographers. Employment opportunities arose for scientists with these disciplines, in oil and gas companies, engineering contractors, and companies specializing in data acquisition, analysis, consultancy, and numerical modeling.

The move into more hostile and deepwater locations, which are typically not well understood, was a major driver for expansion of the geoscience industry and for technological advances in instrumentation and systems capable of working in extreme conditions and depths. This was particularly true for the metocean industry. In deep water, the impact of environmental forces becomes increasingly important for rig selection, platform concept selection, platform design, installation, and planning operations, all of which carry significant cost and safety implications (Shaw, 2003). Ocean currents create a particular challenge. Unlike coastal locations where they are dominated by tidal and wind-driven circulations, in deep water, the oceanic circulation is dominant and is more difficult to predict and characterize. Ocean currents' importance is also elevated as they will have a major impact on the design and performance of the mooring systems, risers, and tendons throughout the whole water column. Therefore, for most deepwater developments, it is necessary to acquire current data through the water column for at least 1 year. This requirement has created an industry of specialist contractors who developed capabilities to acquire long-term current data sets. The data sets acquired are then used to characterize the environment statistically and to undertake Extreme Value Analysis, to predict by extrapolation the 100-year return period parameters (more recently extrapolations out to 1000 years are commonplace).

Influences of contractual clauses on the performance of projects

Unlike the scientific community, the oil industry is driven by economic factors, the ultimate objective is to execute efficient exploration activities, and on the completion of a successful drilling operation, to produce first oil as soon as possible. To achieve this, contractors are required to be operationally responsive and accept the normal contracting and safety requirements of the industry. As timelines for projects are typically compressed, delivery of data on time is key. To ensure this urgency is transferred to the contractors who collected data, oil companies will typically apply clauses that reduce payment if the project is not started and completed on time (liquidated damages) and data return clauses, whereby delivery of incomplete data sets will result in reduced payment. This has several consequences. Firstly, contractors must develop project management skills and strict technical and operational procedures to ensure timely delivery of completed data sets and avoid the financial penalties of liquidated damages and data return clauses. Additionally, this has a trickle-down impact on the instrument manufacturers to ensure their instrumentation provides data of the highest quality and instruments have a very high level of reliability.

As with many industries, the importance of health, safely, and the environment (HSE) has become an increasingly important area of focus in the oil and gas industry and is strictly enforced. Today oil and gas companies require their contractors to have well-developed HSE systems that are compliant with their own internal systems and will audit them on a routine basis. In many cases having an approved HSE system is a prerequisite to being allowed to bid on contracts. This has resulted in the development of detailed safety procedures for all aspects of a contractor's activities and includes the requirement for the workforce to undertake formal HSE training before they are able to go offshore. Safety analyses are conducted on nearly all aspects work, and appropriate procedures are developed. This includes procedures for: the handling and servicing of sensors and instrumentation, working offshore, working on vessels, deployment and recovery of scientific instrumentation, and personal protective equipment (PPE). HSE policies have also influenced technology development. A key metric used in the industry is personnel exposure hours. The number of hours estimated for a project is one of the evaluation criteria in bid selection, and therefore reducing it will enhance the likelihood of being selected for the project.

Contractors will therefore work hard to reduce exposure hours. This can be achieved by minimizing the number of crew on a vessel and relying on shore-based remote operation centers, who support the offshore crews undertaking data quality assurance and quality control, processing data, and supporting crews in planning future operations. Further reductions can be achieved using autonomous underwater and surface systems for data acquisition.

Driving innovation in the workforce

While the oil industry is primarily interested in site-specific characterization of the environment, several alternative approaches can be adopted. In event that there is a need to improve understanding of a particular phenomenon, a regional interest in characterizing conditions or a technology that would have common value across the industry, then a group of oil and gas companies may form a consortium and undertake the study as a Joint Industry Project (JIP). As JIPs tend to have a greater research-based emphasis, in many cases these studies involve scientists from academic institutions as well as contractors. This engagement with the scientific community has created opportunities for not only for the research scientist to gain an awareness of the application of science in the oil and gas industry, but also leverage their expertise into other areas, and in some cases to move into the industrial sector as consultants or small businesses, consistent with the business model of the new blue economy. Two of the longest standing JIPs are the Deepstar Consortium (https://www.thedeepstar.org/program-structure), which is focused on its members' technology needs, and the Climatology and Simulation of Eddies/Eddy Joint Industry Project (CASE-EJIP), which is a consortium of oil companies whose focus is on the Gulf of Mexico. For more than two decades, Deepstar and CASE-EJIP have funded sensor research and development, data acquisition, and analytical and modeling studies, engaging scientists and technologists from government, academia, and industry.

In addition to JIPs, the counter can arise, whereby a product can be generated by a contractor or consultant that has value to a wide variety of oil companies. This has been prevalent in the seismic industry that undertakes multiclient acquisition programs, or in some cases independent speculative programs, with the sales of data to oil companies after completion of the project. This approach has become extremely popular for hindcasting winds and waves in many of the offshore oil basins around the world.

Where there is seen to be common need to improve scientific understanding of the environment, studies may be funded by government agencies. In the United States, these studies are conducted by the Department of the Interior, Bureau of Ocean energy Management, under its Environmental Studies Program (ESP, see Brown, this volume). Research areas covered by this program include physical oceanography, atmospheric sciences, biology, protected species, social sciences and economics, submerged cultural resources, and environmental fates and effects. The program has been operational since 1973 and has provided more than $1B in research funds. The ESP projects typically will have a significant scientific requirement and will require the acquisition of data on the Outer Continental Shelf. As a result of this, in many cases the studies are undertaken by consortia of academic and research institutes in collaboration with industry partners, in a similar manner to JIPs. These projects provide an excellent opportunity for cross-fertilization of the scientific approach and operational data acquisition across the two communities.

More novel approaches of the industry to technological develop have also been adopted such as the Shell Ocean Discovery XPRIZE. This presented a challenge to the ocean community to acquire high-resolution hydrographic data at a remote deepwater location, using only autonomous vehicles. The program was further sponsored by the US National Oceanic and Atmospheric Administration (NOAA) who provided an additional bonus prize for detecting and autonomously tracking a chemical or biological signal to its source and by industry and institutional partners. Participants in the challenge included students at varying stages of their careers, who in many cases collaborated with the engineering, sensor technology, and service industry companies.

Sharing data with the scientific community

The availability of data is an important consideration in the build-out of the new blue economy. Over the last two decades, the oil and gas industry has relaxed its position on the confidentiality of data, as offshore operators and contractors have become more engaged with the ocean observing community. This was first seen in Europe, with the System of Industry Metocean data for the Offshore and Research Communities (SIMORC), which commenced in 2007 (Schaap, 2007). This program provides scientists and engineers with an overview of, and access to, an increasing volume of metocean data sets collected by the offshore energy industry. Major offshore energy companies are participating via the International Association of Oil & Gas Producers (IOGP) and its IOGP Metocean Committee. At present, SIMORC covers more than 7700 data sets from Shell, Total, and BP, covering more than 3000 years of observations of winds, waves, currents, and sea levels. The SIMORC system is operational to serve users in identifying and gaining access to data sets and extends its coverage regularly with additional data sets from oil and gas companies (http://www.simorc.com/).

More recently, a similar program has been undertaken in the United States. The Gulf of Mexico offshore oil and gas industry has collected an immense amount of physical oceanographic data over the last three decades, particularly in deep water. The Gulf of Mexico Coastal Ocean Observing System Regional Association (GCOOS-RA) has gained access to many of these data sets from individual companies, contractors, and from JIPs. The acquired data sets have been fully quality controlled to US Integrated Ocean Observing System (IOOS, see Gouldman, this volume) standards and metadata appropriately supplemented. The data sets are available online through the GCOOS GulfHub data portal (https://gulfhub.gcoos.org/). At the time of writing, more than 4600 unique physical oceanographic data sets are available on the GulfHub ERDDAP server. Of these, more than 2900 are industry proprietary data sets from deep-water moorings, offshore oil rigs, production platforms, and vessels. It is anticipated that more than 5000 data sets will be available upon completion of the project (van Smirren et al., 2020). This work is supported with grants from the National Academies of Sciences, Engineering, and Medicine, as part of the Understanding the Gulf Ocean Systems initiative (https://www.nationalacademies.org/gulf/fellowships-and-grants/understanding-gulf-ocean-systems).

The need for training

The oil and gas industry continues to work closely with universities and community colleges to assist with the development of curricula focused on operational needs, enhancing the opportunities for students after completing their studies. An example of this is the Marine Technology program at Cape Fear Community College (http://catalog.cfcc.edu/preview_program.php?catoid=9&poid=1976&returnto=475), which is focused on providing academic and practical skills necessary for a career in marine scientific support. For hydrography, the University of Southern Mississippi Master of Science degree in hydrographic science (https://www.usm.edu/graduate-programs/hydrographic-science.php) meets the Category A requirements of the Féderation Internationale des Géométres, International Hydrographic Organization, and International Cartographic Association (FIG/IHO/ICA) International Board of Standards and Competence for Hydrographic Surveyors and Nautical Cartographers. More recently, Texas A&M has developed the Master of Ocean Science and Technology (https://ocean.tamu.edu/graduate-students/master-of-ocean-science-technology/index.html). This degree is focused on preparing students for careers in ocean sciences and particularly ocean observing (see also Van Sumeren, and Beaird et al., this volume).

These academic degrees provide a sound foundation to the student; however, as they transition into the workplace, there is a requirement for further training to develop specific expertise to work in the industrial ocean sciences environment. Much of this is achieved through mentoring and on-the-job training; however, for some of the larger organizations, more formal training programs are available. For example, Fugro, a major geoscience contractor to the oil and gas industry, has two academies located in Plymouth, UK, and Abu Dhabi in the UAE. These training facilities provide internal and external training on hydrographic surveying and operation of remotely operated vehicles (ROVs) and include theoretical and practical hands-on training on fully equipped coastal survey vessels.

Oil and gas companies and their contractors are important supporters and sponsors to many marine industry groups including the Society of Underwater Technology (SUT) headquartered in the United Kingdom. SUT's close relationship with the oil and gas industry has led to establishment of Branch offices in 12 of the key operating locations of the industry, including, for example: Aberdeen, Perth, Houston, and Rio de Janeiro. These branches provide short introductory training courses, designed for new recruits into the industry. Currently offered courses include the Subsea Awareness Course, Metocean Awareness Course, and the Integrated Marine Site Characterization Course.

In 2002, the Institute of Marine Engineering, Science and Technology (IMarEST), a UK-based international professional body, introduced two new professional accreditations: the Chartered Marine Scientist (CMarSci) and Chartered Marine Technologist (CMarTech). Note that in the United Kingdom, "chartered professionals" are like "certified professionals" in the United States (see MacKenzie, and Hall, this volume). CMarSci and CMartTech are valuable professional recognition for marine scientists and technologists, signaling professionalism and deep expertise, and can accelerate career progression, as well as providing recruiters with confidence that the capabilities of candidates are proven. In 2019, SUT partnered with IMarEST and the Marine Technology Society (MTS) to offer marine technology professionals the opportunity to seek the title of CMarTech in the United States.

Developing workforce opportunities

Since the middle of the last century, the energy industry has been one of the major providers of opportunities to marine scientists and technicians wishing to pursue a career in industry, with both technical and business career paths. The skill sets and expertise developed in support of oil and gas exploration are equally applicable in many growing markets that will emerge with the new blue economy. As oil and gas companies transition into the offshore renewables markets (see Geerlofs and Moreno, this volume), engineering and scientific consultancies are also making the transition, and the blend of work conducted is shifting away from oil and gas dominance, to more balanced portfolios of projects. Accelerated growth in US offshore wind sector is anticipated and, as with the oil and gas industry, offshore wind projects are trending toward locations further from land and in deeper water, which in many cases requires floating wind turbines mounted on moored semisubmersible platforms, to which much of the expertise developed in the oil and gas industry is directly applicable.

Specialist consultancies are also translating operational capabilities developed in the oil and gas industry, to the ocean observing community. In the United States, the IOOS recently awarded a project for the installation and maintenance of a coastal HF Radar system network, which monitors the ocean current off the coast of Louisiana to Fugro, who partnered with the University of Southern Mississippi, representing the first occasion a commercial organization is operating HF radars in the United States. Experiences gained in the oil and gas industry managing large quantities of data and delivery tools such as data portals have also allowed companies to develop specialist data management skills that are equally applicable in the ocean observing community. For example, several data portals developed by the RPS Group are currently provided to IOOS and its regional associations.

In closing, the oil and gas sector has been the biggest commercial driver for the marine sciences industry, creating career opportunities for marine scientists and technologists for more than half a century and driving operational professionalism and many technical developments. Although there is increasing societal pressure to move to clean energy solutions, it is probable that oil and gas will remain an important market for the ocean science professionals that represent the new blue economy for the next few decades. The emergence of the offshore wind industry has provided an opportunity for the marine science industry to develop a more diversified portfolio, which is less dependent on a single market sector. Nascent markets in hydrokinetic, wave, tidal, and ocean thermal energy, marine aquaculture, and deep-sea mining will create new opportunities. Finally, the growing awareness of the importance of the ocean in the understanding of climate change will increase the demand for comprehensive operational ocean observing systems and a workforce to operate them.

References

Morton, Q., 2016. Beyond the Sight of Land. GEOExPro.

Schaap, D.M.A., 2007. SIMORC - System of Industry Metocean Data for the Offshore and Research Communities, OCEANS 2007 – Europe. Aberdeen, UK.

Shaw, C., 2003. Offshore industry requirements and recent metocean technology developments, WMO. Advances in the Applied. Marine Climatology 223–229. JCOMM Technical Report No. 13.

van Smirren, J.R., Kirkpatrick, B., Gayanilo, F., Knee, K., Magnell, B., Williams, B., DiMarco, S., 2020. GulfHub: Leveraging Industry-Academic Partnerships to Expand Physical Oceanographic Data Availability in the Gulf of Mexico, Oceans 2020 - Singapore – U.S. Gulf Coast.

Health

CHAPTER 12

Oceans and Human Health and the New Blue Economy

Paul A. Sandifer[1,2], Paula Keener[2,3], Geoffrey I. Scott[2], Dwayne E. Porter[2]
[1]Center for Coastal Environmental and Human Health, College of Charleston, Charleston, SC, United States; [2]Center for Oceans and Human Health and Climate Change Interactions, Arnold School of Public Health, University of South Carolina, Columbia, SC, United States; [3]Global Ocean Visions, LLC, Charleston, SC, United States

Introduction

The field of Oceans and Human Health (OHH) is a "metadiscipline." As such, it "provides an overarching framework for adopting and incorporating knowledge across many fields of study ... to yield insights and perspectives not achievable with any one alone" (Mihelcic et al., 2003). OHH combines elements of oceanography, marine biology, ecology, biomedical science, environmental health science, medicine, public health, social sciences, economics, and communication science. Initially, OHH programs focused primarily on illnesses associated with exposure to ocean, coastal, and Great Lakes waters, such as those caused by harmful algal blooms (HABs) and *Vibrio* bacteria, as well as a few health benefits including potential new pharmaceuticals. While these foci remain, OHH programs have expanded to encompass health effects from other freshwater bodies and reported health-promoting effects of "blue" spaces (coastal areas, rivers, or landscapes involving water features) (Grellier et al., 2017; Cracknell, 2019).

While precursors for OHH programs have been around for a long time, as a recognized field, OHH is just a little less than 20 years old (Sandifer et al., 2012, 2013). Named OHH programs got underway in 2004, supported by the National Institute of Environmental Health Sciences (NIEHS), the National Science Foundation (NSF), and the National Oceanic and Atmospheric Administration (NOAA). Dr. Rita R. Colwell (see Colwell interview, this volume) was Director of the NSF when decisions were made for NSF to support OHH work. She was an early supporter and has continued to champion OHH work. The final report of the US Commission on Ocean Policy, published in 2004, devoted a full

Preparing a Workforce for the New Blue Economy
ISBN 978-0-12-821431-2
https://doi.org/10.1016/B978-0-12-821431-2.00057-3

chapter to the OHH theme, and Congress passed a specific OHH Act in 2004 (United States Commission on Ocean Policy, 2004). Following development in the United States, scientists in Europe, particularly in the United Kingdom, rapidly developed similar OHH-related research activities (Depledge et al., 2013, 2019; European Marine Board, 2013; Fleming et al., 2014, 2018) and are making great strides. OHH Gordon Research Conferences in 2008, 2010, 2012, and 2014 cemented the status of OHH as a widely recognized focus area of interdisciplinary research. Currently, US OHH centers are located at Bowling Green State University, Florida Gulf Coast University, University of South Carolina, and Woods Hole Oceanographic Institution, with a European center at the University of Exeter in the United Kingdom.

The OHH centers foster collaboration among oceanographers, marine biologists, biochemists, pharmacological scientists, environmental modelers, public health practitioners, physicians, social scientists, coastal managers, urban and land use planners, private businesses, community representatives, and others to coordinate and integrate research efforts. Community participatory approaches are followed, and results are incorporated into products and tools to increase understanding of health risks and benefits associated with aquatic environments and communicate this information in ways that protect and enhance public health.

This chapter explores OHH topics ranging from ecological forecasting for HABs and *Vibrios* to plastics in the marine environment, coastal influences on wellness, a human health observing system and touches on a few of the myriad opportunities for OHH programs to contribute to the new blue economy and vice versa through data streams, information products and services, and new and developing technologies.

Ecological forecasting

Ecological forecasts (EFs) are "intended to provide early warnings of possible effects of ecosystem changes on ecosystem health (including humans) with sufficient lead time to allow for corrective or mitigative actions" (Allen et al., 2015). NOAA, with numerous partners, is a leader in this arena, contributing a number of prototypes and a few operational EF applications. Two of the most important EFs for OHH are for pathogenic *Vibrio* bacteria and HABs. Used together with other information and analyses, these EFs provide the basic elements for an ocean health early warning system (Fig. 12.1).

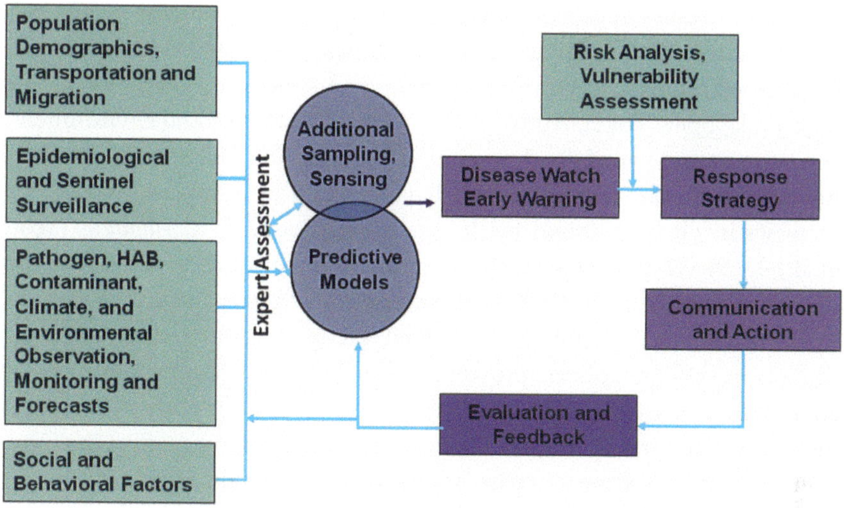

Figure 12.1 Schematic diagram of major components of, and information flows in, a hypothetical ocean health early warning system. *(Modified from Trtanj, J.M., Davis, J. R., 2002. Climate variability information as an adaptation tool: the case of public health. Presentation at Energy Modeling Forum Workshop on Adapting to Climate Change and Variability. Snowmass Conference Series: Climate Impacts and Integrated Assessment (CCI-IA) (Stanford University), Snowmass Colorado)) and previously published in Sandifer, P.A., Collier, T.K, Trtanj, J.M. 2012. Coastal ecosystems, sustainable management. Encycl. Sustain. Sci. & Technol. 4, 2258–2270).*

Vibrio bacteria

Vibrios

While there are ~1400 known human pathogens, they comprise <1% of Earth's microbial species (Microbiology by Numbers, 2011). Among these are *V. cholerae*, the causative agent of cholera; *V. parahaemolyticus*, the leading cause of gastrointestinal illnesses associated with seafood consumption in the United States; and *V. vulnificus*, the primary agent of wound infections resulting from contact with water containing high concentrations of the bacterium (Centers for Disease Control and Prevention, 2016). All of these can cause serious and sometimes fatal illnesses. While we focus here on *V. parahaemolyticus* and *V. vulnificus*, it is important to note that Rita Colwell was the first scientist to link global climate with increases in *Vibrio* (*V. cholerae*) infections (Colwell, 1996; Vezzulli et al., 2016), and she and her colleagues pioneered early predictive tools for cholera that helped inform later development of *Vibrio* forecast models.

Both *V. parahaemolyticus* and *V. vulnificus* occur naturally in brackish coastal waters worldwide and are not directly associated with pollution. Because of their life cycle characteristics and genetic makeup, *Vibrios* are highly adaptable (Baker-Austin et al., 2010). Globally, coastal environments are being transformed by burgeoning urbanization, increasing development, and climate change, which have resulted in significant worldwide increases in illnesses caused by *V. parahaemolyticus* and *V. vulnificus* (Scott et al., 2019). Water temperature is a primary driver of abundance for both species in most coastal areas, with several other water quality parameters (e.g., salinity, pH, turbidity, dissolved oxygen) associated to differing degrees in different places. However, rising seawater temperatures associated with climate change have been identified as important causes of increased temporal and spatial exposure to *Vibrios*, with illnesses now occurring during a greater part of the year (Spring through Fall) and at much higher latitudes in both the northern and southern hemispheres, including Alaska and the North Sea (Baker-Austin et al., 2012; Scott et al., 2019). The estimated number of *Vibrio*-caused illnesses in the United States has increased from around 8000 cases in 2004 to >80,000 cases in 2020 (Centers for Disease Control and Prevention, 2019). In addition, the majority of *Vibrios* are highly antibiotic-resistant (Scott et al., 2016), increasing the potential for serious illnesses. Baker-Austin et al. (2008, 2009) reported that > 98% of *V. parahaemolyticus* and *V. vulnificus* in coastal waters of the southeastern United States were antibiotic-resistant.

Concern about the increased rate of antibiotic resistance is compounded by high population growth in many coastal regions, especially among senior citizens with underlying health conditions. Individuals with chronic health conditions, particularly compromised immune systems and liver disease, are at increased risk for serious infection by *Vibrios*, especially V. *vulnificus* (Kuo et al., 2007). United States coastal states tend to have a disproportionately larger senior citizen population compared to more inland states (National Oceanic and Atmospheric Administration, 2013). Florida has the highest rates of *Vibrio* illness in the United States and *V. vulnificus* has replaced *V. parahaemolyticus* as the most commonly reported *Vibrio* there, as wound infections have replaced shellfish consumption as the most common exposure route (Weiss et al., 2011). Similarly, the growth of coastal urban areas in China is more than three times the national rate of population growth (McGranahan et al., 2007; Neumann et al., 2015).

Vibrio forecasting systems and the data streams that support them are expected to provide substantial new economic opportunities for private business and nonprofits to develop new tools and communicate predictions.

While forecasts are not yet nearly as mature as are weather and climate predictions, considerable progress has been made on region-by-region and area-by-area bases. Distributions and abundance of *Vibrios* are driven in part by water quality parameters such as temperature, salinity, and pH. Routinely collected water quality data are used in experimental forecasts for the North Sea (Oberbeckmann et al., 2012), Chesapeake Bay and the Mid-Atlantic (Randa et al., 2004; Davis et al., 2019; DeLuca et al., 2020), New England (Hartwick et al., 2019), Southeastern United States (Hsieh et al., 2008; Wetz et al., 2008; Froelich et al., 2013; Deeb et al., 2018), the Gulf of Mexico (Lipp et al., 2001), and the Pacific Northwest (Paranjpye et al., 2015). Notably, much of the necessary water quality data are already being collected in publicly supported environmental programs (see, for example, Gouldman, this volume), from both satellite and other remote-sensing sources and in situ sampling. Nevertheless, more robust temporal and spatial data coverage at finer scales is required to fill gaps in understanding and modeling of *Vibrio* abundance and virulence, particularly in response to ongoing climate change. This information is necessary to provide early warning indicators and routine, accurate forecasts for managing shellfish harvests, and potentially dangerous contact exposures. Current OHH research also includes the development of ensemble models analogous to those used to forecast hurricanes. Opportunities for the new blue economy are enormous, ranging from development of sensors, to software, to individual and ensemble forecast products, and many others.

Harmful algal blooms

HABs occur when marine and freshwater algae and cyanobacteria, which are present naturally in aquatic environments, grow beyond normal concentrations, fouling the water and often producing some of the most potent biological toxins known (Backer et al., 2015). Globally, there is substantial evidence that HAB occurrences are increasing, which may in part be related to climate change (Paerl and Otten, 2016; Berdalet et al., 2017; Gobler et al., 2017; Wells et al., 2020). Increasing population growth and use of nitrogen and phosphorus in agricultural fertilizers and industrial applications also likely play roles. Paerl et al. (2018) reported that human-caused nutrient over-enrichment, along with increasing global temperatures and occurrence of extreme hydrologic events (tropical storms, floods, and droughts), is accelerating eutrophication and promoting expansion of HABs across the freshwater-to-marine boundaries within estuaries and coastal waters.

Human exposure to toxins produced by marine and freshwater HABs may occur via seafood consumption, contact recreation, and inhalation of aerosolized toxins (Berdalet et al., 2015). Toxins produced by freshwater HABs may also contaminate freshwater fish and invertebrates and drinking water as well as produce contact recreation hazards to swimmers. Freshwater cyanobacteria produce more than 100 toxins, which can affect several human organ systems and cause wide-ranging and serious health effects, including death (Massey et al., 2018). A major example of impacts to drinking water is the 2014 "do not drink" advisory for the city of Toledo, OH, when a massive cyanobacterial bloom of *Microcystis* formed near its potable water intake on Lake Erie and contaminated the water with microcystin, a potent liver toxin (Bullerjahn et al., 2016). The advisory lasted about 3 days and affected ~400,000 water residential users. A smaller Canadian community was later affected similarly. Recent OHH research has shown that microcystin toxin may increase inflammatory response, adversely affecting the liver, kidney, gut microbiome, and ovary, and may increase insulin resistance in mammalian toxicity studies conducted at environmentally relevant concentrations (Sarkar et al., 2019a,b; Wang et al., 2020).

The Working Group on Harmful Algal Bloom Dynamics, established by the International Council for the Exploration of the Seas and the UN Intergovernmental Oceanographic Commission, has created the international Harmful Algal Bloom Environmental Data Base (HAEDAT). The highest numbers of HAB events in this database have been reported from Europe, North America, Japan, and China (Fig. 12.2).

In the United States, the Centers for Disease Control and Prevention (CDC) collected data on HAB events and health effects from 2007 to 2011 and reported 2201 HAB events, of which 96.7% occurred in freshwater and 2.8% in brackish and marine waters (Backer et al., 2015). However, the majority (57.4%) of reported illnesses in the United States were from marine HABs affecting the safety of seafood (Backer et al., 2015). Identified freshwater HABs accounted for 12.2% of all reported and identified HAB-related illnesses (Backer et al., 2015), and these may be increasing in importance.

DeFlorio-Barker et al. (2018) estimated that 4 billion surface water recreation events occur in fresh and marine waters annually in the United States, resulting in a projected 90 million illnesses, including from pollution-sourced

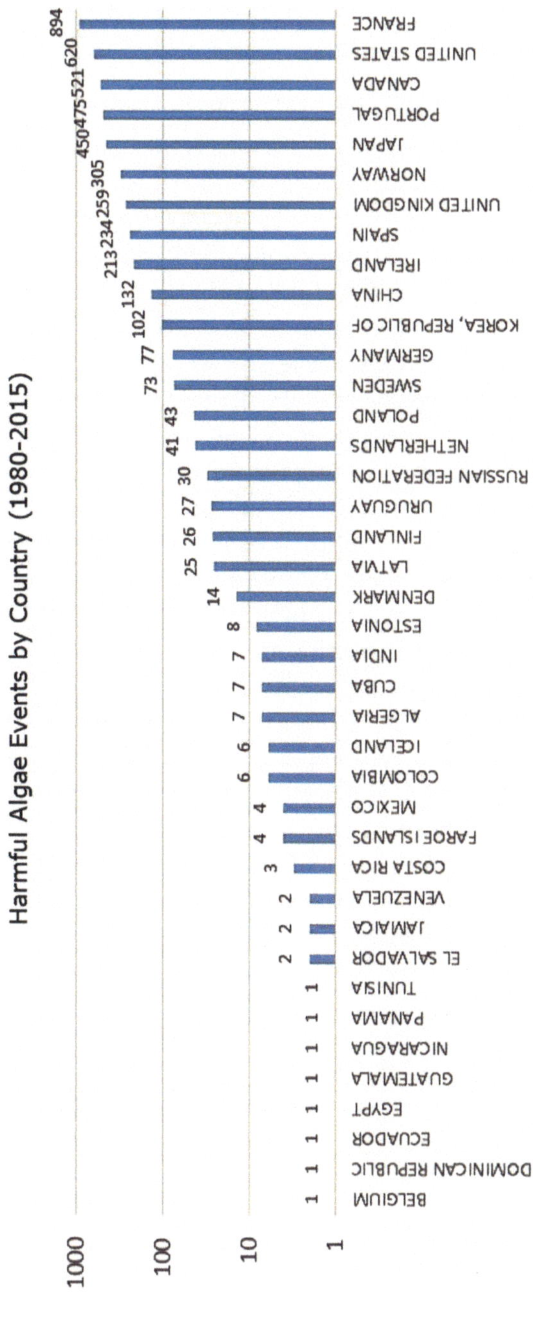

Figure 12.2 HAEDAT Global database of HAB events, 1980–2015, by location.

contaminants as well as HABs and *Vibrios*, and with associated costs of \$2.2–3.7 billion. Health costs specifically due to marine-borne pathogens and HABs were estimated to be ~ \$900 million annually (Ralston et al., 2011).

Development of HAB forecasts has long been a cornerstone EF activity in NOAA's portfolio, with major efforts initially focused on the Florida red tide organism, *Karenia brevis*, and later for other species in other regions. Currently, HAB monitoring and prediction activities are underway in the Gulf of Mexico, the Gulf of Maine, Alaska, the Pacific Northwest, parts of California, the Great Lakes, and the Caribbean (Allen et al., 2015; Anderson et al., 2019). Similar efforts are underway in the European Union, the United Kingdom, Asia, and elsewhere (Anderson et al., 2019). A broad range of technologies enhance model development and the accuracy, timeliness, and utility of predictions. These include remote sensing via satellites and other devices, in situ collection of water quality data and samples using buoys, boats, ships, drones, autonomous, and manually operated underwater devices, "smart" samplers such as the Environmental Sample Processor that can detect DNA signatures in situ (see Curry and Ausubel, this volume), advanced optical instruments, handheld units and kits, and a wide variety of biochemical and molecular analytical equipment (Bullerjahn et al., 2016; Gilbert et al., 2018; Anderson et al., 2019; Stauffer et al., 2019; Vander Woude et al., 2019). Monitoring of nitrogen and phosphorus inputs is particularly important for cyano-bloom modeling in the Great Lakes (Bullerjahn et al., 2016). HABs are a growing problem around the world, with particular dangers to potable and recreational waters. Similar to the situation with *Vibrios*, the critical need to develop better technological and communication tools to enhance HAB forecasts provides significant private-sector economic opportunities.

Plastics

The first United Nation's World Ocean Assessment (United Nations, 2016) addressed marine litter and microplastics and identified a significant "gap in information needed to evaluate impacts of marine debris on coastal and marine species, habitats, economic health, human health and safety, and social values." While this gap remains, public interests and scientific research on ocean plastics, microplastics (particles <5 mm in size), and nanoplastics (<1 μ in size) have increased considerably, with over 1400 species documented as having been affected by marine litter (Claro et al., 2019) and >100 species carrying body burdens of ingested plastics (Schandera, 2020). Adverse effects of plastics include entanglement and ingestion by marine organisms, physical alteration of marine habitats, damage to benthic organisms from shallow waters to the deep ocean, alteration of regional biodiversity, release of toxic chemicals, and transport of invasive species. Numerous species, particularly filter-feeding molluscan shellfish, are known to uptake micro- and nanoplastic particles. Seafood consumption, along with other ingestion and inhalation sources, provides avenues for human uptake (Smith et al., 2018; Yong et al., 2020). While research using the mouse as a model for mammalian systems found that plastic particles can have negative effects such as inflammation, chemical toxicity, and disruption of gut microbiota, major health threats to humans at this time are not known (Smith et al., 2018; Jiang et al., 2020; Yong et al., 2020). However, microplastic particles have now been found in the human placenta (Ragusa et al., 2021), and it is clear that there are significant gaps in our knowledge about how continuing and increasing exposures to plastic particles from the environment may affect human health over the long term.

Collection and availability of data on plastics in the ocean, new and enhanced technologies for remote and automated sensing, use of sophisticated laboratory instrumentation, implementation of ecosystem assessments, and deeper understanding of mechanisms of harmful effects have increased in recent years (Evereart et al., 2018; Galgani et al., 2021), but much remains to be done (see Smail, this volume). Solving the ocean plastics problem presents an unprecedented opportunity for growing new business activities through data integration; technology development; design and application of new source identification, control, and cleanup measures; development of exposure diagnostic tools; and standardization of metrics. International needs, opportunities, and best practices for public–private collaboration to support the new blue economy through solutions to the wicked problem of plastic pollution are highlighted in two new seminal publications, Pew Charitable Trusts and SYSTEMIQ (2020) and Karasik et al. (2020).

Seafood aquaculture

Sustainable seafood or "blue food," whether wild caught or cultured, is a key topic in three recent reports from the High Level Panel for a Sustainable Ocean Economy (https://www.oceanpanel.org/). Costello et al. (2019) note a potential sixfold increase in blue food with more effective management of wild fisheries coupled with sustainable aquaculture practices, resulting in health benefits to the ocean. Hoegh-Guldberg et al. (2019) present a comprehensive, quantitative analysis of potential ocean-based solutions to five urgent climate change problems, including greenhouse gas reduction. One solution is to shift human diets toward seafood, whether from wild-capture or aquaculture sources. Acknowledging data limitations and caveats, Konar and Ding (2020) estimate that the benefits of substantially increasing the proportion of sustainably produced ocean protein in human diets would exceed associated costs by a factor of 10. Similarly, Troell et al. (2019) note that by 2050, the world population will need to rely on the ocean for approximately double of what it does today in edible protein. A push to sustainably grow much more blue food to meet nutritional demands of the planet's growing population and as a means of mitigating climate change offers tremendous investment potential for the new blue economy (see, also, Rasso, this volume).

In the United States, a recent Executive Order (EO) on Promoting American Seafood Competitiveness and Economic Development (Executive Order, https://www.whitehouse.gov/presidential-actions/executive-order-promoting-american-seafood-competitiveness-economic-growth/) has a goal to provide environmentally safe and sustainable seafood. The EO calls for removing barriers to aquaculture permitting including for finfish aquaculture in marine and coastal waters within the territorial sea and Exclusive Economic Zone (EEZ), identifying locations suitable for commercial aquaculture, and updating a National Aquaculture Development Plan. The EO also establishes an Interagency Seafood Trade Task Force to advance fair and reciprocal trade in seafood products.

Achieving sustainable aquaculture practices requires innovation in policy and management as well as in research and development, coupled with advances in environmental monitoring and assessment. Progress in these areas will result in vast volumes of data at every level of process and supply chains, from identifying suitable aquaculture sites to regulatory and permitting requirements, laboratory testing and trials, feed specifications, species interactions, breeding, animal health, transportation, packaging, marketing, consumer perception and response, and more. How these data will be collected, aggregated, modeled, interpreted, communicated, and used to inform decisions at every level provides significant opportunities for growing the new blue economy.

Private sector innovation in digitizing aquaculture is of paramount importance. The use of artificial intelligence, machine learning, robotics, computer visioning technologies, sophisticated sensors, and augmented reality, to name a few, is critical to bringing sustainable aquaculture to scales necessary to meet projected global demands. The New England Ocean Cluster (https://www.newenglandoceancluster.com/) notes that transformations-system (T-systems) changes in the seafood sector are also necessary to scale blue food production to required production levels. Additionally, a Blue Food Assessment is underway by an international team of interdisciplinary researchers to identify gaps in understanding the role of aquatic food in the global food system, with goals of informing and advancing policy changes and highlighting the importance of blue food on the global food policy agenda (Friends of Ocean Action, 2020). However, as we seek to scale blue food production to meet global growing population needs and grow a new blue economy, it is essential that we remain mindful of the critical importance of biodiversity for human health and well-being and the necessity for human society to exist within the safe operating space of the biosphere (Dasgupta, 2021). Sustaining natural ocean biodiversity must be considered a prerequisite for large-scale development of ocean- and coastal-based aquaculture.

Coastal wellness tourism

In the United States, 90% of an annual \$3.5 trillion in health care expenditures are related to chronic diseases and mental health issues (https://www.cdc.gov/chronicdisease/about/costs/index.htm). Depression alone accounts for an economic burden of > \$200B/year (Greenberg et al., 2015), and obesity is an even more significant factor. Obesity is associated with numerous negative health outcomes, including cardiovascular disease, diabetes, cancers, and millions of deaths annually (GBD 2015 Obesity Collaborators, 2017). Lack of physical activity at almost all ages is a major driver of obesity and other chronic diseases (Booth et al., 2012; Gonzalez et al., 2017).

Over the past few decades, health-related tourism, including medical tourism that combines various forms of treatment with some tourist activities, has increased substantially (Hyder et al., 2019). Large-scale wellness tourism is a more recent trend and tends to focus on relaxation and restoration-promoting activities and places and "travel associated with the pursuit of maintaining or enhancing one's personal well-being" (Rubinstein, 2020). Both wellness and medical tourism are expanding globally, with wellness tourism growing at 5%–10% per year, twice as rapidly as medical and overall tourism (Page et al., 2017; Rubinstein, 2020). The UN estimates that there will be more than 2 billion people 60 years of age and older by 2050, comprising 22% of the global population. Older people are increasingly taking advantage of better health, strong financial resources, and excellent quality of life and life expectancy and are participating in tourism in rising numbers (Patterson and Baldera, 2020). Older travelers tend to have broad interests, particularly in tourism that combines interesting places with opportunities for education and health-promoting activities.

As Jarratt and Davies (2019) noted, "No form of leisure is more significant in scale or economic value than coastal tourism." While some salutary effects of coastal exposures have been recognized for many years, there is now an accumulation of scientific evidence that living near or visiting coastal and other "blue spaces" is associated with better mental and in some cases physical health (Depledge and Bird, 2009; White et al., 2010; Gascon et al., 2017; Grellier et al., 2017; Garrett et al., 2019; Britton et al., 2020; Chen and Yuan, 2020). Much of this work has been conducted in the United Kingdom and in parts of Europe, where it is an important component of European OHH research (bluehealth2020.eu). Most beneficial effects attributed to coastal exposures relate to heightened feelings of restoration, reductions in depression, anxiety, other related psychological conditions, and increases in physical activity (the "blue gym" effect), the most important of which is walking. Coastal wellness tourism could capitalize on the uplifting, restorative influences of coastal environments coupled with their depression-diminishing effects and opportunities for healthful outdoor physical and enriching activities, such as walking and wildlife viewing.

An important new blue economy opportunity is the potential to use data and information streams originating from OHH research to "promote well-being as a destination resource" (Page et al., 2017), particularly to older adults and those who may be seeking restoration, lessened anxiety, and life-affirming experiences following the COVID-19 pandemic. Related opportunities lie in innovative uses of natured-based and combinations of "gray" and "green" infrastructure in designing and implementing coastal restoration projects (Sutton-Grier and Sandifer, 2019; Sutton-Grier et al., 2018). Nature-based solutions could help protect blue spaces from climate change impacts, increase coastal resiliency, and enhance health-promoting characteristics and attractiveness to visitors. Exploration of the data-driven "ecological prototype" approach to green design and construction (Hensel, 2020), with inclusion of emerging information on mechanisms of action for health-promoting effects in design criteria for "blue space" infrastructure, may provide additional opportunities for new blue economy businesses. Appropriate development of wellness tourism could also benefit the well-being of local hospitality workforces and tourism-focused businesses and communities (Page et al., 2017), in part by improving local healthcare capacity.

OHH and communities

Coastal communities are facing increasing public health threats as a result of climate change. In addition to human health impacts, climate change increases food insecurity, worsens air pollution, reduces access to clean water, displaces people from their homes, and causes widespread social and economic disruption within both urban and rural coastal communities (Rudolph et al., 2018). These threats affect the health and safety of the coastal communities, as well as their long-term social cohesion and economic stability. What makes a coastal community vulnerable to these threats and pressures, and their capacity to rebound after disrupting events, is their ability to adapt and become more resilient in the face of change (Lam et al., 2016). Community resilience depends on numerous factors, including employment, crime rates, income, race, education, the strength and embeddedness of social networks, community value cohesion, faith-based and/or nonprofit organizations, property values, wealth generation, and municipal financial soundness (Aldrich, 2012; Demiroz and Hu, 2014; Kim and Marcouiller, 2016).

Inequities and climate change are deeply interconnected in coastal communities that are underserved or subjected to environmental injustices; these are described as Environmental Justice (EJ) communities. EJ refers to efforts to "document and redress the disproportionate environmental burdens associated with social inequalities" (Chakraborty et al., 2016). Initially, EJ focused on pollution-based problems, particularly in minority and other socially and economically deprived communities, but it has since expanded to include many other forms of social inequalities such as health disparities and effects of disasters and climate change.

The same physical, social, economic, and lack of health services environments that are associated with poor health outcomes for coastal EJ communities also increase exposure and vulnerability to health impacts of climate change. EJ communities are often historically disenfranchised, seemingly lacking the political and economic status to ensure that decision-makers take their perspectives, needs, and ideas fully into account (Rudolph et al., 2018). Community-managed or community-based resiliency and adaptation efforts are less prevalent in coastal EJ communities (Birkmann and von Teichman, 2010; Priscoli and Stakhiv, 2015). It is critical that investors in the new blue economy and OHH initiatives partner with community leaders to develop and implement community engagement plans for coastal community partners, particularly in EJ communities. Such plans should embrace community-based participatory research (CBPR) principles and protocols and engage leaders from within and across participating new blue economy sectors and local coastal communities. CPBR involves "systematic inquiry, with participation of those affected by the issue being studied, for the purposes of education and taking action or effecting social change" (Green et al., 1995 as cited in Vasquez et al., 2006). It is based on the concepts that the public has a right to participate in research that directly affects it and that the "collective intelligence" that results from researchers, industry leaders, community leaders, and individuals working together will result in better public health outcomes (Friedman et al., 2015). Core principles for conducting CPBR are well established, including for public health work (Israel et al., 1998; Friedman et al., 2015). One approach for implementing CPBR within the OHH structure is based on a framework being used successfully at the OHH Center at the University of South Carolina (Sandifer et al., 2020b). Specific to partnering among EJ communities, OHH initiatives and industries and businesses of the new ocean economy, this approach involves the following three primary elements:

(1) Assess and strengthen the target community's knowledge base, environmental literacy, and resiliency focused on community and public health impacts resulting from natural and human-caused threats and impacts. (**Learn!**)

(2) Build the community's capacity to co-develop research, economic, and education designs to reduce the health impacts of disasters and improve public health through community engagement and information sharing (**Leverage!**).

(3) Develop and implement community-engaged best management practices and tools that include preventive and proactive approaches to mitigating health impacts from disasters, including for the most vulnerable populations, those in low socioeconomic brackets, and in EJ communities. (**Lead!**)

Referred to as L!^{3}, this approach recognizes that engaging community members and OHH scientists in joint research teams from the outset is essential (Kaiser et al., 2019) and includes a focus on EJ communities.

Income, employment, and job security are drivers of public health, and in particular of mental health, that affect quality of life of individuals and families in coastal communities. Murphy and Athanasou (1999) reported job loss and unemployment as significant contributors to stress, whereas finding a new job and sustained employment was associated with stress reduction. In coastal EJ communities facing economic hardships and job instability, OHH and the new blue economy represent excellent opportunities to support development of degree- and certificate-based education and training programs to meet the need for skilled workers necessary to staff new and expanded sustainable aquaculture facilities, coastal tourism and wellness businesses, port facilities, renewable offshore energy initiatives, and coastal and offshore observing and monitoring programs as well as other existing and developing sectors of the new blue economy. Bringing together business leaders with OHH scientific and technical experts and leaders of underserved communities, local school districts, technical schools, colleges, and universities to address training and education needs of the new blue economy would be a first step in building a well-trained local workforce that could support healthier, more economically robust coastal communities while providing substantial economic and social returns on investment.

Need for a human health observing system

The ocean and coastal environments are on the frontline for effects of global change. Ocean waters are sinks for excess heat and CO_2, which in turn are major drivers of sea level rise, extreme weather events, changing precipitation patterns, and ocean acidification. The greatest number of people reside along coastal margins (National Oceanic and Atmospheric Administration, 2013), and these areas contain very high levels of built infrastructure investments and many of the ecological resources in grave danger from global climate change. Despite warnings of climate-related dangers to coastal communities and people, no trend toward retreat or even reduced population growth in these areas has been observed yet, and instead increased population growth is anticipated (National Oceanic and Atmospheric Administration, 2013). In addition, most of the largest cities in the United States and globally are located on coasts, and their populations are projected to increase by 25% by 2050 (Aerts et al., 2014). Many coastal communities and areas are already experiencing significant impacts from sea level rise and other aspects of climate change (Sweet et al., 2019; Sinay and Carter, 2020) and more will in the future (Neumann et al., 2015). Climate change factors also can exacerbate effects of ocean and Great Lakes health threats associated with HABs and water-borne disease.

In addition to further development of ecological forecasting, there is a substantial need for advancements in monitoring of human health indicators through the development of human health observing or surveillance systems. An innovative approach to this, specifically targeted toward disaster-prone regions such as the United States Gulf of Mexico coastline, is the Community Health Observing System framework designed by Sandifer et al. (2020a,b). This first-of-its-kind observing system framework builds on existing ongoing national health databases and adds region-specific longitudinal cohort studies that can track health effects of exposures, whether to environmental disasters, infectious disease such as COVID-19, HAB toxins, or positive health effects of coastal residence and visiting. Implementation of the Gulf of Mexico Community Health Observing System and replication in other areas would be a major step toward improving disaster health care planning and response, describing acute, chronic, and poorly known adverse health effects of disasters and climate change, and understanding how coastal exposures may enhance health. They also could improve targeting of investments for preparedness, response, and prevention, reduce health care expenditures, and perhaps provide early warnings of impending health crises. Data streams produced by large-scale health observing systems would be massive and would create huge "big data" mining opportunities for the new blue economy and other economic sectors.

Conclusions

The UN has proclaimed 2021—2030 as the Decade of Ocean Science for Sustainable Development (Decade) (see Ryabinin, this volume), largely in response to findings of the World Ocean Assessment I that the health of the world ocean is continuing to decline and urgent action is needed to reverse continued damage to the Earth system that supports humankind. The health of the ocean and that of humans (i.e., OHH) are the drivers of the decade. As researchers, business and government leaders, philanthropists, and civil society come together post-COVID-19 to determine how to best support the UN's Sustainable Development Goals through the decade, OHH programs and activities should be considered foundational. Such considerations could establish the basis for a Global Ocean Health—Human Health Observing System that could monitor threats to environmental and human health, provide essential data for health-related forecasts, and document changes, whether improvements or declines, in critical ocean/human health metrics so that appropriate actions could be taken. Academic centers of excellence such as the OHH centers are expected to play catalytic and transformational roles in efforts to grow the new blue economy.

Acknowledgments

This work was supported in part by the National Institute of Environmental Health Sciences of the National Institutes of Health under Award Number P01ES028942 to the University of South Carolina and through a subaward to the College of Charleston. The content is solely the responsibility of the authors and does not necessarily represent the official views of the National Institutes of Health, the University of South Carolina, or the College of Charleston.

References

Aerts, J.C.J.H., Botzen, W.J.W., Emmanuel, K., Ling, N., de Moel, H., Michael-Kerjan, E.O., 2014. Evaluating flood resilience strategies for coastal megacities. Science 344, 473—475.

Aldrich, D.P., 2012. Building Resilience: Social Capital in Post-disaster Recovery. University of Chicago Press, Chicago.

Allen, A.L., Brown, C.W., Lewitus, A.J., Sandifer, P.A., 2015. The roles of emerging technology and modeling in operational ecological forecasting at NOAA. Mar. Technol. Soc. J. 49 (2), 193—203.

Anderson, C.R., Berdalet, E., Kudela, R.M., Cusack, C.R., Silke, J., O'Rourke, E., Dugan, D., McCammon, M., Newton, J.A., Kirkpatrick, B., Hubbard, K., Morrell, J., 2019. Scaling up from regional case studies for a global harmful algal bloom observing system. Front. Mar. Sci. 6, 250. https://doi.org/10.3389/fmars.2019.00250.

Backer, L.C., Manassaram-Baptiste, D., LePrell, R., Bolton, B., 2015. Cyanobacteria and algae blooms: review of health and environmental data from the harmful algal bloom-related illness surveillance system (HABISS) 2007—2011. Toxins 7, 1048—1064. https://doi.org/10.3390/toxins7041048.

Baker-Austin, C., McArthur, J.V., Tuckfield, R.C., Najarro, M., Lindell, A.H., Gooch, J., Stepanauskas, R., 2008. Antibiotic resistance in the shellfish pathogen *Vibrio parahaemolyticus* isolated from the coastal water and sediment of Georgia and South Carolina, USA. J. Food Protect. 71 (12), 2552—2558.

Baker-Austin, C., McArthur, J.V., Lindell, A.H., Wright, M.S., Tuckfield, R.C., Gooch, J., Warner, L., Oliver, J., Stepanauskas, R., 2009. Multi-site analysis reveals widespread antibiotic resistance in the marine pathogen *Vibrio vulnificus*. Microb. Ecol. 57, 151—159. https://doi.org/10.1007/s00248-008-9413-8.

Baker-Austin, C., Stockley, L., Rangdale, R., Martinez-Urtaza, J., 2010. Environmental occurrence and clinical impacts of *Vibrio vulnificus* and *Vibrio parahaemolyticus*: a European perspective. Environ. Micro. Rep. 2 (1), 7—18.

Baker-Austin, C., Trinanes, J.A., Taylor, N.G.H., Hartnell, R., Siitonen, A., Martinez-Urtaza, J., 2012. Emerging *Vibrio* risk at high latitudes in response to ocean warming. Nat. Clim. Change 3, 73–77.

Berdalet, E., Fleming, L.E., Gowen, R., Davidson, K., Hess, P., Backer, L.C., Moore, S.K., Hoagland, P., Enevoldsen, H., 2015. Marine harmful algae blooms, human health and well-being: challenges and opportunities in the 21st century. J. Mar. Biol. Assoc. U. K. https://doi.org/10.1017/S0025315415001733.

Berdalet, E., Tester, P.A., Chinain, M., Fraga, S., Lemée, R., Litaker, W., Penna, A., Usup, G., Vila, M., Zingone, A., 2017. Harmful algal blooms in benthic systems: recent progress and future research. Oceanography 30 (1), 36–45. https://doi.org/10.5670/oceanog.2017.108.

Birkmann, J., von Teichman, K., 2010. Integrating disaster risk reduction and climate change adaptation: key challenges — scales, knowledge, and norms. Sustain. Sci. 5 (2), 171–184. https://doi.org/10.1007/s11625-010-0108-y.

Booth, F.W., Roberts, C.K., Laye, M.J., 2012. Lack of exercise is a major cause of chronic diseases. Compr. Physiol. 2 (2), 1143–1211.

Britton, E., Kindermann, G., Domegan, C., Carlin, C., 2020. Blue care: a systematic review of blue space interventions for health and wellbeing. Health Promot. Int. 35 (1), 50–69. https://doi.org/10.1093/heapro/day103.

Bullerjahn, G.S., McKay, R.M., Davis, T.W., Baker, D.B., Boyer, G.L., D'Anglada, L.V., Doucette, G.J., Ho, J.C., Irwin, E.G., Kling, C.L., Kudela, R.M., Kurmayer, R., Michalak, A.M., Ortiz, J.D., Otten, T.G., Paerl, H.W., Qin, B., Sohngen, B.L., Stumpf, R.P., Visser, P.M., Wilhelm, S.W., 2016. Global solutions to regional problems: collecting global expertise to address the problem of harmful cyanobacterial blooms: a Lake Erie case study. Harmful Algae 54, 223–238.

Centers for Disease Control and Prevention, 2016. National Enteric Disease Surveillance: COVIS Annual Summary, 2014. Centers of Disease Control and Prevention, p. 10. CS265270-A.

Centers for Disease Control and Prevention, 2019. Vibrio parahaemolyticus. Centers for Disease Control and Prevention Web Site, Atlanta, GA. https://www.cdc.gov/.

Chakraborty, J., Collins, T., Grineski, S., 2016. Environmental justice research: contemporary issues and emerging topics. Int. J. Environ. Res. Publ. Health 13 (11), 1072.

Chen, Y., Yuan, Y., 2020. The neighborhood effect of exposure to blue space on elderly individuals' mental health: a case study in Guangzhou, China. Health Place 63, 102348. https://doi.org/10.1016/j.healthplace.2020.10234.

Claro, F., Fossi, M.C., Ioakeimidis, C., Baini, M., Lusher, A.L., McFee, W., McIntosh, R.R., Pelamatti, T., Sorce, M., Galgani, F., Hardesty, B.D., 2019. Tools and constraints in monitoring interactions between marine litter and megafauna: insights from case studies around the world. Mar. Pollut. Bull. 141, 147–160.

Colwell, R.R., 1996. Global climate and infectious disease: the cholera paradigm. Science 724, 2025–2031.

Costello, C., Cao, L., Gelcich, S., 2019. The Future of Food from the Sea. World Resources Institute, Washington, DC. Available online at: www.oceanpanel.org/future-food-sea.

Cracknell, D., 2019. By the Sea: The Therapeutic Benefits of Being in, on and by the Water. Aster, a division of Octopus Publishing Group, Ltd., London, UK.

Dasgupta, P., 2021. The economics of biodiversity: the Dasgupta review. HM Treasury, London 606. https://www.gov.uk/government/publications/final-report-the-economics-of-biodiversity-the-dasgupta-review. (Accessed 10 February 2021).

Davis, B.J.K., Jacobs, J.M., Zaitchik, B., DePaola, A., Curriero, F.C., 2019. Operational in-situ prediction and forecast models for *Vibrio parahaemolyticus* in the Chesapeake Bay are attainable from lagged water quality measurements. Appl. Environ. Microbiol. https://doi.org/10.1128/AEM.01007-19.

Deeb, R., Tufford, D., Scott, G.I., Moore, J.G., Dow, K., 2018. Impact of climate change on *Vibrio vulnificus* abundance and exposure risk. Estuar. Coast 41 (8), 2289–2303.

DeFlorio-Barker, S., Wing, C., Jones, R.M., Dorevitch, S., 2018. Estimate of incidence and cost of recreational waterborne illness on United States surface waters. Environ. Health 17, 3. https://doi.org/10.1186/s12940-017-0347-9.

DeLuca, N.M., Zaitchik, B.F., Guikeman, S.D., Jacobs, J.M., Davis, B.J.K., Curriero, F.C., 2020. Evaluation of remotely sensed prediction and forecast models for *Vibrio parahaemolyticus* in the Chesapeake Bay. Remote Sens. Environ. 250, 112016. https://doi.org/10.1016/j.rse.2020.112016.

Demiroz, F., Hu, Q., 2014. The role of nonprofits and civil society in post-disaster recovery and development. In: Kapucu, N., Liou, K.T. (Eds.), Disaster and Development: Examining Global Issues and Cases. Springer, Cham, Switzerland, pp. 317–330.

Depledge, M.H., Bird, J.W., 2009. The Blue Gym: health and wellbeing from our coasts. Mar. Pollut. Bull. 58 (7), 47–948. https://doi.org/10.1016/j.marpolbul.2009.04.019.

Depledge, M.H., Harvey, A.J., Brozonlee, C., Frost, M., Moore, M.N., Fleming, L.E., 2013. Changing view of the interconnections between the ocean and human health in Europe. Microb. Ecol. 65, 852–859.

Depledge, M.H., White, M.P., Macock, B., Fleming, L.E., 2019. Our future health and wellbeing depend on the oceans. BMJ 366, 14671. https://doi.org/10.1136/bmj.1461.

European Marine Board, 2013. Linking Oceans and Human Health: A Strategic Research Priority for Europe. Position paper for the European Marine Board, Ostend, Belgium.

Evereart, G., De Rijcke, M., Mees, J., Janssen, C., Van Cauwenberghe, L., Koelmons, A., Vandegehuchte, M.B., 2018. Risk assessment of microplastics in the ocean: modelling approach and first conclusions. Environ. Pollut. 242, 1930–1938.

Fleming, L.E., McDonough, N., Austen, M., Mee, L., Moore, M., Hess, P., Depledge, M.H., Philippart, K., Bradbrook, P., Smalley, A., 2014. Oceans and human health: a rising tide of challenges and opportunities for Europe. Mar. Environ. Res. 99, 16–19.

Fleming, L.E., Leonardi, G.S., White, M.P., Medlock, J., Alcock, I., Macintyre, H.L., Maguire, K., Nichols, G., Wheeler, B.W., Morris, G., Taylor, T., Hemming, D., Iacono, G.L., Gillingham, E.L., Hansford, K.M., Heaviside, C., Bone, A., Duarte-Davidson, R., 2018. Beyond climate change and health: integrating broader environmental change and natural environments for public health protection and promotion in the UK. Atmosphere 9, 245. https://doi.org/10.3390/atmos9070245.

Friedman, D.B., Toumey, C., Porter, D.E., Hong, J., Scott, G.I., Lead, J.R., 2015. Communicating with the public about environmental health risks: a community-engaged approach to dialogue about metal speciation and toxicity. Environ. Int. 74, 9–12.

Friends of Ocean Action, 2020. http://www3.weforum.org/docs/WEF_FOA_2pp_FOODSECURITY_9_2019.pdf.

Froelich, B., Bowen, J., Gonzalez, R., Snedeker, A., Noble, R., 2013. Mechanistic and statistical models of total *Vibrio* abundance in the Neuse River Estuary. Water Res. 42, 5783–5793.

Galgani, F., Bebianno, M., Bondareff, J., Griffiths, H., Hassellov, M., Loakeimidis, C., Jambeck, J., Keener, P., Lana, F., Makarenko, I., Rochman, C., Schuyler, Q., Sobral, P., Topouzelis, K., Vethaak, D., Vlahos, P., Wang, J., Weis, J., 2021. Marine litter. New York. In: Chapter 12 of the UN/regular Process/World Ocean Assessment 2 (Changes in Inputs and Distribution of Solid Waste in the Marine Environment/Marine Litter). United Nations, p. 15 (in press).

Garrett, J.K., Clitherow, T.J., White, M.P., Wheeler, B.W., Fleming, L.E., 2019. Coastal proximity and mental health among urban adults in England: the moderating effect of household income. Health Place 59. https://doi.org/10.1016/j.healthplace.2019.102200.

Gascon, M., Zijlema, W., Vert, C., White, M.P., Nieuwenhuijsen, M.J., 2017. Outdoor blue spaces, human health and well-being: a systematic review of quantitative studies. Int. J. Hyg Environ. Health 220 (8), 1207–1221. https://doi.org/10.1016/j.ijheh.2017.08.004.

GBD 2015 Obesity Collaborators, 2017. Health effects of overweight and obesity in 195 countries over 25 years. N. Engl. J. Med. 377 (1), 13–27. https://doi.org/10.1056/NEJMoa1614362.

Gilbert, P.M., Pitcher, G.C., Bernard, S., Li, M., 2018. Chapter 18. Advancements and continuing challenges of emerging technologies and tools for detecting harmful algal blooms, their antecedent conditions and toxins, and applications in predictive models. In: Gilbert, P.M., et al. (Eds.), Global Ecology and Oceanography of Harmful Algal Blooms, Ecological Studies, 232. Springer International Publishing AG, pp. 339–351.

Gobler, C.J., Doherty, O.M., Hattenrath-Lehmann, T.K., Griffith, A.W., Kang, Y., Litaker, R.W., 2017. Ocean warming since 1982 has expanded the niche of toxic algal blooms in the North Atlantic and North Pacific oceans. Proc. Natl. Acad. Sci. U. S. A. 114 (19), 4975–4980.

Gonzalez, K., Fuentes, J., Marquez, L.M., 2017. Physical inactivity, sedentary behavior and chronic diseases. Korean J. Fam. Med. 38, 111–115.

Green, L.W., Daniel, M., George, A.M., Frankish, C.J., Herbert, C., Bowie, W.R., et al., 1995. Study of Participatory Research in Health Promotion: Review and Recommendations for the Development of Participatory Research in Health Promotion in Canada. R Soc Canada.

Greenberg, P.E., Founier, A.-A., Sisitsky, T., Pike, C., Kessler, R.C., 2015. The economic burden of adults with major depressive disorder in the United States (2005 and 2010). J. Clin. Psychiatr. 76 (2), 155–162.

Grellier, J., White, M.P., Albin, M., Bell, S., Elliott, L.R., Gascón, M., Gualdi, S., Mancini, L., Nieuwenhuijsen, M., Sarigiannis, D., van den Bosch, M., Wolf, T., Wuijts, S., Fleming, L.E., 2017. BlueHealth: a study programme protocol for mapping and quantifying the potential benefits to public health and well-being from Europe's blue spaces. BMJ Open 7 (6). https://doi.org/10.1136/bmjopen-2017-01618.

Hartwick, M.A., Urquhart, E.A., Whistler, C.A., Cooper, V.S., Naumova, E.N., Jones, S.H., 2019. Forecasting seasonal *Vibrio parahaemolyticus* concentrations in New England shellfish. Int. J. Environ. Res. Publ. Health 16, 4341. https://doi.org/10.3390/ijerph16224341.

Hensel, D.S., 2020. Ecological prototypes: initiating design innovation in green construction. Sustainability 12, 5865. https://doi.org/10.3390/su12145865.

Hoegh-Guldberg, O., Caldeira, K., Chopin, T., Gaines, S., Haugan, P., Herner, M., Howard, J., Konar, M., Krause-Jensen, D., Lindstad, E., Lovelock, E., Michelin, M., Nielsen, F.G., Northrop, E., Parker, R., Roy, J., Smith, T., Some, S., Tydmers, P., 2019. The Ocean as a Solution to Climate Change: Five Opportunities for Action. Report. World Resources Institute, Washington, DC. Available online at: http://www.oceanpanel.org/climate.

Hsieh, J.L., Fries, J.S., Noble, R.T., 2008. Dynamics and predictive modelling of *Vibrio* spp. in the Neuse River Estuary, North Carolina, USA. Environ. Microbiol. 10 (1), 57–64.

Hyder, A.S., Rydback, M., Borg, E., Osarenkhoe, A., 2019. Medical tourism in emerging markets: the role of trust, networks, and word of mouth. Health Market. Q. 36 (3), 203–2010. https://doi.org/10.1080/07359683.2019.1618008.

Israel, B.A., Schulz, A.J., Parker, E.A., Beckera, A.B., 1998. Review of community-based research: assessing partnership approaches to improve public health. Annu. Rev. Public Health 19 (1), 173–202.

Jarratt, D., Davies, N.J., 2019. Planning for climate change impacts: coastal tourism destination resilience policies. Tour. Plan. & Dev. 17 (4) https://doi.org/10.1008/21568316.2019.1667681.

Jiang, B., Kauffman, A.E., Li, L., McFee, W., Cai, B., Weinstein, J., Lead, J.R., Chatterjee, S., Scott, G.I., Xiao, S., 2020. Health impacts of environmental contamination of micro- and nanoplastics: a review. Environ. Health Prev. Med. 25, 29. https://doi.org/10.1186/s12199-020-00870-9.

Kaiser, B.A., Hoeberechts, M., Maxwell, K.H., Eerkes-Medrano, L., Hilmi, N., Safa, A., Horbel, A.S.C, Juniper, S.K., Roughan, M., Theux Lowen, L.N., Short, K., Paruru, D., 2019. The importance of connected ocean monitoring knowledge systems and communities. Front. Mar. Sci. 6, 309. https://doi.org/10.3389/fmars.2019.00309.

Karasik, R., Vegh, T., Diana, Z., Bering, J., Caldas, J., Pickle, A., Rittschof, D., Virdin, J., 2020. 20 Years of Government Responses to the Global Plastics Pollution Problem — The Plastics Policy Inventory. Nicholas Institute for Environmental Policy Solutions, 311p.

Kim, H., Marcouiller, D.W., 2016. Natural disaster response, community resilience, and economic capacity: a case study of coastal Florida. Soc. Nat. Resour. 29 (8), 81—997. https://doi.org/10.1080/08941920.2015.1080336.

Konar, M., Ding, H., 2020. A Sustainable Ocean Economy for 2050 - It's Costs and Benefits. Commissioned by High Level Panel for A Sustainable Ocean Economy, 62p. https://oceanpanel.org/sites/default/files/2020-07/Ocean%20Panel_Economic%20Analysis_FINAL.pdf.

Kuo, Y.-L., Shieh, S.-J., Chiu, H.-Y., Lee, J.W., 2007. Necrotizing fasciitis caused by *Vibrio vulnificus*: epidemiology, clinical findings, treatment and prevention. Eur. J. Clin. Microbiol. Infect. Dis. 26, 785—792.

Lam, N.S.N., Reams, M., Li, K., Mata, L.P., 2016. Measuring community resilience to coastal hazards along the Northern Gulf of Mexico. Nat. Hazards Rev. 17 (1), 04015013 https://doi.org/10.1061/(ASCE)NH.1527-6996.0000193.

Lipp, E., Rodriguez-Palacios, C., Rose, J.B., 2001. Occurrence and distribution of the human pathogen *Vibrio vulnificus* in a subtropical Gulf of Mexico estuary. Hydrobiologia 460, 165—173.

Massey, I.Y., Yang, F., Ding, Z., Yang, S., Guo, J., Tezi, C., Al-Osman, M., Kamegni, R.B., Zeng, W., 2018. Exposure routes and health effects of microcystins on animals and humans: a mini-review. Toxicon 151, 156=162.

McGranahan, G., Balk, D., Anderson, B., 2007. The rising tide: assessing the risks of climate change and human settlements in low elevation coastal zones. Environ. Urban. 19 (1), 17—37.

Microbiology by numbers. Nat. Rev. Microbiol. 9, 2011, 628. https://doi.org/10.1038/nrmicro2644.

Mihelcic, J.R., Crittenden, J.C., Small, M.J., Shonnard, D.R., Hokanson, D.R., Zhang, Q., Chen, H., Sorby, S.A., James, V.U., Sutherland, I.W., Schnoor, J.L., 2003. Sustainability science and engineering: the emergence of a new metadiscipline. Environ. Sci. Technol. 37, 5314—5324.

Murphy, G.C., Athanasou, J.A., 1999. The effect of unemployment on mental health. J. Occup. Organ. Psychol. 72, 83—99. https://doi.org/10.1348/096317999166518.

National Oceanic and Atmospheric Administration, 2013. National Coastal Population Report: Population Trends from 1970 to 2020. NOAA, Silver Spring, MD, p. 22.

Neumann, B., Vafeidis, A.T., Zimmermann, J., Nicholls, R.J., 2015. Future coastal population growth and exposure to sea level rise and coastal flooding - a global assessment. PLoS One 10 (6), e0131375.

Oberbeckmann, S., Fuchs, B.M., Meiners, M., Wichels, A., Wiltshire, K.H., Gerdts, G., 2012. Seasonal dynamics and modeling of a *Vibrio* community in coastal waters of the

North Sea. Microb. Ecol. 63 (3), 543–551. https://doi.org/10.1007/s00248-011-9990-9.

Paerl, H.W., Otten, T.G., 2016. Duelling 'CyanoHABs': unravelling the environmental drivers controlling dominance and succession among diazotrophic and non-N2-fixing harmful cyanobacteria. Environ. Microbiol. 18, 316–324.

Paerl, H.W., Otten, T.G., Kudela, R., 2018. Mitigating the expansion of harmful algal blooms across the freshwater-to-marine continuum. Environ. Sci. & Technol. 52, 5519–5529.

Page, S.J., Hartwell, H., Johns, N., Fyall, A., Ldking, A., Hemongway, A., 2017. Wellness tourism and small business development in a UK coastal resort: public engagement in practice. J. Tour. Manag. 60, 466–477.

Paranjpye, R.N., Nilsson, W., Liermann, M., Hilborn, E.D., 2015. Environmental influences on the seasonal distribution of *Vibrio parahaemolyticus* in the Pacific Northwest of the USA. Microb. Ecol. 91 (12) https://doi.org/10.1093/femsec/fiv121.

Patterson, I., Baldera, A., 2020. Continuing and emerging trends of senior tourism: a review of the literature. Popul. Ageing 13, 385–399. https://doi.org/10.1007/s12062-018-9228-4.

Pew Charitable Trusts and SYSTEMIQ, 2020. Breaking the Plastic Wave – A Comprehensive Assessment of Pathways towards Stopping Ocean Plastics Pollution. Pew Charitable Trust, 78p.

Priscoli, J.D., Stakhiv, E.Z., 2015. Water-related disaster risk reduction (DRR) management in the United States: floods and storm surges. Water Pol. 17, 58–88.

Ragusa, A., Svelato, C., Santacroce, C., Catalano, P., Notarstefano, V., Carnevali, O., Papa, F., Rongioletti, M., Baiocco, F., Draghi, S., D'Amore, E., Rinaldi, D., Matta, M., Giorgini, E., 2021. Plasticenta: first evidenc of microplastics in human placenta. Environ. Int. 146, 106274 https://doi.org/10.1016/j/envint.2020.106274.

Ralston, E.P., Kite-Powell, H., Beet, A., 2011. An estimate of the cost of acute health effects from food- and water-borne marine pathogens in the USA. J. Water Health 9 (4), 680–694. https://doi.org/10.2166/wh.2011.157.

Randa, M.A., Polz, M.F., Lim, E., 2004. Effects of temperature and salinity on *Vibrio vulnificus* population dynamics as assessed by quantitative PCR. Appl. Environ. Microbiol. 70 (9), 5469–5476.

Rubinstein, P., 2020. How the Wellness Industry Is Taking over Travel. https://www.bbc.com/worklife/article/20200203-how-the-wellness-industry-is-taking-over-travel.

Rudolph, L., Harrison, C., Buckley, L., North, S., 2018. Climate Change, Health, and Equity: A Guide for Local Health Departments. Public Health Institute and American Public Health Association, Oakland, CA and Washington D.C.

Sandifer, P.A., Collier, T.K., Trtanj, J.M., 2012. Coastal ecosystems, sustainable management. Encycl. of Sustain. Sci. & Technol. 4, 2258–2270.

Sandifer, P.A., Trtanj, J.M., Collier, T.K., 2013. A perspective on the history and evolution of an oceans and human health "metadiscipline" in the USA. Microb. Ecol. 65, 880–888.

Sandifer, P., Knapp, L., Lichtveld, M., Manley, R., Abramson, D., Caffey, R., Cochran, D., Collier, T., Ebi, K., Engel, L., Farrington, J., Finucane, M., Hale, C., Halpern, D., Harville, E., Hart, L., Hswen, Y., Kirkpatrick, B., McEwen, B., Morris, G., Orbach, R., Palinkas, L., Partyka, M., Porter, D., Prather, A., Rowles, T., Scott, G., Seeman, T., Solo-Gabriel, H., Svendsen, E., Tincher, T., Trtanj, H., Walker, A.H., Yehuda, R., Yip, F., Yoskowitz, D., Singer, B., 2020a. Framework for a community health observing system for the Gulf of Mexico region: preparing for future disasates. Front. Pub. Health 8, 578463.

Sandifer, P.A., Knapp, L., Lichtveld, M., Manley, R., Abramson, D., Caffey, R., Cochran, D., Collier, T., Ebi, K., Engel, L., Farrington, J., Finucane, M., Hale, C.,

Halpern, D., Harville, E., Hart, L., Hswen, Y., Kirkpatrick, B., McEwen, B., Morris, G., Orbach, R., Palinkas, L., Partyka, M., Porter, D., Prather, A., Rowles, T., Scott, G., Seeman, T., Solo-Gabriel, H., Svendsen, E., Tincher, T., Trtanj, H., Walker, A.H., Yehuda, R., Yip, F., Yoskowitz, D., Singer, B., 2020b. A conceptual framework for a community health observing system for the Gulf of Mexico region. In: Final Project Report and Supporting Materials. Mississippi-Alabama Sea Grant Consortium Technical Report MASGP-20-026, National Sea Grant Library No. GOMSG-T-20-001, 101pp.

Sarkar, S., Alhassona, F., Kimono, D., Albadrani, M., Seth, R.K., Xiao, S., Porter, D.E., Scott, G.I., Brooks, B., Nagarkatti, M., Nagarkatti, P., Chatterjee, S., 2019a. Microcystin exposure worsens nonalcoholic fatty liver disease associated ectopic glomerular toxicity via NOX-2-MIR21 axis. Environ. Toxicol. Pharmacol. 73, 103281S.

Sarkar, S., Kimono, D., Albadrani, M., Seth, R.K., Busbee, P., Alghetaa, H., Porter, D.E., Scott, G.I., Brooks, B., Nagarkatti, M., Nagarkatti, P., Chatterjee, S., 2019b. Environmental microcystin targets the microbiome and increases the risk of intestinal inflammatory pathology via NOX2 in underlying murine model of Nonalcoholic Fatty Liver Disease. Sci. Rep. 9, 8742.

Schandera, L.M., 2020. Investigating Microplastics and Their Presence in South Carolina. Thesis, Masters in Earth and Environmental Resource Management. School of the Earth, Ocean, and Environment, University of South Carolina, Columbia, SC, 22pp.

Scott, G.I., Porter, D.E., Norman, S., Scott, C.H., Uyaguari, M., Maruya, K.A., Weisberg, S.B., Fulton, M.H., Wirth, E.F., Moore, J., Pennington, P.L., Schlenck, D., Denslow, N.D., Cobb, G., 2016. Antibiotics as CECs: an overview of the hazards posed by antibiotics and antibiotic resistance. Front. Mar. Sci. 3 (24), 1–24.

Scott, C.H., Horton, C., Brett, C., Pipes, S., Tufford, D., Sandifer, P.A., DeLorenzo, M., Pennington, P.L., Porter, D.E., Ek, C., Norman, R.S., Scott, G.I., 2019. *Vibrio* bacteria in aquatic ecosystems and effects of climate change on antibiotic resistance: an increasing global threat. In: Maurice, P., Bailey, C. (Eds.), Wiley Water Encyclopedia, Human Dimensions, Water, Society, and Law. Wiley Publishers, NewYork, NY. https://doi.org/10.1002/9781119300762.wsts0226: 37pp.

Sinay, L., Carter, R.W., 2020. Climate change adaptations for coastal communities and local governments. Climate 8, 7. https://doi.org/10.3390/cli8010007.

Smith, M., Love, D.C., Neff, R.A., 2018. Microplastics in seafood and the implications for human health. Curr. Environ. Health Rep. 5, 375–386.

Stauffer, B.A., Bowers, H.A., Buckley, E., Davis, T.W., Johengen, J.H., Kudela, R., McManus, M.A., Purcell, H., Smith, G.J., Vander Woude, A., Tamburri, M.N., 2019. Considerations in harmful algal bloom research and monitoring: perspectives from a consensus-building workshop and technology development. Front. Mar. Sci. 6, 399. https://doi.org/10.3389/fmars.2019.00399.

Sutton-Grier, A.E., Gittman, R.K., Arkema, K.K., Bennett, R.O., Benoit, J., Blitch, S., Burks-Copes, K.A., Colden, A., Dausman, A., DeAngelis, B.M., Hughes, A.R., Scyphers, S.B., Grabowski, J.H., 2018. Investing in natural and nature-based infrastructure: building better along our coasts. Sustainability 10, 523. https://doi.org/10.3390/su10020253.

Sutton-Grier, A.E., Sandifer, P.A., 2019. Conservation of wetlands and other coastal ecosystems: a commentary on their value to protect biodiversity, reduce disaster impacts, and promote human health and well-being. Wetlands 39, 1295–1302. https://doi.prg/10.1007/s13157-018-1039-0.

Sweet, W., Dusek, G., March, D., Carbin, G., Marra, J., 2019. 2018 State of High Tide Flooding with a 2019 Outlook. NOAA Technical Report NOS CO-OPS 090, p. 31.

Troell, M., Jonell, M., Crona, B., 2019. The role of seafood in sustainable and healthy diets. EAT-Lancet Comm. Rep. 24, Stockholm Resilience Centre. https://eatforum.org/content/uploads/2019/11/Seafood_Scoping_Report_EAT-Lancet.pdf.

United Nations, 2016. First Global Integrated Marine Assessment (First World Ocean Assessment). United Nations. In: https://www.un.org/Depts/los/global_reporting/WOA_RegProcess.htm.

United States Commission on Ocean Policy, 2004. An Ocean Blueprint for the 21st Century. Final report of the U.S. Commission on Ocean Policy, Washington, DC, 455pp + appendices.

Vander Woude, A., Ruberg, S., Johengen, T., Miller, R., Stuart, D., 2019. Spatial and temporal scales of variability of cyanobacterial harmful algal blooms form NOAA GLERL airborne hyperspectral imagery. J. Great Lake. Res. 45, 536–546. https://doi.org/10.1016/j,glr.2019.02.006.

Vásquez, V.B., Minkler, M., Shepard, P., 2006. Promoting environmental health policy through community based participatory research: a case study from Harlem, New York. J. Urban Health 83 (1), 101–110.

Vezzulli, L., Grande, C., Helaouet, P., Edwards, M., Hofle, M.G., Brettar, I., Colwell, R.R., Pruzzo, C., 2016. Climate influence on *Vibrio* and associated human diseases during the past half-century in the coastal North Atlantic. Proceedings of the National Academy of Science, USA 113, D5062–E5071. National Academies of Science. www.pnas.org/cgi/doi/10.1073/pnas.1609157113. E5062-E5071.

Wang, Y., Xua, J., Stanley, J.E., Xu, M., Brooks, B.W., Scott, G.I., Chatterjee, S., Zhang, Q., Zelinski, M.B., Xiao, S., 2020. A closed vitrification system enables a murine ovarian follicle bank for high throughput ovotoxicity screening, which identifies endocrine disrupting. Reprod. Toxicol. 93, 118–130.

Weiss, K.E., Hammond, R.M., Hutchinson, R., Blackmore, C.G.M., 2011. *Vibrio* illness in Florida, 1998–2007. Epidemiol. Infect. 139, 591–598. https://doi.org/10.1017/S0950268810001354.

Wells, M.L., Karlson, A., Wulff, R.M., Kudela, C., Trick, V., Asnaghi, B., Cochlan, W., Davidson, Rijcke, De M., Dutkiewicz, S., Hallegraeff, G., Flynn, K.J., Legrand, C., Paerl, H.W., Silke, J., Suikkanen, S., Thompson, P., Trainer, V.L., 2020. Future HAB science: directions and challenges in a changing climate. Harmful Algae 91, 101632. https://doi.org/10.1016/j.hal.2019.101632.

Wetz, J.J., Blackwood, A.D., Fries, J.S., Wiliams, Z.F., Noble, R.T., 2008. Trends in total *Vibrio* spp. and *Vibrio vulnificus* concentrations in the eutrophic Neuse River Estuary, North Carolina, during storm events. Aquat. Microb. Ecol. 53, 141–149.

White, M., Smith, A., Humphreys, K., Pahl, S., Snelling, D., Depledge, M., 2010. Blue space: the importance of water for preference, affect, and restorativeness ratings of natural and built scenes. J. Environ. Psychol. 30, 482–493.

Yong, C.Q.Y., Valiyaveetill, S., Tang, B.L., 2020. Toxicity of microplastics and nanoplastics in mammalian systems. Int. J. Environ. Res. Publ. Health 17, 1509. https://doi.org/10.3390/ijerph17511509.

Interview: Human health and cholera

Rita Colwell

Recent technological advancements have improved our understanding of the ocean and our ability to measure it. How does this impact your research on oceans and human health?

Technological advancements have been incredible considering the genomic revolution entered rather late to the marine sphere. For example, technological developments in laboratories on the West Coast (MBARI, UW, OSU) now include devices that extract DNA in situ. This is an amazing development because it allows us to analyze microorganisms as a complete population, including viruses.

Viruses, bacteria, fungi and plankton populations can be detected and studied in situ rather than as individual species in a laboratory, and those organisms we can't culture have turned out to be the most interesting.

The analytical tools change our understanding of how populations function. And also allow us to understand the entire food chain in a detailed, functional way, which we could not have done without these developments. So as a biologist, my view is that we are a long way from the biological oceanography that I was taught as a graduate student at the University of Washington 50 years ago.

The technological revolution enables us to understand important aspects, namely that marine microorganisms are essentially machines that protect the climate and play a role in human health. Our blue planet depends on these microbial populations in a significant way.

We find bias in previous assumptions. For example, at the turn of the 20th century, it was assumed that fish and plankton, because they can be observed by the naked eye, are dominant in transformation of energy. However, with improved microscopic techniques, bacteria are now known to be important. In addition, there are more viruses in a milliliter of seawater than bacteria. We may find that fragments of nucleic acids in seawater are incorporated in ways that we had not understood.

It's significant that life on our planet is so interrelated, including all forms of life.

There is so much more to explore in the ocean. It is fortunate we haven't been able to routinely reach the deepest parts of the world's ocean because the ocean depths have not been exploited or as destructively polluted by human activity as the coastal shores. Microbial life traces to prehistoric times; not just to the 13 billion years from the origin of the universe, but ca. 500 million years ago when life originally evolved. Thus, technical advancements allow study of life in all stages and are truly extraordinary.

As an eminent scientist, what policies would you like to see change, or need to occur, to ensure responsible development and use of water resources and to engineer a better future? Is it at the policy level or is it more at the personal use level?

The oceans cannot continue to be used as a waste receptor. Policy changes are needed, such as support for building or retrofitting sewage treatment plants as bio reactors to convert and recycle waste products. We have the technology to do this, yet at present, sewage treatment plants release methane into chimneys to be burned off when it could be recycled to power and light cities and provide energy.

And the ocean, at the rate that the ocean is being polluted means we're not going to have fish to feed the projected global population of 10 billion people. Ocean policies have to be holistic and integrative. Every day that we construct a better, more holistic understanding of the oceans, terrestrial sphere, and the atmosphere we can sustain the incredible, integrated life support system that keeps us all alive and healthy. Social and global political policies must reflect such understanding.

What changes are needed in education and training to prepare the workforce of tomorrow?

Science must be taught better and more effectively. Science should be taught as discovery. Science is not memorization. It's truly discovery. Producing larger textbooks does not help. It makes more sense to develop creative experiments for kids that demonstrate phenomena they can understand and build upon.

And we need to use the best of all of the brains. We have to come to an understanding that to survive as a civilization, we must harvest the talent of all the humans, regardless of gender, race or ethnicity And it is critical to overcome biases, learn to communicate, and conduct interdisciplinary research. Communication of science to the public, especially about research findings, is critical.

All of the issues that you've raised are massive, significant, and critical. If there were one or two things that you could do, that you have the power or the authority and responsibility to change immediately, what would be those first things that you would change?

The first action I would take is re-institute a program I initiated at the National Science Foundation, the GK-12 Program. It was for graduate students in scientific and engineering fields, overseen by their professors, to spend 20 h a week in K-12 classrooms, assisting K-12 teachers. The graduate students worked with teachers to create and implement science lesson plans and activities. This helped teachers become more science literate and improve their science teaching skills. It helped graduate students become better communicators and educators. And helped K-12 students by providing them with more robust science learning experiences and allowed them to interact with near-peer instructors.

Many of the lesson plans and activities were augmented to be more discovery focused rather than memorization or worksheet-based experiences. I visited GK-12 schools where the programs were ingenious. The kids were having fun and at the same time they were learning science. I remember visiting the GK-12 program at Ohio State where the program focus that day was on aerodynamics. The kids were in the gym and given airplanes, paper clips and weights. The student challenge was to design a plane and measure how far the plane would fly. It was amusing to watch. Of course, the first reaction of the children was to load the wings and body of the plane with weights. When they flew the plane, it flew barely a foot and crashed to the floor. The students repeated the design, discovering the balance needed for the plane to fly. They measured distance and time, hence calculated speed. It was enjoyable for them to learn otherwise complicated concepts. The graduate students enjoyed the teaching experience. It was highly instructive for the students, the graduate students, and the K-12 teachers.

What do you see as the role of government agencies in preparing the workforce of tomorrow for the new blue economy?

I was Director of the NSF when the anthrax bio-terrorism attack occurred. Representatives from seventeen governmental agencies comprised an advisory committee that functioned for six years. We called ourselves the "National Genome Sciences Interagency Coordinating Committee." We met every Friday afternoon for an hour. I chaired the committee which comprised agency representatives who were highly enough ranking in their agencies that they had budgets, could make decisions, and could work with direct knowledge and approval of their agency head.

Together we successfully tracked the source of the anthrax, using genome sequencing. It was an example of interagency collaboration and cooperation and working together to generate results with lasting impact. However, success required breaking down barriers to communication, collaboration, and cooperation among the agencies, especially since no single agency could do all that was needed to meet the challenge.

If governmental agencies work together to address societal issues, for example, grand science or engineering challenges, or training the workforce of tomorrow, it could create a path for solving problems and preparing the next generation for the jobs and skills of the future. Ocean-related issues, such as protecting the ocean for sustainable use, will require the contribution of many disciplines and participants from all nations. If governments could work together, set policy and pathways for action, preparing the workforce of tomorrow would be a partnership among nations.

How important is professional certification (e.g., certified marine technologist) for the workforce of the future in your field?

Professional certification is an issue that is frequently discussed by the marine sciences community. The question usually ends up as a question of the value of such an effort.

The American Society of Microbiology provides certification for clinical microbiologists and for environmental biologists. A certification for a biological oceanographer in partnership with the American Society for Microbiology would be practical.

Similarly, certification for marine technology by the Marine Technology Society would be useful, as would collaboration with oceanographic institutions and oceanographers serving on boards of certification units.

Certification of marine technologists should build on the strength of the disciplines and strengthen interdisciplinary certification.

We've touched on technology, policy, and workforce development. The one thing we didn't touch on is resources

I've just completed my 10th year as chairman of the $500 million 10-year program, Gulf of Mexico Research Initiative. It's been hugely successful. There were several objectives, one of which was capacity building in the Gulf of Mexico.

The program focused on addressing oil spills, using the best physics, chemistry, biology, engineering, social, and behavioral sciences, and economics to address the societal problem of how to extract energy safely and effectively. And to address it in a way that is environmentally protective, socially constructive, and economically rewarding. The study was done over a 10-year period, a time frame necessary for environmental research.

Another example is the NSF Long-Term Ecological Research (LTER) program. The LTER stations can run 20 years or longer to collect environmental data and measure effects. A long-term time frame, substantial investment, and a variety of technologies are necessary for environmental, including climate, research. In the Gulf of Mexico, data were collected from sensors installed on a variety of platforms, e.g., buoys, underwater robots, airplanes, satellites, and ships, again different from the oceanography taught when I was a graduate student. Now it is possible to tap into machine learning and pattern recognition. One of the developments from the Gulf of Mexico research initiative was establishment of a working database. The collected data were formatted to be readily converted for use in the future, an example of a model for an holistic environmental program.

In retrospect, is there anything that you would have done differently in designing that program now that you have a chance to look back at it?

Yes. Because we were responding to the immediate crisis following the Deepwater Horizon tragedy in the Gulf of Mexico, we were essentially building an airplane while we were flying it.

Social, behavioral, economic, and health impacts of the crisis were studied, which had never been done before at such a scale. Interaction with nature requires more than just collecting samples to analyze in the laboratory. Humans are impacted profoundly by such violent interactions with nature and those interactions need to be studied at a scale over a long term. If we had a better understanding of the complexity at the front end of the initiative, perhaps we would have structured the program differently to include greater emphasis on the social, psychological, and behavioral impacts of the disaster.

You've mentioned that everybody needs to be working together. With improvements to technologies for monitoring and prediction capabilities, how can that assist Non-Governmental Agencies (NGOs)? Specifically how is that working to further your work on cholera and disseminating that information to NGOs?

Let's put it this way. Forty years ago, I had the brashness to propose that cholera was a naturally occurring aquatic bacterium. Well, that hypothesis was not well received by the medical community.

Then I had the audacity 20 years ago to report a multitude of viable microorganisms exist in the environment that are alive but not readily cultured in a laboratory.

We now know that the causative agent of cholera is in fact a globally distributed aquatic bacterium and, like many such bacteria, can exist in a viable but not culturable state, a type of dormancy. My students and I had also determined that zooplankton, particularly copepods, are the major host, or vector for Vibrios. Using Landsat sensors, we were able to determine a relationship of chlorophyll, sea surface temperature, sea surface height, with epidemics of cholera in Bangladesh.

We developed a model 21 years ago and have continually worked to improve that model using data obtained from satellite sensors. Now we are able to predict four weeks in advance when, where, and how high the risk of an outbreak of cholera in countries like Bangladesh, Yemen, and Mozambique.

A study conducted in Yemen with the British aid agency successfully predicted where the greatest risk of cholera would be located geographically, and physicians and medical supplies were deployed to those sites. As a result, the impact of the cholera outbreak was reduced.

We now work with the British humanitarian team, UNICEF, and NASA, and provide monthly predictions of cholera risk. That is exciting because it demonstrates satellites can provide a powerful public health tool. And to support NGO workers, we are developing training modules for how to interpret the data.

We are working with NGOs to collect water samples to ground-truth the data. When an outbreak is massive, it is almost too late to collect water samples to detect the causative strains of the cholera outbreak. We need to test water samples at the time of prediction of risk of an outbreak, and determine when and where to vaccinate to minimize the impact of exposure to the unsafe water. That's the ultimate objective, namely to apply real time data and address a potential health crisis before it occurs.

Author biography

Dr. Rita Colwell is Distinguished University Professor at the University of Maryland at College Park and Johns Hopkins Bloomberg School of Public Health and the Founder and Chairman of CosmosID, Inc. Her interests are focused on global infectious diseases, water, and health.

Dr. Colwell served as 11th Director of the National Science Foundation and Cochair of the Committee on Science, National Science and Technology Council.

Dr. Colwell holds advisory positions in the US Government, nonprofit science policy organizations, and private foundations. She authored or coauthored 20 books and more than 800 scientific publications.

Dr. Colwell served as Chairman of the Board of Governors of the American Academy of Microbiology, President of the American Association for the Advancement of Science, Washington Academy of Sciences, American Society for Microbiology, Sigma Xi National Science Honorary Society, International Union of Microbiological Societies, and American Institute of Biological Sciences. Dr. Colwell is a member of the US National Academy of Sciences, Royal Swedish Academy of Sciences, Stockholm, Royal Society of Canada, Royal Irish Academy, Indian Academy of Science, Bangladesh Academy of Science, American Academy of Arts and Sciences, and American Philosophical Society.

Dr. Colwell has been awarded 63 honorary degrees from institutions of higher education and is the recipient of the Order of the Rising Sun, Gold and Silver Star and the International Prize for Biology, bestowed by the Emperor of Japan, the 2006 National Medal of Science awarded by the President of the United States, the 2010 Stockholm Water Prize awarded by the King of Sweden, and the Lee Kuan Yew Water Prize of 2018. A geological site in Antarctica, Colwell Massif, has been named in recognition of her work in the Polar Regions.

OP/ED: Harmful algal blooms (HABs)

Barb Kirkpatrick

As a harmful algal bloom scientist for the last two decades, I participate in many community engagement sessions. One of the questions I am asked regularly is if Harmful Algal Blooms (HABs) are getting worse. Wearing my cautious scientist's hat, I respond that we think so—but changing sampling strategies and inconsistent sampling makes it a bit difficult to say for sure. What we do know for sure is that (1) the more people that move to our coastal communities, the more people are impacted by the blooms; and (2) the more improved tools we have to detect and track blooms, the more we find them, which can give us the impression that they are more frequent and/or larger in size.

Several large events have occurred in the last few years that have certainly increased public awareness of HABS. The Toledo water crisis in 2014 received widespread national attention when the residents of Toledo were warned not to drink their tap water due to an HAB on Lake Erie (the city's source for drinking water). The California warm water "blob" event occurred from 2013 through 2015 and wreaked havoc on the ecosystem including coral bleaching and depleting the fisheries. And the 2017–18

HAB event in Florida impacted over 400 miles of coastline, causing massive fish, sea turtle, and marine mammal mortality and causing toxic air in coastal communities. All three of these events received national media attention.

Let's return to the first part of my response when folks ask about HABs getting worse. In my home state of Florida, people continue to move into the state at 900 people a day (https://wusfnews.wusf.usf.edu/post/floridas-population-growing-900-people-day-report-says). Where are folks living? Seventy six point five percent (76.5%) live in a coastal area (https://coast.noaa.gov/states/florida.html) or put another way, of the 19.6 million people living in Florida, about 15 million live in coastal portions of the state. So yes indeed, more people are being impacted by HABs by the loss of recreational or commercial fishing, shellfish harvesting, poor water and air quality, and human health concerns. Residents and visitors alike have a growing demand to have improved information about these blooms on both spatial and temporal scales.

So let's examine the second part of my response—that with new HAB detection and tracking tools, we find blooms more often. Satellite imagery for HAB events has improved markedly over the last 15–20 years. Often seen as THE answer to HAB detection and tracking, they unfortunately are still unable to see through cloud cover (which happens frequently even in the "Sunshine State") and also have difficulty in shallow, near shore environments. The traditional method for HAB detection is through microscopic cell enumeration of individual water samples—an extremely accurate technique, which requires a highly skilled microscopist. I take a minute to note that we will always need these folks and it appears to be a field falling into disrepair. However, many new tools and instruments are in use and/or being developed in the HAB research and management community that use spectral absorption, high and low-resolution cameras, and DNA or RNA probes—to name a few (see Ausubel and Curry, this volume). Presently, these new sensors are mostly land or ship-based and/or moored. I suspect that many of these sensors will become part of the payload of autonomous underwater vehicles (AUVs) in the near future. Artificial intelligence is also being employed by these various techniques.

But managing HABs is not just a biology challenge. The chemistry and physics in the water column are also important parameters needed for detecting and tracking blooms. Many of the physical parameters (i.e., temperature, salinity, and density) have been measured in situ for years, but not at the spatial and temporal scales needed for HAB bloom management.

Many chemical parameters still rely on obtaining water samples and analyzing them back in a laboratory setting. New chemical sensors that analyze and transmit in real time are emerging. Once these chemical and physical measurements are made in real time, they must be integrated into the HAB observations for improved detection, tracking, and forecasting.

As coastal populations grow and citizens and visitors demand improved HAB reporting and prediction, a new workforce will be needed to deploy and maintain HAB sensors in the nearshore and offshore environment. Also needed will be technicians and specialists trained in deploying, piloting, and recovering AUVs. Data managers will be needed to curate data, archive data, and work with forecasters and communication specialists to display data in meaningful ways for a variety of users (i.e., the beachgoer may not need the same information and display as a resource manager). Several universities have identified the need for this workforce and offer programs in ocean technology and ocean observing. I would suggest that we also need to reach out to our technical schools and encourage them to do the same. Similar to weather observing, forecasting, and communications activities that help to mitigate harmful impacts from weather events, HAB impact mitigation will require a cadre of trained specialists.

HAB organisms have been around since the dinosaurs and are most likely going to do just fine with our changing ocean climate; perhaps flourish. The rapid growth in our coastal communities will also impact our oceans and the demand for clean, safe, beautiful coastal environments will continue to grow. We need to meet these changes by developing a well-trained HAB workforce to meet these demands. Our coastal economies will demand nothing less.

Barbara Kirkpatrick, EdD

Executive Director, GCOOS

Former COchair and Ex-officio member of the National HAB Committee

Member, Florida HAB Taskforce

Author biography

Barbara Kirkpatrick is the Executive Director for the Gulf of Mexico Coastal Ocean Observation System (GCOOS). She has over 35 years of experience in human and environmental epidemiology and started her career as a Respiratory Care Supervisor at Duke University Medical Center before receiving a Master's Degree in Health Occupations Education at

North Carolina State University and a Doctorate in Educational Leadership from the University of Sarasota.

Kirkpatrick joined Mote Marine Laboratory and shifted her research to environmental human health, particularly the respiratory effects linked to harmful algal blooms. While there she led an 11-year study to scientifically document the impacts of red tide on humans—particularly those who have chronic respiratory diseases.

Now, as Executive Director of GCOOS, she has been instrumental in broadening the scope of the ocean monitoring organization to include biological aspects of ocean monitoring—particularly monitoring for toxic algal blooms and marine animal movements.

CHAPTER 13

Biological information for the new blue economy and the emerging role of eDNA

Alan Curry, Jesse H. Ausubel
The Rockefeller University, New York, NY, United States

Introduction

From microbes to mammals, near shore to mid-ocean, and seafloor to seabirds, humans want and need to know about ocean life. Obvious benefits have derived from more accurate means of locating high-value wild fish for food or for protection in the interests of recreation and conservation. Fishers and dive shop operators may use the same information for opposite purposes. Surveyors have traditionally monitored sea life by observing seafood markets and trawl nets, by diving with goggles and clipboards, and more recently by deploying sonars and cameras, and sieving bits of extracellular DNA shed in seawater.

In earlier times and still today, blue life has other direct uses in addition to those as a recreational resource or as seafood for human consumption. Fish meal and oil provide feed for other forms of animal life, including carnivorous fishes raised in farms. For thousands of years, bones and shells, famously of turtles, became buttons, combs, and other household products. Shells also became ornaments and jewelry. Live animals enter the aquarium trade. Codliver oil promised good health before the word nutraceutical became popular, and other parts and forms of marine life, including algae, become pharmaceuticals and cosmetics. Accurate information about the diversity, distribution, and especially the abundance of marine life obviously has value. Humanity has misjudged abundance and exceeded sustainable exploitation of marine life many times, from abalone and cod to turtles and whales (Caswell et al., 2020).

Let's step back for a moment from catching, farming, and processing of fish and other sea life, feeding and lodging coastal tourists and visitors plus their boats, whale watching, and scuba diving. Consider five additional

Preparing a Workforce for the New Blue Economy
ISBN 978-0-12-821431-2
https://doi.org/10.1016/B978-0-12-821431-2.00030-5

aspects of the new blue economy and examples of their needs for biological information (Jolly et al., 2019).

1. Transferring goods (surface and pipelines), power and information (cables), and people. Operators of ships want to avoid collisions with marine mammals and areas where they may disturb spawning. They want to avoid biofouling and processes that shorten the lifetime of their capital.
2. Building boats, piers, and wharves; dredging channels; constructing barrages and new islands; reconstructing beaches. Builders of boats want to operate quietly and with bow wave pressure that minimizes harm. Port operators want to dredge at times of acceptable environmental impact. Builders of new islands and beaches want to create favorable habitat and not destroy it.
3. Pumping oil and gas, extracting wind and wave energy, and mining minerals; desalinating sea water. Operators of extractive industries want timely knowledge of their environmental impact, including spills and accidents. They want to design over a life cycle so that aged structures may be welcomed as artificial reefs.
4. Disposing of wastes (sewage, brines, plastics, and radioactive and other hazardous materials; decommissioning platforms). Individuals, companies, and municipalities want to create safe and harmless waste streams that will not accumulate dangerously in the tissues of marine animals. They want to monitor microbes, including viruses, in aquatic environments that can give early warning of threats to the health of humans and other animals.
5. Exploring and monitoring for safety, research, and prediction; mapping and hydrography. Researchers studying changes in ocean temperature or the geology of the seafloor may use active acoustics, pinging and listening for echoes and attenuations; they should conduct their experiments at frequencies and intensities that do not cause acute or chronic harm to marine life. Mappers and explorers should search for uncharted or poorly charted ecologically and biologically significant areas. Researchers who monitor the oceans for algal blooms that may be part of natural cycles also want to warn of harm to other marine life and to spur timely beach closures. Researchers exploring new or unusual species may supply valuable model organisms for science, as the squid axon has been for more than a century.

Clearly, needs for blue biological information already range widely, and we have dreams about new value from biomedical prospecting in the ocean. Medical and biological reasons to prospect for chemicals in the ocean do not hide in darkness. Animals such as starfish can regenerate limbs. Marine species ranging from sturgeon and whales to clams and corals live for centuries. Some animals, including eels, can generate strong electric currents. Obviously, all the marine animals, as well as algae (seaweeds) and the marine plants, thrive in salty water. Surely chemicals, compounds, and structures abound in the oceans that could make humans healthier, more youthful, and more beautiful. Regeneration, longevity, resistance to cancers, tolerance to heat and cold and to salinity and alkalinity, absence of seasonal affective disorder, and resistance to fatigue entice bioprospecting.

So far, the outcomes of bioprospecting are meager, a relatively small number of large-molecule drugs for blood cancers. Currently, US regulators have approved about a dozen compounds, and four more have entered Phase III trials, 13 Phase II, 7 Phase I (Midwestern University, 2020). What about the future of marine biopropecting?

The R&D enterprise of the pharmaceutical industry, not to mention the cosmetic industry, is immense. The size of the global pharmaceutical industry investment in R&D, about $180 billion USD in 2018 (Makulic, 2020), tempts a belief that onshore labs are doing everything in the search for designer drugs. However, quite possibly forms of marine life can serve as filters or prefilters. While we can prospect directly for compounds, a more important strategy may be to sequence parts of the genome of the organisms whose functions impress us or undertake other kinds of genetic and molecular studies. Then we may learn of proteins or other molecules and their shapes and other attributes that contribute to the impressive functions.

When a marine product becomes lucrative, we fear the problems of scaling up to meet large and potentially global markets, what Jouffray et al. (2020) call the "Blue Acceleration." The future wealth of the oceans cannot come from direct processing of sea cucumbers or starfish for their precious essences. The scale of life in the oceans could not sustain the sponge industry, nor the demand for tortoiseshell. Moreover, many desirable processes surely involve several genes and their products. While such polygenic products may also scale better through onshore synthesis, crucial clues live in the oceans. Science has barely begun to look at the 250,000 known species of marine life to classify them systematically by useful attributes.

The restraint or regulation of commercial activities in the ocean is premised on reducing collateral damage to marine life and other noncommercial interests (protection for the sake of protection) by shipping, dredging, drilling, waste disposal, research, and fishing itself. Regulations mandate monitoring of pollution levels (e.g., bacterial concentrations at popular beaches, leaks at oil platforms, plastic flotsam) and consequent effects on sea life (e.g., injuries to whales, coral bleaching). While these monitoring activities restrain trade or tax the new blue economy, they are needed to avoid situations where profits are improperly privatized and high costs externalized. An industry of environmental consultancies exists to carry out these studies.

Environmental DNA: new mode of monitoring marine life

As we have suggested, means of collecting biological information are changing, and those entering the field or remaining in it may soon find themselves with a different workload and working environment. What has been a mostly manual undertaking in the salt spray of the field may increasingly resemble space physics, where engineers design and launch probes that feed data to deskbound analysts. To share the example with which we are most familiar, the processing and profitable analysis of extracellular or environmental DNA (eDNA) requires different skills than might be found aboard a typical trawl vessel. Turning deckhands and experts who gauge the age of a fish from its ear bone into crew members of teams alongside molecular biologists and data scientists will challenge the maritime community.

Monitoring the location and movement of marine wildlife has been cumbersome and costly. The advent of techniques to isolate, amplify, and analyze eDNA can reduce the cost in time and money for oil and gas operators to monitor their operations, port operators to dredge, fisheries to open and close, coastal developers to assure the public that their works operate responsibly, researchers to explore and track biological changes, and governments to nominate areas for protection and restoration and to evaluate their success (Stoeckle and Ausubel, 2019).

Typically, in the United States and many other nations, maritime operators contract out evaluations of the impact of their activities on wildlife to an environmental engineering or consulting firm. In some cases, these consultants gather original data (from aerial surveys, for instance). In others, they compile and synthesize data from public agencies. These public agencies (arms of NOAA, for example), rely on reporting from commercial fishing vessels through vessel monitoring systems (VMS) or vessel trip reports (VTR) and "fisheries-independent" data from trawl surveys by governments, universities, and other nonprofits. Trawls have been a mainstay of ocean-life monitoring since at least the 1960s; NOAA's

Northeast Fisheries Science Center began theirs in 1963. The spatial coverage of the resulting data—when extrapolated with extensive modeling—is excellent, but the spatial resolution is poor, typically displayed in blocks of 100 square miles (260 square kilometers) apiece. Genomics offer the eventual prospect of precision, accuracy, and coverage, with the information available anytime, anywhere, and fast, a trio emblematic of the new blue economy.

In the interim, significant opportunities exist to augment trawls with water sampling and eDNA analysis, including when trawls are canceled or reduced owing to storms, equipment failures, costs, or crises such as COVID-19. eDNA promises much greater temporal resolution and precision due to the short time and therefore short distance over which shed DNA is thought to degrade and due to the detection of DNA from fish that would slip through or otherwise evade a trawl net. A time series for 11 species off the coast of New Jersey monitored simply by collecting 1 liter of water from the shore matches well with costly trawls of tens of millions of liters (Fig. 13.1).

Accordingly, many universities and government organizations have begun incorporating eDNA collection and analysis into their oceanographic research, and companies (Jonah Ventures, Mr. DNA, NatureMetrics, Spygen) have begun selling eDNA testing as a service. Typically, the customer mails in a filter through which a small quantity of freshwater or saltwater has been passed, DNA is then extracted from sediment on the filter, and the DNA is then analyzed in a standard next-generation-sequencing and bioinformatics pipeline. DNA identifications are usually based on existing reference sequences (such as those of the mitochondrial

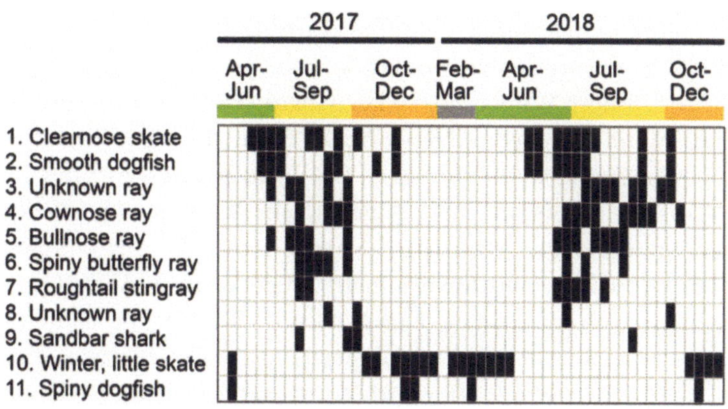

Figure 13.1 Fish species detected through water sampling (eDNA) at Barnegat Light, NJ, by month of appearance (Stoeckle et al., 2020).

12S rRNA gene) in GenBank. In the continental US exclusive economic zone (EEZ) relatively good coverage (>80%) of resident fish and marine mammals makes this process straightforward. eDNA collection can quickly identify a single species of interest (e.g., a marketable fish or an invasive species) with single-species assays (Stoeckle et al., 2018) or search for a class of life (e.g., vertebrates or molluscs) with metabarcoding (Stoeckle et al., 2017) or support certain customer tasks: to characterize completely a parcel of ocean over an appropriate period (e.g., a port over several seasons).

Actual collection and filtration of water still demand labor, including dipping liter containers at the surface, or Nansen/Niskin bottles at depth. Expansion of eDNA's use will require a streamlining of these "field services," particularly to characterize larger spatial areas or volumes of water. While eDNA produces excellent data on the presence and probably on abundance for a specific location, depth, time, and cost (an estimated \$25–50/sample), traditional means complement it with valuable information on age structure and health of sampled populations.

eDNA need not process the volumes of water currently flowing through trawl nets around the world. Work underway will clarify how near to one another in space and time these water samples must be taken to characterize accurately a given surface area or parcel of ocean—at least comparably to trawls or sufficient to satisfy reporting regulations as currently worded. Trawls have a very high throughput of water and must actually catch the fish; eDNA must simply capture the shed genetic material trailing a fish or school of fish. Currently sampling is strictly on a point-by-point basis, with more expansive extrapolations or assumptions necessary (vs. trawling) to deem an area of ocean properly "surveyed." In comparisons, eDNA captured 90% of the species netted by a trawl and detected more unique species than the paired trawls.

A New York–area energy project was the subject of the report "New York State Offshore Wind Master Plan: Charting a Course to 2400 MW of Offshore Wind Energy" (New York State Energy Research and Development Authority NYSERDA, 2017). Its fish and fisheries component, Chapter J (Energy and Environment Engineering, 2017), was prepared for the New York State Energy Research and Development Authority by Ecology and Environment Engineering (NASDAQ: EEI), a \$50 million/year consultancy based outside Buffalo. In addition, the companies who ultimately intend to operate in the six lease areas (Equinor, Orsted, Deepwater Wind, Vineyard Wind, and US Wind) will contract their own studies. At present, we are aware of only one company that has delivered end-to-end fish stock assessment by eDNA using existing technologies: Battelle, whose team members collected and analyzed eDNA from deep water off the North Slope of Alaska for Shell (Stever, 2015).

Indeed, to advance use of biological information in the new blue economy, the oil and gas industry helped inaugurate the International Workshop on Environmental Genomics in 2016. As a member of its trade group, the International Association of Oil and Gas Producers formed an industry subgroup, the International Consortium on Environmental Genomics (ICE-G), which launched in 2019 the Environmental Genomics Research Joint Industry Program (JIP) to pool resources of energy companies (including Chevron, Eni, Equinor, ExxonMobil, Hess, Shell, and Total) to develop protocols and encourage establishment of regulations governing eDNA for environmental monitoring during 2020–23 (IOGP, 2020). The JIP considers proposals on eDNA topics applicable to all phases of the offshore energy industry, from exploration to decommissioning.

While eDNA research has advanced rapidly over the past decade (Tabarlet, 2018), eDNA is in its infancy as a tool for regulators and commercial entities of the new blue economy. Its value proposition (efficiency + precision) awaits regulatory validation and broad commercial endorsement. The cost of incumbent technology such as conventional trawl surveys is somewhat obscured in the budgets of public agencies, or within the broad offerings of environmental consultants, making market-sizing and efficiency comparisons difficult. NOAA's "Fisheries Data Collections, Surveys, and Assessments" budget was $155 million in FY18, while their fishermen's contingency fund had $349 million set aside in FY 18 to compensate fishermen for the effects of oil and gas activity in the US EEZ. The National Marine Fisheries Service manages 474 marine and anadromous (river dwelling) fish stocks (including 230 with commercial value) within the US EEZ as well as invertebrates, sea turtles, marine mammals, and other marine and coastal species and their habitats, all of whom are assessed by what NOAA calls its ABC Method: parameters including abundance, biology, and catch are fed into a statistical model, which produces a stock assessment.

eDNA's employment in these stock assessments would require fewer personnel and less time at sea and necessitate different or retrained personnel and new kinds of equipment on shore. Firms including Biomeme, ANDE, Minion, and Illumina, and counterparts in Europe and Asia, are introducing portable DNA preparation and analysis hardware. As an emerging component of the new blue economy, we could envision a potential commercial operator developing a strategy using ships of opportunity—akin to the VMS and VTR fisheries reporting that already exists—to collect, filter, and store seawater eDNA samples over broad ocean routes to determine cost, reliability, chain of custody, and repeatability of data gathering, analysis, transmission, and archiving.

Companies could cater to monitoring needs with relatively localized operations at sea ("genomic weather stations") and perhaps even provide basin-scale (eventually global-scale) maps and databases with near-real-time updates of the presence and abundance of key marine species of scientific, ecological, and commercial value. Successful market penetration will benefit from advances in water handling at larger scales (tens and hundreds of liters), in some ways reminiscent of the continuous plankton recorders that have been employed since the mid-20th century, or even autonomous water collection and filtration, as demonstrated by MBARI and Proactive. New regulations must explicitly allow or require eDNA monitoring (vs. trawling) or establish that the two methods are comparable in the context of environmental impact submissions. Public attitudes may also play a part, as eDNA helpfully avoids the bycatch and habitat destruction of trawling.

Eventually one could imagine an efficient, end-to-end application of eDNA, if the water collection can be automated and the isolation, amplification, sequencing, and bioinformatics performed by lower-cost, "one-touch" machines that do not require PhD-level personnel in shore-based scientific facilities. Priorities to realize the potential of eDNA for the new blue economy include:

1. Building up the reference library of species.
2. Developing the technology of remote sampling (e.g., drone and marine robots that could collect and filter water and save the filters).
3. Integrating, and speeding and lowering the cost, of automated processing of samples (e.g., development of a single "machine" that could trap sediment containing DNA, extract the DNA from the sample, do the further steps such as sequencing, and then email the sequence file to the user).
4. Improving bioinformatic and big data storage in relation to eDNA; e.g., further improve the relevance and ease of use of programs such as the DNA Subway of Cold Spring Harbor Laboratory (2020) and also develop sites or enterprises that would aggregate and store eDNA information and results so that these could be integrated to reveal larger patterns over space and time.

Summary

eDNA and companion developments in genomics can transform biological information for microbes to mammals, near shore to mid-ocean, and seafloor to sea birds. Combined with advances in tagging and tracking and in acoustics, we can foresee in a decade or two more reliable, more comprehensive, and more timely biological information about the new blue economy. This information will tell us about ecosystems as well as species. Concurrently we hope that researchers unlock some of the secrets of marine life for regenerative medicine and other fields where breakthroughs would bring enormous benefit. The key is not ships and other expensive platforms, but autonomy, sensors, analytics, and the other technologies that characterize the emergent new blue economy.

References

Caswell, B.A., Klein, E.S., Alleway, H.K., et al., 2020. Something old, something new: historical perspectives provide lessons for blue growth agendas. Fish Fish. 21 (4), 774–796. https://doi.org/10.1111/faf.12460.

Cold Spring Harbor Laboratory, 2020. DNA Subway: Fast Track to Gene Annotation and Genome Analysis [Data Set]. https://dnasubway.cyverse.org/.

Energy and Environment Engineering., P.C, 2017. New York State Offshore Wind Master Plan: Fish And Fisheries Study. NYSERDA Report 17-25j. Final report prepared for New York State Energy Research and Development Authority (NYRSDA). https://www.nyserda.ny.gov/-/media/Files/Publications/Research/Biomass-Solar-Wind/Master-Plan/17-25j-Fish-and-Fisheries-Study.pdf.

IOGP, International Association of Oil & Gas Producers, 2020. The Environmental Genomics Research Program for Oil & Gas Industry Applications. http://www.iogp-edna.org/.

Jolly, C., Jolliffe, J., Stevens, B., Hoffman, J., 2019. Rethinking Innovation for a Sustainable Ocean Economy. OECD.

Jouffray, J.B., Blasiak, R., Norström, A.V., Österblom, H., Nyström, M., 2020. The blue acceleration: the trajectory of human expansion into the ocean. One Earth 2 (1), 43–54.

Makulic, M., March 2020. R&D Spending Share of Top Pharmaceutical Companies 2018 and 2024. https://www.statista.com/statistics/309471/randd-spending-share-of-top-pharmaceutical-companies/.

Midwestern University, June 2020. Clinical Pipeline: Marine Pharmacology. https://www.midwestern.edu/departments/marinepharmacology/clinical-pipeline.xml.

New York State Energy Research and Development Authority (NYRSDA), 2017. New York State Offshore Wind Master Plan. NYSERDA. https://www.nyserda.ny.gov/All-Programs/Programs/Offshore-Wind/About-Offshore-Wind/Master-Plan.

Stever, B., February 3, 2015. Genetic Flotsam Offers Clues to Ocean Biodiversity. Scientific American. https://www.scientificamerican.com/article/genetic-flotsam-offers-clues-to-ocean-biodiversity/.

Stoeckle, M., Ausubel, J., 2019. The eDNA Revolution. Sea Technol. 60 (6), 7.
Stoeckle, M.Y., Mishu, M.D., Charlop-Powers, Z., 2020. Improved environmental DNA reference library detects overlooked marine fishes in New Jersey, United States. Front. Mar. Sci. 7, 226.
Stoeckle, M.Y., Mishu, M.D., Charlop-Powers, Z., 2018. GoFish: a versatile nested PCR strategy for environmental DNA assays for marine vertebrates. PLoS One 13 (12).
Stoeckle, M.Y., Soboleva, L., Charlop-Powers, Z., 2017. Aquatic environmental DNA detects seasonal fish abundance and habitat preference in an urban estuary. PLoS One 12 (4).
Taberlet, P., 2018. Environmental DNA: For Research and Biodiversity Monitoring. Oxford U Press.

National security

CHAPTER 14

National security and the new blue economy

Thomas G. Drake[a]
U.S. Navy Office of Naval Research, Arlington, VA, United States

The new blue economy is foundational to unfettered access to the global ocean commons and confidence in the nation's security. The global oceans are essential providers of prosperity not only for the United States, but also for its allies, partners, and indeed, all nations, maritime or landlocked. The United States has enjoyed several decades of maritime dominance in the global oceans, enabling tremendous economic benefits from access to ocean transportation and communications routes, fisheries, opportunities for tourism, among many other benefits. This article will focus on a narrow subset of the elements of national security that depend critically on the new blue economy, namely, the ability of the US Navy and Marine Corps in the Department of Defense and the US Coast Guard in the Department of Homeland Security, the *Naval Service,* to leverage superior knowledge of the ocean environment to create advantage in the array of maritime mission needs that arise across the security spectrum of day-to-day competition, crisis, and conflict.

The national security implications of superior environmental knowledge—as characterized by the new blue economy—extend far beyond the ability to predict, for example, the intensity and path of a typhoon in the Western Pacific. The requirements for environmental prediction drive a vast spectrum of scientific and engineering efforts: new understanding of the ocean, advancements in high-performance computing that enable numerical forecasting, and observational capabilities and technologies that often spin off new opportunities for economic growth. Predicting the maritime environment ever further into the future continues to be a major goal of the US government, which works in partnership across multiple agencies to leverage expertise and common agency aspirations. The principal federal drivers in the operational realm are the Naval Meteorology and Oceanography Command (NMOC) and the National Oceanic and Atmospheric Administration (NOAA).

[a] The view presented is that of the author and does not necessarily represent the views of the Department of Defense or the Office of Naval Research.

Preparing a Workforce for the New Blue Economy
ISBN 978-0-12-821431-2
https://doi.org/10.1016/B978-0-12-821431-2.00045-7

The NMOC's oceanographic mission requirements are daunting: to predict at any point on the globe those ocean variables that impact naval operations. These critical parameters include the height and period of ocean waves, ocean currents, water temperature and optical characteristics, and the conditions that control acoustic propagation throughout the full water column, such as vertical and horizontal density structure and seafloor properties (Burnett et al., 2014). All of these variables in the ocean state are of course inextricably coupled to the atmosphere, the cryosphere, and the Earth's terrestrial environment, and the ability to forecast these conditions into the future with any useful skill requires a holistic understanding of the Earth system. NOAA's mission adds the prediction of space weather and climate change, as well as conservation and management of coastal marine ecosystems and resources, among other duties. Space weather is critical to national security, impacting radio communications, satellites including Global Positioning Satellites, and the nation's power distribution network. While the NMOC focus is on providing short- and medium-range forecasts with lead times of up to a month, fully leveraging the environment for national advantage necessarily extends to the gray area between weather and climate timescales. The undisputed impacts of climate change on the blue economy are discussed in detail elsewhere in this volume.

Astounding advances in environmental prediction skill have occurred over the last century (see Chassignet, this volume, for example). The hurricane that flattened Galveston in 1900 was almost wholly unexpected. We have come a very long way. Though the practice of predicting the weather from first principles is decades old, weather prediction has recently become a highly lucrative business built on decades of government-supported science. Specialized forecasts are now tailored for specific crops, ski areas, and other targeted applications, many driven by the insurance industry. Predicting the details of ocean circulation, on the other hand, remains less amenable to the precise spatial and temporal predictions we have now come to expect from weather forecasters, for several reasons. Perhaps the greatest difficulty is in adequately sampling the essential ocean variables to initiate, constrain, and validate the numerical models used for ocean prediction. While aspects of the Earth's atmosphere can be globally sampled by satellites at useful spatial and temporal scales, no such global observations are possible for the ocean interior. The demand for such ocean observations sparked the revolution in ocean technologies described further below, and in turn, the development of the new blue economy.

Computer models for ocean circulation, based on the same Navier–Stokes equations that describe the dynamics of the atmosphere, started a meteoric rise in complexity and capability in the 1980s, riding on the coattails of Moore's Law (Semtner, 1986). A concomitant increase in observations for the models has not, however, been easily accomplished (see Gouldman, and Watson-Wright and Snelgrove, this volume). Shipboard observations are too sparse by several orders of magnitude; other economical sources of observations are needed. While drifting ocean sensors have provided observations of the sea surface in quantity since the 1970s (leveraging the new availability of such satellite communication systems as Iridium, for example), the ocean interior remains poorly sampled. Perry and Rudnick (2003) provide excellent descriptions of early technologies and their development. Ocean models are capable in theory of providing predictions with considerable spatial and temporal resolution, but the promise of high-fidelity "ocean weather" forecasts is significantly hampered by the lack of sufficient observations of the ocean to initialize the models. It is difficult to predict what the ocean will be doing tomorrow if you don't have enough observations to know what it looks like today (see Reiss, this volume).

Both Navy and NOAA use the Hybrid Coordinate Ocean Model (HYCOM) for global ocean prediction. The system provides 4-day forecasts of the ocean state with a temporal resolution of 3 h and is updated daily (https://www.ncdc.noaa.gov/data-access/model-data/model-datasets/navoceano-hycom-glb). These global simulations have a horizontal resolution of 1/12th degree (roughly 9 km at the equator) and variable resolution in the vertical, which is interpolated onto 40 standard depth levels. The vertical resolution is relatively higher near the sea surface, ranging from a few to tens of meters, but the deep ocean is quite poorly resolved, with the deepest layers having resolutions up to several hundreds of meters. At present, these horizontal and vertical spatial resolutions are relatively high for an operational global model, and models used for climate applications are typically coarser. However, to enable the use of these ocean predictions for maritime naval operations, finer resolutions are required. Because computational limitations preclude running the global models at the necessary horizontal resolution for many naval applications, the Navy also runs a variety of regional models at higher resolutions and builds specialized models on demand for specific operational needs. These models are "nested" within the coarser models, taking their initial and boundary conditions from them, and provide more specific information for naval applications.

The ocean is but one component of the complex physical Earth system, and the skill of numerical ocean predictions is often dependent on the coupled interactions with the other domains. This leads to the development of Earth system models that incorporate the various elements of weather and climate—ocean, atmosphere, land surface, and sea ice, and sometimes distinct dynamic models for features such as ocean surface gravity waves and swell. Depending on their intended use, Earth systems models typically incorporate two or more of these components into a more complex prediction system. Along with other federal agencies, the Navy is working toward an interoperable Earth System Prediction Capability (ESPC) that promises to provide environmental forecasts with greater skill, higher fidelity, and longer lead times. Of course, this comes at a cost, not only in the model complexity, but also in the computational power required to operate these models. For this reason, the Navy maintains a suite of coupled models ranging from regional models such as the Coupled Ocean Atmosphere Mesoscale Prediction System (COAMPS) that incorporate the ocean, waves, and atmosphere, to the global ESPC model that includes both sea ice and land surface components as well. Typically, when predicting environmental conditions for a specific region, as the required forecast lead times get longer, there is a need to consider both larger spatial areas and more components of the Earth system in order to have any chance of a useful forecast. While the ocean evolves more slowly than the atmosphere, the synoptic forecast skill for variables of interest tends to diminish after a week or two, depending on the ocean features and area being considered, so medium- and long-range forecasts are often probabilistic, incorporating ensembles of model runs to account for a variety of uncertainties. While exploiting these probabilistic predictions can sometimes present a problem for operational decision-making, they also put additional strain on the computational requirements to run the coupled prediction systems. Advancements in high-performance computing, as well as the development of more efficient algorithms for running the component numerical models of the ESPC, will be needed to enable better ocean forecasts for the blue economy.

All ocean models are data-starved. Observations of the ocean are required to initialize the numerical models and keep them "on track," which is done through the assimilation of data into the model state. Navy Global HYCOM uses the Navy Coupled Ocean Data Assimilation (NCODA) system to assimilate satellite altimeter observations of anomalies in the sea surface height, as well as vertical profiles of ocean variables from

ARGO floats, moored buoys, and other available observations. This system, based on a three-dimensional variational data assimilation approach, blends these near-real-time observations of the ocean with the model's ocean state to create the initial conditions for the numerical forecasts. More advanced data assimilation techniques, such as those incorporating adjoint or inverse models in four-dimensional assimilations schemes (4DVAR), promise to extract even more information from the existing observations to initialize the predictive models and improve predictions.

By far, most of the available observations of the global ocean are at the sea surface. Global observations of ocean currents, surface gravity waves, and sea surface temperatures are now widely available, and this data helps make the numerical predictions of the dynamics at the ocean surface sufficiently skillful for many naval operations. Optimal routing models for ship transit are in common use (though such tragedies as the recent loss of the El Faro still occur), as are models for tracking oil spills (see Walker, this volume), though the largest uncertainties are in the subsurface ocean currents. Improving the predictive skill for the interior flow will enable a variety of useful applications, including the ability to optimally route undersea vehicles to exploit favorable subsurface currents, but these efforts are still developing. While the analysis of surface currents demonstrates that there are useful two-dimensional structures on many spatial scales—from the mesoscale meanderings of western boundary currents such as the Kuroshio to the presence of Lagrangian coherent structures at smaller scales—understanding the evolution of three-dimensional ocean flows is still a question for the research community. Efforts to develop advanced mathematical techniques, such the extension of dynamical systems theory into three dimensions to understand how the surface currents interact with the interior ocean, hold promise for future naval applications, but are still in their infancy at present (e.g., Peacock and Haller, 2013).

Sampling the ocean interior requires new approaches, and a team at the Massachusetts Institute of Technology (MIT) supported by the Office of Naval Research and the NOAA Sea Grant Program jump-started the development of new ocean sensing platforms to fill the observational gaps and empower the new blue economy (Buderi, 2013). These technologies, detailed further below, are now widely expected to herald the Navy's newest ocean-going platforms, joining aircraft carriers, destroyers, and submarines: a family of powered unmanned undersea vehicles that will in the future augment the Navy's manned platforms in much the same way that unmanned aerial systems have supplemented the US Air Force's manned airplanes.

The delineation between remotely operated vehicles and autonomous ones is blurry. The energy industry and the Navy both have long histories of using remotely operated vehicles for a variety of applications. Autonomous vehicles, operating and *making decisions* without human input, are still a substantial work in progress (see Manley, this volume). A spectrum of approaches from human-in-the-loop to human-on-the-loop to completely autonomous operations will be the norm for some time in the future, as risks and rewards for fully autonomous behaviors are enabled by technical advance and evaluated from economic, political, legal, and in the defense realm, ethical perspectives.

Fixed moorings for subsurface ocean sampling have long been an arrow in the oceanography quiver. These Eulerian measurements, fixed in space, produce useful time series of observations. Profiling sensors able to transit from the sea surface to depth and back, gathering such critical observations as temperature, conductivity (salinity), and pressure (depth), were enabled by the development of underwater gliders driven by highly efficient buoyancy engines. These Lagrangian sensors, moving with the flow, provide a new perspective on ocean processes and require new mathematical techniques for interpretation. The ability to change the bulk density and thus buoyancy of a vehicle from negative to positive with very little expenditure of energy provided the opportunity to deploy so-called gliders: person-portable, hydrodynamically shaped housings containing oceanographic sensors and communications capabilities that execute sawtooth profiles in the ocean and transmit data to overhead satellites when at the sea surface. The larger Liberdade class of "flying wing" gliders have high lift-to-drag ratio wing design, providing long distance, high-speed travel with larger payloads. Such platforms are quieter than most powered vehicles and have been successfully used to acoustically monitor marine mammals (Olmstead et al., 2010).

While the energy efficiency of gliders and Lagrangian sensing platforms commonly provide months to years of operation, powered vehicles with much more extensive capabilities for ocean mapping and a variety of Naval missions are readily available and rapidly improving. Commercial systems capable of autonomous mapping operations at 6000 m are common, and such improvements as synthetic aperture sonar systems enabled by relatively inexpensive inertial motion sensors and highly accurate clocks offer increasingly high-resolution swath mapping. A great deal of science and technology remains to develop the autonomous behaviors necessary for complex unattended operations and to combat corrosion and biofouling that such systems must address on their own for long-term deployments.

Autonomous surface vessels are rapidly proving their utility in both commercial and military applications. For instance, the ability to map the seafloor in shallow coastal waters, economically inefficient for ship operations, can now be accomplished semiautonomously by such companies as Saildrone (https://www.saildrone.com/technology) using wind- and solar-powered sailboats controlled by remote operators. Such technologies are expected to enable seafloor mapping of the entire US Exclusive Economic Zone in a decade.

As autonomous ocean platforms and sensors continue to develop new operational capabilities and longer endurance, there will be a need to enable the exchange of information both among these nodes and back to shore. The delivery of the observations from these platforms back to NMOC for assimilation into the predictive models in near real time is critical for improving the ocean forecasts. Similarly, these forecasts can be very useful for optimal routing, sampling, and decision-making on board the autonomous platforms themselves, requiring a capability for two-way sharing of information. However, often communications at sea are bandwidth-limited, even for high-powered surface vessels, let alone submarines or power-limited UxVs. For this reason, for many maritime applications, running local environmental forecasts aboard these platforms themselves would enable the exploitation of the operational environment without requiring the transmission of large data sets from land to sea, taxing the existing communications infrastructure. Ever-increasing low-power computational capabilities will enable the development of ocean models that can be run on the platforms themselves, incorporating local observations and making forecasts for the local ocean environment that can be exploited for more efficient operations. Research is ongoing to explore the development of new predictive capabilities, such as reduced-order ocean models designed for low-power GPU computing systems that will enable platform-specific ocean forecasts to be generated directly on a variety of unmanned autonomous platforms themselves. This will significantly reduce the need for high-bandwidth communications to reach back to shore for useful environmental knowledge.

National maritime security and, in turn, the new blue economy depend on detailed knowledge of and ability to predict the Earth's environment from space down to the sea surface, throughout the water column and deep into the seabed. While the Navy, Marines, and Coast Guard are not profit-seeking, they persistently seek a competitive edge through superior use of information, precisely like other participants in the new blue economy.

At present the volume of ocean information is too low and the cost to obtain it is too high. Although the opacity of the ocean to many sensing modalities presents daunting challenges, new technologies and approaches for characterizing the ocean and optimizing sampling strategies promise to dramatically expand the availability of inexpensive observations. With such observations, we will soon realize the potential for forecasts of the ocean "weather" to become as familiar as the nightly weather report.

References

Buderi, R., 2013. Naval Innovation for the 21st Century: The Office of Naval Research since the End of the Cold War. Naval Institute Press, p. 273.

Burnett, W., Harper, S., Preller, R., Jacobs, G., LaCroix, K., 2014. Overview of operational ocean forecasting in the US Navy: past, present, and future. Oceanography 27 (3), 24–31. https://doi.org/10.5670/oceanog.2014.65.

Olmstead, T., Roch, M.A., Hursky, P., Porter, M.B., Klinck, H., Mellinger, D.K., Helble, T., Wiggins, S.M., D'Spain, G.L., Hildebrand, J.A., 2010. Autonomous underwater glider based embedded real-time marine mammal detection and classification. J. Acoust. Soc. Am. 123 (3), 1971.

Peacock, T., Haller, G., 2013. Lagrangian coherent structures: the hidden skeleton of fluid flows. Phys. Today 66, 41–44.

Perry, M.J., Rudnick, D.L., 2003. Observing the ocean with autonomous and Lagrangian platforms and sensors: the role of ALPS in sustained ocean observing systems. Oceanography 16 (4), 31–36. https://doi.org/10.5670/oceanog.2003.06.

Semtner, A.J., 1986. History and methodology of modelling the circulation of the world ocean. In: O'Brien, J.J. (Ed.), Advanced Physical Oceanographic Numerical Modelling. NATO ASI Series (Series C: Mathematical and Physical Sciences), vol. 186. Springer, Dordrecht. https://doi.org/10.1007/978-94-017-0627-8_3.

Environment

CHAPTER 15

The new blue economy in the American Arctic: challenge and opportunity in the 21st century

Tim Gallaudet, Alexandra Skrivanek, Nicholas I. Flocken
United States National Oceanic and Atmospheric Administration, Washington, DC, United States

Arctic change and challenge

During the annual meeting of the American Geophysical Union (AGU) in December of 2019, I spoke at the press release of the 2019 NOAA Arctic Report Card (ARC) (Richter-Menge et al., 2019). The ARC has been produced annually since 2006, and it documents the physical, biological, and geochemical changes and impacts in the region due to rising air temperatures, declining sea ice, and warming waters.

The impacts have been significant. The warming in the Bering Sea, for example, has caused commercially important fish species (Pacific cod and pollock) to expand north approximately 500–1,000 km over a period of less than 12 months in 2017. Up to 15% of the US stock of pollock migrates across the US–Russia Exclusive Economic Zone (EEZ) boundary. Pollock comprises the largest commercial fishery by volume in the United States. Pacific cod biomass in the Gulf of Alaska declined by 58% between 2015 and 2017 due to multiple marine heat waves, forcing this $85M (2018 wholesale value) federal fishery to close in the Gulf of Alaska in 2020 to support marine mammal protection measures.

Other impacts have been equally dramatic. Declining sea ice and melting permafrost are increasing coastal erosion and reducing habitat for endangered and threatened species and species important for Alaskan subsistence harvesting. Invasive species are also threatening native wildlife in Alaska, and thawing permafrost is impacting critical infrastructure. These changes and others are thoroughly documented in all of the NOAA ARCs.

Then there is the global COVID-19 pandemic, which has put the brakes on a booming American blue economy. How are we to address all of these challenges and achieve the potential of the new blue economy in the Arctic of the 21st century? As put forward elsewhere in this book, the

Preparing a Workforce for the New Blue Economy
ISBN 978-0-12-821431-2
https://doi.org/10.1016/B978-0-12-821431-2.00039-1

answer is people, policies, and products that better help us adapt. In the following, I will describe how NOAA is doing this in the American Arctic of the Alaska region (as defined by the Arctic Research and Policy Act (Arctic Research and Policy Act of, 1984)). For a broader treatment of Arctic change, the recent report by the Congressional Research Service is a useful resource (O'Rourke et al., 2020).

NOAA in Alaska

NOAA has a permanent presence in Alaska and makes periodic scientific and survey expeditions with partners in the Alaskan Arctic, Arctic Ocean, and Arctic coastal areas. Our 600+ people in Alaska perform our mission of science, service, and stewardship for the very purpose of helping the region adapt to challenging changes described earlier. This dedicated workforce and their products and services are essential to providing the data, information, and knowledge to support the new blue economy in the Arctic through advancing the understanding, health, and sustainable use of the natural environment in the region.

Weather, water, ice, climate, and space

NOAA's National Weather Service (NWS) has three forecast offices in Anchorage, Fairbanks, and Juneau. The NWS Weather Forecast Offices, in collaboration with NWS National Centers, issue weather forecasts and warnings for some of the most extreme conditions in the country. With small private and commercial aircraft providing much of the state's transportation, aviation forecasts are at a premium. During the spring, ice breakup and river water level prediction provided by the Alaska Pacific River Forecast Center are critical to the safety of communities that depend on the water resources comprising 40% of the nation's total capacity, including an estimated 12,000 rivers and 3 million lakes (Fig. 15.1). The US National Ice Center, a multiagency partnership between NOAA (NWS), the US Navy, and the US Coast Guard, monitors sea ice conditions and movement in support of marine navigation and subsistence living. Also important are space weather prediction, which is vital to the protection of the nation's electrical and telecommunication infrastructure, and tsunami detection and warnings of the National Tsunami Warning Center (NTWC) in Palmer, Alaska. These lifesaving data and services are enabled by a multibillion-dollar weather data, observation and satellite infrastructure—feeding the new blue economy—that includes NOAA's

Figure 15.1 NWS expanded the use of drones in 2020 to support water level monitoring along streams and rivers and also began expanding staff capability to conduct more UxS missions.

National Environmental Satellite, Data, and Information Service (NESDIS) Command and Data Acquisition Station in Fairbanks (Fig. 15.2) and our Office of Oceanic and Atmospheric Research (OAR) Barrow Atmospheric Baseline Observatory in Utqiagvik (Fig. 15.3).

Oceanography and hydrography

NOAA's oceanographic and hydrographic activities in the American Arctic are supported by NOAA Ships Fairweather and Rainier. Scientists from across the agency, including the Pacific Marine Environmental Laboratory (PMEL) and the Earth System Research Laboratory (ESRL), regularly conduct field campaigns with partners from academia, international institutions, and NGOs to map, characterize, and explore the seafloor and ocean volume and coastal zone in the region. Specific areas of focus that support our mission include ocean circulation, ocean acoustics, marine chemistry, air—sea interaction, marine meteorology, and coastal processes.

Figure 15.2 NOAA's Fairbanks command and data acquisition station. *(photo: NOAA.)*

Figure 15.3 Barrow atmospheric baseline observatory in Utqiagvik.

Fisheries and ecosystem science and management

NOAA's scientific study in Alaska underpins the conservation of the natural resources and sustainable management of the fisheries in the region. NOAA Fisheries' Alaska Fisheries Science Center and Alaska Regional Office include teams of expert scientists, managers, and professionals. Along with NOAA Ship Oscar Dyson, NOAA-supported Alaska Sea Grant, the National Centers for Coastal Ocean Science's

Kasitsna Bay Lab, and the Office of Response and Restoration, these fine public servants conduct fisheries surveys and stock assessments, habitat conservation, protecting endangered and threatened species, marine debris removal, and oil spill response.

The new blue economy in the American Arctic

In addition to NOAA's workforce and products, our policies also ensure that those in the American Arctic thrive. Our remarkable advances in science and technology (S&T), regulatory efficiency, and innovative partnerships are creating the foundation for the necessary recovery and eventual renewal of a blue economy throughout the United States and the Alaskan Arctic, as well as the data and information that serve as the foundation for the new blue economy.

Science and technology

Emerging science and technology (S&T) is helping us turn the corner in our economic recovery. NOAA has identified 6 S&T focus areas that are enabling transformational improvements in our performance and efficiency. These include the fields of Autonomous and Uncrewed Systems (UxS), Artificial Intelligence (AI), Omics, Data, Cloud, and Crowdsourcing and Citizen Science (National Oceanic and Atmospheric Administration [NOAA], 2020a). We have developed strategies and 5-year implementation plans are nearing completion for each to guide the dramatic expansion of their application to every NOAA mission area. Further, I describe how they are supporting the new blue economy in the American Arctic.

The Unified Forecast System

NOAA is leading the development of the Unified Forecast System (UFS), a community based, coupled Earth modeling system that is the source of our operational numerical weather prediction forecasts. UFS numerical applications span from local to global domains and predictive timescales from subhourly analyses to seasonal predictions. The system includes the Finite Volume Cubed Sphere (FV3) dynamical core, and cloud computing will assume an increasing role in UFS improvements going forward (NOAA, 2019). A key component of NOAA's effort to advance our world leadership in Earth system prediction, the UFS is especially important in enabling accurate, 2-week predictions of Hurricane Force Aleutian low-pressure systems by NWS's Ocean Prediction Center. There are on average over 40 of these systems in the North Pacific annually, and they are extremely hazardous to our fishing fleet, commercial shipping, and the safety of coastal communities in the American Arctic.

Marine transportation

A study by the Committee on the Marine Transportation System (CMTS) projects that by 2025, vessel traffic through the Bering Sea could increase by 50%–200% relative to 2014 levels (U.S. CMTS, 2019). To prepare for this expected growth, as well as the growth in maritime commerce across the country, the Administration released a Presidential Memorandum (PM) on Ocean Mapping the US EEZ and the Shoreline and Nearshore of Alaska (Trump, 2019). NOAA is the lead in implementing this PM, and it has done so by spearheading development of a National Strategy for Mapping, Characterizing, and Exploring the US EEZ (Ocean Policy Committee [OPC], 2020) and an Alaskan Coastal Mapping Strategy (Mapping the Coast of Alaska, 2020). Specific actions that are already underway include leveraging UxS, AI, and partnerships (NOAA Office of Coast Survey, 2020; Fig. 15.4).

Figure 15.4 NOAA tested the Benthic Explorer and Navigator (BEN) ASV from NOAA Ship Fairweather in 2018 off Point Hope Alaska and will expand mapping operations using a variety of UxS in the Alaskan Arctic over the next decade. *(Photo by Christina Belton, NOAA.)*

Ocean exploration and characterization

In addition to ocean mapping for marine transportation, ocean exploration and characterization are important activities directed in Trump (2019) and OPC (2020) that support fisheries, offshore energy, critical minerals, and bioprospecting for pharmaceuticals. At a White House Summit in November of 2019, the topic of Ocean S&T Partnerships was explored, discussed, and promoted (Droegemeier and Neumayr, 2019). Since then, NOAA has formalized partnerships with Caladan Oceanic, Vulcan, Schmidt Ocean Institute, OceanX, Ocean Infinity, Viking Cruise Lines, Scripps Institution of Oceanography, and the Navy—all for the purposes of advancing our ocean exploration and characterization capabilities through advanced S&T (NOAA, 2020b–e; NOAA, 2020g–h; Boyle, 2020; NOAA Oceanic and Atmospheric Research, 2020).

Seafood competitiveness

The pandemic has presented challenges to our fisheries, notably with the cancellation of our fishery independent surveys. As with our ocean mapping efforts, we leveraged UxS to conduct our critical pollock survey (NOAA, 2020f, Fig. 15.5). We will build on this innovation in the future, including machine learning applications that we also have demonstrated in Alaska (NOAA Alaska Fisheries Science Center, 2018, Fig. 15.6).

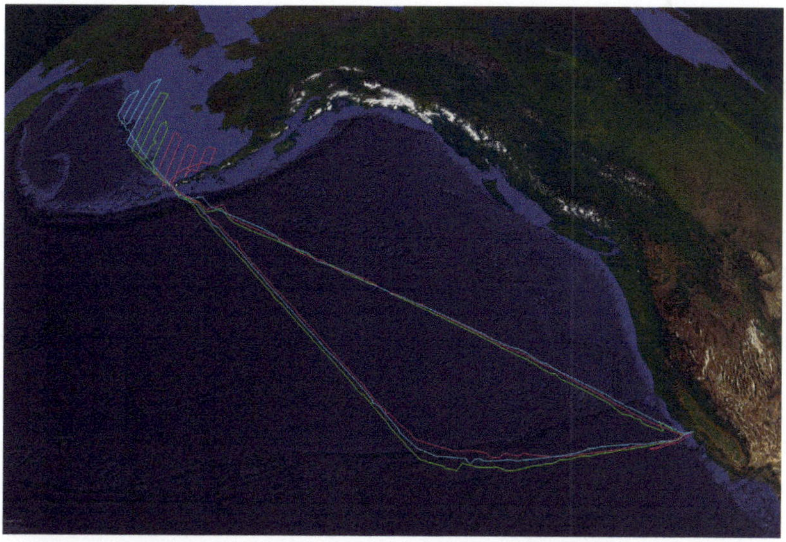

Figure 15.5 To mitigate the fisheries survey collection gap in the American Arctic due to COVID-19, NOAA leveraged a 5-year partnership with Saildrone to perform our Alaskan pollock survey in 2020 with UxSs along the *track lines* shown here. *(Source: https://www.saildrone.com/news/usv-complete-milestone-alaska-pollock-surveyhttps://www.fisheries.noaa.gov/feature-story/ocean-going-robots-effective-surveying-pollock.)*

Figure 15.6 NOAA's Alaska Fisheries Science Center has pioneered the application of machine learning to fisheries-independent stock assessments in the Gulf of Alaska. Here is an example of automated bycatch detection for the Halibut fishery. We are presently expanding this to the Bering Sea in support of the new blue economy in the American Arctic. *(photo: NOAA.)*

As the seafood sector recovers, it is important to note that economic benefits of NOAA Fisheries in Alaska to the American blue economy have been enormous: 101,000 jobs and $13.8B in economic output. Alaska seafood exports total $3.0–3.4B annually, while the first wholesale value of Alaska fisheries is $4.7B with an ex-vessel value of $2.0B (McDowell Group, 2020). These remarkable figures are the result of advances in fisheries science, dynamic fisheries management, and a diversified approach to seafood production. To continue these advances, the White House released the Executive Order on Seafood Competitiveness and Economic growth in May of 2020 (Trump, 2020). The actions directed by this EO are well underway and include (a) removing barriers to American fishing (b) combating illegal, unreported, and unregulated (IUU) fishing, (c) removing barriers to aquaculture permitting, (d) implementing Aquaculture Opportunity Areas, and (e) improving transparency for aquaculture. For the American Arctic in Alaska, actions (a) and (b) are critical, and advancing aquaculture actions could significantly aid this promising industry. While finfish farming is prohibited by state law, shellfish and macroalgae mariculture has great appeal and is promoted through the Alaska Mariculture Development Plan (Alaska Mariculture Task Force [MTF], 2018). Although Alaskan mariculture is presently limited to sub-Arctic regions of the state, the vision of a $100-million-dollar industry in 20 years implies that this sector will depend critically on the knowledge produced by the new blue economy in the American Arctic going forward.

The power of partnerships

The full potential for NOAA's contributions to the new blue economy in the American Arctic is being realized through partnerships. Innovative collaboration with the private sector and philanthropies, native communities, the interagency, academic institutions, and priority international allies is enabling rapid advances in our science, service, and stewardship.

Public–private partnerships

As mentioned previously regarding Ocean Exploration and the 2019 White House Summit, NOAA has prioritized ocean S&T partnerships to advance our mission. A terrific example involves the Alaska Fisheries Science Center's collaboration with commercial fisherman to improve stock assessments in the Gulf of Alaska (NOAA Alaska Fisheries Science Center, 2020). As portions of the fishery migrate northward to the Arctic, we will apply Omics, UxS, and AI/ML to further advance public–private partnerships like this (MTF, 2018).

Academic, interagency, and international collaboration

Because the geophysical and geopolitical environment of the Arctic is interconnected both within the region and with the rest of the globe, advancing the abundance and availability of data provided by the new blue economy in the American Arctic requires collaboration with a wide range of partners. International agreements with international partners are vital to NOAA's ability to monitor the Arctic with satellites and in situ measurements enabling us to better monitor harmful algal bloom events, ocean color, the ice edge, know Arctic weather conditions, monitor hurricane force winds offshore, and predict ice movement. NOAA works with multiple universities, agencies, and international experts across the Arctic region for joint research and operations. Examples include the US Navy's ICEX (Fig. 15.7), NSF-funded scientific expeditions on the USCG Cutter

Figure 15.7 NOAA's weather, ice, and ocean climatology, observations, and forecasts are used in the planning and execution of the US Navy's biennial ICEX in the Arctic. *(photo: Michael B. Zingaro, US Navy.)*

Figure 15.8 NOAA scientists, with US university, interagency, and international partners embarked onboard the icebreaker research vessel Polarstern as part of the MOSAiC expedition near the North Pole. *(photo: Alfred Wegener Institute.)*

Healy, and the Multidisciplinary Drifting Observatory for the Study of Arctic Climate (MOSAiC) expedition (Fig. 15.8). Operational ice and iceberg analyses and forecasts are also closely coordinated and shared within the US and Canadian North American Ice Service (NAIS) and the several national ice centers that make up the International Ice Charting Working Group (IICWG). The Interagency Arctic Research Policy Committee (IARPC) coordinates these efforts within the United States, while several working groups under the Arctic Council do so for the member and observer countries. NOAA leads several of these working groups, and we colead the US delegation to the biennial Arctic Science Ministerial meetings. The scientific research promoted through these activities is essential to the sustainable development and use of the natural resources in the region. A positive example acknowledging this was the Agreement to Prevent Unregulated High Seas Fisheries in the Central Arctic Ocean.

Native Alaskan outreach and support

The perspectives and contributions of Native Alaskan and tribal communities are invaluable to the new blue economy in the American Arctic (similarly, see the contribution from Noelani Puniwai, in this volume, regarding the application of traditional ecological knowledge to the new blue economy). NOAA consults with tribes and native corporations about Federal actions that may affect tribal governments and their members. Habitat, fisheries, and marine mammals are the topics we consult and engage on routinely, and we maintain a close relationship with the Alaska Federation of Natives and other Alaska Native Organizations and tribes. Additionally, NOAA advocates for Aboriginal Subsistence Whaling (ASW) quotas of the Bering-Chukchi-Beaufort stock of bowhead whales to the International Whaling Commission (IWC) each year. To expand these collaborations, NOAA has recommended that the IARPC establish a Tribal Communications Coordinator position, which would ensure a whole of government approach in our scientific support to native communities and their economies.

Arctic opportunity and optimism in the 21st century

During the AGU Ocean Sciences Meeting in February of 2020, I was a panel member addressing the interagency follow-up to the previously mentioned November 2019 White House Summit on Ocean S&T Partnerships. I acknowledged the challenges associated with changes in the world's oceans, but I also described the growing confidence among ocean stakeholders in the potential that technology, innovation, and partnerships can improve the understanding, health, and sustainable use of our oceans.

A terrific example of this is the Ocean Optimism Campaign of Dr. Elin Kelsey, Dr. Nancy Knowlton, and Elisabeth Whitebread (Ocean Optimism, 2017). NOAA's contributions of data, information, and knowledge in the Alaskan Arctic are representative of the kind of positive approach we are taking toward mitigating the challenges in all of our nation's oceans, coasts, and Great Lakes, to recover from the COVID-19-associated economic downturn, and to further empower the American new blue economy throughout the 21st century.

References

Alaska Mariculture Task Force, March 23, 2018. Alaska Mariculture Development Plan. Alaska Fisheries Development Foundation, Inc. https://www.afdf.org/wp-content/uploads/Alaska-Mariculture-Development-Plan_v2018-06-29_FINAL_digital.pdf.

Arctic Research and Policy Act of 1984, July 31, 1984. Public Law 98-373. 98 Stat. 1242. https://www.congress.gov/bill/98th-congress/senate-bill/373.

Boyle, A., April 21, 2020. NOAA Makes a Pact with Vulcan to Deepen Collaboration on Ocean Science. GeekWire. https://www.geekwire.com/2020/noaa-makes-pact-vulcan-deepen-collaboration-ocean-science/.

Droegemeier, K., Neumayr, M., November 13, 2019. Droegemeier and Neumayr: Why Trump's Making Ocean Exploration a Top Priority. Fox Business. https://www.foxbusiness.com/lifestyle/kelvin-droegemeier-why-trumps-making-exploring-unknown-ocean-top-priority.

Mapping the Coast of Alaska, June 2020. A 10-Year Strategy in Support of the United States Economy, Security, and Environment. U.S. Department of Commerce, National Oceanic and Atmospheric Administration. https://iocm.noaa.gov/about/documents/strategic-plans/alaska-mapping-strategy-june2020.pdf.

McDowell Group., January 2020. The Economic Value of Alaska's Seafood Industry. Alaska Seafood Marketing Institute. https://uploads.alaskaseafood.org/2020/01/McDowell-Group_ASMI-Economic-Impacts-Report-JAN-2020.pdf.

National Oceanic and Atmospheric Administration, December 19, 2019. Cloud Platforms Unleash Full Potential of NOAA's Environmental Data. U.S. Department of Commerce, National Oceanic and Atmospheric Administration. https://www.noaa.gov/media-release/cloud-platforms-unleash-full-potential-of-noaa-s-environmental-data.

National Oceanic and Atmospheric Administration, 2020a. NOAA Science and Technology Focus Areas. U.S. Department of Commerce, National Oceanic and Atmospheric Administration. https://nrc.noaa.gov/NOAA-Science-Technology-Focus-Areas.

National Oceanic and Atmospheric Administration, January 9, 2020b. NOAA Teams with Ocean Infinity to Advance New Tools for Ocean Exploration and Mapping. U.S. Department of Commerce, National Oceanic and Atmospheric Administration. https://www.noaa.gov/media-release/noaa-teams-with-ocean-infinity-to-advance-new-tools-for-ocean-exploration-and-mapping.

National Oceanic and Atmospheric Administration, February 11, 2020c. OceanX and NOAA Join to Advance Ocean Exploration and Mapping. U.S. Department of Commerce, National Oceanic and Atmospheric Administration. https://www.noaa.gov/media-release/oceanx-and-noaa-join-to-advance-ocean-exploration-and-mapping.

National Oceanic and Atmospheric Administration, April 7, 2020d. NOAA Teams with Pioneering Explorer to Understand and Map the Deepest Parts of the Ocean. U.S.

Department of Commerce, National Oceanic and Atmospheric Administration. https://www.noaa.gov/media-release/noaa-teams-with-pioneering-explorer-to-understand-and-map-deepest-parts-of-ocean.

National Oceanic and Atmospheric Administration, June 9, 2020e. NOAA, Schmidt Ocean Institute Team to Explore and Map the Ocean. U.S. Department of Commerce, National Oceanic and Atmospheric Administration. https://www.noaa.gov/media-release/noaa-schmidt-ocean-institute-team-to-explore-and-map-ocean.

National Oceanic and Atmospheric Administration, June 19, 2020f. NOAA Ramps up Use of Drones to Collect Fish, Seafloor and Weather Data. U.S. Department of Commerce, National Oceanic and Atmospheric Administration. https://www.noaa.gov/stories/noaa-ramps-up-use-of-drones-to-collect-fish-seafloor-and-weather-data.

National Oceanic and Atmospheric Administration, July 22, 2020g. NOAA Partners with Scripps Institution of Oceanography on Unmanned Systems. U.S. Department of Commerce, National Oceanic and Atmospheric Administration. https://www.noaa.gov/media-release/noaa-partners-with-scripps-institution-of-oceanography-on-unmanned-systems.

National Oceanic and Atmospheric Administration, August 4, 2020h. NOAA, U.S. Navy Will Increase Nation's Unmanned Maritime Systems Operations. U.S. Department of Commerce, National Oceanic and Atmospheric Administration. https://www.noaa.gov/media-release/noaa-us-navy-will-increase-nation-s-unmanned-maritime-systems-operations.

National Oceanic and Atmospheric Administration Alaska Fisheries Science Center, May 31, 2018. Developing Machine Vision to Collect More Timely Fisheries Data. U.S. Department of Commerce, National Oceanic and Atmospheric Administration National Marine Fisheries Service. https://www.fisheries.noaa.gov/feature-story/developing-machine-vision-collect-more-timely-fisheries-data.

National Oceanic and Atmospheric Administration Alaska Fisheries Science Center, July 14, 2020. Fishermen and Scientists Pioneer Cooperative Rockfish Survey. U.S. Department of Commerce, National Oceanic and Atmospheric Administration National Marine Fisheries Service. https://www.fisheries.noaa.gov/feature-story/fishermen-and-scientists-pioneer-cooperative-rockfish-survey.

National Oceanic and Atmospheric Administration Oceanic and Atmospheric Research, January 16, 2020. NOAA Teams up with Viking to Conduct and Share Science Aboard New Great Lakes Expedition Voyages. U.S. Department of Commerce, National Oceanic and Atmospheric Administration. https://research.noaa.gov/article/ArtMID/587/ArticleID/2580/NOAA-teams-up-with-Viking-to-conduct-and-share-science-aboard-new-Great-Lakes-expedition-voyages.

National Oceanic and Atmospheric Administration Office of Coast Survey, July 2020. Mapping U.S. Marine and Great Lakes Waters: Office of Coast Survey Contributions to a National Ocean Mapping Strategy. U.S. Department of Commerce, National Oceanic and Atmospheric Administration National Ocean Service. https://nauticalcharts.noaa.gov/learn/docs/hydrographic-surveying/ocs-ocean-mapping-strategy.pdf.

Ocean Optimism, 2017. Inspiration. http://www.oceanoptimism.org/inspiration/.

Ocean Science and Technology Subcommittee of the Ocean Policy Committee, June 2020. National Strategy for Mapping, Characterizing, and Exploring the U.S. Exclusive Economic Zone. White House Press Office. https://www.whitehouse.gov/wp-content/uploads/2020/01/20200611-FINAL-STRATEGY-NOMEC-Sec.-2.pdf.

O'Rourke, R., Comay, L.B., Folger, P., Frittelli, J., Humphries, M., Leggett, J.A., Ramseur, J.L., Sheikh, P.A., Upton, H.F., September 10, 2020. Changes in the Arctic: Background And Issues for Congress (CRS Report No. R41153). Congressional Research Service. https://fas.org/sgp/crs/misc/R41153.pdf.

Richter-Menge, J., Druckenmiller, M.L., Jeffries, M. (Eds.), December 2019. Arctic Report Card 2019. U.S. Department of Commerce, National Oceanic and Atmospheric Administration. https://www.arctic.noaa.gov/Report-Card.

Trump, D., November 19, 2019. Memorandum on Ocean Mapping of the United States Exclusive Economic Zone and the Shoreline and Nearshore of Alaska. White House Press Office. https://www.whitehouse.gov/presidential-actions/memorandum-ocean-mapping-united-states-exclusive-economic-zone-shoreline-nearshore-alaska/.

Trump, D., May 7, 2020. Executive Order on Promoting American Seafood Competitiveness and Economic Growth. White House Press Office. https://www.whitehouse.gov/presidential-actions/executive-order-promoting-american-seafood-competitiveness-economic-growth/.

U.S. Committee on the Marine Transportation System, September 2019. A Ten-Year Projection of Maritime Activity In the U.S. Arctic Region, 2020–2030. https://www.cmts.gov/downloads/CMTS_2019_Arctic_Vessel_Projection_Report.pdf.

Interview: Marine litter

Emily Smail

What are the biggest challenges (technological, policy, logistical, etc.) in marine litter today?

A major challenge for policies related to marine litter is a lack of baseline data. Policymakers need baseline data in order to develop targeted policies and to determine if policies implemented to stem the input of litter and plastics into the environment actually are having impact.

On the policy side, the lack of a clear understanding on the impacts of marine litter is another challenge. It is difficult to drive public support for policy without demonstrating the harm of marine litter. What is the harm of ingestion of microplastics? Similarly, there are significant challenges associated with the recycling system and policies related to packaging materials.

Regarding logistics of data collection on marine litter, people (research scientists, governments, citizen science programs) are using different methods for data collection so it becomes difficult to compare results. Uniformity in data collection and open data sharing would be helpful. Many technologies for monitoring and mitigating marine litter are still in the research and development phase. As these technologies advance, information about marine litter will be more accessible. Significant investment in these technologies is needed in order to move forward.

What are some examples of applications of technologies to current problems in addressing marine litter today and what we could do to entice the future workforce to be interested in solving these problems with the technology of today and tomorrow?

Identifying the types of plastics using hyperspectral imaging through remote sensing and in water probes are examples of interesting applications of technologies to the problem of understanding the distribution of marine litter. To prevent marine litter, new materials technologies are needed, especially in areas such as fishing net design and packaging manufacture.

The future professionals who are interested in stemming the flow of marine litter should also consider looking into applications of technology to waste management. For example, areas that are in need of improved waste management can be identified by applying various remote sensing and modeling technologies to the identification of dump sites and estimation of waste leakage. There is also a developing market for the application of technologies such as artificial intelligence for waste sorting and recycling.

An important aspect of drawing in the future workforce is tied to getting visibility for applications of new technologies. When new applications for ocean cleanup make the news, it helps attract a potential new workforce. People who are interested in technology should also be encouraged to explore opportunities to apply technology to solving sustainability challenges such as marine litter and waste management.

What are some examples of training/education that are the most effective for the future workforce in reducing marine litter?

Some of the emerging programs at community colleges, such as the efforts in San Diego tied to the work of TMA Blue Tech, are encouraging (see Jones and Fischel, this volume). At the higher education level, the efforts of the Partnership for Observation of the Global Ocean (POGO) (https://www.ocean-partners.org/) and the Global Partnership for Marine Litter (http://marinelitternetwork.com/the-partnership/) have provided some useful training programs.

Better communication strategies for reaching "the masses" to have them better understand the issue and potential careers in various aspects of the marine litter life cycle will be important to the development of the future workforce. Campaigns at schools and universities to make students aware of this as an issue and apply their problem-solving skills to coming up with local solutions are also useful.

How relevant/important is the "data revolution" (i.e., "big data," "cloud computing," etc.) to the marine litter community?

The data revolution is very important for determining baselines and establishing interoperable data sets to enable impactful decisions. Innovative processing of data is also needed to combine various types of data, such as environmental data and socioeconomic data, in order to understand the scope and impacts of the marine litter issue and how impactful efforts to collect and contain litter are. Modeling capability is also crucial for prediction and analysis of concentrations and transport of marine litter.

There is also a challenge in integrating data on marine litter with other data, such as environmental data and economic data. As mentioned previously, artificial intelligence/machine learning is showing promise in analyzing some types of marine litter data and information.

We also need to work with countries, such as small island states, which lack access to data collection equipment and network connections to understand the scale of the issue globally.

What levels of education/training are most needed in the workforce of the future (high school, community college, college, advanced degrees, etc.)?

The job market is currently flooded with Ph.D.s so I would encourage the future work force to explore technical training programs and interdisciplinary programs. The marine litter workforce of the future will need people from multiple disciplines, not just marine scientists, but also engineers, computer scientists, and policy experts to work together in interdisciplinary teams.

We should get away from traditional thinking about education and look to innovative approaches, including coding training programs in the military, for example. Fellowships and cross-training programs, such as the Knauss Marine Policy Fellowship (https://seagrant.noaa.gov/Knauss-Fellowship-Program), are also great ways to support the development of a diverse workforce.

How important is professional certification (e.g., certified marine technologist) for the workforce of the future in your field?

I could see data collection certifications being helpful for providing assurance to decision and policymakers that the data they are using to develop or implement a policy is reliable.

It could be helpful to ramp up the needed workforce quickly to tackle this issue as more technician-level workers are needed, and this could be accomplished through community colleges.

From the data and information side, having coders and programmers be part of the solution, with awareness created through a certificate approach, could also move the field forward.

Author biography

Emily Smail is the Executive Director of the GEO Blue Planet Initiative and a Senior Faculty Specialist at the NOAA—University of Maryland Cooperative Institute for Climate and Satellites. She also serves as the cochair of the GEO AquaWatch Initiative's outreach and user engagement working group and supports outreach and education efforts for the NOAA CoastWatch/OceanWatch program. Previously, Dr. Smail worked in informal science education at the Waikiki Aquarium and policy in the US Senate through the Knauss Marine Policy Fellowship Program. She received a B.S. in Biology from the Pennsylvania State University and a Ph.D. in Biology from the University of Southern California where her research focused on water quality and marine biogeochemistry.

Interview: Entrepreneur utilizing ocean data and drones for ocean surveys

Oscar Garcia-Pineda

Recent technological advancements have improved our understanding of the ocean, and our ability to measure and observe it—how does this impact your business model?

In my area of work of remote sensing and detection of oil spills, there is a constant challenge of bringing new discoveries to respond more practically and efficiently to oil spills. The challenge is to apply the science in a practical way so it can be used by the community on a regular basis. Recent developments in new tools have simplified our processing workflows, making it much simpler to obtain actionable results from field data.

What are the major technological limitations to your business?

The technology for surveillance, response, monitoring, or assessment oil spills is well identified. Technologies from in-situ sensors to spacecraft are used world over for the remote sensing of spills. These technologies have improved over the years and will continue to be refined and enhanced. Currently, responders understand well the advantages of choosing one tool or the other depending on the case at hand. From passive sensors to active sensors, we understand how much information we can extract or what can we expect from a diverse suite of data collection. There is no one single sensor or platform capable of resolving all the needs for assessing the magnitude of a spill; therefore, a combination of these technologies is always needed. An ideal sensor/platform (or a combination of them) would have the capacity to detect not only the presence/absence of an oil spill on real time and under all conditions, but also the thickness of the floating layers, which is an important factor for planning remediation operations. This is a limitation as this technology does not exist, yet!

In parallel, current limitations in oil spill analysis relate to our ability to adapt our policies and laws to the use of newer technologies. A good example of this are drones. This type of technology can be harnessed for the monitoring of our environment; however, sometimes regulations impinge our use of this technology. In the case of environmental monitoring specifically, for example, it would be tremendously helpful and much more efficient to be able to fly drones at a higher altitudes, at night, beyond line of sight, or under several other circumstances in which current policies prohibit. While these policies made sense in years past, when drone technologies were in their infancy, now that drone safety has matured, policies should adapt to these changes. So, our limitations are related to adapting to changing technologies rather than inherent limitations of our tools.

What are some examples of applications of technologies to current problems and what we could do to entice the future workforce to be interested in solving these problems with the technology of today and tomorrow?

From my point of view, for the earth-ocean sciences, recent developments in drone technology have been as revolutionary as the invention of the internet or the cellular phone. Being able to observe, to monitor, and to protect the environment so efficiently has transformed the way we operate. Technology is increasingly affordable, and the barrier of entry is very low: recreational use does not require operating licenses in many situations, and many entry-level models can be operated from a smartphone. The current stage of drones (or more properly called Unmanned Aircraft Vehicles—UAVs) involves trained pilots and data analysts; however, the future of this technology in parallel with artificial intelligence points toward automatization of surveillance tasks in which UAVs drones not only operate autonomously, but also detect and alert automatically in real time. Not only from the air, but in parallel, we will foresee the introduction of swarms of Autonomous Underwater Vehicles (AUVs) or Autonomous Surface Vehicles (ASVs) monitoring our beaches, coastal, and marine environments with operational alert systems for multiple natural or unnatural occurrences (not only for Oil Spills!)

In growing your business, what skill sets will the workforce need?

Generally, a proficiency in Geographical Information Systems (GIS) knowledge is required but an increasingly necessary skill set is programming. Many of our day-to-day operations require repetitive tasks, which can be easily automated with some knowledge of basic coding. A crucial skill, however, is the ability to adapt to new technologies; it is crucial to ensure we are always providing as many services as possible.

Oscar Garcia biography

Dr. Garcia earned his Ph.D. in Coastal and Marine System Science Program from Texas A&M University-Corpus Christi, Texas. He currently owns and operates Water Mapping, a consultant group focused on geological exploration and coastal and marine environmental monitoring using state-of-the-art technologies including satellite imagery and geophysical data followed by aerial and on-site observations to provide spatial and temporal analysis for coastal monitoring, oil exploration, and oil spills.

CHAPTER 16

Tsunami hazard and the new blue economy

Vasily Titov
Pacific Marine Environmental Laboratory, United States National Oceanic and Atmospheric Administration, Seattle, WA, United States

Introduction

Tsunamis, along with other marine hazards, threaten world coastlines that are the foundation of any sea-focused strategy. Ports and increasingly dense coastal population are under increasing pressure of finding effective solutions for reducing the tsunami risk. Developing successful strategies against tsunami threat, however, presents some unique challenges.

On the one hand, tsunamis are one of the most devastating natural disasters. The December 26, 2004 Indian Ocean tsunami is one of the deadliest natural catastrophes in history, while the March 11, 2011 Tohoku tsunami is officially the costliest natural disaster of all time (Kajitani et al., 2013, Fig. 16.1). The global risks of geophysical phenomena such as damaging earthquakes and tsunamis are now listed as third highest in terms of likelihood and as the top fifth in terms of impact (World Economic Forum, 2019). And the risk is growing; in the third millennium, tsunamis have become the deadliest natural disasters, while they ranked only fifth previously.

On the other hand, tsunamis are relatively infrequent events, imposing a distinctive challenge for creating sustainable hazard mitigation solutions. While about 10 tsunami events occur per year in the world oceans (based on the 20th century statistics—the most reliable part of tsunami historical records), only a small portion creates severe destruction and loss of lives, and even fewer, one in a decade on average, would be catastrophic, reverberating around the world with news of death and destruction. Hence, a particular coastline may not see strong tsunamis for generations, which may only increase the risks for an unprepared coastal population. The vivid illustration of the destructive power of such forgotten threat was the devastation of the unprepared Indian Ocean coastlines during the December 26, 2004 Indian Ocean tsunami.

Preparing a Workforce for the New Blue Economy
ISBN 978-0-12-821431-2
https://doi.org/10.1016/B978-0-12-821431-2.00007-X

Figure 16.1 Inundation of the March 11, 2011 Tohoku tsunami on the Sendai region near Sendai airport.

As the memories of the 2004 Indian Ocean and the 2011 Tohoku tsunami catastrophes fade, the support for publicly financed warning and mitigation systems may become increasingly challenging. In the near future, the US coastal population and infrastructure may become more vulnerable to that (semi-)forgotten hazard, similar to the Indian Ocean coastlines.

This paper explores the development history, existing capabilities, and emerging technologies for tsunami warning systems and mitigation strategies. Analysis of the historical trends may provide insight for future solutions and associated requirements for a new level of tsunami resilience. The new blue economy can provide the need and the opportunities not only to maintain the current status, but also to bring the tsunami warning services to a new level of effectiveness by providing economic incentives for new high-tech and high-efficiency tsunami services.

Evolution of tsunami warning technology: successes and challenges

Tsunamis have been documented to impact human coastal developments from the very beginning of the modern civilization. Historical records provide evidence of impressive tsunamis dated back to 2000 BCE, some of which may have even altered the course of history, like the 1755 Lisbon tsunami that may have contributed to the demise of the Portuguese empire. But it was only less than 100 years ago, when the idea of tsunami warning services first emerged. Japan, as the most tsunami-affected country in the world, was the first to recognize the value of the tsunami warning after repeated devastations of the Sanriku coasts by the 1886 and 1933 tsunamis. The Japanese government proposed a list of tsunami hazard mitigation measures shortly after 1933 and established a government-run tsunami warning operation 8 years later in 1941 (Shuto and Fujima, 2009). The other tsunami-prone countries of the Pacific followed with their own national tsunami warning systems after major tsunamis affecting their countries' coastlines. All systems used the same basic principles as the first Japanese warning method, utilizing the earthquake data to infer the tsunami potential and verifying the tsunami impact with coastal observations (Bernard, 2006; Bernard and Titov, 2015). It was not until the devastating 2004 Indian Ocean and the 2011 Tohoku tsunamis that the major shift of the tsunami warning methodology started to occur, fueled by new real-time measurements and new model forecast advances. However, over 30 significant events that occurred since the 2011 Tohoku revealed many existing gaps in the existing tsunami warning and hazard mitigation strategies. Most of these recent events that affected populated coastlines resulted in casualties, even at areas where newly established tsunami warning services were implemented. The death toll of over 2000 from just two tsunamis in Indonesia in 2018, where the new tsunami warning system had been operational since 2008 (Lauterjung and Letz, 2017), demonstrated that much more need to be done to make the tsunami warning systems robust and effective. The analysis of success and challenges of the past tsunami warning developments may identify opportunities for the future of tsunami warning.

Saving lives requires quick reaction and prediction accuracy of the tsunami warning systems to provide more time for coastal population to react to the warnings and more accurate information to guide actions. The speed of modern warning systems for initial tsunami warnings has improved about 10-fold since the over-30-minute warnings of the initial Japan system of 1941 (Shuto and Fujima, 2009). The Japan Meteorological Agency (JMA) and US National Oceanic and Atmospheric Administration (NOAA) both are issuing their first tsunami information in less than 2–3 min after an earthquake most of the time. That initial warning reaction time may be close to the physical limit of how fast the data from large earthquakes can be obtained; so it may be unrealistic to expect much faster initial source assessment.

The need for the fast reaction time of the modern tsunami warnings has imposed utmost importance on the accuracy of the tsunami warning. Underwarning and overwarning have become the major challenges for the modern tsunami warning systems.

"False alarms," when coastal evacuation is called for a nonflooding tsunami, had become an especially acute problem for Hawaii in the previous century. Public trust in the tsunami warning system started to erode when 15 out of 20 warnings since the system was established were perceived as unnecessary (Bernard, 2006). The public has become more tolerant to "false alarms" after seeing the horrors of the 2004 Indian Ocean catastrophe documented in photographs and videos, and after witnessing the real-time broadcast of the destruction of the Sendai coast during the 2011 Tohoku tsunami. The economic impacts of unnecessary evacuations, however, are not that forgiving and will only grow with rising coastal population and commerce. The 1986 estimate of the $40M cost of unnecessary Hawaii evacuation after an M7.7 Alaska earthquake (Bernard, 2006) is over $100M in today's dollars without accounting for infrastructure growth on Hawaiian coasts. With memories of the large tsunamis fading, the new generation will quickly develop zero tolerance for tsunami false alarms.

The underwarning for tsunamis may lead to even more disastrous consequences. During the 2011 Tohoku tsunami, JMA issued the first tsunami warning in just 3 min after the onset of the earthquake. Unfortunately, this first warning underestimated the earthquake energy by a factor of 10 (Tsushima et al., 2011; Ohta et al., 2012; Koshimura and Shuto, 2015) simply because of the limited data for the initial assessment. As a result, the initial wave amplitude forecast—thought to be a conservative estimate—was for 3–6-meter tsunami at nearby coasts, which many residents ignored relying on 10 meter-high seawalls protecting most settlements. The actual tsunami ran up to 40 m along coastlines near the tsunami source and consistently over 10 m for most of the impacted shores, easily overtopping and destroying most seawalls, killing many who were deceived by the benign first warning. The correct assessment of the earthquake and tsunami energy was realized only days later (Ohta et al., 2012). A clear lesson of the 2011 Tohoku tsunami is that the tsunami warning should not only be fast but has to be accurate.

Successes

The Intergovernmental Oceanographic Commission (IOC) facilitated major expansion of the tsunami warning services to global level from pre-2004 Pacific-centric service by coordinating multiple national tsunami warning systems, hazard mitigation programs, and data sharing. All member states of the IOC benefit from this new distributed global tsunami warning system. All partners contribute to the system with data and services and benefit with better forecasts based on more complete data and improved accuracy.

One of the most striking successes for the global tsunami warning system development is the implementation of the real-time deep-ocean tsunami observation network across the world oceans. Deep-Ocean Assessment and Reporting for Tsunamis (DART) was developed in the 1990s (Milburn et al., 1996) with the goal to reduce rampant "false alarms" of the tsunami warning system at the time. But it was the Indian Ocean tsunami that emphasized the value of the DART data for improving overall tsunami warning accuracy to achieve actionable tsunami forecast capability. The United States spearheaded the development and initial deployment of the DART system in the interest of the US tsunami warning system, so the initial DART array was US-sponsored and owned. However, international partners immediately saw the benefits of shared data for improving the national and global tsunami warnings. Presently, almost half of the DART array is maintained by international partners—a notable example of international cooperation and cost sharing among nations for the mutual benefit of improved tsunami warning capabilities.

The 2011 Tohoku tsunami clearly demonstrated that much more data are needed for tsunami forecasting. Japan is now deploying a vast ocean-bottom cabled network of pressure gauges to supplement DART data for local tsunami hazard assessment, with the United States, Canada, and other nations following suit with their own cabled observatories.

The most obvious implication of the new data capability and new international integration of the tsunami warning systems is the drastic reduction of tsunami "false alarms," especially for the so-called far-field events, when the source of a tsunami is more than 1000 km away from the coastlines to warn. The Hawaiian Islands placed in the middle of the Pacific are a good example of the beneficiary of that improvement. Tsunamis from most of the hazardous sources for Hawaii would be "intercepted" by several existing DART systems and with enough time for analysis, a data-based forecast can be obtained well before the evacuation decision needs to be made.

Challenges

Despite recent expansion of the tsunami detection network offshore and along world ocean coastlines, the tsunami forecast is still very much a data-poor process. With possibly one geographic exception along the East Coast of Japan, the tsunami models have to provide forecasts based on one, two, or at the very best three short time series from nearby offshore sea-level gauges. Increasing the density of the tsunami detectors may be the most straightforward way to improve tsunami forecast accuracy.

Tsunami warning for the near field, when waves hit coastlines closest to the tsunami source, is a much more difficult problem. The 2011 Tohoku tsunami disaster clearly demonstrated this near-field warning challenge. Part of the difficulty is the same lack of real-time data for accurate and timely forecasts. Japan spared no expanses addressing this real-time data issue by building extensive network of cabled gauges to provide timely data for fast local tsunami forecasting. The modeling part of the forecast may also need to be redesigned for quick utilization of massive real-time data assimilation.

Future needs and possible solutions

Why do we need improved warning?

A relevant question that should be asked before making a case for improvement of any system is the following: Is there a need for further improvement besides pure technological advancement opportunities? In the case of the tsunami warning, saving human lives, reducing property and economic damage are the main goals and the main drivers for tsunami warning system developments. All these goals have financial implications that could be used as a metric for assessing general risk, tsunami warning system efficiency, and the economic incentives of tsunami warning improvements.

The US Tsunami Hazard Mitigation effort has prepared Tsunami Hazard Assessments for US coastlines based on available historical data (Dunbar and Weaver, 2015). The purpose of this report was to develop priorities for the US Tsunami Program. The report found that tsunami hazard is high or very high for most of the United States and territories except the US Atlantic Coast, Gulf Coast, and Alaska Arctic Coast. However, the Atlantic Coast assessment did not include coastal hazards from meteotsunamis.

Tsunami economic impact assessments, while rare, are also available in the literature, ranging from historical event impact evaluations to economic analyses based on model scenarios. The assessments of economic impact for the 2011 Tohoku tsunami are one of the most complete studies, with staggering results. Direct economic impact from the tsunami attack is estimated in the range from $211B to $230B. Additional indirect impacts include measurable losses of GDP in the following year due to decline of economic activities, which adds tens of billions of dollars to the direct impact (Kajitani et al., 2013). These astronomical numbers illustrate the tsunami destruction potential of the level of national security threat. Several model studies for the US coastlines may also provide more localized statistics for assessing tsunami warning value for specific communities or coastal facilities. Borrero et al. (2005) estimated impact from strong local tsunami flooding for the port of Los Angeles and Long Beach. The economic impact ranges from $7B to $40B depending on the level of damage to the port and the speed of recovery. Rose et al. (2016) performed a more comprehensive study of economic impact to California by a potential tsunami from Alaska. The study estimates $5.9B of total economic damage including port disruption, property damage, and evacuation cost. Most of the economic losses are due to tsunami impact in ports. One of the interesting conclusions of the study is that costs of evacuation ($2.8B) are similar to those from the physical damage and port disruption ($3.1B). This California evacuation put the 1986 estimate of economic losses from the "false alarm" in Hawaii of about $100K into perspective. Another conclusion of the study was that resilience measures may reduce the impact by 80%–85%. Among various economic resilience measures (which may be costly), the effective tsunami warning and mitigation measures may be one of the least expensive. Just shortening or eliminating unnecessary evacuation with specific actionable forecast capability may reduce the direct economic impact by half.

With high tsunami hazard along most of the US coasts and potential economic impacts in billions of dollars from even mild tsunami impact, the case is vividly clear that concerted efforts are needed for closing the existing gaps in tsunami warning to mitigate the tsunami risk for the US coastal communities and facilities.

How: paradigm shift in warning operations from "detection and warning" to "data assessment and forecast".

To provide the next level of tsunami resilience, the tsunami warning operations may need to shift the paradigm from "Detection and Warning" to "Data Assessment and Forecast." Many pieces of technology to enable that shift are already in place, but new approaches may be needed to translate these technologies into actionable products.

Tsunami warning operations and mitigation measures around the world have implicitly inherited the general strategy from the very first tsunami warning ideas that came from the original 1941 Japanese tsunami warning procedures: a handful of experts analyze earthquake and limited tsunami data, make decisions about a possible tsunami threat, and if deemed necessary, disseminate a tsunami warning along the established channels to local managers and then to the public (Shuto and Fujima, 2009). Nowadays, the data are received through satellite instead of telephone lines, decisions are made with the help of computer programs instead of using manual analysis of paper charts, warning messages are distributed via Internet and satellites instead of radio and telephone, but the inherited structure of the centralized warning centers is still in place. Every part of the procedure works faster, but the structure itself may not allow for the flexibility that new volumes of data, new methods of data analysis, new computer model expertise, and new dissemination options may allow.

While complex computer simulations are used for tsunami analysis at many tsunami warning centers, most of the public and event emergency managers see just text messages as tsunami warning products. The weather forecasts, for example, use text messages also, but forecasts are also available in a variety of different specialized products. Just like in weather forecasting, the tsunami warning operations can use a distributed model when many functions of the tsunami assessment can be done not in one central location but are delegated to other experts. Data analysis, forecast distribution, and specialized products can all be done outside of the core tsunami warning structure to improve flexibility and ultimately making more actionable and useable products. The structure of distributed responsibilities for data analysis and forecast products is not a purely theoretical concept. Several examples in tsunami warning operations illustrate the usability of such an approach.

DART network development and IOC tide gage network data service are examples of different countries sharing responsibilities for providing real-time data for tsunami assessments and at the same time receiving benefits of such distributed services. Modeling forecast expertise has been distributed in many different ways. Science conferences and articles are the more traditional scientific way of expertise sharing. A more practical approach was initiated as the TIME (Tsunami Inundation Modeling

Exchange) project of Tohoku University, as part of the International Decade for Natural Disaster Reduction back in the 1990s, which transferred the expertise of Japanese tsunami modeling to 22 countries for the development of inundation maps (Shuto and Fujima, 2009). More recently, the integrated and Internet-connected tsunami modeling capability provided by the ComMIT system (Titov et al., 2011) was established as IOC-sponsored activity. It has now transformed into a very successful self-sustainable program with an established training program and hundreds of participants from over 30 countries. This Internet-enabled modeling platform can be easily developed into a shared distributed forecast capability using an Internet aggregator system such as the Tweb tool (Bernard and Titov, 2015). Accurate digital bathymetry is an example of the essential data for tsunami forecast.

The Internet, cloud computing, big data, and the abundance of individual communication devices among populations may provide unique opportunities to develop fast automated tsunami forecasts and distribution systems as a unified capability. The future generation and a big part of the current population do not consume information from centralized sources such as radio, TV, and even the Internet via computers; these are becoming an obsolete way of obtaining information. Personal communication devices may not only give immediate access to forecasts and warnings, they may also provide immediate feedback information for warning and forecast updates. There are plenty of ways to get different slices of weather forecasts with graphics and easy-to-understand graphical summaries via a myriad of apps on one's phone, but it is still just a couple of sources of text message for a tsunami warning. The elements are in place to develop actionable easy-to-understand graphic forecast products that can be delivered directly to needed populations or clients quickly. The science of such products is continuously evolving and may provide better timing and better accuracy for the forecast (Bernard and Titov, 2015).

The future tsunami forecast may not need a centralized clearing house for data and models. The fast-responding service may rather be a facilitated distributed system where information is shared via a complex data network automatically assimilated into distributed forecast nodes (Fig. 16.2) with a complex hierarchy of selecting "authoritative" forecasts, streaming user-tailored products to all requested customers simultaneously. Fast reaction times for tsunami forecasts demand a high level of automation. Examples of

Figure 16.2 Conceptual diagram illustrating tsunami forecast capabilities with a distributed forecast system to be shared via a congregate system. An example of the 2011 Tohoku tsunami is used for the simulation. The global view shows the "authoritative" source propagation scenario that is used for local flooding forecasts. Each color-coded *circle* indicates the level of flooding threat for the location (green—no flood, yellow—medium coastal threat, red—high threat of flooding), forecasted in real time. Inserts show examples of locally computed flooding predictions for two sites with a confirmation photograph at Sendai.

such automated data analysis systems exist already—and they are not all in the financial sector. ShakeAlert, for example, uses real-time earthquake data assessments for shaking prediction when there are only seconds available to analyze huge amounts of data (Given et al., 2018). Imagine what can be done when minutes and hours are available for a tsunami forecast! With the future abundance of real-time data, the accuracy may be benchmarked in real time with forecast updates provided immediately and seamlessly.

The next-generation tsunami forecast system is not going to appear magically out of an imaginary blueprint or theoretical exercise. As this paper has shown, pieces of such a system already exist and some more are under development. While the government plays an important driving role in putting these pieces together into new system development for tsunami warning, it is the demand and partnership from participants of the new blue economy that are needed to make this development of the new tsunami forecast capability successful and sustainable.

Acknowledgments

The author acknowledges the dedicated research support of NOAA Pacific Marine Environmental Laboratory scientists and engineers that made this research possible (PMEL contribution number 5168) and would like to thank the staff of the NOAA Center for Tsunami Research for multiple scientific contributions described in this paper. The scientific results and conclusions, as well as any views or opinions expressed herein, are those of the author and do not necessarily reflect the views of NOAA or the Department of Commerce.

References

Bernard, E., et al., 2006. Tsunami: Scientific frontiers, mitigation, forecasting, and policy implications. Phil. Trans. Roy. Soc. Lon. A 365 (1845), 1989–2007. https://doi.org/10.1098/rsta.2006.1809.

Bernard, E., Titov, V., 2015. Evolution of tsunami warning systems and products. Phil. Trans. R. Soc. A 373, 20140371. https://doi.org/10.1098/rsta.2014.0371.

Borrero, J., Cho, S., Moore, J.E., Richardson, H., Synolakis, C., 2005. Could it happen here? Civ. Eng. 75 (4), 54–65, 133.

Dunbar, P.K., Weaver, C.S., 2015. United States and Territories National Tsunami Hazard Assessment: Historical Record and Sources for Waves– Update. NOAA Tsunami Program Special Report. NOAA, USGS. https://nws.weather.gov/nthmp/documents/Tsunami_Assessment_2016Update.pdf.

Given, D.D., Allen, R.M., Baltay, A.S., Bodin, P., Cochran, E.S., Creager, K., de Groot, R.M., Gee, L.S., Hauksson, E., Heaton, T.H., Hellweg, M., Murray, J.R., Thomas, V.I., Toomey, D., Yelin, T.S., 2018. Revised Technical Implementation Plan for the ShakeAlert System—an Earthquake Early Warning System for the West Coast of the United States: U.S. Geological Survey Open-File Report 2018–1155. https://doi.org/10.3133/ofr20181155, 42p.

Kajitani, Y., Chang, S.E., Tatano, H., 2013. Economic impacts of the 2011 Tohoku-Oki earthquake and tsunami. Earthq. Spectra 29, 457–478. https://doi.org/10.1193/1.4000108.

Koshimura, S., Shuto, N., 2015. Response to the 2011 great East Japan earthquake and tsunami disaster. Phil. Trans. R. Soc. A 373. https://doi.org/10.1098/rsta.2014.0373, 37320140373.

Lauterjung, J., Letz, H. (Eds.), 2017. 10 Years Indonesian Tsunami Early Warning System: Experiences, Lessons Learned and Outlook. GFZ German Research Centre for Geosciences, Potsdam. https://doi.org/10.2312/GFZ.7.1.2017.001, 68p.

Milburn, H.B., Nakamura, A.I., Gonzalez, F.I., 1996. Real-time tsunami reporting from the deep ocean. In: OCEANS 96 MTS/IEEE Conference Proceedings. The Coastal Ocean - Prospects for the 21st Century, Fort Lauderdale, FL, USA, vol. 1, pp. 390–394. https://doi.org/10.1109/OCEANS.1996.572778.

Ohta, Y., Kobayashi, T., Tsushima, H., et al., 2012. Quasi real-time fault model estimation for near-field tsunami forecasting based on RTK-GPS analysis: application to the 2011 Tohoku-Oki earthquake (Mw 9.0). J. Geophys. Res. 117, B02311. https://doi.org/10.1029/2011JB008750.

Rose, A., Wing, I.S., Wei, D., Wein, A., 2016. Economic impacts of a California tsunami. Nat. Hazards Rev. 17 (2). https://doi.org/10.1061/(ASCE)NH.1527-6996.0000212.

Shuto, N., Fujima, K., 2009. A short history of tsunami research and countermeasures in Japan. Proc. Jpn. Acad. Ser. B 85. https://doi.org/10.2183/pjab.85.267.

Titov, V.V., Moore, C., Greenslade, D.J.M., Pattiaratchi, C., Badal, R., Synolakis, C.E., Kânoğlu, U., 2011. A new tool for inundation modeling: community modeling interface for tsunamis (ComMIT). Pure Appl. Geophys. 168 (11), 2121–2131. https://doi.org/10.1007/s00024-011-0292-4.

Tsushima, H., Hirata, K., Hayashi, Y., et al., 2011. Near-field tsunami forecasting using offshore tsunami data from the 2011 off the Pacific coast of Tohoku earthquake. Earth Planets Space 63, 56. https://doi.org/10.5047/eps.2011.06.052.

World Economic Forum, 2019. The Global Risks Report 2019. https://www.weforum.org/reports/the-global-risks-report-2019.

CHAPTER 17

Rising sea level

John Englander
Rising Seas Institute, The Explorers Club, Boca Raton, FL, United States

With the 1°C of ocean warming already observed and the current projections for 1−3°C additional, it is now urgent that we consider the effect of sea level being meters higher than it has been during all of human civilization. The geologic record shows that as recently as 11,000 years ago, sea level was rising more than 4 m in a century, a rate that would challenge almost every coastal area (Lambeck et al., 2014). From the perspective of the new blue economy, it is essential to understand the forces and realistic ranges of sea-level change in the coming century, both globally and locally. This chapter will cover rising sea level from four aspects:

- Historic sea-level change, for context.
- Primary factors for sea-level change, globally and locally, including potential abrupt change as well as the related phenomena for short-duration flooding.
- Realistic scenarios for future sea level rise (SLR) on the timescale of both the next three decades (mid-century) and 100 years as useful frames of reference relevant to the typical design life for assets and infrastructure.
- Perspective for the economic impacts, opportunities, and risks that may be affected by realistic scenarios for SLR in the new blue economy.

Historic ranges of sea level

Of course, the height of the ocean determines the location of the global coastline. Since both have been relatively stable throughout all of human civilization, that fact has seemed almost theoretical. Our common perception is that the current sea level and shoreline are "normal." Sea level has hardly changed during the last 2000 years, the common era of human history. Over long periods of Earth history, however, sea level has changed greatly, even without human influence. Ordinarily our expectations of the future are shaped by the current trends or the historical record, perhaps looking back several decades or even a century. Unfortunately, both current trends and the recent past fail to prepare us for the profound change posed by rising sea level in the decades ahead.

Preparing a Workforce for the New Blue Economy
ISBN 978-0-12-821431-2
https://doi.org/10.1016/B978-0-12-821431-2.00031-7

In all recorded human history, roughly 8000 years, sea level and shoreline have changed little. Looking back 20,000 years, however, the geologic record is clear. Global mean sea level (GMSL) was 120 m below present and at times changed abruptly (Lambeck et al., 2014). To help visualize that vertical scale 120 m is roughly equivalent to 30 floors of a commercial building. As sea level rises, coastlines move inland. Anyone looking ahead to the new blue economy of the coming decades will want to understand the nonlinear and potentially abrupt future changes in global sea level.

For the last 6000 years or so, essentially the span of human civilization, sea level has been in a plateau phase, with relatively little change. Prior to that, coming out of the last "ice age," sea level rose at an average rate of a meter a century. That rate would be extremely disruptive in our modern world of heavily developed coastlines. Far more extreme, in one 500-year period, some 14,000 years ago, referred to as Meltwater Pulse 1A, sea level rose at rates up to 4.7 m per century (Lambeck et al., 2014).

Putting SLR into the context of the ice age cycles enables us to see the larger pattern and direction of sea-level change. The 400,000-year historical graph from my book, *High Tide on Main Street*, shown further, provides perspective (Englander, 2013). Four ice-age warming and cooling cycles are shown. In common usage, "ice ages" refers to the glacial maxima when ice sheets were at their greatest volume. The most recent occurred 22,000 years ago. This article will use that common terminology, but should note that in geologic terms, we are still technically in an ice age, since there is multiyear ice at the poles.

During the last warm period, 122,000 years ago, referred to as the Eemian, the high-water mark was 7 m (more than 20 feet) above current sea levels (Hoffman et al., 2017). That was the *interglacial*, the previous period of warm temperature between the "ice ages," with reduced ice sheets and higher sea level, obviously a natural cycle. The chart just below shows 400,000 years of sea level in blue, global average temperature in red, and global average carbon dioxide (CO_2) in green. The data are a composite from several sources assembled by Drs. James E. Hansen and Makiko Sato, of Columbia University's Earth Institute. (https://data.giss.nasa.gov/gistemp/ and https://www.sciencedirect.com/science/article/abs/pii/S0012821X02009500)

This chart illustrates several important points:

- Over the long term, atmospheric carbon dioxide, global average temperature, and sea level move in tight synchronization, with their peaks lining up.

- The recent ice age cycles have been a recurring natural cycle, roughly every 100,000 years. The pattern follows the Milankovitch Cycles, largely determined by the changing elliptical orbit of the Earth around the Sun and the inclination, which varies the amount of heat energy we receive. (Muller and MacDonald, 1997) Though four cycles are shown here to provide some detail, the general pattern of cycles approximately every 100,000 years has existed for the last two and a half million years of the Pleistocene.
- The 100,000 year cycle can be broken up into roughly 20,000 years of warming and rising seas and 80,000 years of cooling and falling sea level.
- During the last two million years, there has been an upper and lower limit to global sea level, temperature, and CO_2. Sea level cycles up and down approximately 120 m, global average temperature rises and falls 5°C, and CO_2 ranges between 180 and 280 parts per million (PPM).
- Now, CO_2 has spiked upward, appearing as a vertical line on the chart, passing the 400 PPM level. This level has not been reached for millions of years. The rapid increase in CO_2 correlates closely with the burning of fossil fuels. In 2019, we passed a new record with CO_2 at 415 PPM, 40% above the upper bound of 280 PPM that has prevailed for millions of years (For the current level, see: https://www.esrl.noaa.gov/gmd/ccgg/trends/).
- There is a lag time for temperature to rise with carbon dioxide and a further lag time for the ice to melt, causing sea level to rise. Yet we can already see the evidence of warming temperature and rising sea level, as the three lines proceed to maintain their long-term synchronization. Lag times will be covered further in my third section about realistic future scenarios.
- Dividing the range of sea-level change by the range of temperature change, you can easily estimate the ratio of SLR to long-term global temperature change. It works out to approximately 20 m per degree Celsius or about 35 feet per degree Fahrenheit. The question of how quickly the ice can melt is uncertain. Regardless, the natural correlation of the three parameters makes clear that sea level will be significantly higher, even if we ceased all CO_2 emissions immediately.

In accordance with the *natural* cycle of the ice ages, we should be in the early stage of a period of 80,000 years of cooling temperature and falling sea level with the ice sheets slowly increasing in size. In fact, the reason that sea level seemed rather stable for the last 5000 years or so is that it was at the turning point following the *upward* phase and prior to the *downward* phase. Looking at the pattern, we can see that the current warming is a break from the natural patterns of the past.

Even at the upper limit of the roughly 120 m of sea-level variation during the ice ages (Pleistocene), there remains enough ice to raise global sea level a further 65 m if it were all to melt (Allison et al., 2009). For the decades and century ahead, anticipating the realistic range of future SLR will be important for many aspects of the new blue economy.

Primary factors for SLR, plus short-duration flooding

Changes in sea level, both global and regional, are surprisingly complex (Church et al., 2010). Regardless, in a given location, the big changes of multiple meters in base ocean height are primarily determined by the amount of ice that exists on land in the form of ice sheets and glaciers. This varies over centuries and millennia depending on global temperatures. Thus, changing sea level is largely caused by long-term changes to global average temperature, commonly referred to as climate change.

Over the last century, the thermal expansion of seawater has been a significant contributor to global sea-level change, accounting for about half of the 18–20 cm (7–8 inches) of rise (USGCRP, 2017). An article by Widlansky et al. (2020)"Increase in sea level variability with ocean warming associated with the nonlinear thermal expansion of seawater" points to increasing effects from thermal expansion. Nonetheless, as this century progresses, the amount of sea-level increase from melting ice on land will very likely be far greater, making thermal expansion a relatively lesser issue (Hall et al., 2019).

Just two places, Antarctica and Greenland, hold 98% of potential sea-level rise. If they were to melt completely, global average sea level would be more than 60 m (~200 feet) higher. To be clear, that is not even considered remotely possible in the next few centuries. That leaves the important question, however, of just how much higher it could get. Mostly that depends on how warm the world gets. At present, no one can know for certain since the amount of energy that we will produce and the means of production (e.g., fossil fuel, renewable energy, nuclear) are subject to many future policies and variables—the "unknown knowns."

Higher sea level is often confused with beach erosion or flood events that may last for a few hours or days. Storms, heavy rain, and downhill runoff can be devastating and deadly, but those localized or regional floods are temporary. As the water recedes, it is generally possible to recover and rebuild. Rising sea level also raises the height and extent of flooding during these short-duration floods in coastal areas.

Extreme high tides are also associated with higher sea level. Of course, tides are caused by the gravitational pull of the sun and the moon and are quite predictable. Tide heights vary daily, with extra high tides at the full moon each month and extreme high tides, "spring tides" (or perigean tides) now colloquially called "king tides." They follow an 18.6-year lunar orbital period, sometimes described as a 19-year tidal epoch. As global sea level rises, it is often noticed first during the extreme high tides.

In addition to the impact of global sea-level change, there are regional and local variations in sea level. Perhaps most important is what is described as vertical land movement. In many places the land is moving down or up, usually referred to as subsidence or uplift. One adds to the effect of SLR, the other subtracts. Presently, the most extreme case of subsidence can be seen in Jakarta, Indonesia. https://www.bbc.com/news/world-asia-44636934. They have had more than 3 m (~10 feet) of subsidence in the last few decades. There, the cause is compacting of the sediments coming out of a dozen or so rivers, plus the effect of pumping groundwater. Partly due to their unsolvable vulnerability to rising seas and flooding compounded by sinking land, in May 2019, the Prime Minister announced plans to relocate their capital from Jakarta. Similar problems have plagued New Orleans, Tokyo, and Venice, Italy.

Realistic scenarios for SLR: decades and centuries

Though many people believe and remain hopeful that significant SLR can still be avoided through rapid reduction of greenhouse gases, or other technologic miracles, there is nothing realistic to support that salvation. With the excess heat already stored in the ocean, basic physics makes rather clear that there will be less ice on the planet, resulting in higher global sea level, following the correlation in Fig. 17.1. The most recent literature dealing with sophisticated thermal analysis affirms that more than 90% of the energy trapped in the atmosphere by greenhouse gases is stored in the sea (Wu et al., 2020).

A July 2020 article "Delayed emergence of a global temperature response after emission mitigation" (Samset et al., 2020) makes clear that even the most aggressive plans to reduce greenhouse gases will take many decades to have any noticeable effect and affirmed studies such as Marotzke (2018) that

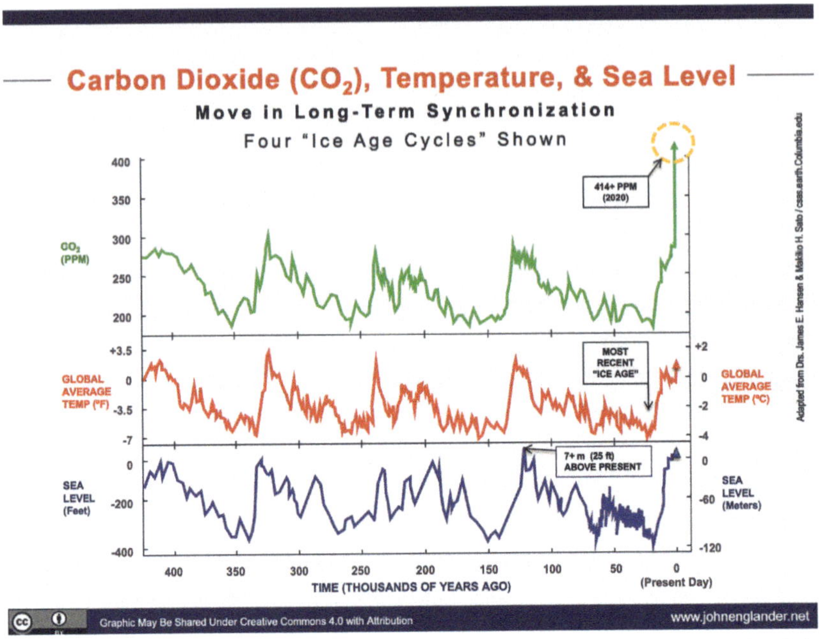

Figure 17.1 Four hundred thousand year record of carbon dioxide (CO_2), global average temperature, and sea level. *(Graphic by John Englander; Data from Hansen & Sato, www.https://csas.earth.columbia.edu/.)*

warming will continue for decades, even with the most aggressive proposals being advanced. Another 2020 article "The hysteresis of the Antarctic Ice Sheet" (Garbe et al., 2020) finds that the rate of SLR associated with warming global temperature increases sharply. Their work says that we should expect 1.3 m of SLR per degree of warming, for the first 2° of warming. That includes the degree we have already taken onboard. The rate almost doubles to 2.4 m per *degree*, for warming between 2 and 6°C. It increases to 10 m per degree, for the range 6–10°C of warming. They are not able to specify the number of decades for the adjustment to occur.

As can be seen in that chart earlier, the rate of sea-level rise has changed abruptly in the past. These abrupt nonlinear changes (called "meltwater pulses") can happen again. Until recently, 1 m of GMSL rise was generally thought to be the worst-case scenario this century. Partly that is due to a misreading of the eminent and authoritative Fifth Assessment Report of the Intergovernmental Panel on Climate Change (IPCC, 2014), which presented global mean sea-level projections in accordance with their four hypothetical scenarios, ranging from RCP 2.6 to 8.5. Few noticed an anomaly in summary table 13.5 therein and graphic 13.10. Even in the most extreme scenario, RCP 8.5, the highest increase in sea level by the end of this century was 0.98 m. Closer examination showed that the maximum contribution from Antarctica under that scenario was a mere 15 cm. To put that in perspective, RCP 8.5 stipulates that the planet will be 4.3°C above the baseline. That tiny glacial contribution from Antarctica at that temperature level is incongruous, if not ridiculous. That IPCC report does explain what is lacking from the sea-level projections in extensive footnotes, describing the "uncertainties." Unfortunately, nearly all readers simply read the topline figure of approximately a meter, not realizing that it effectively omitted Antarctic contribution. To be crystal clear, Antarctic holds some 86% of potential sea-level rise and is showing rapid signs of melting and destabilization. To be fair to the IPCC, they have to satisfy strict protocols that are difficult to change in order to preserve consistency with the previous three decades of their important work. In my book, *Moving to Higher Ground*, I use the term "Antarctic asterisk" to describe this burying of the big potential contribution from Antarctica level rise, essentially as a footnote (Englander, 2021).

The problem is that given the unprecedented increase in ocean heat, the thickness of the ice sheets, and the recent observations of ice sheet and glacial melting in the Polar Regions, it is not possible to predict precisely the future rate of SLR.(Hansen et al., 2016) While the models are getting more sophisticated, they are nowhere near the sophistication and accuracy of forecasting for routine weather and major events such as cyclones. Although we can't accurately predict what the future rate of sea-level rise will be, we do know the rate of rise is accelerating (Bamber et al., 2019).

Over the last decade, other projections for SLR beyond the IPCC have been repeatedly increased. At present, roughly 2 m (~7 feet) is considered a reasonable upper boundary for SLR by the end of this century (Sriver et al., 2018). "Characterizing uncertain sea-level rise projections to support investment decisions" analyzed 16 projections for SLR this century in comparison to the IPCC reports. It showed a range of 2–3 m in sharp contrast to the IPCC's worst case of less than a meter. There is increasing recognition that the IPCC characterization of future SLR may well be understated. "Twenty-first century sea-level rise could exceed IPCC projections for strong warming futures" is but one example (Siegert et al., 2020). NOAA's technical report "Global and regional sea level rise scenarios for the United States." (Sweet and Kopp, 2017) presents an extreme

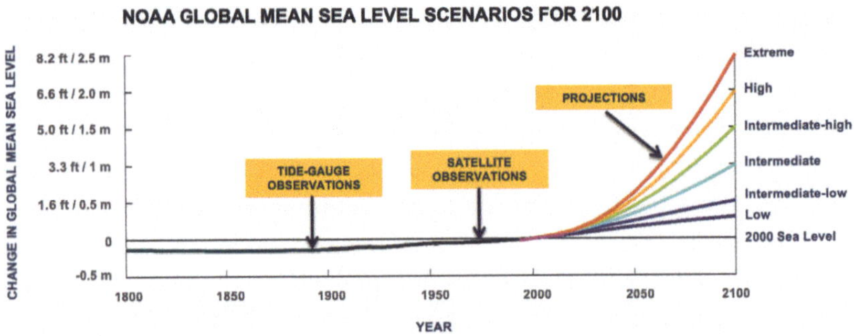

Figure 17.2 Six different scenarios or projections for SLR are depicted. Note their smoothness and parallel trajectory, which can be quite misleading. It illustrates the great diversity of SLR projections and the limitations to precise long-term forecasting SLR. Rising sea level will not likely follow any smooth trajectory, but rather will have some abrupt changes as has happened in the past. Melting and collapse of the great ice sheets on Antarctica and Greenland defies long term prediction. (Author's re-creation of Figure 6, in above cited NOAA publication (Sweet and Kopp, 2017).

scenario of 8 feet (~2.5 m) of possible SLR by 2100 as shown in Fig. 17.2 but notes that even that is not a true upper limit.

Considerable SLR of as much as a meter by mid-century and more than 2 m by end of century is now a realistic possibility. In the next century, rates could well accelerate further. It is a time of unprecedented change. All coastal areas will be affected. The blue economy will be insulated from some effects out on the blue ocean, but in the coastal, estuarine, and tidal riverine environments, rising sea level promises to be full of disruption—and opportunity. The following graph shows the change in global sea level since 1992 when satellites enabled far more precise measurements than the previous tide gauges allowed.

In Fig. 17.3, three segments of the overall slope are evident, with the slope of each shown in millimeters (mm) per year. Over the full data set of the 140 years, the rate of GMSL rose at the rate of 1.7 mm per year. Starting in 1993 with more precise satellite-based measurements, the rate has been 3.3 mm per year. In the most recent decade, the rate is 4.5 mm a year. The rate of rise is clearly increasing. The key question to answer is what the future rate of SLR will be. Unfortunately, it is not possible to forecast accurately how the ice sheets, miles thick on Greenland and Antarctica will collapse.

One of the problems is that scientific models for higher sea level are assumed to be the basis for safe engineering. Whether one is designing a

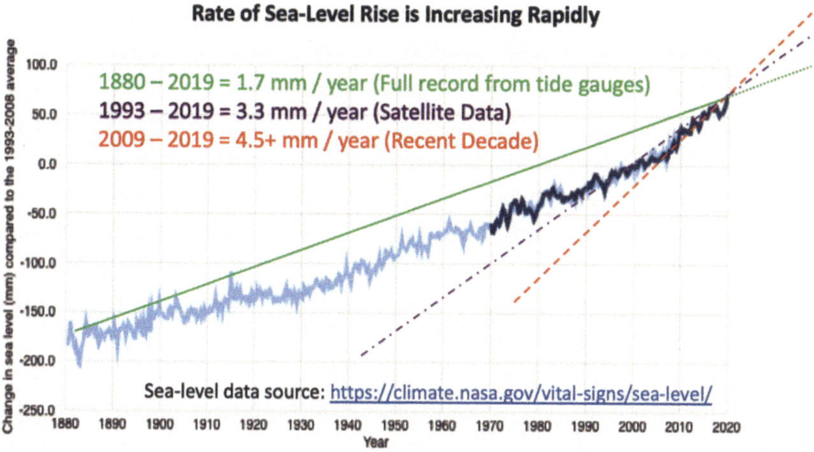

Figure 17.3 Precision sea-level records started in 1992 with satellite-based measurements. Comparison of the data trend line since 1880, to the full satellite record since 1992, and the most recent decade shows that the rate of rise is accelerating sharply. Author's trendlines on sea level overlay on data source: https://climate.nasa.gov/vital-signs/sea-level/.

bridge or an airplane wing, it is always understood that there needs to be a margin of safety to cover the underlying force or risk. Whether one looks at the IPCC's portrayal of 98 cm as the worst case or the NOAA projection of 2.5 m, those are heights of future water level. Designs, engineering, zoning, buildings, and infrastructure should be well above the estimated water level, so as to stay *high and dry* for the design life of the structure. In 2019, the UK-based Institution of Mechanical Engineers (IMechE) did a year-long review of the issue. Their report "Rising Seas: The Engineering Challenge" takes a far-sighted approach and serves as a good example. It endorses a "9 box matrix," to incorporate a margin of safety, looking at three different time horizons, 30, 50, and 100 years. The report is available for download, https://www.imeche.org/policy-and-press/reports/detail/rising-seas-the-engineering-challenge (In full disclosure, I was a subject matter expert for them throughout the process and developed the 9-box matrix).

SLR economic impacts, risks, and opportunities for new blue economy

In the 21st century, we are faced with an entirely new situation. Rising sea levels will change every marine and brackish coastline in the world, extending far up estuaries to cities not normally thought of as "on the ocean." Washington, D.C., London, Sacramento, and Hartford, Connecticut are just a few examples of places that will be changed as sea level rises to new heights. Rising sea levels will cause essentially permanent flooding of land and worsening short-term flood events in coastal areas all over the world. Even small amounts of higher base sea level can have severe effects. Just an additional few centimeters (an inch) can cause the water to overflow a building's lowest opening, a seawall, levee, or critical piece of infrastructure. A 1-foot vertical rise can move the shoreline hundreds of feet inland, depending on the steepness of the elevation rise.

More accurate, abundant, and accessible data and information about future water depth will be central to the development of the new blue economy. Failure to see what lies ahead could lead to expensive design and engineering errors and exposure. On the positive side, those that have the perspective to gather data about the changing baseline of sea level may have a strategic advantage. Realizing that SLR could be meters higher this century, with potentially abrupt change, can yield more resilient and adaptive designs. This is particularly important since facilities and infrastructure typically have a design life of 50 or 100 years and often function even longer.

Rising sea level will have tremendous impacts financially as vast areas of low-lying coastal real estate, infrastructure, and buildings become more vulnerable to routine flooding and eventual submergence. In turn, those assets secure all manner of financial instruments from mortgages, to bonds, publicly traded shares, and complex financial derivatives. There are two special aspects to this vulnerability. First, as the sea level gets higher, expanding short-term flood areas and pushing the shoreline closer to properties, their values will be impacted even *before* actual submergence. Second, compared to all other risks and devaluations, the write-downs of asset values for SLR are essentially permanent, never to recover. In economic terms, the depreciation and write-off of those vast low-lying coastal assets will have a tremendous negative impact as they go "underwater" both literally and figuratively. However, there will very likely be a positive economic impact this century from the adaptation to rising sea level.

The positive value from engineering, designing, and constructing the places that can operate safely in the new environment of higher sea level will have a much greater value than the write-off of those largely depreciated structures. Most of the current efforts are directed at solving modest amounts of flooding, but with little idea where the long term is headed.

That shortcoming presents opportunities for those that can "see over the horizon" (Englander, 2021).

Examples of relevant areas of work and study in the new blue economy related to SLR:

- Increased ship-based, ship-supported, and remote sensing observations of ice sheets and glaciers in the Polar Regions.
- A new look at data gathering and databasing recognizing the changing baseline of sea level, which, up until now, was assumed to be rather static.
- Refined bathymetry taking into account accelerating SLR, as well as factors such as the lunar nodal cycle, and local vertical land movement (subsidence/uplift).
- Advanced vulnerability assessments for coastal assets, infrastructure, communities, and supply chains as flood levels rise and coastlines change more than in all of recorded human civilization.
- New approaches for adaptive engineering and construction in low-lying coastal areas recognizing more severe flood events, in addition to higher base sea level. Assets on floats, even jack-up structures, will have more merit in a world of dynamic SLR.
- Refined modeling, but recognizing the limitations of models, due to unpredictable nonlinear and abrupt changes in the cryosphere.
- In addition to the physical impacts of rising base sea level, there are unprecedented ecological and ecosystem effects from changes to larval estuarine breeding grounds to coral reefs and algae blooms that warrant study and broader planning.

To conclude, slowing the rate of SLR is possible if we get very aggressive with all the efforts to reduce carbon dioxide and other greenhouse gases, possibly combined with efforts to remove the CO_2 already in the atmosphere and efforts to limit solar radiation by various schemes.

Yet, regardless of all those efforts, vast quantities of glaciers and ice sheets will continue to melt at somewhat unpredictable rates. Sooner or later, that will cause sea level to rise meters higher, something never experienced in human civilization. Like no other challenge in history, we now have the facts in front of us to realize the world is going to change. The sooner we accept that and begin to design a future world for generations to come, the better.

Looking "over the horizon" to see what is heading toward us makes it possible to design our coastal environment so that it is more adaptable and durable, likely yielding a better return on investment. For those designing, engineering, building, and financing in the coastal zone for the coming decades, it is essential to have an understanding of the profound change to sea level and shoreline that is just beginning. Rising and accelerating sea level will add complexity and challenge to the already dynamic maritime environment. Rising sea level will also create new opportunities. In fact, the necessity to adapt to unstoppable and accelerating SLR may become the greatest engine for economic growth this century.

References

Allison, I., et al., 2009. Ice sheet mass balance and sea level. Antarct. Sci. 21 (05), 413–426. http://journals.cambridge.org/abstract_S0954102009990137.

Bamber, J., et al., June 4, 2019. Ice sheet contributions to future sea-level rise from structured expert judgment. Proc. Natl. Acad. Sci. U. S. A. 116 (23), 11195–11200. https://doi.org/10.1073/pnas.1817205116. First published May 20, 2019.

Church, J., Woodworth, P., Aarup, T., Wilson, S., 2010. Understanding Sea-Level Rise and Variability. Wiley-Blackwell. https://doi.org/10.1002/9781444323276.

Englander, J., 2013. High Tide on Main Street: Rising Sea Level and the Coming Coastal Crisis, second ed. The Science Bookshelf, Boca Raton.

Englander, J., 2021. Moving to Higher Ground: Rising Sea Level and the Path Forward. The Science Bookshelf, Boca Raton.

Garbe, J., Torsten, A., Levermann, A., Donges, J., Winkelmann, R., 2020. The hysteresis of the Antarctic ice sheet. Nature 585, 538–544. https://doi.org/10.1038/s41586-020-2727-5.

Hall, J., Weaver, C., Obeysekera, J., Crowell, M., Horton, R., Kopp, E., 2019. J. Coast. Manag. 127–150. https://doi.org/10.1080/08920753.2019.1551012.

Hansen, J., Sato, M., et al., 2016. Ice melt, sea level rise and superstorms: evidence from paleoclimate data, climate modeling, and modern observations that 2°C global warming could be dangerous. Atmos. Chem. Phys. 16, 3761–3812. https://doi.org/10.5194/acp-16-3761-2016.

Hoffman, J.S., et al., January 20, 2017. Regional and global sea-surface temperatures during the last interglaciation. Science 276–279. https://science.sciencemag.org/content/355/6322/276.

IPCC, 2014. Climate Change, 2014: Synthesis Report. Contribution of Working Groups I, Fifth Assessment Report of the Intergovernmental Panel on Climate Change. IPCC, Geneva.

Lambeck, K., Rouby, H., Purcell, A., Sun, Y., Sambridge, M., October 2014. Sea level and ice volume since the glacial maximum. Proc. Natl. Acad. Sci. U. S. A. 111 (43), 15296–15303. https://doi.org/10.1073/pnas.1411762111.

Marotzke, J., 2018. Quantifying the Irreducible Uncertainty in Near-term Climate Projections. Wiley Interdisciplinary Reviews. https://doi.org/10.1002/wcc.563.

Muller, R., MacDonald, G., 1997. Glacial cycles and astronomical forcing. Science 215. https://doi.org/10.1126/science.277.5323.215.

Samset, B.H., Fuglestvedt, J., Lund, M., 2020. Delayed emergence of a global temperature response after emission mitigation. Nat. Commun. 11, 3261. https://doi.org/10.1038/s41467-020-17001-1.

Siegert, Martin, Alley, Richard, Rignot, Eric, Englander, John, Corell, Robert, et al., 2020. Twenty-first century sea-level rise could exceed IPCC projections for strong-warming futures. One Earth December 18, 691–703. https://doi.org/10.1016/j.oneear.2020.11.002. In press.

Sriver, R.L., et al., 2018. Characterizing Uncertain Sea-Level Rise Projections to Support Investment Decisions. https://doi.org/10.1371/journal.pone.0190641.
Sweet, W., Kopp, R., 2017. Global and Regional Sea Level Rise Scenarios for the United States. NOAA Technical Report NOS CO-OPS 083.
USGCRP, 2017. Chapter 12: Sea Level Rise. Climate Science Special Report. Retrieved 12-27-2018 from: science2017.globalchange.gov.
Widlansky, M.J., Long, X., Schloesser, F., 2020. Increase in sea level variability with ocean warming associated with the nonlinear thermal expansion of seawater. Commun. & Earth Environ. 1, 9. https://doi.org/10.1038/s43247-020-0008-8.
Wu, W., Zhan, Z., Pen, S., Ni, S., Caliles, J., September 18, 2020. Seismic ocean thermometry. Science 369 (6510), 1510–1515. https://doi.org/10.1126/science.abb9519.

Interview: Surfing

Ben Freeston

How relevant/important is the "data revolution" (i.e., "big data," "cloud computing," etc.) to Surfline?

I think in some senses we were ahead of the curve here, we've been using large data sets for predictive modeling long before the new machine learning paradigm evolved and as a web-based business working in the cloud in one form or another has been part of our DNA for the last 10–15 years. There's no doubt that being there that long and being disciplined in our data archiving has given us a significant advantage now the data revolution is in full swing and we see new data techniques as key to our business future.

What are the biggest challenges (technological, policy, logistical, etc.) to your business?

Historically a big challenge was marrying technology designed for research, academia, or enterprise use with consumer requirements. Machine learning has democratized the process of building models to predict for very specific outputs and is increasingly aiding our ability to deliver exactly what our customers want.

What are some examples of applications of technologies to current problems of the industry today and what we could do to entice the future workforce to be interested in solving these problems with the technology of today and tomorrow?

One significant challenge for our business has been getting decent ground truth data for the sort of predictive models we've wanted to build. We're solving that now by applying machine learning to camera video to create novel data sets. In the present that gives us a set of actionable real-time data for customers, but it also paves the way for new, detailed, long-term data sets that can be used to build better models in the future. Right now, data science is perhaps 80% the business of sourcing, cleaning, and preparing data with just 20% of time spent actually building predictive models from that. The hope for the next generation to enter the workforce is that foundation is then well laid, and they can work with large, well-organized data layers to make the next round of discoveries. That said the space has changed so fast in the last few years, it's almost impossible to predict how it'll develop moving forward in detail—that alone should attract inquisitive and intelligent people into this kind of role.

What are some examples of training/education that are the most effective for the future workforce for your business?

Strides in data science and machine learning are best made when domain knowledge is married with data science skill. On the domain knowledge side, we still rely on traditional institutions and processes, typically looking for graduates with ocean science degrees. On the data science side, the space is moving so fast that we're more interested in aptitude, typically demonstrated by a background in a math-heavy discipline coupled with a demonstrable ability to self-teach. Here we're increasingly using online learning platforms at a junior level and pulling insight straight from published research at a more senior level.

What levels of education/training are most needed in the workforce of the future (high school, community college, college, advanced degrees, etc.)?

We're open-minded—an advanced degree is certainly a marker of a certain kind of aptitude, but increasingly we see candidates using a self-teaching route to this level with online learning platforms, typically as the equivalent of a postgraduate qualification in a more traditional setting. In the fastest-moving data and engineering fields, this is sometimes the only way to be truly current, or at least will be an ongoing requirement for professional development.

Even outside those data science roles a comfort and familiarity with using data to drive decision-making are increasingly crucial in product management, marketing, and other business roles, the earlier this kind of competency is taught as part of the education process, the better for us as a company.

Ben Freeston biography

Ben is responsible for product development and innovation at Surfline.com. As the founder of Magicseaweed, Ben spent almost two decades playing at the intersection of data and product in surfing, becoming an evangelist for the power of new data science techniques to help deliver the next generation of surf zone modeling. He lives in California with his wife and three children and still tries to surf everyday, regardless of the conditions.

CHAPTER 18

The importance of seafloor mapping to the new blue economy: a private sector perspective

David Millar
Government Accounts Director, Americas at Fugro, Frederick, MD, United States

The new blue economy is knowledge-based, using scientific data and information about the sea not just for the benefit of material wealth, but also to identify and inspire solutions to environmental and societal challenges (Spinrad, this volume). As the world's leading Geo-data specialist with decades of experience working in ocean and coastal environments, Fugro is leading the offshore survey industry in support of this new business model. Through integrated data acquisition, analysis, and advice, we are collaborating with clients, technology partners, and science-based organizations to improve ocean knowledge for a safe and livable world.

You can't manage what you don't know

Each year the world ocean contributes trillions of dollars to the global economy, feeding billions of people and providing sources of recreation, maritime commerce, energy, and medicine to the planet. It also supplies more than half of the oxygen in the air we breathe, helps regulate climate, and influences the weather. At the same time, more than 600 million people live in coastal areas that are less than 10 m above sea level, meaning these communities are particularly vulnerable to coastal risks, amplified by a changing climate. The balance between a healthy ocean, a healthy population, and a strong economy is driving the push toward a new blue economy. But before a sustainable path can be forged, we must first fill known data gaps in ocean science and seafloor mapping (Fig. 18.1).

The value of a wholly mapped ocean

We tend to think of oceans as vast bodies of water that extend to unknown depths, providing transportation on the water surface and fisheries within

Preparing a Workforce for the New Blue Economy
ISBN 978-0-12-821431-2
https://doi.org/10.1016/B978-0-12-821431-2.00020-2

Figure 18.1 Turbines in an offshore wind farm that are increasingly providing energy to the planet and contributing to the New Blue Economy. *(Image courtesy of Fugro.)*

the water column. In reality, the oceans are floored with solid earth that is as dynamic and topographically diverse as the land masses that envelope them.

To date, only 20% of the global seafloor is mapped using direct, modern survey methods. Instead, the bathymetry that exists for most of the world's oceans is based on satellite altimetry, a technique that produces derived bathymetry of limited resolution and accuracy. These methods yield horizontal resolutions of up to 13,000 m and can have vertical errors up to ±1900 m. While suitable for studying regional geology and physiography, these data do not support studies needed to conduct ocean resource assessments, perform environmental studies, establish risk mitigation plans, or support a new blue economy (Fig. 18.2).

Without such basic knowledge as water depth, seabed characterization, and water column properties, we now find ourselves in a situation where much of the ocean is seriously degraded. We can and should do better.

Modern maps for a sustainable future

Except in shallow, clear, coastal waters, modern seafloor mapping is most efficiently accomplished using multibeam echosounder technology. Today's multibeam systems are operated from surface vessels or underwater vehicles, generating acoustic waves that are capable of penetrating kilometers of water column, meaning that highly accurate and resolute

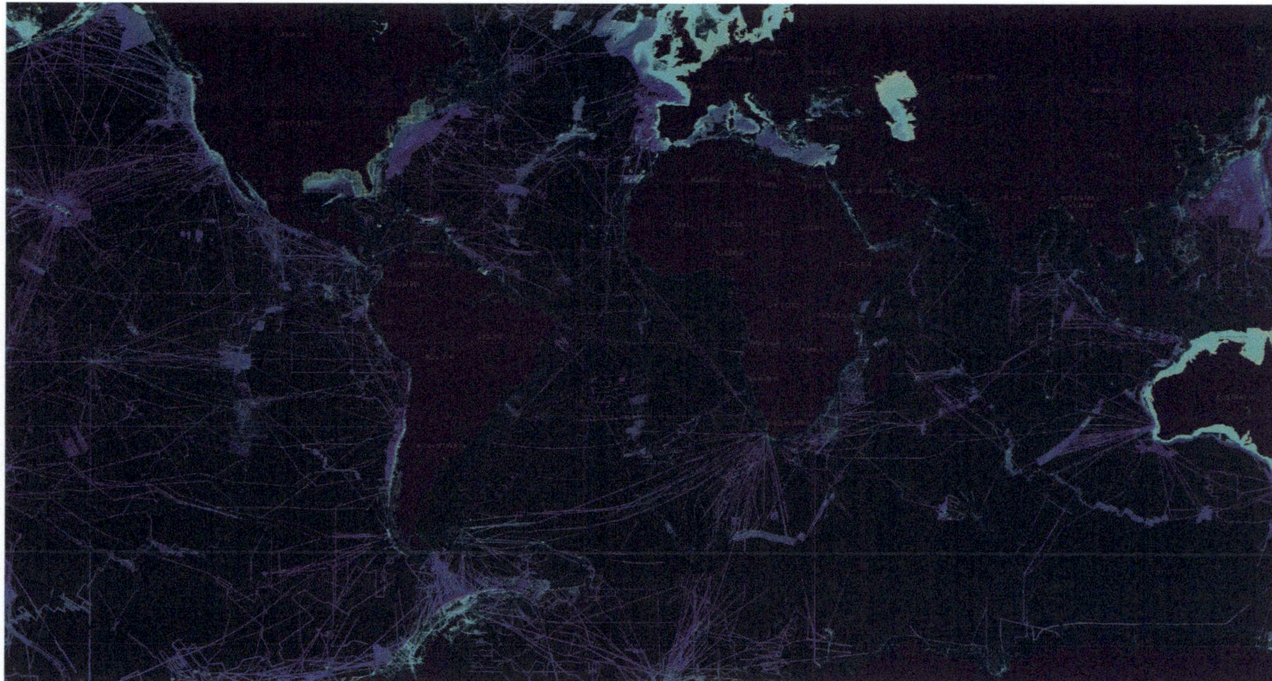

Figure 18.2 Map showing the limited extent of global bathymetry coverage from direct, modern survey methods (GEBCO 2020 grid and TID). *(Image courtesy of Vicki Ferrini - Seabed 2030.)*

mapping data can be obtained in all areas of the world oceans, including its deepest canyons. For example, where bathymetry derived from satellite altimetry may not have even detected a large seamount, data obtained from multibeam systems is capable of distinguishing shipwrecks, resources, and biology located on that previously unknown seamount.

Multibeam surveys provide additional benefits beyond bathymetry. By taking advantage of the acoustic waves' reflected energy, it is possible to interpret information about the physical characteristics of the seafloor, providing a first-order proxy of sediment type and seafloor roughness. Remote seafloor characterization capabilities through analysis of the angular dependence of the backscatter response are now used to understand the lithologic composition of seafloor sediment and to map the spatial distribution of seafloor habitats.

More recently, multibeam systems have been designed to record the time series for each beam providing backscatter amplitude measurements of the water column. This acoustic sensing of mid-water scattering properties serves multiple purposes. In addition to conducting fish stock and biomass assessments, this technique can also be applied to mapping suspended sediment from dredging operations and detecting mid-water bubble plumes from natural or human-made seeps. The latter capability has important implications for quantifying seafloor methane flux. Mid-water backscatter can also detect free gas release from the seafloor that expels clusters of oil and gas bubbles hundreds of meters into the water column. These data allow for efficient mapping and localization of seafloor hydrocarbon seeps, fish schools, underwater cultural heritage, algae and seagrass, and internal ocean waves.

This availability of water column and seafloor backscatter alongside traditional bathymetry represents a diversification of data that will help fuel the new blue economy. By increasing the number of user groups who can derive information and knowledge from a single survey mission, modern multibeam data sets are accommodating more efficient and sustainable development opportunities.

Map once, use many times

Given the expanded capabilities of multibeam sonars, multibeam surveys should serve multiple stakeholders and purposes. The same bathymetry data used to guide safe navigation can also help improve weather forecasts, predict climate change, model ocean garbage gyres, and protect coastal communities from flood, storm, tsunami, and marine geologic risks.

Bathymetry, along with backscatter, also supports responsible development and protection of marine resources. Recent offshore wind development activity along the US Atlantic Coast serves as a timely example. Fugro has been active in several lease development projects, performing multibeam surveys as a critical input for feasibility assessments, environmental impact studies, and regulatory compliance. As these projects progress, the same multibeam data used during these early studies can be applied to inform a wide range of design, engineering, and construction parameters.

Continual technology advancement

Seafloor mapping solutions and technologies are now advancing at an unprecedented pace, driven primarily by innovations in core technologies, such as remote operations, autonomous vehicles and robots, and analytics and cloud automation. While these core technologies have been used in seafloor mapping solutions for a while, the pace, breadth, and sophistication of their application are now very high and will ultimately provide safer, more efficient, more cost-effective, and more sustainable solutions. In addition, these innovations and technology developments are making higher volumes of higher-quality data more accessible, which again will help drive the new blue economy (Fig. 18.3).

Remote operations

Fugro has been using remote technologies in offshore operations for some time, but the relevance and value of these innovations were highlighted

Figure 18.3 Fugro's Blue Essence USV, which is being developed via a partnership between Fugro and SEA-KIT. *(Image courtesy of Fugro.)*

anew during the recent COVID-19 outbreak and associated lockdown. Advances in communications, telemetry, and data compression have all contributed to the development of remote command and control technology, which greatly reduces the number of field staff required on a project, instead allowing qualified staff to conduct their work from a shore-based office environment.

Fugro first introduced remote operations in 2013 in the United States to provide remote navigation and positioning in support of rig moves. Since then, the scope and scale of Fugro's remote services offering have grown appreciably, with remote operations centers now located around the globe supporting a wide range of projects. Fugro personnel are now providing positioning and navigation services, operating multibeam and other mapping sensors, processing mapping data, and even piloting remotely operated vehicles (ROVs) from our remote operations centers. With fewer field staff required to execute these projects, Fugro can acquire and deliver data faster, with improved operational safety, increased business continuity, and a significantly reduced carbon footprint. Remote operations also facilitate increased data availability and access. In theory, these innovations result in faster data delivery to the public, ultimately helping to drive the new blue economy.

Autonomous vehicle and robotic technologies

The innovation and development around remote operations have helped accelerate the innovation and development of autonomous and uncrewed platforms. These advancements are occurring in the water with autonomous underwater vehicles (AUVs), on the water with uncrewed surface vehicles (USVs), and in the air with uncrewed aerial vehicles (UAVs). While each platform supports different aspects of ocean mapping, they all share similar technology goals and challenges. In a mapping context, AUVs support subsea multibeam applications, USVs support surface multibeam applications, and UAVs support coastal lidar applications. Sensor miniaturization, power, endurance, and collision avoidance are important aspects to the development of all three of these platform types. Because AUVs typically operate in deeper water near the seafloor, they do not necessarily have the same traffic management, collision avoidance, and regulatory requirements as UAVs and USVs. In many respects, AUV technology is more mature and has been operating without human intervention for years. With respect to UAVs and USVs, however, the ultimate goal is a platform that can operate autonomously without human intervention. Until the collision

Figure 18.4 Fugro's Blue Shadow class of USVs have been designed for coastal hydrography. *(Image courtesy of Fugro.)*

avoidance technology matures and a regulatory regime is established, uncrewed operations will still require some level of supervision.

Fugro has been offering AUV-based high-resolution mapping solutions for nearly two decades. As mentioned earlier, this technology tends to be relatively mature compared to its USV and UAV counterparts, though there have still been significant recent advancements in terms of power, endurance, and sensor payload. When it comes to aerial mapping and remote sensing, the latest generation bathymetric lidar technology for shallow water coastal mapping applications is now UAV-capable with approximately 5 h of operational endurance. In areas where water clarity does not support bathymetric lidar, a new class of USVs is providing a cost-effective force multiplier capability to increase productivity. Lastly, the latest generation of "over the horizon" USVs is capable of full ocean depth mapping, while hosting an eROV or AUV for remote and uncrewed high-resolution mapping, exploration, and/or inspection applications (Fig. 18.4).

Whether by AUV, UAV, or USV platforms, all autonomous seafloor mapping solutions acquire data in a safer, more cost-effective, and more sustainable way. As a result, we will see more data acquisition platforms collecting a much higher volume of higher-quality data than has historically been possible, innovations that will again further spur the new blue economy (Fig. 18.5).

Artificial intelligence and cloud processing

Advances in artificial intelligence (AI) and cloud processing are also contributing to data acquisition and processing improvements. For example,

Figure 18.5 SEA-KIT USV preparing to receive a HUGIN ASV. *(Image courtesy of Fugro.)*

AI is critical to the route optimization, dynamic line planning, and collision avoidance of autonomous and uncrewed survey platforms. Both also allow data to be pulled from acquisition platforms, whether crewed or uncrewed, in near real time, for use by shore-based clients, processors, analysts, and consultants. Lastly, AI and cloud processing are now being used in cleaning, processing, and running analytics on ocean mapping data sets.

A changing workforce

The demographic of the ocean mapping community is diversifying and becoming more inclusive. This change is enabled not only by the above technology developments but also by the broader application of ocean mapping data to ocean science in general.

Moving offshore work onshore

Traditionally, ocean mapping data were acquired by specialists working on survey vessels and processed by different specialists back in shore-based offices and laboratories. Often, those working in the field also had to double as software programmers and/or electronic technicians to ensure that all systems and processes functioned properly during lengthy deployments. The offshore role was not particularly attractive to those needing or wanting to stay close to home to raise a family or develop other interests beyond the at-sea experience. Today, the ability to manage these operations remotely is making it possible to merge these two distinct roles into one, with specialists participating in all aspects of a project from the safety and comfort of a shore-based office.

Expanding beyond traditional skill sets

As seafloor mapping systems and processes become more sophisticated, we can expect less emphasis on data collection and data processing. Some expertise will still be required in these areas, of course, but where that is the case, ocean mappers with skills in autonomy, robotics, and AI will become increasingly important and relevant. So now, instead of creating expertise that combines ocean mapping with computer programming and/or electronics technology for shipboard positions, there will be increasing demand for expertise that combines ocean mapping with remote operations, autonomy, robotics, and AI for shore-based positions (see Van Sumeren, this volume).

Similarly, instead of creating expertise that is focused solely on processing data, there will be a demand for extracting knowledge of automatically processed data and applying it to other ocean science, social science, and economic science disciplines. Some experts will still be required to quality assurance and control (QA/QC) automated processes but, in general, the workforce will be able to migrate from processing data to applying data in meaningful ways. Further, because seafloor mapping is so important to many of society's challenges and opportunities, expertise that connects seafloor mapping with other skills and training will play a larger role in ocean science and policy.

Putting it all together: a global call to action

Technology is on our side as we prepare for the new blue economy, but time, unfortunately, is not. That's why Fugro is actively involved in two closely related initiatives: The Nippon Foundation-GEBCO Seabed 2030 Project and the United Nations Decade of Ocean Science for Sustainable Development (see Ryabinin, this volume and McLean, this volume). Both programs are working to close known gaps in ocean science data, including ocean mapping, and both require cross-sectoral, interdisciplinary collaboration and partnerships. Committed to the cause, Fugro has participated in the programs' planning stages, facilitated the contribution of existing data, contributed new "in-transit" data, and collaborated with clients and partners to advance new and innovative approaches to large-scale seafloor mapping that will help accelerate the pace and value of regional datasets.

As the technology developments described here are scaled up and provide greater access to a larger volumes of more diverse and higher-quality ocean mapping data, a more diverse workforce with new skills will translate this data into information and knowledge to help drive the new blue economy. The ocean is a shared resource, and it will take all of us working in the marine community to safeguard its future. Becoming part of the global push to improve ocean knowledge now, via seafloor mapping, will help ensure a sustainable ocean economy for decades to come.

SECTION IV

Mechanisms for growth

The New Blue Economy will grow organically through traditional entrepreneurship and with the advent of new mechanisms to facilitate networking, tradecraft, and collaborative development. Three different examples are presented here. A very new concept for public support of economic development is presented by **Irene Penesis and John Whittington. Michael Jones and Melissa Fischel** present the role of a BlueTech Cluster in Workforce Development and Economic Development originating in an historical stronghold for marine economy, Southern California. **Laurie Jugan and Anna Kate Baygents** describe the collaboration between state economic development activities and individual companies forming the Marine Industries Science and Technology Cluster in Mississippi. The Canadian Centre for Ocean Ventures and Entrepreneurship, COVE, serves an important incubation/acceleration function in Halifax and is presented by **Jim Hanlon and Sherry Scully**. Finally, two contributors specifically address issues in technology development that will serve the New Blue Economy. **Justin Manley's** interview provides a vision for future industry trends. **Peter Fietzek's** op-ed discusses the value of and potential for science and industry collaboration.

CHAPTER 19

Australia's Blue Economy Cooperative Research Centre

Irene Penesis[1,2], John Whittington[1]
[1]Blue Economy Cooperative Research Centre, Newnham, TAS, Australia; [2]Australian Maritime College, University of Tasmania, Launceston, TAS, Australia

Introduction

The Australian blue economy

The Australian blue economy is made up of well-established industries (e.g., marine & coastal tourism, oil & gas, fisheries & inshore aquaculture, shipping & ports, marine observations) and emerging industries (e.g., offshore renewable Energy, offshore aquaculture, ecosystem services, marine biotechnology). Australia's National Marine Science Plan (2015–2025) (National Marine Science Committee, 2015) predicts the value of Australia's blue economy to be AU $100 billion annually by 2025. The growth in the blue economy is necessarily reliant on the emergence of the new blue economy (NBE), as defined, and described in this book (see Spinrad and Rayner, this volume). Without the innovations resulting from the NBE, our traditional blue economy will not meet its predicted potential.

Australia's opportunities to grow the blue economy are vast. Australia's Exclusive Economic Zone (EEZ), the third largest in the world, is made up of 8.2 million square kilometers off Australia and its remote offshore territories and 2 million square kilometers off the Australian Antarctic Territory. The EEZ includes tropical, subtropical, temperate, and Antarctic environments capable of supporting diverse aquaculture while also being home to some of the world's best ocean energy resources.

Under Australia's federated system of government, State Governments have jurisdiction over inshore waters to 3 nautical miles, and the Australian Government has jurisdiction from State waters out to 200 nautical miles offshore, the limit of the EEZ. With multiple jurisdictions responsible for administration of Australia's oceans, there is a complex, and at times inconsistent, arrangement of planning and regulatory controls. In addition, there are vast areas of Australia's ocean territory without the appropriate regulatory frameworks to support the development of emerging blue economy industries.

Preparing a Workforce for the New Blue Economy
ISBN 978-0-12-821431-2
https://doi.org/10.1016/B978-0-12-821431-2.00043-3

The focus of this paper is to outline the opportunities that the NBE can make possible in Australia's emerging blue economy industries; offshore aquaculture and offshore renewable energy, and the benefits through integration, and how industry, government, and academia have partnered to form the Blue Economy Cooperative Research Center (CRC) to realize these opportunities.

The NBE is implicit in the research and training focus of the Blue Economy CRC. Our vision of seeing aquaculture and renewable energy moving further offshore is predicated on the NBE. That is, the development of new technologies—autonomous systems, sensors, numerical models, advances in cloud computing—will unlock our ability to measure, model, and predict our oceans, fish stocks, and structures and generate the data needed to inform future application, decision-making, and addresses challenges in real time. Hence, the development of the NBE (and application of the knowledge-based products generated) will allow traditional blue economy industries to transition into new environments and sustainably grow in ways that are not currently possible.

Offshore aquaculture

Australia has a global reputation for quality, safe, and sustainable seafood and management of its marine natural resources and biodiversity. Australia's fisheries though are relatively small by world standards with Australia importing 72% of its seafood. While the majority of Australia's wild fish stocks are classified as sustainable (Stewardson et al., 2018), there is limited capacity to sustainably increase wild capture seafood, and any significant growth in sustainable seafood production will need to come from aquaculture.

Australia has an existing mature inshore aquaculture industry based on salmonids, mussels, oysters, pearl, tuna, abalone, and prawns servicing domestic and international consumers. Aquaculture produces 36% of the total volume of Australian fisheries production (ABARES, 2018). This contrasts with global marine aquaculture, which currently produces 16% of global marine fisheries and aquaculture production (Costello et al., 2020).

Australian aquaculture's gross value of production of AU $1.42 billion (ABARES, 2018) is dominated by Tasmania's salmonid industry. With a gross value of AU $838 million in 2017–18 (DPIPWE, 2018), it has doubled in value in the last decade (ABARES, 2018). Australia's National Aquaculture Strategy (DAWR, 2017) has a target to increase gross value of aquaculture production to AU $2 billion by 2027, and Tasmania has similar target to double the value of salmonid production by 2030. Major expansion in salmonid aquaculture is currently constrained by the limited availability of nearshore production sites and the increasingly strong competition with other sectors for access to inshore waters including from wild fishers, navigation, conservation, and recreational users.

In contrast to the Australian salmonid industry, which is well established and globally competitive, the Australian seaweed industry is very small, with production valued at AU $3 million, with the majority of this value from collection and processing of storm cast kelp in Tasmania. AgriFutures Australia (2020) recently released an Australian Seaweed Industry Blueprint that outlines plans for AU $1.5 billion Australian seaweed industry that could employ 9000 people; however, Australia currently has no commercial-scale seaweed ocean farms, and growth will rely significantly on reformed policy and regulation to allow for ocean cultivation of native seaweeds in offshore zones, and the formation of a dedicated research and development plan. Integrated multi-trophic aquaculture (IMTA) systems could combine the culture of finfish with seaweeds and mussels that can extract organic and inorganic nutrients coming from the fish (Chopin, 2006). This presents the opportunity to maximize the economic benefits for the producers by diversifying the products, and it has the potential to improve social acceptance of aquaculture (Hughes et al., 2016; Chopin et al., 2008).

Meeting the ambitious goals for increasing aquaculture production, whether finfish, seaweed, or bivalves, will require data to inform marine spatial planning, new offshore farming systems, and new regulatory systems to manage them. While offshore aquaculture is increasingly recognized as the solution, there remains a lack of knowledge and technology for operating in this new environment, and the data required to inform technological solutions, operations and production constraints, economic assessments, and fit-for-purpose policy and regulation, to support industry expansion into offshore waters.

The performance of existing inshore fish pens and their suitability for offshore/exposed environments will need to be understood. Environmental data and new sensors will be required to monitor the behavior of these floating structures to identify the requirements for improved application for different operating envelopes and failure modes. Salmon production biology challenges in offshore environments will also need to be understood such as climate change impacts (e.g., thermal, acidity and dissolved oxygen stressors); increased energy expenditure in high energy wave and current environments; effective feeding in complex water systems; assessing and managing general animal health and well-being. Hence, innovation in new technologies such as autonomous technologies (drones, ASVs, AUVs), sensors, acoustics, and eDNA will provide the necessary observational data to inform fish stock and growth assessments, feeding, water quality, temperature, oxygen, predator interactions, and harmful algal blooms.

Offshore renewable energy

Australia has some of the world's best offshore wind, wave, and tidal resources, which have enormous potential to deliver energy to support offshore operations and exportable renewable energy such as hydrogen. Adjacent to its coasts, Australia (and particularly Tasmania) possesses immense high-quality, temporally consistent offshore wind resources accessible to existing offshore technologies, with a generation potential of approximately 25 PWh/yr that would be equivalent to that of Europe, United States, Japan, and United Kingdom combined (Eurek et al., 2017). In March 2019, Australia approved its first ever offshore wind project, "The Star of the South," which will also be Australia's largest electricity project with a reportedly AU $8bn (US $5.4bn) investment. This wind farm proposed by Copenhagen Infrastructure Partners (CIP) off the southern coast of Gippsland in Victoria hosts a 2.2 GW of offshore wind capacity (Buckley and Shah, 2019). The southern coastline of Australia also has a great wave resource. Estimates suggest that wave energy could contribute up to 11% of Australia's energy (enough to power a city the size of Melbourne) by 2050, making it a strong contender in Australia's renewable energy mix (Hemer et al., 2016). Preliminary assessments of Australia's tidal resource show that energy availability is less widespread than Australia's wave resource, but several localized regions exhibit promising resource potential; however, the magnitude of this resource is still being quantified (Hemer et al., 2019; Penesis et al., 2020).

The main challenge for the marine renewable energy sector is the development of devices for offshore conditions that demonstrate commercial competitiveness and can achieve meaningful scale of deployment. In addition, the regulatory frameworks in Australia to facilitate renewable energy projects offshore are immature. Enabling our transition to offshore renewable energy generation will require improvement in technology for measuring, modeling, and predicting the ocean (see Spinrad, this volume). Tailored forecasting and predicting of waves, currents, wind, and interactions will inform device design and infrastructure resilience, energy generation and power take-off optimisation, support dynamic software models to accurately determine time-varied inputs/outputs from various renewable devices and support microgrid management, and in particular address challenges in real time during extreme events or when environmental (e.g., marine mammal and bird) interactions occur.

In the context of marine energy converters and microgrid architectures, hydrogen is expected to play a critical role offshore as it is the only foreseeably commercially viable option for long-term management (storing, transforming, distributing) of variable electricity in the offshore marine environment, where it is available in excess of either onsite operational requirements or export to the grid. As fuel hydrogen can be used for powering offshore vessels and other equipment, while offering a new high-value, high-volume market for marine device developers via exportable renewable energy and chemicals, with no port constraints. Hydrogen is gaining a reputation as a flexible, safe, transportable, and storable fuel. Overall, hydrogen is a key "energy carrier" that can enable offshore renewables to compete commercially. It will also enable access for aquaculture to low-cost electricity and energy storage offshore, reducing exposure to diesel prices while decarbonizing the overall supply chain and delivering synergy with oxygen, ozone, and freshwater supply.

Developing the emerging offshore renewable energy industry in Australia offers opportunities to contribute to carbon mitigation measures for other offshore industries that are currently heavily reliant on fossil fuels to power their operations. The Tasmanian salmon industry is an example of an offshore industry seeking to decarbonize operations. The energy-intensive resource (power, freshwater, fuel, and potentially heat and oxygen) demands of the aquaculture industry span a range of operations, including, for example, lighting, pumps, feeding, aeration, desalination, cleaning, refrigeration, monitoring and support (barge) operations, among others. Different aquaculture operators will have alternative optimal power solutions to meet their requirements. Time-series demand profiles will be required for major elements of typical offshore/high energy aquaculture facilities. Total annual fossil fuel-based energy use across the Tasmanian salmon industry is estimated to be 125 GWh per annum to produce 63,000 Head-On-Gutted (HOG) tonne of salmon annually. Moving further offshore will significantly increase aquaculture's diesel energy demands given the need to sever the current ties to land-based power and the increased offshore transport requirements. A shift to renewable energy powered, off-grid energy solutions for the offshore aquaculture sector has the benefit of displacing CO_2 emissions (anticipated to be more than 70 kT

CO_2-e), reducing noise and diesel costs as well as the risk of diesel transport and contamination in more exposed operating conditions.

Opportunities through integration

To date, in Australia, the aquaculture and renewable energy sectors have operated independently of each other. There has been limited appreciation of the cobenefits that the different sectors could derive from integration and limited understanding of the accelerated development that collaborations across the sectors could bring to make offshore developments a commercial reality. While moving offshore offers a substantial increase in available sites and a potential step change in the scale of production, the cost of offshore operations is much larger. Offshore engineering will be central to this emergence, leveraging experience in the design, installation, and operation of maritime using infrastructure in very exposed, offshore environments safely.

Research and development projects in Europe have demonstrated that sharing costs through innovative and multiple use of offshore platforms can lead to improved economic outcomes for multiple sectors. Through collaboration, these sectors can work together to address the challenges of offshore operations and explore colocation benefits including shared operations and maintenance (Buck et al., 2010), mooring systems (Connolly and Hall, 2019) construction costs, resources and logistics (Dalton et al., 2019), leading to increased production, spatial efficiency, and hence value generation (Griffin et al., 2015). There has been a growth in international projects promoting the integration of several sectors and to move offshore and the benefits of colocation (Campbell, 2016; Aquatera Ltd., 2014; MARIBE), which are exploring the development of modular multiuse floating offshore platforms, which would be suitable for transport, energy, aquaculture, and leisure (Abhinav et al., 2020). The TROPOS concept integrates offshore culture of finfish and microalgae with offshore wind and solar energy generation; however, this concept has not been tested yet in a commercial setting (Papandroulakis et al., 2017). There are some initiatives growing seaweeds in the vicinity of offshore wind turbines, but the energy produced has not been used to support the culture activity itself (Buck and Langan, 2017).

Making offshore/high-energy aquaculture farms and renewable energy systems safe to operate and financially viable means a step change in the adoption of NBE technologies. For example, traditional aquaculture farms are labor-intensive environments with staff requiring transport, accommodation, protection from the elements, and a safe working environment. Recent developments in remote feeding, supported by machine learning, are changing the way farms operate and making operation in offshore environments safer and more efficient.

Operational decisions for offshore aquaculture and energy generation require accurate and reliable knowledge of current conditions and the ability to model and predict system changes in real (or near-real) time. There is an extensive array of model analysis tools and forecasting systems (e.g., provided through Australia's CSIRO and Bureau of Meteorology), advances in telemetered 4G instrumentation, remote sensing technology, and emerging artificial intelligence (AI) methods with the potential to provide useful input to site selection, design optimization, feed optimization, environmental prediction, and tactical operational decisions and management strategies that can significantly reduce operational risks or comprise operations in offshore environments.

The Blue Economy CRC is working with the industry to develop and integrate this information into a knowledge platform to facilitate Australia's traditional blue economy industries move offshore to meet the growing demand for sustainable seafood and other marine products at a low carbon energy cost, through innovative offshore facilities such as multiuse platforms and colocation.

An example of how these synergies through integration could emerge is shown in Fig. 19.1 and includes:

- The repurposing of existing offshore floating structures can provide a central "hub" for offshore operations, such as this oil and gas accommodation barge.
- Technical innovation in new designs, materials, and structures will help to prevent damage to plant and equipment, such as this wave attenuation barrier that can dampen incident wave energy.
- Process automation using robotics will remove people out of harm's way by undertaking operational, monitoring, and maintenance tasks.
- Next-generation ocean remote sensing will provide in situ ocean observing data, support ocean modeling and forecasting, and provide real-time monitoring data to inform operations.
- Intensive finfish production in specially adapted offshore pens and production of invertebrates such as abalone, lobster, and oysters raised in floating artificial reefs may also feature, with large-scale kelp cultivation used to bioremediate point source discharge of nutrients, capture carbon, and also to help dampen wave action.

Figure 19.1 An example of the integration of the aquaculture, offshore renewable energy, and maritime engineering sectors to deliver innovation in sustainable offshore aquaculture and renewable energy systems. *(Image provided by Blue Economy CRC.)*

- Supply chain logistics change quite dramatically far offshore, presenting opportunities to explore "vertical integration" at sea to reduce operating costs—bringing hatchery, juvenile rearing, seafood processing, and even cold storage to an offshore base to reduce cost and complexity.
- Capturing renewable energy offshore through floating wind turbines, tidal, or wave energy converters means that there will be immediate savings in diesel generation, but these energy sources can also be used to power vessels, drones, cold storage, lighting, and desalination or oxygenation.
- Energy, delivered via a microgrid, can be stored in various forms for short- or long-term use or even exported.

The Blue Economy CRC

The Australian Government's CRC Program supports industry-led collaborations between industry, researchers, and the community. It is a proven model for linking researchers with industry to focus on research and development that is geared toward end use and commercialization.

Established in 2019, the AU $329 million Blue Economy CRC over a 10-year life brings together 40 industry, government, and research partners from nine countries with expertise in maritime engineering, aquaculture, renewable energy, environmental management, and policy and planning.

Moving colocated seafood and energy production offshore into new and energetic operating environments requires production systems that can withstand both regular and extreme weather events, while being safely and economically managed. These will be capital-intensive systems with NBE technologies that are evolving at a rapid pace. The integration of the offshore aquaculture, renewable energy, and maritime engineering research sectors into one center is deliberate; they are naturally synergistic and provide the opportunity to share critical knowledge and infrastructure, delivering productivity and operational cost benefits. Without the Blue Economy CRC, the participating industry partners could not assemble the critical mass of knowledge, resources, and infrastructure required to address the research and development challenge of working offshore alone and to harness the opportunities provided by the NBE.

These emerging offshore industries require a policy and regulatory environment that provides the confidence to make long-term investments, society will require confidence that they will be demonstrably environmentally sustainable and socially responsible, and existing users of the ocean will expect to continue to enjoy their current access and opportunities. The Blue Economy CRC is working with Governments to develop contemporary planning, management, and monitoring frameworks that provide confidence to society while facilitating investment in a sustainable blue economy.

The Blue Economy CRC's research program has been developed through an iterative dialogue with industry and government to ensure it is focused on the challenges faced by industries in achieving scale and commercial success in the offshore environment. Key activities are focused around five integrated user-defined research programs:

- The Offshore Engineering and Technology program will generate the infrastructure that supports the development of offshore systems. It brings together industrial engineering expertise to collaborate with the aquaculture and offshore renewable energy sectors to build the required infrastructure for integrated offshore operations. Intellectual Property will emerge in the design of sea-pen infrastructure, support systems for operating (e.g., anchoring devices), innovative maintenance technologies (e.g., anticorrosive or antifouling devices), and monitoring (e.g., advanced materials for longevity and structural reliability; in-built sensors in composite materials to detect fatigue in offshore platforms). Commercial prototypes will be developed for monitoring and maintenance using robotics, artificial intelligence, integrated sensors, and real-time visualisation.

- The Seafood and Marine Products program will develop offshore aquaculture systems that provide viable and sustainable growth opportunities for this sector. Commercialization opportunities include novel aquaculture system designs for emerging species in collaboration with offshore engineering program, and new seafood products, as well as the development of supply chain aquaculture activities (e.g., platform-based hatcheries and processing). Identification and development of premium export products and new export markets will ensure the expectations of high-end users are met.
- The Offshore Renewable Energy Systems program will support offshore aquaculture through supplies of lower-cost energy and ancillary products (oxygen and freshwater) and to contribute to the cost of offshore infrastructure through the development of exportable energy carriers (e.g., hydrogen). Commercialization opportunities include the design and development of renewable energy conversion devices; optimal offshore storage solutions and export products and microgrid architecture solutions and control systems for intelligent management of integrated end-user demands. ORES will also focus on essential resources such as freshwater (via desalination) and oxygen (for hatchery and fish culture).
- The Sustainable Offshore Developments program will profile and advocate for the regulatory frameworks that will provide confidence for aquaculture and renewable energy industry to invest and for the public to be confident that offshore developments operate to the highest environmental standards for sustainability and ecosystem integrity.
- The Environment and Ecosystems program will evaluate the environmental footprint of the infrastructure, culture systems, and energy generating devices. The program connects with the Sustainable Offshore Developments program to develop management systems to monitor environmental impact and interactions with other sectors and with offshore engineering, aquaculture, and renewable energy programs to monitor the impacts of the environment on health, maintenance, and performance of species, infrastructure, and devices, respectively. Novel monitoring systems including models and user interfaces to deliver real-time data and information for use by government, industry, and the public will be developed.

Accelerating the move offshore will require the delivery of targeted education and training to develop a skilled workforce. Currently the offshore aquaculture, renewable energy, and the NBE industries as new and emerging sectors in the Australian economy have poorly developed training protocols and a limited skill base for future workforce development. The Blue Economy CRC will focus on higher education programs to train a cohort of specialist scientists, engineers, social scientists, lawyers, and economists with detailed cross-discipline knowledge to work in the NBE. Higher degree by research students (e.g., PhD students) will be linked to industry-driven research topics and will, as part of their candidacy, be embedded with partners to help ensure they are work-ready and are connected to potential employers.

Developments in offshore aquaculture and marine renewable energy systems will be of significant interest to many sectors of society and sustained long-term investment in these industries will require facilitative government policies and a supportive community. A key role of the Blue Economy CRC will be the development and dissemination of knowledge, tools, and advice to assist Governments, industries, and the community to make, understand, and then monitor and assess the impacts and competing trade-offs of developments in the offshore environment.

The Blue Economy CRC has commenced the development of an electronic, publicly available, knowledge platform that brings together industry practice and research knowledge about the integration through colocation of offshore aquaculture and offshore renewable energy. It will describe and evaluate existing technologies, solutions, and trends, which will identify the major research challenges and opportunities. To develop the platform, the Blue Economy CRC has commenced 17 short-term scoping study projects to address knowledge gaps in areas such as:

- Key challenges for offshore high-energy salmon and kelp aquaculture production.
- Integrated offshore aquaculture and renewable energy infrastructure, mooring, and vessel designs.
- Marine energy conversion technologies suited to offshore conditions that support energy export (such as hydrogen) and storage, as well as aquaculture applications.
- Understanding the operational and energy requirements of offshore aquaculture systems.
- Robust site selection procedures and environmental and operational monitoring strategies including the application of remote and autonomous monitoring technologies.
- Decision support tools for identifying trade-offs and synergies among emerging blue industries and human activities.

- Economic assessment frameworks for the blue economy.
- Ethical, policy, and regulatory frameworks for Australia's emerging blue economy and systems for their integration.

The scoping study projects will be completed in late 2020 and form part of the CRC's foundational knowledge platform. The knowledge platform's contemporary information will be used to drive the direction of the Blue Economy CRC's ongoing research investment.

Conclusions

The formation of the Blue Economy CRC is timely; the Australian aquaculture sector is moving offshore into deeper waters and more exposed environments and is seeking additional capability to support this transition. Simultaneously, offshore renewable energy is gaining attention as a prospective and important energy alternative with unique opportunities and niche markets to service, but the sector is fragmented and needs resources to help its establishment. The offshore engineering sector is facing a period of cyclical downturn and needs to find new ways to repurpose infrastructure and apply its knowledge and experience to emerging sectors.

These sectors have traditionally been disconnected. On the current development trajectory, there was little appreciation of any new cooperative relationships and the benefits that the different sectors could derive from integration or an understanding of accelerated development that collaborations and knowledge sharing across the sectors can bring to help make scaled offshore development a reality. The footprints of offshore industries can overlap with each other and with sensitive environments, potentially creating trade-offs among the triple-bottom-line of social, economic, and environmental outcomes, hence requiring innovations and strategies generated by the NBE, to inform these emerging blue economy sectors.

Australia is well placed to become a leader in quality seafood, marine products, energy production, and NBE technologies for both domestic and international markets. Importantly, the investment from our Partners and the programs 10-year time period will enable the Blue Economy CRC to manage risks effectively and demonstrate the viability of integration for enhanced production. Underpinned by the NBE, this will support a "step change" in productivity and scaled operation of Australia's seafood production, as well as support the reduction in the operating costs of offshore renewable energy deployments. The Blue Economy CRC will facilitate much needed device demonstration and optimization and test new products while contributing to Australia's carbon mitigation commitments. Our conservative economic impact modeling suggests that through this CRC over a billion dollars of economic benefit will be returned to our partners and the broader community over the next 15 years.

Acknowledgments

The authors acknowledge the financial support of the Blue Economy CRC, established and supported under the Australian Government's CRC Program, grant number CRC-20180101. The CRC Program supports industry-led collaborations between industry, researchers, and the community. The authors also wish to acknowledge the support of all of their Partners (and their staff), Board, and Research Program Leadership Team for their ongoing commitment and collaboration in the development of the Blue Economy CRC.

References

ABARES, 2018. Australian Fisheries and Aquaculture Statistics. Australian Government Department of Agriculture and Water Resources.

Abhinav, K.A., Collu, M., Benjamins, S., Huiwen, C., Hughes, A., Jiang, B., Jude, S., Leithead, W.E., Lin, C., Liu, H., Recalde-Camacho, L., Serpetti, N., Sun, K., Wilson, B., Yue, H., Zhou, B.-Z., 2020. Offshore multi-purpose platforms for a Blue Growth: a technological, environmental and socio-economic review. Sci. Total Environ. 734, 138256. https://doi.org/10.1016/j.scitotenv.2020.138256.

AgriFutures Australia, 2020. Australian Seaweed Industry Blueprint-A Blueprint for Growth. Accessed from: https://www.agrifutures.com.au/wp-content/uploads/2020/09/20-072.pdf.

Buck, B.H., Langan, R., 2017. In: Buck, B., Langan, R. (Eds.), Aquaculture Perspective of Multi-Use Sites in the Open Ocean: The Untapped Potential for Marine Resources in the Anthropocene. Springer International Publishing.

Buck, B.H., Ebeling, M.W., Michler-Cieluch, T., 2010. Mussel cultivation as a co-use in offshore wind farms: potential and economic feasibility. Aquacult. Econ. Manag. 14 (4), 255–281.

Buckley, T., Shah, K., 2019. Australia's First Offshore Wind Project a Step in the Right Direction. The Institute for Energy Economics and Financial Analysis. Accessed from: https://ieefa.org/wp-content/uploads/2019/11/Australias-First-Offshore-Wind-Project-a-Step-in-the-Right-Direction_Nov-2019.pdf.

Campbell, D.J., 2016. WaveNET - the road to commercialisation. In: Proceedings of the 3rd Asian Wave & Tidal Energy Conference, Singapore, pp. 59–62.

Chopin, T., Robinson, S.M.C., Troell, M., Neori, A., Buschmann, A.H., Fang, J., 2008. Ecological engineering in aquaculture: towards a better waste management in western world mariculture. Ecol. Eng. 2463–2475. Accessed from: http://www.sciencedirect.com/science/article/pii/B9780080454054000653.

Chopin, T., 2006. Integrated Multi-Trophic Aquaculture: What it Is, and Why You Should care…'and Don't Confuse it with Polyculture. Aquaculture North America, p. 4.

Connolly, P., Hall, M., 2019. Comparison of pilot-scale floating offshore wind farms with shared moorings. Ocean Eng. 171, 172–180.

Costello, C., Cao, L., Gelcich, S., et al., 2020. The future of food from the sea. Nature 588, 95–100. https://doi.org/10.1038/s41586-020-2616-y.

Dalton, G., et al., 2019. Feasibility of investment in Blue Growth multiple-use of space and multi-use platform projects; results of a novel assessment approach and case studies. Renew. Sustain. Energy Rev. 107, 338–359.

DAWR, 2017. Australia's National Aquaculture Strategy, Department of Agriculture and Water Resources 2017, National Aquaculture Strategy, Canberra, August. CC BY 4.0. Access from: https://www.agriculture.gov.au/sites/default/files/sitecollectiondocuments/fisheries/aquaculture/national-aquaculture-strategy.pdf.

DPIPWE, 2018. Tasmanian Agri-Food Scorecard 2017-18, AgriGrowth Tasmania. Department of Primary Industries, Parks, Water and Environment. Accessed from: https://dpipwe.tas.gov.au/Documents/Agri-Food%20SCORECARD%202017-18.pdf

Eurek, et al., 2017. An improved global wind resource estimate for integrated assessment models. Energy Econ. 64, 552–567. https://doi.org/10.1016/j.eneco.2016.11.015.

Griffin, R., Buck, B., Krause, G., 2015. Private incentives for the emergence of co-production of offshore wind energy and mussel aquaculture. Aquaculture 436, 80–89.

Hemer, M.A., Zieger, S., Durrant, T., O'Grady, J., Hoeke, R.K., McInnes, K.L., Rosebrock, U., 2016. A Revised Assessment of Australia's National Wave Energy Resource. Renewable Energy, pp. 1–23. Accessed from: https://doi.org/10.1016/j.renene.2016.08.039.

Hemer, M.A., Herzfeld, M., Rizwi, F., Rosebrock, U., 2019. Towards characterisation of the Australian national tidal power resource. In: Proceedings of the 13th European Wave and Tidal Energy Conference (EWTEC2019), Napoli, Italy.

Hughes, A.D., Corner, R.A., Cocchi, M., Alexander, K.A., Freeman, S., Angel, D., Chiantore, M., Gunning, D., Maguire, J., Mendoza Beltran, A., Guinée, J., Ferreira, J., Ferreira, R., Rebours, C., Kletou, D., 2016. Beyond Fish Monoculture - Developing Integrated Multi-Trophic Aquaculture in Europe.

MARIBE, Unlocking the potential of multi-use of space and multi-use platforms (accessed April 28, 2018). URL http://maribe.eu/wp-content/uploads/2016/09/maribe-booklet. pdf.

National Marine Science Committee, 2015. National Marine Science Plan: Driving the Development of Australia's Blue Economy. Access from: https://www.marinescience.net.au/.

Papandroulakis, N., Thomsen, C., Mintenbeck, K., Mayorga, P., Hernández-Brito, J., José, 2017. 'The EU-Project "TROPOS"'. In: Buck, B.H., Langan, R. (Eds.), Aquaculture Perspective of Multi-Use Sites in the Open Ocean, pp. 355–374.

Penesis, I., Hemer, M., Cossu, R., Nader, J.R., Marsh, P., Couzi, C., Hayward, J., Sayeef, S., Osman, P., Rosebrock, U., Grinham. A., Herzfeld, M. and Griffin, D., 2020. Tidal Energy in Australia: Assessing Resource and Feasibility in Australia's Future Energy Mix. Final Report. Australian Maritime College, University of Tasmania. Accessed from: https://arena.gov.au/assets/2020/12/tidal-energy-in-australia.pdf.

Stewardson, C., Andrews, J., Ashby, C., Haddon, M., Hartmann, K., Hone, P., Horvat, P., Klemke, J., Mayfield, S., Roelofs, A., Sainsbury, K., Saunders, T., Stewart, J., Nicol, S., Wise, B. (Eds.), 2018. Status of Australian Fish Stocks Reports 2018, Fisheries Research and Development Corporation, Canberra. Accessed from: https://fish.gov.au/about/foreword.

Case study: The new blue economy: People, products, and policies

Michael B. Jones, Melissa Fischel

The role of a BlueTech cluster in workforce development and economic development

Background on clusters and TMA BlueTech

The term "business cluster"—sometimes called an industry cluster or competitive cluster—is a concept popularized by Prof. Michael Porter at Harvard (Porter, 1998). It is typically understood to mean a geographical region where a critical mass of capabilities and companies has developed resulting in faster regional development and a sustainable competitive advantage over other areas.

TMA BlueTech (TMA) began as part of a 501(c) (3) educational nonprofit in January 2007 and was recognized as an "organized cluster" in 2009. In 2013, TMA established a 501(c) (6) nonprofit to act as a membership-based, industry association focused on BlueTech. The original entity was renamed TMA Foundation (TMAF) and continued the focus on education and research. TMA has a geographic center (i.e., a "geographic cluster") as it was started to organize the BlueTech community in San Diego county, California. TMA has spread from beyond the original regional focus as it added international activities and members. In addition, TMA is a "sectoral cluster," which aggregates stakeholders in common sectors, in this case ocean and water sectors.

TMA's definition of the Blue Economy includes ocean, fresh water/ wastewater—"everything wet"—so it includes traditional Blue industries (e.g., fishing, boat and ship building, marinas, and other "visible" industries in ports), BlueTech, and supporting industries. BlueTech is the innovation segment of the Blue Economy, which typically includes the fastest-growing companies doing business globally. Most clusters focus on either oceantech or watertech. TMA believes the two belong together because water takes many forms in its natural cycle and TMA member companies sell across the water spectrum.

The Mission Statement of TMA is "Promoting Sustainable, Science-Based Ocean & Water Industries." Its tagline is "Promoting BlueTech & Blue Jobs," which reflects its role as an industry association interested in creating good-paying Blue Jobs with its members and regional, national, and international partners.

The creation of organized Blue Economy and BlueTech clusters globally is important to develop strong, regional Blue Voices region by region ("*Think global, act regional*"). These coordinated Blue Voices form through a process of education and alignment of interests across three major Triple Helix stakeholder groups—academia/education; industry; and policy (government, NGOs, etc.)—and are critical to pursue sustainable, science-based usage of the ocean and water resources.

TMA is a leading voice for the Blue Economy and BlueTech internationally. TMA conceived and helped launch the BlueTech Cluster Alliance (BTCA) in January 2017 to engender cross-cluster/multilateral collaboration for academia and business. As of January 2021, BTCA has 10 leading BlueTech cluster partners in eight countries (Canada, France, Ireland, Norway, Portugal, Spain, the United Kingdom, and United States) and will continue to expand. BTCA partners benefit from the sharing of best practices and issues of common interest including education, workforce development, and capacity building for Blue Economy clusters in developing countries.

TMA has been a multiyear Strategic Partner for maritime technology with the ITA (International Trade Administration) in the US Department of Commerce. It coauthored *The Ocean Enterprise: A study of US business activity in ocean measurement, observation and forecasting* (ERISS & TMA BlueTech, 2015)—the first national-scale assessment of the US ocean enterprise for NOAA/US IOOS—(see contributions by Rayner and Gouldman, this volume). TMA was chosen by ITA to organize the first ever US Maritime Technology Export Initiative, a funding program to help innovative, BlueTech SMEs (small-to-medium enterprises) from across the United States to compete in international markets by participating in select trade shows and follow-on missions globally. TMA's annual BlueTech Week held each November in San Diego attracts hundreds of high-level executives and policymakers across the Triple-Helix from around the world to convene, share best practices, and explore ways to collaborate during seven events over 5 days. BTW2019 participants came from 18 countries.

TMA has long acted like a "virtual incubator/accelerator" and has helped support numerous SMEs (small and medium-sized enterprises) with business development, industry connections, and marketing support. In October 2017, TMA launched a physical incubator in conjunction with the San Diego office of Denton's (the world's largest law firm), the City of San Diego, and Scripps Institution of Oceanography. The BlueTech Incubator is a physical incubator and a first step toward a larger facility on or close to the water where companies can do R&D and testing.

TMA focused on education and workforce development

A commitment to the development of human resources is important for the long-term success of a regional cluster. According to Prof. Porter, "Outstanding talent is a scarce resource in any nation. A nation's success largely depends on the types of education its talented people choose, where they choose to work, and their commitment and effort" (Porter, 1990).

Regions such as San Diego (and most port cities) are costly to move to and live in, which makes it critical to retain local talent by educating and training the entrepreneurs and workers of the future Millennials and Post-Millennials are recognized as being purpose-driven generations. The sustainable, science-based nature of the Blue Economy is a good fit for employees looking for purpose through their work. Blue Economy companies should work to attract talent from the younger generation to mitigate the aging demographic of many blue economy employees.

There are multiple ways to attract new talent:

- Create visibility and awareness for the Blue Economy/BlueTech;
- Demonstrate the size and growth of the ocean and water/wastewater market(s) with data and studies to highlight career opportunities in this field for students and their parents;
- Enable "storytelling" platforms to connect emotionally to students about the sustainable usage of ocean and water/wastewater resources as a moral imperative and as a future career;
- Emphasize the "Circular Economy" as an economic system aimed at eliminating waste, reducing pressure on the environment, stimulating innovation, and creation of new, good-paying jobs;
- Establish BlueTech Centers of Excellence and retool existing institutions to provide hands-on learning and regional training to develop and retain talent; and
- Work with traditional educational systems and workforce development agencies to become committed stakeholders to these sectors.

To attract talent to BlueTech, TMA has an extensive educational and workforce development outreach effort in conjunction with many partners to leverage limited resources to attract and train talent. Activities include:

- BlueSTEM Career Pathway: In Fall 2017, San Diego Unified School District (SDUSD) in collaboration with TMA launched a BlueSTEM Career Pathway pilot program at a Middle School and a High School. Based on the success and popularity of that pilot, SDUSD subsequently announced it would develop Blue Economy curriculum to use as a unifying BlueTech "theme" across the entire 105,000 K-12 student district from Transitional Kindergarten through 12th grade. Teachers are developing ocean and water-related curriculum to match California next-generation teaching standards. TMA in collaboration with members and partners enhances the career pathway with postsecondary opportunities, including internships, externships for teachers, company site visits, and industry speakers. The goal is to align industry needs with student experience. Students will learn "ocean and water literacy" as part of a comprehensive program that also introduces them to an array of good-paying careers:

- Blue Economy Career Pathway Videos: In Fall 2016, TMA was awarded funds through the SoCal Promise Grant project to focus on education programs in the Blue Economy, including career pathways, job websites, outreach, and videos. Up to 16 career pathway videos have been or are in development to highlight good-paying blue-collar and white-collar jobs in the Blue Economy;
- Blue Economy Education Collaborative (BEEC): TMA organized a Blue Economy Education Collaborative to catalogue individual programs, create regional awareness, and promote collaboration among organizations across San Diego county interested in Blue Economy and BlueTech (ocean and water-related) education and workforce development (over 40 organizations identified);
- Blue Economy Immersion Program: In conjunction with members and partners TMA runs 3-day summer Blue Economy immersion programs providing HS students insight into the Blue Economy, jobs, and hands-on skills;
- Company site visits: TMA organizes visits to regional companies for economic development and elected officials, nonprofits, teachers, workforce development leaders, and others to gain firsthand knowledge into the Blue Economy;
- Research: TMA cosponsored the first ever *San Diego Maritime Industry Report 2012* to begin to identify, categorize, and size the industry in the region. In November 2016, the San Diego Workforce Partnership published a *Blue Economy Labor Market Analysis* coauthored by TMAF. The report identified current labor needs of the Blue Economy and identified actions needed to strengthen the labor pool;
- SeaPerch chapter and Porpoise Robotics: In Spring 2016, TMA assumed responsibility for the San Diego SeaPerch chapter, which is part of the national SeaPerch program supported by the US Office of Naval Research (ONR). There is an annual regional SeaPerch Regional Competition that ties into a national effort. The SeaPerch program is being coordinated with Porpoise Robotics and other more advanced programs to create a continuum of robotics learning for students;
- STEM Leadership Series: TMA cosponsors a STEAM Leadership Series event each year on the USS Midway Museum for 300+ high school students from SDUSD focused on BlueTech careers;

- Summer Internships: TMA Foundation has received grant funding for student stipends so low-income high school students can take advantage of summer internship opportunities with TMA member companies; and
- University Outreach: TMA collaborates with colleges and universities in the region to connect industry with academia, educators, and students. As examples:
  - MiraCosta College's Technology Career Institute (TCI) developed a BlueTech training program including a Remotely Operated Vehicle (ROV)/Drone technician program that provides operator training for underwater robots and drones;
  - Scripps Institution of Oceanography and TMA collaborate on several programs throughout the year including a cohosted day during BlueTech Week;
  - In partnership with San Diego Community College, TMA members, and SDUSD, TMA is developing a Blue Economy Apprenticeship program; and
- Workforce of Tomorrow track: TMA organizes a "Workforce of Tomorrow" track as part of the annual BlueTech Week (www.bluetechweek.org) each Fall.

While much has been done, there is still so much to do. The industry needs more data and location information in order to understand the size of the market and the impact on our ocean and waterways. In order to accomplish this, we need to develop data literacy programs in our schools. Programs combining ocean literacy and data literacy will unleash the full potential of the blue economy and create a workforce with transferable skills.

Conclusion

Success is a "team sport." It is achieved through the assistance and collaboration of many people and organizations. This is certainly true for cluster development including education and workforce development. TMA has a long history of collaboration nationally and internationally. It is proud to share what it has accomplished and always looks forward to learning from others.

Author biographies

Michael Jones is President of TMA BlueTech (TMA), which is an organized nonprofit industry association founded in 2007. It is active across the "Triple Helix"—academia/education, industry, and policy. Its Mission Statement is "Promoting Sustainable, Science-Based Ocean & Water Industries." Originally focused on San Diego—the largest US BlueTech (ocean and water tech) cluster—it has members across the United States and abroad. TMA is a Strategic Partner with the US Department of Commerce and helped found the BlueTech Cluster Alliance, which includes 10 leading BlueTech clusters from eight countries. Michael has a broad background as a Board member, entrepreneur, investment banker, and Founder/President of TMA. He received his undergraduate degree from the University of Arizona and an M.A. from Johns Hopkins University School of Advanced International Studies (SAIS). His graduate work included a year in Bologna, Italy; a year in Lima, Peru; and a 5-month internship at the EU headquarters in Brussels.

Melissa Fischel is Director of Operations and Workforce Development at TMA BlueTech. Ms. Fischel supports the growth of the organization through programs and initiatives. Melissa has an MBA from University of California, Irvine and a B.A. in Economics from University of California, San Diego. She brings 20+ years of industry experience including a Fortune 100 company, start-ups, and nonprofits.

References

Porter, M.E., March–April 1990. The Competitive Advantage of Nations. Harvard Business Review.

Porter, M.E., November–December 1998. Clusters and the New Economics of Competition. Harvard Business Review.

Case study: Marine Industries Science and Technology (MIST) Cluster

Laurie Jugan, Anna Kate Baygents

Introduction

The Marine Industries Science & Technology (MIST) Cluster (www.mistcluster.org) is a regional group of organizations involved in the development and implementation of applied technologies for operating in, working around, and monitoring the marine and coastal environments. Large and small business, federal, state, and local agencies, universities, and nonprofits through their applied technology efforts support other related marine industries. These include commercial and recreational fishing, shipbuilding, defense, aqua/mariculture, off-shore oil and gas, and environmental restoration and recovery, to name a few.

Supported in part by a US Small Business Administration (SBA) Regional Innovation Cluster contract, the University of Southern Mississippi (www.usm.edu), the Mississippi Enterprise for Technology (www.mset.org), and their cluster program team members provide direct and linking services to assist MIST Cluster members. These services address the 3 C's of Counseling, Capital Access, and Contracting with the government and their prime contractors. This includes helping government agencies and large businesses identify local companies that can assist them in meeting small business goals and introducing them to the innovative culture of these small, high-tech firms.

Background

In 2010, the existence of a maritime technology cluster in the Northern Gulf of Mexico region was recognized by the Mississippi Enterprise for Technology (MSET), a private nonprofit dedicated to technology-based economic development and located at Stennis Space Center (SSC) in South Mississippi. Though the presence of the National Aeronautics and Space Administration (NASA) in this area is well known, the larger presence of marine-related federal agencies and large companies that operate in the area is not widely recognized (https://www.nasa.gov/centers/stennis/pdf/518017main_FS-2011-02-00105-SSC.pdf). The Navy has the largest number of direct and contracted employees at SSC, and the National Oceanic and Atmospheric Administration (NOAA) also has a significant workforce at the center. In addition, a number of companies supporting maritime projects and programs are located along the Mississippi Gulf Coast. There are several large companies supporting the marine-related defense industry in the State, including Lockheed Martin, Northrop Grumman, Science Applications International Corporation (SAIC), Rolls Royce Marine, Leidos, VT Halter Marine, Trinity Yachts, and Huntington Ingalls Shipbuilding—Mississippi's largest private sector employer. However, there was no formal framework in place to quantify, organize, or activate this ecosystem that is supported less by direct commercial ventures and more by federal agencies and programs.

In 2010, at the inaugural *Oceans In Action* Workshop, now the premier Blue Tech conference in the region (https://www.mtsociety.org/oceans-in-action/), presentations were given on defining the local cluster and the benefits a formal cluster could bring to members. During the event, the audience was asked to complete a card with contact information if they were interested in supporting a maritime industry cluster. More than 100 attendees submitted cards. This group, consisting of representatives from federal agencies, the state of Mississippi, the state's university system, and private industry, formed the initial core membership of the cluster. Presentations on the cluster have continued at each *Oceans In Action* Workshop since and now include talks and panels on recent cluster activities.

In 2014, the University of Southern Mississippi (USM) held a Marine Industry Economic Forum to call attention to the significant portion of the State's economy that was attributed to the maritime industry as a whole. Focused on the economics of the industry, the program featured speakers from large and small companies that could quantify their annual revenues, types of jobs, average salaries, supply chains, and impacts on the community. A presentation of data collected through the NOAA National Ocean Economics Program (NOEP, https://oceaneconomics.org/) indicated that the value of Mississippi's maritime-based efforts was estimated at nearly $2B annually with more than 31,000 jobs.

However, the NOEP data did not capture certain aspects of the economy that are attributable to federal jobs, commercial fishing, oil and gas, or ancillary jobs created as a result of each job in the industry. This indicated that the NOEP data only scratched the surface of the portion of Mississippi's GDP attributable to the Blue Economy. To further highlight this discrepancy, Michael Jones from The Maritime Alliance, the BlueTech Cluster in San Diego (now TMA BlueTech, see contribution above in this volume), gave a keynote that provided information on the difficulties of identifying companies in maritime clusters and quantifying their economic impact.

The workshop was the catalyst for two significant follow-on efforts. The first was a capstone project from the USM Master of Science in Economic Development program entitled "Mississippi's Blue Economy: An Analysis of Mississippi's Maritime Commerce" (https://www.usm.edu/sites/default/files/groups/trent-lott-national-center/pdf/mississippis_blue_economy.pdf). For this study, authors used standard economic development data sets to quantify key aspects of the state's maritime industry. Although the data used have known issues when assessing companies doing business within the Blue Economy, one conclusion was, "From the secondary data presented and the industries identified in the analysis, it is clearly demonstrated that there is potentially a largely undefined maritime cluster on the Mississippi Gulf Coast. Only through continuing, sustained efforts to further qualify and quantify its maritime activity can Mississippi move itself into a position of informed action." The report also indicated that the Blue Economy portion of Mississippi's overall economy may be as high as 35%.

The second effort following the USM Marine Industry Economic Forum was the submission of a proposal to formalize the existing cluster in the region to the Small Business Administration's Regional Innovation Clusters (RIC) program (https://www.sba.gov/local-assistance/find/?type=Regional%20Innovation%20Clusters&pageNumber=2). USM, in concert with MSET and leveraging MSET programs, submitted the proposal—using information provided within the Marine Industry Economic Forum and the Mississippi's Blue Economy study—to formalize the MIST Cluster. The contract was awarded to USM in 2015 with a 5-year period of performance. Under the contract, USM and MSET have created the MIST Cluster Program to support maritime companies and strengthen local business flow.

Connections

A primary benefit of MIST Cluster membership is the opportunity to connect with suppliers, customers, consultants, partners, and even competitors. Leveraging existing MSET programs, the MIST Cluster coordinates quarterly Small Business Summits to connect companies with the federal agencies and prime contractors in the region. As mentioned, a high percentage of maritime business along the Mississippi Gulf Coast is associated with federal agencies (https://www.mset.org/wp-content/resources/2010/09/MS-Defense-Related-Blue-Economy_MDDI_Official-Copy.pdf). However, information on business opportunities from these agencies is not widely advertised. The MIST Cluster Summits are designed for speakers from these organizations to define their missions and present upcoming opportunities in terms of Simplified Acquisitions (purchases under $250,000) and upcoming Requests for Proposals (RFPs) for products and/or services needed.

MIST Cluster team members assume the role of understanding the federal contracting landscape in the region and, to an extent, at the national level. As such, with an understanding of a company's capabilities, recommendations can be made on which Small Business Specialists in federal agencies could assist them; upcoming contracts for which they could compete; other companies that could be strategic partners; and training/certifications they may need to increase their offerings.

Networking with other maritime industry clusters throughout the world is another MIST Cluster value-added service. For example, Stennis-based Navy groups began an initiative to seek out foreign technologies in unmanned systems. MSET, through MIST Cluster contacts, spread the word of this initiative to organizations in Canada, the United Kingdom, Portugal, Spain, France, and other nations, as well as to clusters in the United States, resulting in a surge of applicants to this initiative.

Training

Working in the federal contracting space is significantly different than working commercially. Companies must abide by the Federal Acquisition Regulations (FAR, https://www.acquisition.gov/browse/index/far) and the Defense Federal Acquisition Regulations Supplement (DFARS, https://www.acquisition.gov/dfars), which require auditable cost accounting, specified policies and procedures, enhanced cybersecurity, and more. The MIST Cluster has been able to absorb the cost of training in these areas, offering classes at no or low cost to members. Additional training opportunities in other subject matter, such as process improvement, are offered locally and at reasonable costs.

Workforce development

To address workforce development, MIST Cluster works to identify issues in supporting local maritime companies' and organizations' workforce needs. MIST Cluster members serve on several local committees focused on the State's workforce to support the development of new solutions and have participated in efforts to initiate the ACT Workforce Ready Communities Program (https://www.workreadycommunities.org/) throughout Mississippi.

MIST Cluster members engage with local universities and community colleges to assist in the creation and/or promotion of maritime programs. One example is USM's Unmanned Maritime Systems Certificate program (https://www.usm.edu/ocean-science-engineering/unmanned-maritime-systems-ums-certification.php) a 5-week class; the first of its kind to offer a credential in the field. MIST Cluster is working with the Mississippi Gulf Coast Community College and serving on the Transportation and Maritime Advisory Committee, to create programs (https://mgccc.edu/programs/career-technical/) to train the needed workforce in these areas. When a broader reach is needed for support, MIST Cluster collaborates with institutions throughout the northern Gulf of Mexico region.

Partnerships

Because MIST Cluster is comanaged by USM and MSET, capitalizing upon their existing partnerships with other organizations is especially key in growing the cluster. For example, MSET has long served as a partner with NASA in managing leased space at Stennis Space Center, hosting nearly two dozen companies ranging from start-ups to large business anchors. Through a Space Act Agreement and a Use Permit, MSET provides office space and services to technology companies supporting the missions of Stennis tenants, including the Navy and NOAA. Services provided to these companies vary and are customized to their individual needs.

Over the past several years, MSET has developed a trusted relationship with the US Navy, entering into Partnership Intermediary Agreements with the Naval Meteorology and Oceanography Command (NMOC) and the Naval Undersea Warfare Center Division Newport (NUWC). Additionally, MSET works with NOAA's Technology Partnerships Office at SSC and nationally.

Supporting NASA, the Navy, and NOAA, MIST Cluster partners assist in technology transfer, technology demonstrations, Small Business Innovation Research (SBIR)/Small Business Technology Transfer (STTR), and commercialization efforts. Companies are identified that can take federally developed technologies to a commercial market, and new technologies are sought for federal agency missions. SBIR/STTR topics are reviewed, and companies are sent relevant topics for their consideration. USM's Business Innovation and Assistance Center (https://www.usm.edu/vice-president-research/business-and-innovation-assistance-center.php) assists companies in the development of their SBIR/STTR proposals.

In recent years, the frequency of arranging technology demonstrations for federal agencies and larger companies has increased. MIST Cluster partners review emerging technologies and summarize for potential users, then arrange for the demonstration, and assist in promoting follow-on conversations. Through the annual *Oceans In Action* Workshop, MSET and the Marine Technology Society's Gulf Coast Section (https://gulfcoast.mtsociety.org/) work to include new technology developers on the agenda, and in the exhibit rooms, to provide access to federal agency buyers, potential users, and strategic partners. The MIST Cluster focuses on bringing hi-tech companies of interest to the exhibit hall. Additionally, business-to-government and business-to-business meetings are coordinated so companies can explain their technologies in detail to potential federal and commercial customers.

Regional support

The MIST Cluster is active in working at the local, state, regional, national, and international levels to highlight and promote Mississippi's Blue Economy. For example, in 2017, then Governor of Mississippi Phil Bryant challenged the University of Southern Mississippi by Executive Order to stand up the Governor's Ocean Task Force. Drawing from the Navy's Task Force Ocean (https://www.onr.navy.mil/task-force-ocean), the goal was to create a Master Plan for economic development centering around marine science and technology assets and capabilities within the State. The task force developed nine key recommendations and a roadmap for implementation that is being addressed through several state initiatives and stakeholder efforts.

One recommendation from Governor Bryant's Ocean Task Force (https://www.mset.org/wp-content/resources/2010/09/Governor-Bryants-Ocean-Task-Force_Official-Copy.pdf) is to establish a maritime systems innovation and commercialization center. The Port of Gulfport recently broke ground on The Roger F. Wicker Center for Ocean Enterprise, a multiuse building where federal agencies, academia, and private companies will collaborate on strategic initiatives of national interest. The center, scheduled to open in 2022, will serve as a hub for entrepreneurial activity and another location for MIST Cluster-provided support and programming.

With support from USM's Mississippi Defense Diversification Initiative, MIST Cluster partners have documented various aspects of Mississippi's Blue Economy. These include an assessment of the State's assets supporting the defense-related Blue Economy and unmanned maritime systems (https://www.mset.org/wp-content/resources/2010/09/UMS-in-MS_Official-Copy.pdf). A study of the local technology workforce (https://www.mset.org/wp-content/resources/2015/10/South-MS-STEM-Roadmap.pdf) was supported by the US Economic Development Administration. Through these efforts, awareness of the State's Blue Economy and BlueTech capabilities has been elevated and broadly publicized.

Summary

The MIST Cluster currently has over 150 members and collaborates frequently with federal, state, and local agencies, academia, business resource providers, other clusters, and other entities to spur economic activity in Mississippi's Blue Economy and the Northern Gulf of Mexico region. The MIST Cluster allows technology innovators to come together with industry end users to understand each other's capabilities and needs; to collaborate and expand business opportunities; and to address shared issues.

Plans for cluster hubs across the Mississippi coast are in the works, as well as expanded efforts to support the implementation of Governor Bryant's Ocean Task Force recommendations. In support of a knowledge-based Blue Economy (i.e., the New Blue Economy), the Cluster and local partners are formulating ways to address current and future challenges—contests to support the handling of big data and its uses; outreach programs to engage the public and students and to emphasize the issues associated with our oceans and waterways; and dissemination of critical information through organized events. Through continued strategic partnerships and concerted efforts, The MIST Cluster is poised to catalyze the expansion of Mississippi's Blue Economy and bring greater awareness to the State's immense BlueTech capabilities.

Author biographies

Laurie A. Jugan

Laurie Jugan is a consultant supporting the Mississippi Enterprise for Technology (MSET) and the Marine Industries Science and Technology (MIST) Cluster, formalized under a Small Business Administration contract with the University of Southern Mississippi. For these two efforts, she serves as the Program Director, creating and implementing programs supporting small businesses and business interactions with organizations with promising partnership opportunities. Prior to this, she supported a technology-based small company, providing technical and administrative services under Navy and NOAA contracts. Ms. Jugan holds Master's degrees in Oceanography and Business Management and is an active member of the Marine Technology Society.

Anna Kate Baygents

Anna Kate Baygents served as the Project Manager at MSET, supporting the MIST Cluster to seek and assist companies interested in doing business with or locating at Stennis Space Center and along the Mississippi Gulf Coast. She served as Project Manager for the Governor Bryant's Ocean Task Force, focusing on the Blue Economy. She assisted in a number of documents describing and quantifying Blue Economy efforts along the Mississippi coast and in the state. Currently, Anna Kate works as a Program Manager for the Center for International Education at The University of Southern Mississippi.

Case study: Centre for Ocean Ventures and Entrepreneurship (COVE)

Jim Hanlon, Sherry Scully

Introduction

The Centre for Ocean Ventures and Entrepreneurship (COVE) (www.coveocean.com) is an ocean technology hub located on an 8-acre urban oceanfront site in Halifax Canada. COVE's mission is to sustainably grow the ocean economy through collaborative investment in innovative ocean technology. Since opening in October of 2018, COVE has attracted over 65 tenants representing all sectors of the ocean economy and ranging in size from small technology start-ups to innovation teams from large multinational companies.

Key COVE features include private and open office and workshop space, shared meeting and conference spaces, on-site access to machining and fabrication space, dockside testing, and access to research vessels based at

COVE. COVE values its relationships with academic researchers and provides oceanfront laboratory space for Dalhousie University's ocean science program and for the applied ocean research and training programs of the Nova Scotia Community College. The broad mix of tenant types and the shared physical infrastructure of COVE continues to attract top innovation performers.

The Start-up Yard at COVE is a classic start-up incubator that supports the launch of new ocean technology companies. Cohorts of new entrepreneurs compete for a 6-month residency in the COVE Start-up Yard that includes formal training and mentoring, access to financing, and use of the shared facilities at COVE. Most importantly, start-up companies work alongside established tenant companies allowing them to develop valuable partnerships and to receive real-world advice. At any given time, there are about a dozen start-up companies in the Start-up Yard, supported by Innovacorp (www.innovacorp.ca).

In addition to the physical premises, COVE provides three programming elements to support the growth of COVE tenant businesses. They are:

Networking

COVE provides formal and informal opportunities for tenants to interact with each other and with visitors; all designed to encourage more business interactions. These include monthly events called Ocean Connectors. These typically draw about 100 people from industry, government, and academia for 2 h of networking plus a brief speaker presentation on a topic of ocean interest. COVE also organizes smaller lunch and learn sessions on specific topics and hosts larger lectures and workshops in collaboration with partners. COVE staff also provide one-on-one "technology brokering" to connect tenants, partners, and students around commercialization opportunities.

COVE has established a formal partnering relationship with the University of Massachusetts—Dartmouth that facilitates exchanges of market opportunities, two-way access to office space for visiting tenants, and student exchanges between Nova Scotia and Massachusetts. COVE is planning to establish similar formal relationships with a limited number of prominent ocean technology centers around the world.

COVE also has formal partnering agreements with academic research programs including the Ocean Frontier Institute (www.oceanfrontier institute.com) and DeepSense (www.deepsense.ca) hosted at Dalhousie

University and the SeaTAC program of the Nova Scotia Community College. COVE works collaboratively with these programs to provide commercialization opportunities for academics and to provide early access to novel research for industry.

Technology integration projects

COVE staff will work with groups of tenant companies to create new projects that will result in business growth for the tenant companies. Specifically, COVE will convene interested tenants to develop project ideas and will also seed fund feasibility assessments for projects. COVE staff will assist tenant groups in seeking financing for new projects and can provide system engineering and project management as required on a noninterference basis. These projects provide market access for smaller COVE companies and can help to develop working relationships among companies that will continue beyond the specific project.

Workforce development and research

The COVE Workforce Initiative (COVE WI) operates out of COVE and is a project that focuses on primary research and evidence-based program development oriented around developing current and future workforce for the broader ocean industry. Since its inception in 2014, then called the Marine People Partnership, the COVE WI has contributed to national and international literature on issues affecting recruitment, retention, and skill development for ocean careers ranging from shipbuilding and boat building to marine renewable energy to aquaculture and commercial fishing, with particular focus on critical skilled trades roles and emerging entrepreneurs. Early research in the project (Scully, 2015) pointed to several key hurdles that contributed to lower-than-expected interest in the emerging blue economy. Key among them were: low levels of awareness of the industry, dated notions of the breadth and modernity of the industry, and skepticism about the durability of the industry arising from a pervasive narrative that ocean industries are inherently obedient to a boom—bust cycle.

Subsequent research into the first hurdle revealed gaps in awareness of the industry that could be traced back to missed opportunities to formally or informally build interest and curiosity in the ocean industry among Canada's youth. At the start of this project, ocean STEM concepts were, and in many provinces still are, nearly absent from the public-school curricula, and a study of youth in Nova Scotia (Scully, 2016) showed that few young people are interacting with the ocean through extracurricular programs or with their families. As a result, we are graduating youth from our schools who lack a fascination for, or even an awareness of, the countless curiosities that the ocean holds, or the myriad blue economy careers that are potentially available to them.

When building a workforce strategy, it is critical to consider how the emerging cohort perceives the industry of interest—our research revealed that those perceptions were not only negative, but also not well informed. The 2016 student study of more than 14,000 youth showed that half (49.8%) of youth would not even consider a career in the blue economy. Open-text responses (Scully, 2016) from those youth revealed why this was so, with comments that ranged from apathetic and disinterested ("*Because I don't have enough interest in the sea and stuff that has to do with that job*"—grade 7 male), to disparaging (e.g., "*I do not care about the ocean or anything that inhabits it. If jobs in the marine industry is all Nova Scotia has to offer, then I won't be living here*"—grade 8 female), to fearful (*because I don't like water its spooky*—grade 7 male; "*because I am afraid of boats and fish are gross*"—grade 9 female). What most of the student comments revealed was an unambiguous lack of awareness of the blue economy, and a disheartening lack of curiosity about the deep blue mystery that surrounds our country, the origins of industry for Canada's people.

The challenge of low awareness of the blue economy careers is exacerbated by dated notions of what those careers are. The student study showed that where many students were familiar with navy (89%), shipbuilding (64%), and commercial fishing (75%) careers, that significantly fewer were familiar with less traditional roles involving marine robotics (32%), ocean technologists (35%), and marine fitter/welders (16%). And where awareness was high, the prevailing notions of these careers were fixed in a past that characterized them as dirty, low-tech, high-risk jobs that require brawn over brain. Most absent was an awareness of the extent to which these careers have evolved, adopting technology and modern-day practices. Student responses also reflected a general lack of awareness and understanding of the breadth and scope of ocean-oriented careers and of the skills and competencies associated with them, for example; "*because I am interested more in jobs that involve math and science*" (grade 8 female); "*Because I would like to be an engineer when I grow up and I don't really like marine jobs in general*" (grade 6 male). Since this study, much of our programming work at COVE WI has focused on building awareness and shifting the narrative of blue economy careers with information and exposure-building opportunities. This work includes youth-oriented and teacher-oriented programs.

Perhaps the most difficult hurdle to overcome is the lack of confidence in the industry's future that is held by some key influencers—particularly those Maritimers whose lived experience has been one of boom and bust. Student responses provided some insight into the narrative they hear at home about the blue economy, for example, "*A large amount of my family has had a job related to the marine industry so I am aware that it would not be in my best interest to pursue one*" (grade 9 male). Where parents have been shown to be the biggest influencers in student career decisions, this narrative emerges as a priority area of focus for change.

Where our youth are the future of the blue economy, our current workforce is our present. COVE WI also conducts primary research and program creation oriented around building awareness, skill relevance, and engaged retention of this current cohort. We have conducted several studies on this group, including mid-career and senior-level employees to analyze skill relevance and skill gaps in an economy where future skills are shifting the notions of "competent" and "current" (Scully, 2017) COVE WI is leveraging our primary research (Scully, 2019a,b), and leading extant research, to develop professional development opportunities that will best meet the learning and skill requirements of this cohort to help build a current workforce that is agile and future-focused, and that can effectively integrate a diverse new wave of workers into the booming blue economy.

Author biographies

Jim Hanlon

Jim is a 40-year veteran of the ocean tech industry, having worked in product design, marketing, and management for companies in Atlantic Canada, Ontario, and New England. His career has spanned the aerospace and defense sectors as well as the marine environmental monitoring field. He has been an owner in two separate high-tech companies that have successfully grown and been purchased by multinationals. Jim holds an electrical engineering degree from Dalhousie University's Faculty of Engineering and an MBA in marketing from Saint Mary's University in Halifax and is a registered Professional Engineer. In June of 2020, Jim was elected a Fellow of the Canadian Academy of Engineering. He serves on many advisory boards for ocean research and commercialization ventures.

Sherry Scully

Dr. Sherry Scully is CEO/Principal, SES Workforce Research and Consulting, where she leads a research project relating to workforce development across the National marine and ocean industries, to attract, train, and retain highly qualified people to sustain the competitiveness and growth of the Canadian shipbuilding and marine industries. With this focus on current and future workforce, Dr. Scully also conducts projects and programs and develops educational resources for teachers and youth in the k-12 system and leads an annual Ocean Institute for educators in several regions across the country. Sherry holds an interdisciplinary Ph.D. in social psychology and management from Dalhousie University, as well as an MBA with a concentration in change strategy, and two degrees in Education.

References

Scully, S., 2015. The Challenges, Needs and Opportunities for Strategic Workforce Development in the Greater Marine Industry. Institute for Ocean Research Enterprise. https://coveocean.com/projects/marine-people-partnership.

Scully, S., 2016. Student Intentions and Perceptions Study: Report of Findings, Analysis and Recommendations. Institute for Ocean Research Enterprise.

Scully, S., 2017. Ocean Technology Sector Study: Competency Assessment and Workforce Study. Centre for Ocean Research Enterprise.

Scully, S., 2019a. The Professional Development Needs of Mid-career and Senior Workforce Across Ocean Industries: Report of research findings and recommendations. Centre for Ocean Research Enterprise.

Scully, S., 2019b. Planned for obsolescence: workforce agility is the new competitive advantage. J. Ocean Technol. 14 (2).

Interview: Industry trends

Justin Manley

As data, data literacy, and the ability to use various technologies are becoming increasingly important across several sectors in the ocean industry, are there any trends in what employers are looking for when hiring?

To be honest there are no definitive trends. My hypothesis is because what we're calling the new blue economy in this conversation is relatively nascent, it's not nearly as dynamic as in other sectors, so in the ocean side I'm not seeing trends. I have yet to come across somebody who's going crazy for a data scientist as with, for example, Wall Street. There are hedge funds who will pay very well for data scientists and skills such as machine learning as well as some autonomy and artificial intelligence. But within ocean-focused companies, it is not something I'm seeing a huge demand for at the moment.

What skill sets will the workforce of the future need?

Broadly speaking, for the entirety of a blue economy or really focusing more on this hypothetical new blue economy, I think we need to look at the flexible skills sets. Talents such as adaptability and an understanding of how software, cloud computing, data science, and machine learning, work generally, even if you can't write the code yourself. How to apply these technologies in creative problem-solving is going to be important. Another key piece of all of these businesses and all of these jobs, business or government or science is the ability to communicate the complexity of what you're doing to those who won't have the same technical or scientific understanding.

What directions are companies pivoting to in order to capitalize on new applications of data?

There are a lot of entrepreneurs and a lot of big businesses watching, thinking, and head-scratching; however, I am seeing very few definitive pivots or new starts where a business is embracing an exclusively data-driven/focused future.

But the flip side of that is a trend that does seem to be working in the traditional ocean data businesses. If you think about someone doing say hydrographic survey or a pipeline survey for an oil company, we've seen an evolution from a very manual, vessel-intensive effort to putting AUVs on ships. The resulting data became more complex and now, for example, one of the successful growing small companies built their own unmanned surface vessel not because they wanted to but because they couldn't buy exactly what they wanted. Their business model has changed, so the business model used to be based on day rates, pay X for the ship, Y for the people, Z for the cost of processing, and your job total ends up being X+Y+Z with negotiation and competition at each phase. A new company, XOCEAN out of Ireland, offers a different model of essentially a flat fee. They say you'd like X square meters surveyed for pipeline purposes? Okay, our price is Y dollars. They've figured out how to use the tools, both the hardware and the software tools to deliver a better value for their customers. The technology was important but so was the approach to market.

This evolution is still project-based ocean data collection and reporting. It is not yet a "wholesale" ocean data approach. So it is not exactly what you're getting at, but it's the kind of transition in business thinking that will continue and will drive the evolution in the success of businesses.

That is a good example. Looking at what's happened over the last 10 years, where IBM bought out the Weather Channel, Monsanto bought out the Climate Corporation, more of a mergers and acquisitions than a pivot, but it is a pivot in the sense that folks were not thinking of IBM investing in the weather industry until they started doing that. Are you seeing any other examples of that magnitude or dimension?

In the ocean space, we continue to see large-scale moves. For example, if we look at Ocean Infinity as a relative newcomer to the industry, they're changing business models. They're coming in and stating, it's a data and uncrewed systems future world. We have the resources and vision to accelerate the process by employing significant capital. There's a little bit of faith in that by increasing the pursuit of collection of data, or the efficiencies there of, that they can grow business, and it appears to be working.

A different kind of example, and again it's not a pivot, is a new company called Scoot Science. They're based out of California and are still under 10 employees. They have developed an analytic offering aimed at the fish farming market, and it's basically an undersea conditions forecast for the next 3–5 days. They sell this to fish farms, and they mostly develop their forecasts from existing data sources. They adapt their offering to their client's available data collection sources.

My understanding is that they have worked with insurance underwriters as part of their market outreach. Think about a fish farm pen full of salmon. You could lose millions of dollars of inventory in a matter of days if you get something wrong. So this data-focused firm is working with both fish farmers and underwriters to demonstrate that their forecasts have value.

So rather than pivoting, what we're seeing is an organic growth within the community but absent the sort of approach by some of the bigger corporations, like the IBM example.

One another example is coming out of Cuna del Mar. This is an organization that benefits from both a philanthropic and market-driven approach. They have put together a series of technology companies under the banner of a subsidiary called InnovaSea. They're not, on a day-to-day business basis, driven by a profit motive. They do aim to be profitable, but they're essentially an investment arm trying to decrease the cost of high technology for fish farming. This has data implications and they are thinking at scale because they are thinking about industrial fish farming globally. So again it's not exactly a pivot but a large influx of capital moving into a big industry driven by a future vision enabled by ocean data.

What levels of education/training are most needed in the workforce of the future (high school, community college, college, advanced degrees, etc.)?

My personal opinion is that skills are going to matter more than degrees. There will always be meaning to say a bachelor's degree as compared to a Ph.D. but for a lot of these jobs, it's going to be about transferable and scalable skill sets.

For example, a worker will need to become more computer-savvy because they may have a digital instrumentation package they have to plug in and understand. They might need to retrieve the historic data produced by a device and, seeing these fluctuations, decide to take certain maintenance steps. The skills to perform this function are important, but do not demand a college degree.

I am very supportive of community colleges and focused education programs that deliver a defined skill set. I could envision the same skills that enable a heating technician to debug and/or maintain a modern heating system being applicable to a fish farm technician. I would look to see this kind of training opportunity grow, as the jobs do.

How important is professional certification (e.g., certified marine technologist) for the workforce of the future in your field?

There will be opportunity for certification to have value. I'm not sure how broad a segment of the workforce that will apply to. Some fields will still value degrees over certifications, but in the end, it's about what can you do. Your drive, ideas, grit, charisma in some cases will be most important. However, I do envision increasing parity between degrees and certificates, as there will be increasing focus on skills and competencies.

And there will be a spectrum of need. For example, somebody could be a premier hydrographer and could have demonstrated this through certification at a number of levels, all of which are narrowly defined in terms of hydrography. But a person who's established certifications in a range of technical areas might be the kind of person you want to hire as your Chief Technology Officer for a start-up or might be the person you want to have as the Director of a major program of a federal agency as opposed to the project manager working on a particular program where you want that deep-dive skill set.

What sectors are doing the best job of positioning their workforce for the challenges of the future?

Most of us have to change what we do over a working career. The marine technology space is pretty diverse. We could be working with weather issues, we could be working with chemistry issues, and the technologies over our careers change relatively quickly.

If we think about a company such as Sea-Bird, the folks who are designing and even manufacturing Sea-Bird's oceanographic instruments, they probably could transition into the biomedical field right now. So I think those kinds of industries or those kinds of skills and employment are well or better positioned for the challenge of the future versus say a person who is an offshore oil roughneck. The future in offshore is wind and that person may have a much harder time transitioning. I suspect the heavy iron side of our industry (or heavy machinery, whatever you choose to call it) is probably less well equipped for the future.

How would you speculate on the fundamental differences in the workforce working the blue economy in the year 2050 compared to the workforce that we've got right now?

I suspect one thing that will make the blue economy workforce really stand out in the future, the one thing we haven't really talked about but we probably should, is global perspective. In 30 years you probably could be a still be a successful data scientist working for Google and your worldview can be shaped by the cubicles and the crowded streets of Silicon Valley and you may not need to have any clue about the rest of the world. I'm not saying you would, but you could perform in the future workforce 30 years from now with a limited regional view. However, working in the blue economy will need an understanding of both the science and at some level the policy and the economic impacts of our planet so I think that's a big one. We inhabit an ocean planet and our blue economy workforce will need a large-scale perspective that crosses geographic, cultural, and political boundaries.

So, I suspect that leads to people who are more creative, more communicative, more open to learning about different points of view. This is another trait that will likely be particularly useful to the blue economy.

I also think our workforce will require strong data literacy and numeracy skills to interpret data and model output.

Any other thoughts you would like to share?

I remain bullish on the vision of this new blue economy. I'm excited about it and yet I still don't know how to make it come true. I'm having more conversations with more people. Some of the start-ups I advise both formally and informally are working on this. I'm excited and energized even if the path is not entirely clear. I wish all my blue economy colleagues fair winds and following seas!

Justin Manley biography

Justin Manley is an innovative technologist and executive with experience in start-up, public corporation, academic, and public sectors. He attended the Massachusetts Institute of Technology and earned a B.S. in History, a B.S. and a M.S. in Ocean Engineering. In addition, he obtained an Executive Certificate in Strategy and Innovation from MIT's Sloan School of Business. Mr. Manley has been working with marine technology and robotics since 1990 and is a recognized leader in unmanned systems development and operations. After professional roles at MIT, supporting NOAA and in the private sector, Mr. Manley founded Just Innovation Inc. in mid-2015 to support clients with a focus on robotics and undersea systems. He is extensively involved in the marine technology profession as a Senior Member of IEEE, a Life Member of the Marine Technology Society (MTS), and a Fellow of the Institute for Marine Engineering Science and Technology (IMAREST). Mr. Manley holds two patents in unmanned systems oversight and security and regularly advises start-ups in the ocean tech sector.

OP/ED: Enhancing ocean observing through collaborations

Peer Fietzek

Growth of the ocean observing industry

The market of ocean observation is changing from a focus on new technology to a focus on information derivation and allocation. Projects and products increasingly involve either single industrial entities or industrial consortia spanning the full value chain of products, systems, data, information, and services and not just focusing on selected components anymore.

The success of the New Blue Economy will depend on the value it generates for its customers and society. In fact, efficient, reliable, and adequate (or high-quality) data acquisition with small spatial footprints and power requirements or, in short, advanced ocean observing technology comprising instrument carriers (platforms) and sensors will be essential for the New Blue Economy businesses utilizing ocean data as the basis for deriving valuable information products. As such a growing New Blue Economy represents a trend positively driving the maturity of the market for marine sensing technology just like and accompanied by the ongoing harmonization, consolidation, and structuring in the scientific sector. Some examples include:

- the Framework for Ocean Observing (FOO) (Lindstrom et al., 2012) and the Essential Ocean Variables (EOVs) (GOOS, 2020a)
- the Global Ocean Observing System (GOOS, 2020b) with its regional OS networks (GOOS, 2020c; EuroGOOS, 2020; IOOS, 2020, and see Gouldman and Ryabinin, this volume)
- the Atlantic Ocean Research Alliance (AORA, 2020) and the All Atlantic Cooperation for Ocean Research and Innovation (AANChOR, 2020)
- scientific networks and "networks of networks" in general
- the trend toward best practices publications (IODE, 2020), open access publications, and FAIR data (Wilkinson et al., 2016).

Reaching further, a momentum initiated by the New Blue Economy might finally become a driver and enabler for market creation, taking ocean observing technology to a higher level of industrialization, beyond the current state of a diverse market with "individual" scientists as the main customers. The New Blue Economy could continue, expand, and eventually achieve for ocean observing what could be observed over the last decades within traditional blue/ocean economies such as fisheries, offshore oil and gas, hydrography, navy or, depending on the sector definition, wastewater. The result should be to reach a level of industrialized products with clear advantages for the users, i.e., significant drops in cost per data point ratios, improved ease of operation, user-friendliness, and maximized reliability, application of common protocols and standardized interfaces. The drop in cost per data point ratios is already seen for several parameters, such as dissolved gas measurements of oxygen or carbon dioxide, and is expected to materialize for others, since many new sensors operate (semi-) continuously providing a large amount of data where beforehand only discrete water samples were possible. In addition, state-of-the art sensor–platform combinations will be deployed for longer missions transmitting or bringing back home much more data leading to a relative increase in data points that is not mirrored equally in an increase in respective sensor and platform prices. In fact, these combinations are more cost-efficient than existing methods, especially if the automated measurement process does not involve human interactions anymore. Speaking about industrialized products, we talk about instruments, sensors, solutions that are specifically designed products targeting certain applications in the context of ocean observation and that explicitly go beyond the state of a "commercialized product" as encountered often in the ocean sensing community. As a result, ease of operation and user-friendliness will benefit. Looking at consumer electronics or the automotive industry provides an exemplary notion of how complex devices can be tweaked for improved operability.

On the entrepreneurial side, an increasing maturity of the market would lead to an enhanced clarity and transparency as well as predictability and planning reliability supporting benefits for the sensor and platform businesses. On the one hand, this would increase the attractivity of the market for a (strengthened) engagement of large industrial companies, and on the other hand, it could stimulate foundation and growth of observation technology companies. The latter would go in line with an increased economic stability and security as well as the opportunity to enhance manufacturing efficiency and implement larger-scale industry processes. It can also be expected that these advantageous conditions of a more matured market will further encourage the incorporation of new technologies and concepts from other industrial areas as additional innovation drivers and positively affect everyday business experiences such as the customer acceptance for industry service engineer hourly rates. More economic value in the field would thus fuel the technological as well as business status-related advancement and innovation that is needed to overcome the challenges related to enhanced and persistent global ocean observing systems, i.e., maximized spatiotemporal coverage, more efficient data collection, conversion of data into widely used products.

Scientists and staff working in the research sector tend to focus on innovation and exploration aspects rather than on long-term, large-scale operational and maintenance work. Furthermore, a clearly enhanced ocean observing environment and matured ocean observation business will generate more operational work to maintain the observing system in place for enhanced data acquisition. Certainly, observation procedures will evolve and improve, but it will be exciting nonetheless to see how far the operational observation sector will be redefined and empowered to cope with these changes of a larger workload. In fact, it will likely be these operational services that will finally deliver tangible value closer to societal needs to the customers, i.e., provide services that can be more easily translated into economic value. These developments on the operational and service side imply that efficiency increases are not just requested on the sensor and platform side, but especially also in the way data are handled, quality controlled, stored, shared, etc., and that more room is given to the data acquisition business model in the sector, i.e., *ocean-observing-as-a-service* (OOAAS) besides *data-as-a-service* (DAAS) or *information-as-a*-service (IAAS) businesses, adding to, but not replacing, the established industry concepts of technology provision or data acquisition enablement. In any case, the operational and service sector for ocean observing will also require its specialized workforce to grow and cope with the market developments.

The Argo Program (Roemmich et al., 2019; Watson-Wright and Snelgrove, this volume) is one example nicely illustrating some of the arguments above and should be kept in mind when reading further; it is and will be a central element and driver for global ocean observing and the science and industry behind it.

Collaboration to support the transition

Fostered collaboration between the industrial and academic (research and education) sectors can stimulate and contribute to the strengthening of the ocean observing sector. The sectors of marine technology and marine sciences and education offer great promise for enhanced collaboration and resulting impact, especially related to monitoring, observation, and platform technology through aspects such as:

- The vessels and other platforms serve as interfaces for the work of different disciplines and offer many joint working opportunities.
- Both sectors are highly internationally organized and various shows and conferences regularly bring in touch the industry with the science representatives.
- Both sectors benefit from the currently growing societal importance and role of ocean observations driven by a growing environmental awareness; c.f. the Fridays for Future movement (https://fridaysforfuture.org/), the Sustainable Development Goals (SDGs) in which the ocean (Claudet et al., 2020; von Schuckmann et al., 2020) and data for formal status and progress assessment of the individual goals (Campbell, 2019) play a central role and the United Nations Decade of Ocean Science for Sustainable Development (the Decade, see Ryabinin, this volume). The latter underlines the truly global extent and ambitions for extended engagement and community building.

Research and development collaborations including educational and training aspects should be intensified. It is important that collaborators contribute according to their professions, strengths, and skills. The partners should be aware of and communicate their mutual and individual interests as the basis to get clear on interfaces and to prepare a win-win setting. This implies reflecting and adequately considering the different motivations to enter collaborations for businesses of different sizes (start-ups, small/medium enterprises, and large companies). Lastly, they should be willing to partner on eye level and have realistic expectations. Within collaborations academic partners can, for example, benefit from the development, engineering, and production well as economic, market, and marketing know-how of the industrial team mates, while industry, for example, values the innovative input on new applications and dissemination through publications, talks, etc (Fietzek, 2020). The private sector is highly qualified to develop technologies for the users. It's important that researchers and technicians come together, find and agree on common technological demands related to ocean observing, and express them in unison to the industrial sector. As part of this process the users should use their rising market power of a growing, more organized, and consolidated customer base as an incentive to leverage the realization of desired products and negotiate favorable contractual conditions.

Science—industry collaboration and private—public partnerships are key to disrupting the chicken-and-egg situation of "there must first be a significant demand for advanced sensors before companies can develop targeted products (i.e., customer push)" vs. "there must first be advanced sensors on offer, before orders can be placed (i.e., technology or manufacturer push)" and support the movement to more industrialized sensing solutions in an advanced market.

Fig. 1 depicts a relationship leading to a mature ocean observing environment capable of significantly increasing output data. Step 1 describes the economic value and societal benefit related to ocean observing as the kick-start to the process and linked to the generation of funding needed for the transition. Effective collaborations are essential at this stage. The current discussion of environmental awareness and sustainability generates a large momentum that science and industry should use to jointly highlight the importance of ocean observing and related societal benefits and economic value. This is a "once in a lifetime opportunity" as often stated in connection with the Decade.

During these promising times related to ocean observing, we should also consider and openly discuss new collaboration models. In this context it is valuable to learn from historical success stories, for example, in fisheries science and the development of echosounder equipment. There is a long history of successful science—industry collaboration in Norway between the Institute of Marine Research in Bergen and the company Simrad, now part of Kongsberg Maritime, dating back as far as the late 1940s (Schwach, 2004). The early days of this partnering reveal contemporary characteristics: scientists described their needs and qualified engineers built the instruments

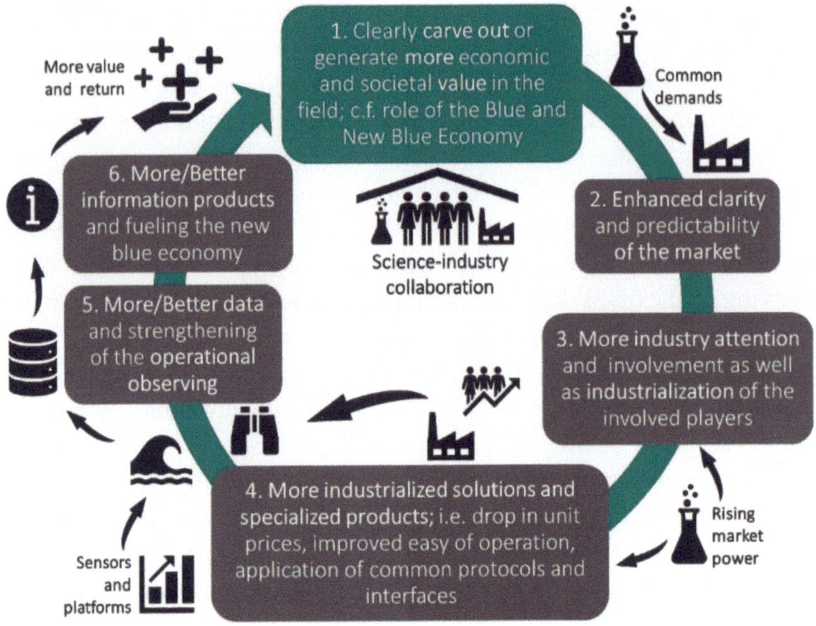

Figure 1 Circular relationship for achieving a growing and more industrialized ocean observing environment benefitting all stakeholders and society.

addressing them, industry benefitted from the input from research institutes by scientists and technicians (an often overlooked group of valuable innovation contributors), importance of good intersectorial connections allowed for efficient and informal ways to discuss problems and suggest improvements. Obviously, there are other contemporary examples of inspiring public–private-partnership activities: the US National Oceanic and Atmospheric Administration (NOAA) collaborated with an exploration company to advance new tools for exploration and mapping (NOAA, 2020a), with a cruise company to conduct and share science (NOAA, 2020b) and with an advanced Unmanned Surface Vehicle (USV) manufacturer to evaluate the suitability of USVs for cost-efficient, large-scale, high-quality oceanic and atmospheric observing (PMEL, 2020). The "collaboration approach to carve out and/or generate value" aspect of step 1 is also beneficial in the sense that difficult assessment questions, concerning, for example, the connection of ocean observing to the exploitation of ocean resources (c.f. mesopelagic fish, deep-sea mining), are automatically approached and discussed by different stakeholders and opened to society thus helping to achieve consensus.

As a result of the enhanced value proposition, the market would become clearer and more predictable (step 2), causing the business environment to industrialize (step 3), resulting in more fit-for-purpose solutions (step 4), a more solid database, and an enhanced operational sector (step 5) and finally support the New Blue Economy (step 6). Step 6 closes the loop, takes us back to step 1, and thus fuels the upward spiral, since a growing New Blue Economy raises the value of ocean observation and provides an economic return to society through taxes, jobs, etc. According to this process, advanced ocean observing is more a profiteer of than a driver for effective science—industry collaboration.

Summary

The New Blue Economy supports the industrial maturation of the ocean observing environment resulting in benefits for the stakeholders such as cost per data ratio reductions, market clarity improvements, and boosted innovation. The business sector of operational observing and business models related to information services will develop further. Enhanced collaboration between the academic and industrial sector represents a means to generate more value in the field, break up the customer push-manufacturer push stasis, and initiate an upward spiral of a growing sector benefiting the New Blue Economy and increasing the return to society. The currently growing environmental awareness and societal importance of ocean observing provide favorable conditions to foster collaborations and discuss new approaches.

References

AANChOR, 2020. Building an All Atlantic Ocean Community. Retrieved March 15, 2021, from: https://allatlanticocean.org/aanchor_csa.

AORA, 2020. Atlantic Ocean Research Alliance. Retrieved March 15, 2021, from: https://allatlanticocean.org/aora_csa.

Campbell, J., 2019. Measuring Progress: Towards Achieving the Environmental Dimension of the SDGs. Nairobi, Kenya. ISBN: 978-807-3750-9. United Nations Environment Programme. Retrieved from: https://www.unenvironment.org/resources/report/measuring-progress-towards-achieving-environmental-dimension-sdgs.

Claudet, J., Bopp, L., Cheung, W.W., Devillers, R., Escobar-Briones, E., Haugan, P., et al., January 24, 2020. A roadmap for using the UN decade of ocean science for sustainable development in support of science, policy, and action. One Earth 2 (1), 34—42. https://doi.org/10.1016/j.oneear.2019.10.012. Perspective.

EuroGOOS, 2020. Regional Operational Oceanographic Systems (ROOS). Retrieved March 15, 2021, from European Global Ocean Observing System: http://eurogoos.eu/regional-operational-oceanographic-systems/.

Fietzek, P., February 17, 2020. Science-industry collaboration in the maritime sector — analytical reflections and experiences from sensor innovation initiatives. In: 2020 Ocean Sciences Meeting, 16-21 February 2020, San Diego, CA USA. https://doi.org/10.1002/essoar.10502302.1.

GOOS, 2020a. Essential Ocean Variables. Retrieved March 15, 2021, from The Global Ocean Observing System - GOOS Framework: https://www.goosocean.org/eov.

GOOS, 2020b. The Global Ocean Observing System. Retrieved March 15, 2021, from: https://www.goosocean.org/.

GOOS, 2020c. GOOS Regional Alliances. Retrieved March 15, 2021, from The Global Ocean Observing System: https://www.goosocean.org/index.php?option=com_content&view=article&id=83&Itemid=121.

IODE, 2020. Ocean Best Practices. Retrieved March 15, 2021, from UNESCO/IOC Project office for International Oceanographic Data and Information Exchange (IODE): https://www.oceanbestpractices.org/.

IOOS, 2020. Regions. Retrieved March 15, 2021, from Integrated Ocean Observing System: https://ioos.noaa.gov/regions/.

Lindstrom, E., Gunn, J., Fischer, A., McCurdy, A., Glover, L.K., 2012. A Framework for Ocean Observing. Intergovernmental Oceanographic Commission. Retrieved from: https://unesdoc.unesco.org/ark:/48223/pf0000211260.

NOAA, January 9, 2020a. NOAA Teams With Ocean Infinity to Advance New Tools for Ocean Exploration and Mapping. Retrieved from News & Features: https://www.noaa.gov/media-release/noaa-teams-with-ocean-infinity-to-advance-new-tools-for-ocean-exploration-and-mapping.

NOAA, January 16, 2020b. NOAA Teams Up With Viking to Conduct and Share Science Aboard New Great Lakes Expedition Voyages. Retrieved from NOAA Research News: https://research.noaa.gov/article/ArtMID/587/ArticleID/2580/NOAA-teams-up-with-Viking-to-conduct-and-share-science-aboard-new-Great-Lakes-expedition-voyages.

PMEL, 2020. Saildrone. Retrieved March 15, 2021, from Ocean Climate Stations - Pacific Marine Environmental Laboratory: https://pmel.noaa.gov/ocs/saildrone.

Roemmich, D., Alford, M.H., Hervé, C., Johnson, K., King, B., Moum, J., et al., August 02, 2019. On the Future of Argo: A Global, Full-Depth, Multi-Disciplinary Array. Front. Mar. Sci. 6 https://doi.org/10.3389/fmars.2019.00439. Article 439.

Schwach, V., 2004. An Eye into the Sea: the Development of Early Fisheries Acoustics in Norway 1935-1960. In: Keuren, D.v., Rozwandowski, H.M. (Eds.), The Machine in Neptune's Garden: Historical Perspectives on Technology in the Marine Environment. Science History Publications/Watson Publishing International/USA, Sagamore Beach, pp. 211—242.

von Schuckmann, K., Holland, E., Haugan, P., Thomson, P., 2020. Ocean science, data, and services for the UN 2030 Sustainable Development Goals. Marine Policy 121 (104154), 1—14. https://doi.org/10.1016/j.marpol.2020.104154.

Wilkinson, M.D., Dumontier, M., Aalbersberg, I.J., Appleton, G., Axton, M., Baak, A., et al., March 15, 2016. The FAIR guiding principles for scientific data management and stewardship. Sci. Data 3, 160018. https://doi.org/10.1038/sdata.2016.18.

SECTION V

Building the workforce

Having defined the rationale and foundational concepts for the New Blue Economy in previous sections, the question remains … how will we ensure that we've got the people and structures to make this happen? In this section we provide perspectives, opinions, opportunities, case studies, and cautionary notes about moving forward with the education and training of the next-generation workforce. **Liesl Hotaling** presents the overall framework for preparing the workforce and the context for subsequent examples. **S. B. Moran** discusses the development of a formal graduate program aimed at bringing ocean expertise and business acumen together more effectively. Similarly, **Nick Beaird et al.** discuss a successful program to train operational oceanographers at Rutgers University. **Deidre Gibson's** case study of personal accomplishment provides an important retrospective that should inform future development. The role of formal training of marine technicians in a rapidly evolving economy and in the context of the articulation between community colleges and 4-year degree programs is discussed by **Hans Van Sumeren.** The role of big data in education—an important consideration in light of expected demand for workers fluent in this area—is expressed in an op-ed by **Randy Kochevar.** **Louisa Koch**'s op-ed provides a perspective on the importance of key skills. In particular consideration of the fact that the next 10 years will be the

UN Decade of Ocean Science for Sustainable Development, we felt it was important to consider the perspectives of those early career ocean professionals who will be rising to leadership positions by 2030; an interview with **Alfredo Giron** provides a firsthand description of some of the challenges that cohort will face. Similarly, the value of traditional ecological knowledge may be considered a *sine qua non* in the build-out of the New Blue Economy, a focus of the interview with **Noelani Puniwai.** Key populations of our community deserve important considerations in how we address the future. We present several op-eds addressing these concerns. The op-ed from **Ben Cuker and Deidre Gibson** frames these challenges in an important historical perspective. Similarly, **Kathy Sullivan's** op-ed is a compelling first-person argument supporting proactive and aggressive attention to diversity, equity, and inclusion. A specific example of how retired/separating military can play an important role in this workforce is discussed in an interview with **Bob Coniglione**. The question of professional certification also bears some treatment in recognition of the expected growth in the workforce for the New Blue Economy. An interview with **Bev MacKenzie** and an op-ed by **Steve Hall** present strong evidence justifying the development of more formal credentialing programs.

CHAPTER 20

Preparing the workforce for the new blue economy

Liesl Hotaling
Eidos Education, Highlands, NJ, United States

Introduction

Opportunities associated with the emerging new blue economy are coupled with the responsibility to use the ocean sustainably, to share knowledge, and to establish effective practices to ensure that professionals working in and developing commercial applications for the sector emphasize sustainability of all ocean resources. Currently many educational preparation and workforce training programs are not structured to generate graduates prepared to meet these opportunities. Which skills will the workforce need to succeed?

Members of the 21st century science, technology, engineering, and mathematics (STEM) workforce will need the abilities of adaptability, complex communication/social skills, nonroutine problem-solving skills, self-management/self-development, and systems thinking (Anderman et al., 2012).

While no student or worker will have the breadth of knowledge, skills, and information alone required to power the new blue economy, members of the workforce will require a transdisciplinary understanding of the ocean and the ability to integrate science concepts, engineering methods, business acumen, and sociopolitical considerations. And directly alongside, and potentially one of the most important skills, the ability to collaborate within a transdisciplinary team. The need for resolving complex ocean issues, while developing whole new ventures related to economic stability, environmental health, and national security will require a collaborative workforce with diverse skills and backgrounds.

These future workers will have a significant impact on the future ocean, helping advance and implement renewable energy sources; generating more robust and accurate data leading to improved predictive capabilities, tools, and services for mineral, marine management, and commercial

Preparing a Workforce for the New Blue Economy
ISBN 978-0-12-821431-2
https://doi.org/10.1016/B978-0-12-821431-2.00015-9

shipping; mitigating coastal damage and severe weather hazards increasing with impacts of climate change; and supporting sustainable food harvesting.

It is estimated by 2030, ocean-focused industries are anticipated to employ approximately 40 million full-time equivalent jobs. That number does not include the collaborative, adjacent industries. The anticipated fastest growth sectors are considered components of the new blue economy including offshore wind energy, marine aquaculture, fish processing, and port activities (OECD, 2016; (OECD report)). The majority of these jobs will require technology-focused skills and abilities. If trends in occupation types continue, there is also an expectation of 8% of jobs available by 2030, which have never existed before and require technology skills (McKinsey Global Institute; (McKinsey & Company,)). The new blue economy will be propelled by the data-enhanced existing ocean sector and the emergence of new transdisciplinary applications and businesses to effectively utilize ocean data and resources.

However, to meet the demand of a workforce possessing the required skills and abilities, changes in primary to secondary educational experiences are needed because given our current US education system, by 2030, less than 50% of students will be prepared for technology-related jobs (McKinsey Global Institute), which could impact the growth and potential of the new blue economy.

Big picture changes needed in education for the 21st century

Improving primary STEM education

Although this chapter is not intended to serve as a comprehensive treatise on STEM education, it is important to shine a spotlight on some existing, underlying issues that will hamper progress in nurturing the needed workforce.

The first is the method of delivery of STEM education in typical K-12 schools. Although improvements are being made in some schools, all too often, science especially is taught more as a history course, memorizing dates and accomplishments, instead of a living, breathing field of study. To gain the necessary STEM skills, all students must be provided with more opportunities to learn practically (Bransford et al., 2000), rather than theoretically—through self-motivated discovery, as opposed to didactic lecturing—and in collaborative learning environments using technology-based problem-solving.

Data-rich, technology-based, collaborative problem-solving STEM activities provide a context for learning, lend themselves to sustained inquiry and revision, and are proven methods for engagement of all learners and instill 21st century workforce skills. However, a large portion of instruction in US classrooms cannot be characterized as such. It is expensive to outfit schools with sorely needed equipment and provide proper professional development for teachers to implement the technology with purpose; however, it is a needed step to prepare our children properly for their future. It is clear that the pathway for unskilled workers is increasingly narrow and unstable. Today, technology skills are not just required for students pursuing degrees in higher education, technology is permeating every facet of our lives and job requirements. Future workplaces will only increase the demand for workers with technology skills. We cannot afford to lose students' interest in STEM any age.

To integrate these data-rich, technology-based, collaborative problem-solving learning experiences into preparation programs, students require computer and Internet access.

Access to connectivity and equipment (or lack thereof)

Just because you are connected does not mean all are connected.

Although schools have increased the number of computers and improved Internet access for students, greater access is still needed for technology to become a reliable learning tool. Studies continue to document that the availability, quality of technology resources, and quality of teacher professional development are distressingly inadequate, and most educators report that their classroom is not the main location in the school where their students use computers. Meaning most schools require students to share, travel within the school to access technology resources, or access the technology resources outside of school. Therefore technology is not integrated as a learning tool into most educational experiences. A national count in the United States of computers in public schools shows a ratio of 3.8:1 (Wells and Lewis, 2006) for the number of students sharing an "instructional computer" with Internet access. Although some school systems have implemented a 1:1 program (one computer to one student), this is not the case in the majority of schools in the United States, nor other affluent countries in the world (Fraillon et al., 2018). In many schools, the ratio from 15 years ago is more often the ratio still found today, and students still needing to transit to computer laboratories in schools.

Meanwhile access at home is dependent on a range of factors. In 2015, in the United States, 94% of children aged 3–18 had a computer and 61% of children aged 3–18 had Internet access. The percentages of children with access at home were higher for children who were older, those whose parents had higher levels of educational attainment, and those whose families had higher incomes. Also, higher percentages of children who were White (66%), Asian (63%), and of two or more races (64%) had home Internet access in 2015 than did Black (53%), Hispanic (52%), and American Indian/Alaska Native children (49%) according to the US Department of Education, National Center for Educational Statistics (2018). Again, this inequity is not just found in the United States, according to the 2018 IEA International Computer and Information Literacy Study Report (Fraillon et al., 2018), patterns of lack of access connected to socioeconomic status, migrant status, and more rural locations ripple through the 12 countries participating in the study.

The two main reasons children aged 3–18 lacked access to the Internet at home were (1) expense and (2) their family was not interested in having it. A higher percentage of suburban students had Internet access at home than students in rural areas, with the largest difference noted in remote rural areas.

The writing of this chapter coincided with the COVID-19 pandemic and the advent of nearly ubiquitous remote learning. This acute shift has not only fully exposed issues related to the digital divide, it has turned the concerning issue into a state of emergency. Lockdowns contributed significantly to lack of access issues as people who cannot afford Internet services at home would regularly use public libraries and other public facilities, which were shuttered, further exacerbating educational divides, as well as economic and healthcare divides (Watts, 2020).

A Pew Research study conducted in April 2020 (Pew Research Center, 2020) found that "36% of low-income parents reported their children were unable to complete their schoolwork" due to lack of Internet connectivity, compared to "14% of middle-income parents and 4% of upper-income parents."

The same Pew survey (N = 4917 US adults) explored attitudes about the "role of government in addressing these issues." Results revealed that "62% do ***not*** think it is the federal government's responsibility to ensure a high-speed internet connection at home." However, when asked about education, "80% think schools have an obligation" to provide connectivity to at least some (low-income) students, while "about one-in-five (19%) say they do not have this responsibility to any students." Clearly, this is a disconnect. Schools are already under tremendous financial pressure, yet the majority of Americans feel that it is a school's responsibility to provide these services, yet also feel the government should not need to pay for the service.

One result of this digital divide forced higher-poverty school districts (47%) to resort to distributing paper packets as the "primary component of distance learning" according to preliminary results from a survey (N = 2500 school district leaders) conducted by the American Institutes for Research (American Institutes for Research, 2020). The same survey found "rural districts were significantly more reliant" on paper packets, with the lack of connectivity contributing to the "COVID-slide."

While organizations donated funds or laptops (new or used), Internet service providers (ISPs) did drop temporary hotspots in densely populated areas, it was a temporary attempt at a solution, but did not help all in need. Americans require permanent solutions, yet little is being proposed. More creative solutions using existing infrastructures should be explored. For example, schools already are paying for Internet access with signal directed "inward." Why not, with additional antennas, during out of school times, direct the signal "outward" to provide service to students living in neighborhoods around the school.

It should be an imperative to generate creative solutions to this issue as connectivity and digital literacy are ever-increasing needs to exist in the modern age and employment. The new blue economy (NBE) will rely on workers to possess technology skills, even jobs requiring a high school diploma. Just as increased data collection abilities have decreased the need for oceanographers to go to sea to collect data and live and work in areas not on the coast, NBE jobs will follow the trend enabling ocean-focused employment opportunities in remote locations, both coastal and inland.

If we were able to quickly resolve access issues, which skills would students need to acquire for workforce preparedness?

Data literacy

The increasing integration of the Internet has already augmented how we operate our lives. We use Internet sources to make decisions as workers, consumers, and citizens. Therefore, developing a student's understanding and faculties for accessing and evaluating the legitimacy of internet sources and overall digital literacy is essential in a child's evolution to an informed citizen and consumer, let alone being prepared for a career in the NBE.

Learning to manipulate data (the use, interpretation, analysis, synthesis, and application) should be a cornerstone in primary school education. Beginning in primary and extending through secondary schooling, students should encounter a range of increasingly difficult learning experiences focused on utilizing increasingly complex data to build skills and comfort levels. Simplified, a learner's experience with data should first begin with orientation (basics of the data set), moving to interpretation (what facts does the data set tell you), and then to synthesis (what do these facts mean). The educator's role is to provide increasingly complex data sets, which lead to the problem-solving of unclear or murky real-world problems and increased learning outcomes (Hotaling et al., 2019).

The need for these data-focused educational experiences is a worldwide issue. According to the 2018 IEA International Computer and Information Literacy Study Report (Fraillon et al., 2018), of the 45,561 grade 8 student participants from 12 countries, approximately 36% were operating at the orientation level, 19% were operating at the interpretation level, and only 2% were operating at the synthesis level. And sadly 18% of students were well "below the lowest proficiency level."

While technology access will remain a challenge for schools in the near future, the pivot point on this issue is supporting teachers using integrated technologies in their teaching. Quality and continued professional development for educators are required to increase their competency and comfort integrating technologies into their teaching. This goes beyond using a SMART Board in their classrooms. This means increasing data literacy skill levels so all teachers achieve the synthesis proficiency level, becoming adept with the use of a variety of technologies (the right tool for the right job), learning classroom management approaches for effective integration of technologies into student learning experiences, and effective and creative methods for assessment of technology-based student and student group work.

Is that all? Nope, one last consideration.

Rethinking mathematics education

To prepare our students for a world with big data, they must not only have access to big data and learn basic data orientation skills, they need to be fluent and acquire data manipulation and synthesis skills. These needs are the fundamental argument for upending the traditional mathematics curriculum (algebra, geometry, calculus) in favor of integrating either Data

Science or Statistics courses to more closely parallel the math skills they will use in their future. Currently only 10% of US high school students take a statistics class, and that Statistics curriculum is mostly theoretical. Other countries changed their mathematics curricula to include more statistics years ago, and in addition to better preparing those students for a world of data use, those countries often outperform the United States in the Program for International Student Assessment (PISA) mathematics scores.

There is hope, just as in the past universities have driven the top-down approach to mathematics curricula. Currently the SAT, the college admissions exam administered by The College Board, is in the process of changing their questions to focus on data literacy and data manipulation with an underlying knowledge of statistics, which might drive a change in mathematics curricula and implementation of data literacy skills across all grades. In a recent iteration of the SAT, 20% of the SAT math questions focus on problem-solving with data analysis, and the reading section includes questions focused on data interpretation and claims reasoning questions, many of which were science data themed.

Students learning and assimilating these skills in primary school would prepare them well with transferable skills for the future workforce. Students could be challenged with examples from ocean observatory time series data to conduct analyses and forecasts or use sparse data sets to conduct exercises in interpolation. Exposure to these data, data manipulation, and general data literacy skills are needed to support future ocean observatories and the NBE.

Primary education summary

The dream scenario of every student with a personal computer experimenting every day in a MakerSpace is unrealistic; however, we can make effective changes to primary education experiences within the existing design constraints.

The first action is to improve the faculties and confidence of technology use with primary educators. Educators need support to learn the effective and practical application of data-rich, technology-based problem-solving in their teaching. We of course want to continue to improve access to quality technologies for our students, but it is also important to maximize the resources already in schools.

The second action is to improve data and STEM literacy by augmenting the STEM curricula to increase the number of collaborative, real-world, problem-solving activities in STEM activities in classrooms. Students need to "see" the application of what they are learning to their lives today or future interests.

The third action is to provide multiple opportunities for students to "see" themselves in a job or career using technology skills. Again, students who drop out of school claim the main reason for leaving is not seeing the connection between what goes on in school and their future.

What can you do? Although many readers of this chapter are not driving national-level education policy, you do have schools near you. Investigate and consider ways to connect teachers to resources your universities, companies, or agencies might offer such as: technology donation programs; summer intern programs for teachers; campus visitation programs to allow students to actually see STEM "in action"; producing STEM career information that spans all education levels (not just graduate-level opportunities); producing STEM career information featuring diverse people so all students can "see" themselves doing the job; or creating programs for your colleagues or undergraduate students to serve as near-peer mentors for local high schools, to name a few ideas.

Secondary education and workforce preparation

Globally, there is already a strong need for skilled people across multiple sectors, including sectors pertaining to the marine industry. There is a rapidly increasing need for skilled workers and interdisciplinary workers with core and marine-related technical competencies or knowledge (data literacy including data processing and management, working with marine platforms and instrumentation) along with the project management skills required for work in the highly complex marine industry ranging from offshore oil extraction to marine aquaculture to storm prediction. To prepare students expeditiously for entry into the workforce, educational experiences must be created with direct input and support from the marine industry while emphasizing the skills needed for immediate employment opportunities.

Looking forward to the NBE in the near future, when new industries are exploding with growth, memberships in professional societies, flexible degree attainment, professional certifications and accreditations will be mechanisms to validate the knowledge and expertise of those working in the field and quickly entrain workers into the workforce.

Again, this chapter is not intended to serve as a comprehensive treatise on NBE workforce preparation; however, the following sections will shine a spotlight on some existing, underlying issues hampering progress in nurturing the needed workforce and potential solutions.

Addressing diversity

According to the world ocean initiative, to date, the ocean industry continues to be one of the male-dominated industries. This is changing; however, accelerated change is needed. Research repeatedly reveals that more diverse and inclusive workplaces lead to better innovation, increase in effective responses to change, and exceeding financial goals.

It is our collective responsibility to change the "face" of the industry (see contribution from Sullivan, this volume). People need to be able to see themselves performing the required skills. This begins at the grade-school level by providing real-world learning examples that are attractive applications of STEM skills for students so they can see the bigger picture as to "why" they need to learn and how that learning will be applied when they graduate. These examples need to include ocean-themed examples, which rarely occurs. Grade-school students need to know there are STEM jobs available for a wide spectrum of education levels. Not everyone with a STEM-related job needs a Ph.D. or many years of schooling, which is a common misconception. Outreach from industry must include diverse faces for students to begin to self-identify with STEM opportunities. Outreach through awareness campaigns to Community Colleges and the Military also requires diverse faces as the racial and ethnic diversity in both populations is much higher than students attending 4-year colleges and universities. Once a more diverse workforce is recruited, significant effort must be continually given to provide inclusive workplace environments through providing nonhostile work environments, equity in pay, creating safe environments for sharing ideas and taking all employee concerns seriously, and providing equitable growth opportunities for all employees.

Many of the suggestions provided in the following sections will be steps in the right direction for addressing the lack of diversity in the ocean industry.

Flexibility in degree attainment

The 21st century has brought a new type of undergraduate student. The majority of undergraduate students are employed full-time while attaining their degree, are 25 years or older, and attain their degree while attending college part-time and less than 20% of undergraduates live on a campus (Le et al., 2019). This is most likely not the person conjured when thinking about an undergraduate and unfortunately does not match well with many existing degree program designs.

With the ever-increasing pulls on our time, the need for flexible and targeted education pathways is also increasing. To be more inclusive, degree-granting organizations need to increase the availability of flexible learning hours for their students and create better "on-ramps and off-ramps" (remediation, transfer, military crosswalks, and leave policies) for students pursuing degrees, and integrate industry standards and workplace competencies into curricula to better match job opportunities.

Degree-granting institutions that have made improvements to their transfer policies including articulation agreements and reverse transfer programs, improvements to their remediation programs, offering more online course options, and improved leave programs are increasing their graduation rates. And the students who benefit overwhelmingly from these changes are community college students.

Community college

Community Colleges are the lifeblood of higher learning. In 2016, nearly half of all enrolled undergraduate students were enrolled at a Community College. Students attending Community Colleges are much more racially and ethnically diverse than students attending 4-year colleges and universities, come from low- to middle-income families, nearly one-third are first generation to attend college, and approximately one-third go on to obtain a bachelor's degree, borrowing less money to do so (Radwin et al., 2018) than students who enroll directly in a 4-year college or university. It is important to note that according to the Science and Engineering Indicators 2020, more than half of all science and engineering graduates attend community colleges at some point in their career. This trend will most likely increase as college costs continue to rise (New America, 2019).

Given that Community Colleges are an entry point for such a large percentage of students, and the generator of so many of the technicians needed to support the NBE endeavors, the stigma that sometimes is associated with Community College graduates must be eliminated. In fact, there are multimillion-dollar campaigns in several states, including California and Washington, to do so. Community Colleges with programs combining technical education and college preparatory programs are more effective in preparing students for immediate employment. These students tend not to require as much on the job training as many programs are informed by industry, so a more inclusive recognition of the practical knowledge and immediate application of skills is a more apt view of including Community College graduates in the workforce.

The collaboration between Community Colleges and industry has many advantages; however, Community Colleges must always keep an eye toward local need and the other to national need. Industry collaborations, partnerships with business incubators and local cluster organizations are essential in identifying local workforce needs. These needs then frame and often augment the curricula offered by the school. These collaborations also result in internship and apprenticeship opportunities for students making them even more prepared to enter the workforce. However, industry collaborations can have some pitfalls if Community Colleges align too closely to one local industry, and skills gained may be less transferable to other job opportunities.

With respect to the NBE, a spectrum of technicians will be needed within the next 10 years. From mechanical, computer, biomedical, and network engineering, to telecommunications, to instrumentation technology to web development, all of these skill sets are applicable to a range of job opportunities within and beyond the NBE. So it is important that the existing technician training programs understand the opportunities within the NBE and offer insight to students through an awareness campaign to enable graduates to apply their skills to the NBE.

Vocational or trade schools

What if a student is just not interested in college? Over the past few decades, there has been a strong push to encourage all high school graduates to pursue a bachelor's degree; however, this is not a solution of best fit for all. Vocational schools are a great option to gain skills quickly and enter the workforce. The shortage of skilled workers mentioned earlier is driving salaries higher for millions of vacant jobs requiring skilled labor. The NBE will be no exception, especially individuals with training in the trade and industry (construction, mechanics and repairs, and precision production) and technical and communications sectors. Over the past decade, an average of 3 million unfilled jobs in the United States are available with average salaries in the $50,000.00 range (U.S. Bureau of Labor Statistics, Occupational Employment Statistics, 2018). This is a gap we cannot continue to ignore.

Different models to address this issue include apprenticeship programs; postsecondary vocational training programs (similar to the US community college model but with emphasis on vocational skills); and the German

Vocational Training System (Spees, 2018), often referred to as the dual training system. This hybrid approach combines the theory provided in traditional university courses and real-world apprenticeships in German companies. Similar to the German model, the P-TECH or Pathways in Technology Early College High School model emerged about a decade ago as a collaboration between IBM and the City University of New York (CUNY) (Spees, 2018). The P-TECH model has been implemented in other locations; however, the model should be considered by more schools to more rapidly prepare a technical workforce.

However, there is another route quickly gaining ground.

The role of certificates and microcredentials as degree substitutes and in life-long learning

Micro-credentials began their ascent about a decade ago in the field of computer programming and continue to rise in popularity and spread to other disciplines. Microcredentials are a method to recognize achievement in successfully demonstrating a very specific skill with a resulting digital badge. Certificates are obtained when a prescribed series of digital badges are collected representing a suite of skills required for the certificate. Microcredentials are low-cost, often online courses that are completed quickly and learners can choose to engage at their own pace, maximizing flexibility for the student.

The NBE could utilize the microcredential infrastructure existing across the world to quickly provide workers with initial job qualifications to fill vacancies for skilled workers, quickly retrain existing workforce, and also utilize the infrastructure to provide supplemental skills to traditional degree holders to create an adaptable workforce with a rich variety of skills.

Microcredentials and certifications are not just for vocational or community college students, the concept is also taking hold at the university level for preparing undergraduates as employers are issuing a clear call to action for universities to integrate their course work to real-world experiences.

Another interesting approach to life-long learning is to create a hybrid microcredential and microinternship program. Experiential learning is provided through the partnerships of colleges and industries, such as the Experiential Network (XN) hosted at Northeastern University. XN provides opportunities for students to participate in 6-week, virtual projects with real companies. These experiences "empower graduate and professional learners to integrate and apply their learning and development in a real-world professional setting" (https://careers.northeastern.edu/students/experiential-network-xn/).

The NBE could bring a diversity of partners, vocational schools, community colleges, universities, small and medium-sized enterprises into flexibly organized networks. Preparing students to work on a range of scientific and technological innovations (e.g., marine robotics and autonomous vehicles, aquaculture, marine renewable energy, biotechnologies, offshore oil and gas) coupled with entrepreneurial or business education basics can leverage skill diversity and resources to benefit all parties with more industry-aligned and competency-focused workers.

The organized networks could also serve each other (articulation agreements) to provide microcredentials contributing to certificates conducted between partners, utilizing each other's resources (laboratory equipment, welding equipment, hydraulics training systems, etc) to quickly prepare workers for the NBE.

Professional certifications or qualifications are also offered through professional societies or other designated agencies. These certifications attest to an individual's proficiency in an industry or profession. Most certifications are intended to be part of an individual's life-long learning path, therefore certifications are normally time-limited with the expectation of evolving changes to the industrial or professional collective knowledge. The ocean sector already has several organizations offering certifications in a number of topics, and this is anticipated to increase in response to the growth and the interdisciplinary nature of the NBE (see contributions from MacKenzie and Hall, this volume).

Professional development for educators

In addition to awareness campaigns, professional development for educators is needed for increased awareness of all the opportunities the NBE has to offer and modifications needed for existing degree programs to be more flexible for adult learners and increase inclusivity for veterans. Also professional development is needed to enhance educators' disciplinary capabilities, teaching skills, understanding of current technologies and practices in industry, and employability skills. Activities typically include workshops, intensive seminars, microcredential programs, industry internships, or a combination.

Military transition assistance program (TAP) and civilian job equivalent assistance

One large group of highly skilled individuals with strong potential to participate in the NBE workforce yet to be discussed are service members transitioning to civilian life. Currently there is an educational divide between many retiring and separating veterans and requirements for jobs and second careers available within the NBE. Many veterans who enlisted in the military and achieved the rank of E5—E9 have spent more than a decade accruing technological skills but often hold a high school diploma or GED and little to no college credit (90%+ of all enlisted military), which serves as an impediment after leaving service and entering the civilian workforce. Yet contemporaneously there is a need for experienced marine professionals with many of the skills these veterans possess.

First, similarly to the Community Colleges, an awareness campaign is needed within the Transition Assistance Program (TAP). TAP provides information, tools, and training services to assist service members with their transition from military to civilian life. They provide information and short courses on preparing for transition to employment, enrolling in courses at colleges and universities, or even starting your own business. The ocean industry and forward looking the NBE would benefit greatly from connecting with TAP personnel to raise awareness of the numbers and types of jobs immediately available and the great potential for entrepreneurs.

Attention should also be given to Military Occupation Codes (MOC) Crosswalk programs. Currently serving military personnel identify all of their experiences, credentials, education, and skill sets obtained while in service and compare them to civilian career opportunities while conducting a gap analysis to understand how to become better prepared for the civilian employment market. Most use the O*NET Interest Profiler to help determine the type of civilian work and occupations of interest. O*NET, just like other similar tools, is online and choice selection is required. If the opportunities are not in the database, they will not be available choices in the programs, so again, an awareness campaign is needed providing a framing of all the opportunities available and simple maps of direct application of existing skills and if needed, simple pathways to obtain other needed skills to prepare many veterans for successful transitions into the NBE.

Secondary education and workforce preparation summary

Globally, there is already a strong need for skilled people across multiple sectors, including sectors pertaining to the marine industry. To prepare students, veterans and retrain workers from other industries for entry into the NBE workforce, we must collectively take proactive steps to create attractive and inclusive opportunities. The NBE workplace must be a diverse and welcoming place of employment, capitalize on skill sets from a range of workers and encourage communications to create interdisciplinary workers, and encourage life-long learning through obtaining micro-credentials and professional certificates to promote personal growth and keep pace with technological changes.

Case study example

To summarize many of the earlier comments and suggestions while offering a concrete example, let's revisit the Portland Grain Shipping example of Optimized Commerce provided by Spinrad in this volume. The commented example will provide specifics as to the educational requirements for the knowledgeable workforce and potential economic development of the NBE.

Optimized commerce

Consider, for example, the commercial shipping activity associated with a port like Portland, Oregon. Portland's main commodity is grain; the US Department of Agriculture values wheat exports from Oregon, Washington, and Idaho at $1.3B, or 40% of US wheat exports annually (https://www.pnwa.net/wp-content/uploads/PNWA_Handout.pdf).

In 2017, the Columbia River saw over 1500 ship visits (NOAA, 2010), many of which transited from Portland, down the Columbia River to the Pacific Ocean (a distance of 105 miles) and then to receiving ports around the world. Given the tidal mode of that part of the world, the transit down the Columbia River from Portland/Vancouver typically takes about one tidal cycle (6–8 h). There are five bridges under which the ships must pass, and a shipping channel of typically 42 feet depth is maintained.

The US Pacific Northwest is a region of considerable precipitation, imposing significant hydrological impact on the Columbia River (with the consequent changes in relative buoyancy of the ships depending on the periodic injection of fresh water into the system). Given that the goal of the grain exporters is to optimize the load-out of their products, there must be careful consideration given to factors such under-hull clearance, air gap (under bridges), as well as the important factors of surface currents and winds, especially in the broad lower Columbia River. In short, the challenges with determining how much grain to load, when to debark from Portland/Vancouver, and what speeds to use in the transit down the river are significant.

A reliable predictive model of all the environmental factors (tides, currents, winds, waves, salinity, etc.), coupled with a vessel-specific load-out model would be of enormous value to the shipping community. And, of course, such a system of models will be unique to each port environment; Portland is different from New Orleans, Baltimore, Boston, etc. Again, such an economic opportunity critically depends on the availability of publicly funded data sources, but can be developed in a tailored, client-specific, fully commercializable manner.

Comments

The earlier example highlights the need for data from coupled systems and data applications needed by a variety of audiences with a spectrum of background content knowledge. Following is a sample of jobs and skills required to meet this challenge:

Job	Role/skills
Scientists • Ocean • Climate • Atmosphere • Agriculture Engineers	Scientists and engineers work together to develop the observation networks, analyze resulting data, work to produce forecasts and other applications of the data sets. These scientists and engineers primarily hold PhDs, therefore years of training and honed technical skills and abilities.
Data modeler, data visualization, data management	These jobs range in technology skills and range in education requirements (A.S.—M.S); however, the ability to work in transdisciplinary teams is essential.
Data translator, data communicator	Translating and communicating data sets to nonexpert users is the lynchpin for utilizing data sources. These jobs require skills in technology, science, engineering, education, and psychology; however, usually degrees (B.S—Ph.D. focused in one or more fields).
User experience (UX) designer	These jobs range in skill specificity from coding to customer facing skills and range in education requirements; however, technology skills and ability to work in transdisciplinary teams are essential.

Job	Role/skills
Sensor developer	These jobs range in technology skills, science, and engineering background, but tend to be transdisciplinary individuals. Education requirements range (A.S.–M.S); however, the ability to work in transdisciplinary teams is essential.
Technicians • Sensor • Platform • Communications	These jobs range in technology skills and science background, but tend to be transdisciplinary individuals. Education requirements range (A.S.–M.S); however, the ability to work in transdisciplinary teams is essential.
Farmer	Data literacy and data management skills to access and apply climate and weather data, respond to field sensor data, optimize growing season, and yield through data applications.
Transportation logistics • Ocean • Rail • Road • Port operations	Port workers and supply chain management (H.S. diploma–Master's) involve a great deal of data management. Data provided by IOOS (see contribution by gouldman, this volume) and observing systems around the world provide the underpinning data for port logistics for port operations, which is coupled with the rail, road, and weather conditions to move goods.
Ship captain Harbor pilot	Navigating a ship from open ocean to the dock depends on having the necessary real-time and forecast environmental information in varying situations. Not only do they need to understand these data, but be prepared to respond quickly to changing situations. Education requirements range (H.S. diploma–Master's); however, require years of on the job experience and training.
Coast guard	Data literacy skills to apply real-time ocean and weather data for search and rescue and other emergency situations.
Harbor police	Data literacy skills to apply real-time ocean and weather data for search and rescue and other emergency situations.

Summary

The tools, techniques, curricula, and postsecondary educational experiences needed for the workforce of the NBE will demand careful attention. Building on mainstream resources and expanding traditional, vocational, hybrid, microcredentials and certificates, and industry-based (e.g., internship) approaches, a new set of skills and experiences must be established quickly to attend to the growing need. A primary focus must be on defining the required skill sets for training this workforce and nurturing the capacity to integrate diverse subject matter and data across different scales—including temporal scales and spatial scales.

Looking forward to the NBE, flexible degree attainment, professional certifications and accreditations will be essential mechanisms to validate the knowledge and expertise of those working in the field and quickly entrain workers into the workforce.

The opportunity in the NBE is infinite. Let's work together now to prepare a workforce ready to take on the challenges and opportunities in the very near future.

References

American Institutes for Research, 2020. The National Survey of Public Education's Response to COVID-19. Retrieves from: https://www.air.org/sites/default/files/National-Survey-on-Public-Educations-Coronavirus-Pandemic-Response-First-Look-July-2020.pdf.

Anderman, E.M., Sinatra, G., Gray, D., 2012. The challenges of teaching and learning about science in the 21st century: exploring the abilities and constraints of adolescent learners. Stud. Sci. Educ. 48 (1), 89–117.

Bransford, J.D., Brown, A.L., Cocking, R.R., 2000. How People Learn: Brain, Mind, Experience, and School. National Academy Press, Washington DC.

Fraillon, J., Ainlet, J., Shulz, W., Friedman, T., Duckworth, D., 2018. This Disconnect Is Leading to a Compounding Workforce Readiness Issue Facing Us in the Very Near Future. Source: U.S. Department of Education, National Center for Education Statistics. Student Access to Digital Learning Resources Outside of the Classroom (NCES 2017-098), Executive Summary.

Hotaling, L., McDonnell, J., Ferraro, C., Florio, K., Lichtenwalner, S., 2019. Educating with data. In: Fauville, G., Payne, D., Marrero, M., Lantz-Andersson, A., Crouch, F. (Eds.), Exemplary Practices in Marine Science Education. Springer International Publishing, 2018.

Le, C., Pisacreta, E.D., Ward, J.D., Margolis, J., December 12, 2019. Expanding Pathways to College Enrollment and Degree Attainment: Policies and Reforms for a Diverse Population. https://doi.org/10.18665/sr.312296.

New America, Fishman, R., Nguyen, S., Acosta, A., Clark, A., 2019. Varying Degrees. Retrieved from:: newamerica.org/education-policy/reports/varying-degrees-2019/.

NOAA, 2010. Estimating Economic Benefits from NOAA PORTS INformation: A Case Study of the Columbia River. NOAA Technical Report, https://tidesandcurrents.noaa.gov/publications/Columbia_River_PORTS_Economic_Study.pdf.

OECD, 2016. The Ocean Economy in 2030. OECD Publishing, Paris. https://doi.org/10.1787/9789264251724-en.

McKinsey & Company, Jobs lost, jobs gained: What the future of work will mean for jobs, skills and wages (https://www.mckinsey.com/featured-insights/future-of-work/jobs-lost-jobs-gained-what-the-future-of-work-will-mean-for-jobs-skills-and-wages).

Pew Research Center, 2020. Special Report: 53% of Americans Say the Internet Has Been Essential During the COVID-19 Outbreak. Retrieved from: https://www.pewresearch.org/internet/2020/04/30/53-of-americans-say-the-internet-has-been-essential-during-the-covid-19-outbreak/.

Radwin, D., Conzelmann, J.G., Nunnery, A., Lacy, T.A., Wu, J., Lew, S., Wine, J., Siegel, P., 2018. 2015–16 National Postsecondary Student Aid Study (NPSAS:16): Student Financial Aid Estimates for 2015–16 (NCES 2018-466). U.S. Department of Education. National Center for Education Statistics, Washington, DC. Retrieved [date] from: http://nces.ed.gov/pubsearch/pubsinfo.asp?pubid=2018466.

Spees, A., 2018. Could Germany's Vocational Education and Training System be a Model for the U.S.? World Education News and Reviews. Retrieved from: https://wenr.wes.org/2018/06/could-germanys-vocational-education-and-training-system-be-a-model-for-the-u-s.

U.S. Bureau of Labor Statistics, Occupational Employment Statistics, 2018. National Occupational Employment and Wage Estimates in the United States. https://www.bls.gov/oes/current/oes_nat.htm.

U.S. Department of Education, National Center for Education Statistics, 2018. Student Access to Digital Learning Resources outside of the Classroom.

Watts, G., 2020. COVID-19 and the digital divide in the UK. Lancet 2 (8), E95–E396. August 1, 2020. Retrieved from: https://www.thelancet.com/journals/landig/article/PIIS2589-7500(20)30169-2/fulltext.

Wells, J., Lewis, L., 2006. Internet Access in U.S. Public Schools and Classrooms: 1994–2005. U.S. Department of Education, National Center for Education Statistics, Washington, DC. http://nces.ed.gov/pubs2007/2007020.pdf.

Further reading

Mistler-Jackson, M., Butler Songer, N., 2000. Student motivation and internet technology: are students empowered to learn science? J. Res. Sci. Teach. 37, 459–479. https://doi.org/10.1002/(SICI)1098-2736(200005)37:5<459::AID-TEA5>3.0.CO;2-C.

National Science Foundation, 2021. Science & Engineering Indicators 2020. Two Volumes. National Science Foundation, Arlington, VA.

The OECD report Rethinking Innovation for a Sustainable Ocean Economy. (https://www.oecd-ilibrary.org/sites/9789264311053-en/index.html?itemId=/content/publication/9789264311053-en).

CHAPTER 21

Workforce development and leadership training for the new blue economy

S. Bradley Moran
College of Fisheries and Ocean Sciences, University of Alaska, Fairbanks, AK, United States

Introduction

With greater societal awareness of environmental issues resulting from climate change, demand is growing for sustainable economic and environmental programs, plans, and budgets at the local, regional, national, and global level. There is also increased recognition of the need to broaden traditional academic training and degree programs to address the consequences of a rapidly changing climate, depletion of natural resources, and related global challenges of food security and human health. Many of these challenges present significant opportunities to a wide range of ocean-related industries, which has led to an increase in workforce demand in the traditional ocean economy, as well as in the emerging sustainable ocean economy—the New Blue Economy (see Kildow, 2021; Spinrad, 2021).

In particular, the new blue economy calls for a broad range of innovative and interdisciplinary education programs. Such programs will require a greater integration of technical, scientific, engineering, data information, and business management skills, including rigorous training in marine science and engineering (see Hotaling, 2021). The need is also growing for leaders with skills in commerce and science to develop sustainable business models that capitalize on emerging market opportunities.

This chapter discusses the need for a broad spectrum of education and training organizations, including technical institutes, community colleges, and universities, to align their educational offerings to build a workforce that meets industry and societal needs and sustainable development goals in the maritime sector. Key challenges and strategies in developing training programs for the new blue workforce are outlined, such as adapting science, technology, engineering, and mathematics (STEM) disciplines to a more integrated curriculum in support of a workforce that requires greater

Preparing a Workforce for the New Blue Economy
ISBN 978-0-12-821431-2
https://doi.org/10.1016/B978-0-12-821431-2.00002-0

technical data and information expertise in marine science and engineering. This is followed by a description of the opportunity to develop new leadership programs, such as the Blue Master of Business Administration (MBA), that combine graduate business training with formal education in marine science, engineering, and technology. Taken together, such innovative academic programs can facilitate greater collaboration between faculty, students, and industry as businesses adapt their management practices to address real-world challenges using sustainable business models as part of the new blue economy. In turn, there is the opportunity to build a more diverse, equitable, and inclusive next-generation maritime workforce.

Workforce needs of the traditional and new blue economy

In describing the diverse workforce needs and opportunities of the blue economy, it is important to recognize and distinguish the traditional blue economy (sometimes referred to as the ocean economy), which includes fishing, tourism and oil and gas industries, and the emerging, sustainable, new blue economy that is driven by data, information, and knowledge, such as ocean engineering and technology, renewable energy, and marine biotechnology. Both sectors are characterized by significant regional and global economic impact (OECD, 2016; Kildow, 2021). The development, application, and commercialization of new data, information, and knowledge to traditional blue economy sectors—such as incorporating fisheries and renewable energy technology—is one of the fastest-growing sectors of the new blue economy. Thus, the new blue economy is a knowledge-based economy that is focused on developing sustainable solutions to meet societal needs and challenges related to human health and food security.

As a regional example, the state of Alaska has more than half of the US coastline, one-third of the US exclusive economic zone, and access to vast aquatic natural resources. Alaska's traditional blue economy includes fisheries with an annual wholesale value over $4 billion, rich deposits of oil and minerals, and a robust maritime tourism industry. The vision for Alaska is that by 2040 the state would grow by 50,000 jobs and $3 billion in wages, approximately equal to Alaska's current oil and gas industry. Globally, the combined traditional and new blue maritime economic sector is currently valued at $1.5 trillion (measured as marine-based industrial contribution to economic output and employment) and is predicted to expand to $3 trillion by 2030 (OECD, 2016). The expansion of the global blue economy and maritime trade that is being driven by established industries and increasingly by technological innovation will spur demand for a skilled workforce to meet the needs of both traditional and new sectors.

Aligning STEM education curriculum to support the new blue economy

As with previous efforts to support the green economy workforce, technical institutes, community colleges, and universities play an important role in educating both the traditional and new blue economy workforce. An important challenge is to develop interdisciplinary STEM curricula with sufficient rigor in marine science, technology, and engineering disciplines at the associate, undergraduate, and graduate degree levels to meet industry needs. An additional challenge is to adapt and deliver these curricula within the current, and often rigid, higher education structure.

In particular, universities may face an internal challenge: whereas the new blue economy workforce requires highly interdisciplinary training, universities tend to be siloed institutions that struggle with academic or financial incentives at the unit level to develop a truly flexible, interdisciplinary curriculum. For example, engineering colleges typically have a course curriculum that adheres to professional engineering accreditation standards, leaving little flexibility in the program for engineering students to take courses external to the college, such as marine-related fields. Further, tuition typically flows to the unit delivering the course, thus creating a financial disincentive for a student's home college (e.g., engineering) to include courses taken in other colleges (e.g., business). This situation may be exacerbated by the fact that the annual financial budget for a given college is built in part from tuition revenue generated by delivery of their own course offerings.

Such structural challenges will require universities to reassess their academic and financial models in ways that promote the development of interdisciplinary programs and curricula in support of the new blue economy. For example, university senior leaders could foster the establishment of interdisciplinary blue economy "grand challenge" instructional programs that cut across the entire institution (Fig. 21.1). New interdisciplinary certificate and degree programs could include marine science, policy, social science, business, economics, engineering, and technology, while leveraging private sector support, public stakeholder engagement, and state and federal economic support. Such a construct should include academic

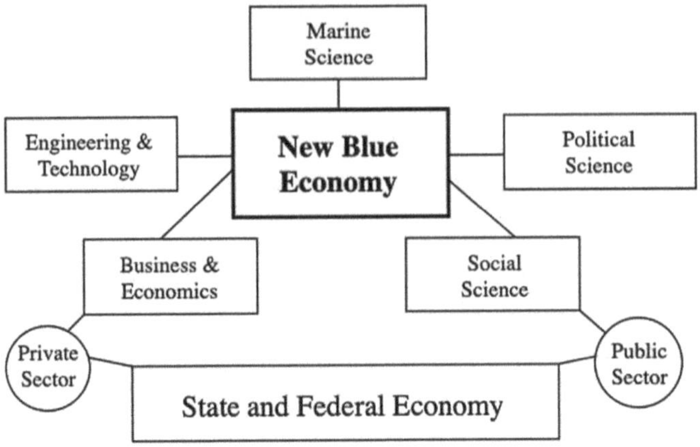

Figure 21.1 A proposed framework for a university interdisciplinary "grand-challenge" instructional program to meet the workforce and leadership needs of the new blue economy.

and research incentives that relate directly to faculty tenure and promotion. Greater financial equity for colleges that jointly offer interdisciplinary academic programs should be established up front, with the goal of encouraging creative ways to incentivize colleges, financially or otherwise, to work collaboratively across disciplines. Addressing these internal challenges and needs a priori would help foster the creative development of new and innovative curricula in support of industry needs in the new blue economy. Also, implicit in the convergence of the different instructional and stakeholder sectors is the opportunity to enhance diversity, equity, and inclusion both within the academic institution and in the emerging workforce (see Sullivan, 2021).

Interestingly, technical and community colleges may have a leg up in adapting curricula to meet the needs of the new blue economy when compared to the more established academic and financial structure of colleges and universities. For one, technical and community colleges tend to be much smaller than universities. This facilitates greater flexibility in course delivery simply due to scale, as well as having reduced, or absence of, rigorous accreditation requirements. Community colleges also typically offer courses more broadly across the institution (see Van Sumeren, 2021), as opposed to university colleges that generally operate with greater autonomy in regard to academics, research activities, and budget. For these reasons, technical and community colleges may be better positioned to develop, market, and implement curricula tailored to meet the technical and operational needs of the new blue economy.

In addition to the challenge of developing new academic programs across disciplines, the successful implementation of these programs will depend significantly on course delivery modes that extend beyond traditional face-to-face course delivery. Specifically, because both established and emerging maritime industries have an existing technical and leadership workforce, new interdisciplinary academic programs will need to shift to online instruction to be accessible and hence successful. Ideally, development of such programs for distance delivery should be accessible through both synchronous and asynchronous online modalities; for example, using a hybrid approach.

Coincidently, at the time of writing this chapter, the global coronavirus pandemic has resulted in a global-scale societal shift to communicating and working online, with many universities having to urgently transition entire academic programs to online delivery. The extent to which the recent and sudden transition to online course delivery remains part of the future of higher education remains uncertain. However, in general, the potential for greater delivery of online programs represents an opportunity for universities to bolster enrollment and tuition revenue while addressing contemporary workforce needs.

With regard to training the new blue economy workforce, a combination of traditional and online instruction modes would help serve the existing workforce as they seek to advance their training and certifications, while also increasing the opportunity to educate the next-generation workforce.

Notwithstanding some of the challenges and opportunities facing higher education institutions, what changes are needed in existing STEM curricula to meet the needs of the new blue economy? While defining individual curricula for technical schools, community colleges, and universities is beyond the scope of this chapter, some broad considerations are worth noting. In this regard, the notion of a blue economy "grand challenge" instruction program provides a useful framework (Fig. 21.1).

First, technical and undergraduate curricula should provide a foundation in STEM fields. This would ideally include courses in marine science and engineering. However, a basic challenge facing many institutions is the lack of both marine science and engineering expertise needed to build on their STEM curricula. A possible solution would be to work across institutions and leverage the expertise of those institutions with marine science faculty; for example, by online instruction. However, developing new academic programs that include cross-institution collaboration may present even greater academic and financial challenges. Depending on a student's specific career goals, additional coursework at the community college or undergraduate level in business administration, economics, and policy would also be recommended. Finally, the increasing need to address societal impacts of climate change as they relate to the marine environment would warrant courses in social sciences and communication as part of a robust blue economy academic program.

Regarding graduate leadership programs to support the new blue economy, it is worth noting that the number of MBA degree programs has grown significantly over the past several decades. However, the recruitment strategy for MBA students has remained largely unchanged; universities generally seek to attract the best students regardless of their undergraduate degree. While many elite MBA programs require prior work experience, the undergraduate area of expertise is often less important. The primary criteria for acceptance to the vast majority of MBA programs are the undergraduate grade point average and the Graduate Management Admission Test (GMAT) score. Further, the curriculum of most MBA programs is generally similar; for example, instruction in economics, accounting, marketing, finance, and organizational behavior. A key difference between various MBA programs is the depth to which these topics are covered. For institutions that offer a 1-year MBA program, the depth of topical coverage is limited; and, unlike 2-year programs, students typically focus on one topical area (e.g., capital markets). Students enrolled in 2-year MBA programs typically pursue additional courses in one or more specialized areas, thereby providing greater depth in multiple areas of business. In both of these types of programs, the curriculum is designed for the student to be hired in an entry- or mid-level management position.

To address the workforce needs of the new blue economy, a number of business schools have embraced corporate social responsibility and sustainability as underlying themes in their MBA programs. For example, there are a number of highly ranked programs in the Aspen Institute Center for Business Education's "green" MBA guide, which rates MBA programs on the extent to which they integrate social and environmental stewardship into the curriculum (Aspen Institute, 2008). However, as with many sustainability initiatives, challenges may exist with regard to programmatic rigor (Hart, 2008), as many schools do not fully integrate sustainability into their curriculum. Rather, sustainability initiatives may be in addition to standard course content. For schools that do integrate sustainability and environmental stewardship into their curriculum, broader strategy elements may become the focus rather than underlying science fundamentals (Bridges and Wilhelm, 2008).

Graduate workforce training in marine science: the Blue MBA

A growing number of business schools offer MBA degrees that include coursework in climate-relevant issues as part of more traditional management curricula. These degree offerings range from corporate social responsibility to more climate-specific areas such as carbon markets, finance, and sustainable supply chain management. Indeed, there is a growing demand for a workforce that can manage and capitalize on the interdependence of financial markets and extreme weather events due to climate change (Duncombe, 2020).

In the evolution of these business school sustainability degree programs, universities have begun to combine marine science and management training. For example, Scripps Institution of Oceanography offers a joint graduate program consisting of a Ph.D. in Oceanography and an MBA offered through the Rady School of Management in the University of California San Diego. The University of Rhode Island developed the first Blue MBA degree specializing in marine science (Moran et al., 2009). While relatively new, the Blue MBA degree provides an opportunity for students to develop skills in management, leadership, and science that can be applied to societal needs and challenges related to the new blue economy. Subsequently, the University of Alaska Fairbanks has developed an online Blue MBA in fisheries and marine science, and the University of South Florida is working to develop a Blue MBA degree. Such blue economy graduate degree programs are designed to equip business leaders with the knowledge and skills to address a broad array of marine environmental challenges, not only in regard to the business and social aspects of environmental stewardship, but in the science of maritime sustainability.

Sustainability MBA programs are generally designed for students who have an interest in learning about broader social, environmental, and economic perspectives as they relate to environmental business challenges. The Blue MBA degree differs from many of these programs because it is intended for students with a science or engineering background. Importantly, the Blue MBA generally requires a foundation in science education and builds on this science base with more traditional business courses that can be applied to marine environmental issues. With the Blue MBA degree, students gain a business understanding of the environmental challenges impacting corporations and a foundation in the scientific principles needed to address these challenges. The Blue MBA program is unique in that it produces students who can either work as part of a senior leadership team as the CENO (Chief Environmental Officer) or function as the CEO of a corporation that offers services and products that provide solutions to marine environmental challenges.

The Blue MBA can foster productive interactions between marine science and business schools by providing a platform for synergies and collaboration. In this way, the Blue MBA is a strategic and mutually beneficial graduate program for universities that have both a college of business and a college of marine science. Defined broadly, the Blue MBA helps to address the need for future business leaders to manage sustainable development goals of corporations focused on the marine environment.

Related graduate programs in climate science and management

The Blue MBA was developed with the vision that future corporate leadership would benefit from individuals who have foundational expertise in marine science and engineering related to the new blue economy. The Blue MBA graduate program is intended primarily for students with undergraduate training in science or engineering. Such coursework provides the necessary rigor and knowledge base to succeed in a blue economy career path that requires both management and marine science and engineering skills. In addition to the Blue MBA and other degree programs offered by business schools that are broadly defined as sustainability MBAs, an increasing number of other graduate programs now support workforce needs of the blue economy.

For example, Scripps Institution of Oceanography recently developed a policy-focused Master of Advanced Studies in Climate Science and Policy. Columbia University offers a Master of Arts in Climate and Society degree program. The Stevens Institute of Technology offers an innovative Master's of Maritime Systems, with six concentration tracks: maritime environment, management, security, and structures; marine transportation; and systems engineering. In the United Kingdom, two graduate degree programs are complementary to the Blue MBA, both focused on carbon management: The Norwich Business School offers an MBA in Strategic Carbon Management, and the School of Geosciences and the Business School of the University of Edinburgh offer a joint Master of Science in Carbon Finance. And, the Imperial College Business School in the Imperial College London offers a Master's of Science in Climate Change, Management, and Finance.

Industry, government, and stakeholder partnerships

The development of training and leadership programs to support the new blue economy workforce creates an opportunity for structured interactions between higher education institutions and industry, local, state, and federal government. For example, the creation of new advisory boards with important industry and government expertise and resources can result in new student internship programs and innovative programs to enhance funding opportunities to support students and research activities. In this way, blue economy workforce development programs can foster relationship building between marine industry and science leaders as they seek to develop future leadership talent.

Supporting the involvement of the private sector, government and stakeholders not only encourages invaluable input and advice on undergraduate and graduate degree curricula; it provides a bridge between business, government, and academia that is of mutual benefit. Establishment of formal advisory councils can also provide needed input in shaping training programs and curricula. Large corporations, small businesses, and local community stakeholders can positively engage with academic institutions to help steward the new blue economy workforce. In these ways, business, government, stakeholders, and academic science leaders can all contribute to and have a vested interest in the development of relevant and successful workforce programs for the new blue economy.

Summary

While it is the responsibility of the government to set policy and regulations to sustain a healthy ocean economy (e.g., Moran, 2014), the business community is increasingly recognized as essential to the development of sustainable and economically viable ways to address the global challenge of climate change. Traditional ocean economy industries and emerging new blue economy sectors require a workforce equipped with the necessary training in ocean sciences, engineering, information technology, and management. The growing demand for a technically competent blue economy workforce is achievable through associate degree and undergraduate training programs in marine engineering and technology. Similarly, advanced degree programs such as the Blue MBA provide needed leadership training that combines technical and science skills with proficiency in business management. Together with innovative, interdisciplinary curricula, and online instruction modalities, such degree programs will benefit individuals seeking careers in the full array of maritime industries. This includes fisheries and aquaculture, tourism, oil and gas recovery, renewable energy, ocean observing, ocean technology and engineering, marine biotechnology, finance, and risk management.

References

Van Sumeren, H.W., 2021. Filling the pipeline: northwestern Michigan College building a new generation marine technician for tomorrow's workforce. In: Spinrad, R., Hotaling, L. (Eds.), Preparing the Workforce for the New Blue Economy. Elsevier.

Aspen Institute (2007–2008) Beyond Grey Pinstripes, 2008. Retrieved September 23 from: https://www.aspeninstitute.org/news/press-release/aspen-institute-releases-green-mba-guide/.

Bridges, C.M., Wilhelm, W.B., 2008. Going beyond green: the "why and how" of integrating sustainability into the marketing curriculum. J. Market. Educ. 30 (1), 33–46.

Duncombe, J., 2020. How financial markets can grow more climate savvy. Eos 101. https://doi.org/10.1029/2020EO143124.

Hart, S., October 13 2008. Sustainability must be integral to schools' DNA. Financ. Times 15.

Hotaling, L., 2021. Preparing the workforce for the new blue economy. In: Spinrad, R., Hotaling, L. (Eds.), Preparing the Workforce for the New Blue Economy. Elsevier.

Kildow, J., 2021. Global expectations for the new blue economy. In: Spinrad, R., Hotaling, L. (Eds.), Preparing the Workforce for the New Blue Economy. Elsevier.

Moran, S.B., 2014. Strengthening America's ocean economy: the national ocean policy. Sea Technol. 55 (1), 24–25.

Moran, S.B., Higgins, M.M., Rosen, D.E., 2009. Educating future business leaders in the strategic management of global change opportunities: the blue MBA. In: Wankel, C., Stoner, J.A.F. (Eds.), Management Education for Global Sustainability. Information Age Publishing, pp. 227–241.

OECD, 2016. The Ocean Economy in 2030. OECD Publishing, Paris. https://doi.org/10.1787/9789264251724-en.

Spinrad, R., 2021. The new blue economy. In: Spinrad, R., Hotaling, L. (Eds.), Preparing the Workforce for the New Blue Economy. Elsevier.

Sullivan, K.D., 2021. The new blue workforce: a case study for diversity, equity and inclusion by design. In: Spinrad, R., Hotaling, L. (Eds.), Preparing the Workforce for the New Blue Economy. Elsevier.

Case study: RUCOOL Operational Oceanography Masters—workforce development case study

Nicholas Beaird, Scott Glenn, Travis Miles, Grace Saba, Josh Kohut, Oscar Schofield

Students graduating from marine science programs today will be critical to the multidisciplinary teams that power the New Blue Economy (NBE). They must be prepared to exploit ever-increasing oceanic observational and forecast data streams and archives and convert those raw data into actionable information (see Hotaling, this volume). Their training will enable the productive and sustainable use and stewardship of ocean resources in the NBE. They will serve a growing global population that is increasingly concentrated at the world's coastlines and will be required to do this while the climate is changing, sea level is rising, the ocean is acidifying, and extreme weather events are increasing in frequency and severity. In an effort to fill a human resource capacity gap, Rutgers University began a new, compact, Master's track in Operational Oceanography (Kohut et al., 2018). The program aims to meet the workforce needs early in the value chain of the emerging NBE through hands-on training in cutting-edge ocean observing data and forecasting technology.

Oceanography as an applied science is not new, it has always been both a research and an applied discipline. The field is always expanding as new technologies spin up to more efficiently provide defense, societal, and economic benefits. These changes impact not why, but how we observe, understand, and forecast the ocean. Autonomous observing systems, satellite remote sensing, cabled and coastal ocean observatories are providing exponentially larger data sets, often in real time. Computational resources and advances in data assimilation enable powerful and accurate forecast models that integrate these complex data streams (see Chassignet, this volume). The NBE will be enabled by how we collect and make available these data sets at the beginning of the value chain.

An efficient NBE requires a workforce that can implement modern computing methods (Oceans of Data Institute, 2015), understand the advantages and limitations of various marine data products, synthesize information, communicate knowledge, and adopt technical advancements. Operational oceanographers must have hands-on experience with cutting-edge ocean observing technology, computing tools (including data formatting, management, manipulation, and visualization), critical thinking, and technical communication that they can apply to new situations. The Rutgers program focuses on practical training in the collection, analysis, and communication of ocean observing system data, focusing on new observing elements critical to the Global Ocean Observing System (GOOS) such as gliders and high-frequency radar (see Gouldman and Ryabinin contributions, this volume). The Master's is built on an understanding that to fill the operational workforce gap, students require hands-on training beyond an undergraduate education, and there is no better place to receive that training than in an operational ocean observatory (Glenn et al., 2015).

The Rutgers Master's track grew out of a workforce needs assessment undertaken by program faculty at an international ocean technology conference. The assessment identified several broad areas of workforce skills required for the NBE (see Van Sumeren, this volume). The core curriculum was designed around these needs with the goal of training job-ready students. Table 1 outlines the workforce needs and describes program

Table 1 Workforce development for NBE capacity building.

Solutions → **Needs ↓**	**Knowledge building**	**Experiential training**	**Integration/synthesis**
Marine domain expertise	*Core courses*: Physical, Biological, and Coastal oceanography	Build core instruments (OpenCTD, OpenROV, fluorometer)	Weekly departmental research seminars
Observing system technology expertise	*Core courses*: Field methods, Integrated Ocean Observing Systems	Field experiences: glider deployments and piloting, HF radar installation and operations, RV Rutgers cruises	Embedded in operating ocean observing system Attend Marine Technology Society meetings and TechSurges
Data literacy	*Core course*: Software Bootcamp	Work with real-time data streams from observations and models: QA/QC, plotting, analysis	Fully data life cycle: Identify, collect, evaluate, analyze, interpret, archive, present, and support decision-making
Ocean model literacy	*Core course*: Operational Modeling and Visualization	Accessing and plotting model output. Run ROMS and WRF models on Rutgers HPC cluster	Analyze data-assimilative model output (DOPPIO) as integrated ocean observation framework
Collaborative skills	Team-building workshop, team-based class projects	Lead/participate in field teams, participate in RUCOOL operations meetings	Contribute to group "grand challenges" e.g., hurricane glider fleet, national HF radar network, Endurance line

Continued

Table 1 Workforce development for NBE capacity building.—cont'd

Solutions → **Needs ↓**	**Knowledge building**	**Experiential training**	**Integration/synthesis**
Communication skills (oral and written)	Core course regular paper and presentations with feedback from cohort and instructors	Fieldwork "cruise reports," experience blog	Thesis, based on MTS extended abstract, MTS style talk/defense
Ready for action	Weekly cohort meetings with survey and discussion of current job announcements; government and industry NBE efforts	Projects and experiences focused on "Real-world" activities. Worked with real data, deployed and managed real field projects	Host/meet with industry and government visitors Visit and attend ocean professional events

implementation strategies to address each. The main workforce requirements include: marine domain expertise, observing system technology expertise, data literacy, ocean model literacy, and collaborative skills. Students need to have general domain knowledge of marine processes and know how to enhance their background knowledge when required. They need to be literate in data handling and processing and know how to advance along with the field. We define ocean model literacy as the ability of students to understand the core concepts of numerical ocean modeling and to be proficient in working with an increasing variety of model output. Our goal is that students will have an understanding of the full life cycle of marine data, from collection, to quality assurance and control (QA/QC), to management, and ultimately, to interpretation and knowledge production. The program aims to make clear to students how each stage of the ocean data life cycle impacts the others.

The program follows a three-component framework to address training for each need. Students engage in ***Knowledge Building*** activities that provide core domain-specific knowledge, typically through coursework. However, coursework alone is not sufficient. Students build lasting competency through active learning, and thus we emphasize ***Experiential Training*** activities to solidify the concepts learned through coursework. These frequent real-world experiential activities take place within an operating ocean observatory housed at the Rutgers Center for Ocean Observing Leadership (RUCOOL) and Department of Marine and Coastal Sciences. Finally, students also engage in ***Integration and Synthesis*** of their coursework and experience. These activities develop the critical thinking and communication skills needed to connect information and ideas and contribute to higher-order function of the workplace.

The Master's track is structured as a compact 15-month program providing thorough but efficient applied training. It provides a concise path for marine science undergraduates to earn a Master's with a single year of additional training (a "4+1" model). It is led by a full-time teaching faculty member and draws from the domain expertise of researchers and applied science staff across the Center. The program begins in summer with an immersive ocean observing experience. Students arrive in late summer and engage in hardware camps focused on core underwater Glider (week 1) and High-Frequency Radar (week 2) technologies, followed by a 2-week Software Bootcamp. The goal of this period is a concentrated introduction to the technologies that form the foundation of the program. It also serves to develop a cohort mentality and to introduce the new students to program faculty and key research and operational personnel that run the local observing system.

Students continue through two semesters of coursework and research, culminating with an applied thesis submission and defense in the second summer. Each semester includes three (3-credit) courses that cover a wide range of ocean observing technologies and include both lecture and hands-on laboratory components. Courses heavily employ active learning pedagogy, engaging students with ongoing operational activities within RUCOOL. Because RUCOOL is deeply engaged with research, industry, and stakeholders including offshore wind development, fisheries, and regional forecasting, the program can build these real-world activities into student experiences. In this way, and through continuing industry needs assessment, the program will be responsive to workforce needs in order to provide students with skills for immediate employment.

Students also participate in independent research (3-credits) during each semester, building toward a thesis submitted during the second summer. Theses are selected from a curated list that have been vetted by faculty for feasibility on a short timeline and that emphasize skill building for marine industries rather than traditional academic research. We envision student training as a "T-shape"—where students gain broad exposure to many aspects of ocean observing, and gain experience in how to develop deeper knowledge of a particular area of interest through their thesis topic. Knowledge building and experiential training give all students broad exposure to the connected marine data life cycle. The thesis provides the opportunity to focus and develop deep expertise in a particular aspect of the program and to practice integration and synthesis. By going through this process in parallel with other students in their cohort, they both learn from others and mentor others in how to enhance their deep background. Theses are formatted as submissions to the MTS/IEEE OES OCEANS meeting, and students are encouraged to submit extended abstracts.

As the program develops, we intend it to evolve to meet emerging needs of the NBE. Regular reevaluations of industry workforce needs will be undertaken, and new skills training incorporated into the curriculum. Course accreditation and evaluation by IMarEST will be sought (see MacKenzie, this volume), in addition to utilizing ongoing efforts by the Marine Technology Society to provide professional certification in the United States. Proposals are underway to fund an "Endurance" glider line offshore of New Jersey. This component of an operational observatory would provide dedicated, continuous, operational training for successive cohorts of students in the Master's track. The program's goal is to utilize the unique operational programs housed at Rutgers to fill an NBE capacity gap

Figure 1 Rutgers Operational Oceanography Master's students lead a glider deployment in 2019. With program faculty guidance, students work as a team to ensure that assets in the water provide data in real time to operational streams (GTS). These students gain hands-on practical skills while participating in "grand challenges" such the creation and maintenance of a sustained observational network to improve US Hurricane forecasting. Synergies between hands-on practical student training and participation in "real-world" actions strongly engage students and increase skill retention.

with relevant, efficient, and hands-on training of a new generation of job-ready, marine-data-literate students (Fig. 1).

Author biography

Nicholas Beaird is a physical oceanographer and an Assistant Teaching Professor at Rutgers University. He completed his Ph.D. at the University of Washington and postdoctoral fellowships and Woods Hole Oceanographic Institution and Oregon State University before joining Rutgers. Along with the other coauthors, he has led the development of the Master's in Operational Oceanography. The authors comprise the Center for Ocean Observing Leadership (RUCOOL) at Rutgers University. RUCOOL creates knowledge of our ocean planet by pushing the limits of science and new technologies while inspiring and teaching future generations of ocean explorers.

References

Building Global Interest in Data Literacy: A Dialogue, Workshop report, October 2015. Oceans of Data Institute and IB.

Glenn, S., Schofield, O., Saba, G., Miles, T., Roarty, H., Kohut, J., 2015. The ocean is our classroom: A 4-year research track for undergraduate exploration research and discovery in oceanography. In: OCEANS'15 MTS/IEEE Washington, pp. 1–5.

Kohut, J., Glenn, S., McDonnell, J., Miles, T., Saba, G., Schofield, O., 2018. Workforce development supporting the blue economy: a master's program of integrated ocean observing at Rutgers University. In: IEEE/MTS Oceans Meeting Proceedings. Charleston, SC.

OP/ED: Overcoming roadblocks and charting a course forward to the new blue economy: an African American woman's perspective on being the first

Deidre M. Gibson

Like many People Of Color (POC) in the marine science field, I did not come to a career in marine science in a straightforward path. Though I enjoyed boating and crabbing with my family while growing up in New Orleans, I never thought about a career in the field of marine or environmental science—I had never seen any one that looked like me, nor had I been exposed to anything like this before. I did not know about POC scientists such as Ernest Evert Just (20th century marine biologist) and Daniel Pauly (preeminent fisheries scientist). When I thought of the image of a scientist, it was not my image.

I was first exposed to marine science while taking undergraduate courses at the University of New Orleans. I was just searching...trying to have some fun, so I took a SCUBA diving class, and that opened my eyes to a lot of different things that I hadn't thought about before. I started doing research and looking at different college guide books. You see, back in the 1980s, there was no Internet access for civilians, so I spent hours on the library floor looking through large books that listed every institution of higher education, in the country. After taking a number of different science classes and doing my own research, I made the decision to pursue marine science as a career and transferred to Shoreline Community College in Seattle. That move was a culture shock. I thought I was going to the enlightened north, but I quickly realized that many of my white classmates had never met a black person before. It was astounding, but at that moment I knew I had to be the best representative for the black race. It was a very heavy burden to carry, but I tried my best at being the first black student in that program. The hardest thing to deal with was the disbelief my classmates and a few professors expressed when I scored better on exams than the white students; they literally asked me if I cheated. I had to develop a thicker skin to ensure that I could endure the subversive racism in this field. After earning both an Associate's degree in science and an Applied Associate's degree in marine biology and oceanography, I transferred to the University of Washington (UW), where I was told that I was the first black undergraduate to enter into their program. Again, astounded! Earning my B.S. in Oceanography in 1991, at UW was a special journey, in itself. The university was like a city, and with the African American population at about 1%, I was always ecstatic to see another black person walking on campus, let alone in a science class. Needless to say, my journey to becoming an oceanographer was a lonely one; one filled with self-doubt and questions about my place in this field.

After earning my undergraduate degree, I moved back home and worked as a research technician at Louisiana University's Marine Consortium for 5 years. I worked for a biological oceanographer, investigating the effects of nutrients from the Mississippi River on the zooplankton community in the Gulf of Mexico. While I found the work meaningful, I found it to be one of the most difficult times of my career. My bosses did not encourage my aspirations. In fact, the white male scientist took the research I conducted as his own and had no intention of giving me coauthorship as he had promised. The white woman scientist I worked for told me and the other white women in her lab that we were "not good enough" to attend graduate school. Fortunately, I did not listen to her or let the white male's betrayal discourage me. I earned my Ph.D. in 2000, in Marine Science from the University of Georgia and conducted my research at the Skidaway Institute of Oceanography in Savannah where I was, again, the first African American to enter the program. There were challenges there, but by then I had developed the thick skin and tools to deal with the constant roadblocks designed to deter my plans. I then went on to do more marine research at two post-doc positions, at Savannah State University and the University of Connecticut-Avery Point, where I received more encouragement than I had in the past.

I began my career at Hampton University in 2002, as a Research Assistant Professor in the National Oceanic and Atmospheric Administration (NOAA) Living Marine Resources Cooperative Science Center. In 2004, I became a tenure-track Assistant Professor in both the Biological Sciences and Marine and Environmental Science departments. In 2012, I received tenure, was promoted to Associate Professor and the Chair of the Department of Marine and Environmental Science. The family atmosphere at Hampton University, and in the Marine and Environmental Science department, along with the ability to work directly with students has afforded a number of African American students the opportunity to pursue higher degrees in the sciences. When I started at Hampton University, many students were not thinking of grad school as an option, but obtaining the grants from the National Science Foundation (NSF) and NOAA has increased their level of interest in research and eventually going on to grad school. Our students have gone on to become Marshall Scholars, graduate students at Yale, University of California—Los Angeles, University of Maryland, Old Dominion University, Virginia Institute of Marine Sciences (VIMS), University of Massachusetts/Boston, and many others. Being the chair of my department allows me to focus on research, writing grants, and directing a number of mentorship and professional development efforts to bring more POC into the marine and environmental sciences.

Because of some of the negative mentorship I have received during my quest in this field, I take my role as a mentor very seriously. When working with programs designed to attract POC to scientific research and meetings, I am surprised by how excited students of color are to see "someone like them" going on to graduate school. As a scientist, I did not get that experience until I began graduate school. Currently, as the only black professor in marine science at Hampton University, I am a mentor to our students. This is a role I take very seriously, because I know how damaging roadblocks can be for students of color pursing their place in the New Blue Economy.

As for my research, a lot of good things are happening for me. I was able to finish my doctorate in only 4 years, an accomplishment made even more special when I got the chance to visit my former boss, who told me that I would never succeed in graduate school. Currently, I am encouraged by the success I have had providing students of color with access to the research sciences. The highlight of my career so far was serving as chief scientist on R/V Savannah, in the summer of 2011, with a "diverse" all-female scientific crew (see Fig. 1) to conduct feeding experiments in order to understand the ecological role and significance of doliolids in continental shelf pelagic ecosystems. The ability to expose my students to research has made the world of difference in our lives. I get joy in watching my students begin the journey of becoming scientists, and the students gain the confidence to leave their undergraduate experience to pursue M.S. and Ph.D.s in this New Blue Economy.

Figure 1 Dr. Gibson (in blue bandana) with her crew of diverse female scientists sampling zooplankton in the Mid-Atlantic Ocean.

Programmatic approaches to bringing diversity and equity to the scientific whelm of the new blue economy

One of the best parts of my job is directing diversity programs to provide the experience and build the confidence in students of color that have aspirations to join this New Blue Economy. Such a program, the Association for the Sciences of Limnology and Oceanography Multicultural Program (ASLOMP) facilitated my own journey for success as an academic in the blue economy (see Cuker, this volume). At the time (c.1997), it was about the only programmatic endeavor directed at promoting diversity in the aquatic sciences. My ASLOMP participation ultimately led to my position at Hampton University, an elected seat on the ASLO Board of Directors, and provided a network for helping to build my career. Fortunately there's a push in marine science now to increase diversity—being involved in these programs has encouraged me to stay in the field and be a part of making that difference. I am most proud of my first NSF grant, the Diversity in Research in Environmental and Marine Sciences (DREAMS) project. The DREAMS I project was a collaboration between Hampton and VIMS, and DREAMS II with VIMS and Elizabeth City State University. These programs jump-started the changed in culture for the Marine and Environmental Science department at Hampton University; the students started to think of themselves as scientists. Now when students are asked to draw an image of a scientist, the images are of them, not Einstein.

In addition to my involvement with DREAMS I and II, I also codirected the Hall-Bonner program (a fellowship program for minority Ph.D. students), I am a codirector and mentor in the ASLO Multicultural Program, and codirected the Mid-Atlantic Center of Ocean Science Education Excellence, and the COSEE-Coastal Trends programs, and now I direct the NOAA Living Marine Resources Cooperative Science Center, and have been funded for over 12 years to conduct NSF funded research studying the ecology of the pelagic tunicate, the Doliolid. These successes have made all of the negative treatment I received, and roadblocks I crossed, to get here, well worth the fight.

The NSF-funded **DREAMS I** program provided comprehensive research training and career preparation to underrepresented undergraduate students throughout the school year. These students also participate in summer research internships and outreach activities. The program supported 14 undergraduate students; three have received their Master's degrees, two received their Ph.D. degrees, two transitioned on to professional schools, and seven are working in the field of marine science or teaching. This program via the DREAMS seminar series impacted over 500 undergraduate students.

The NSF-funded **DREAMS-II program** provided comprehensive research training and career preparation to a cohort of eight under-represented undergraduate students from Hampton University and Elizabeth City State University, who also participate in summer research internships and outreach activities. This program via the DREAMS seminar series has impacted over 400 undergraduate students, aimed to expose Hampton University and ECSU marine science students to research opportunities. While the students took classes, they were also doing small projects on the Hampton campus to get a feel for research and working as a group. That experience prepared them for internship experiences at Virginia Institute of Marine Science and others research institutions. The Hampton University students all transitioned to M.S. programs and completed their degrees.

The NSF-funded **Center for Ocean Science Education Excellence Coastal Trends (COSEE)** The goal of this program was to link scientists, educators, and undergraduate students of color in research and education activities. This program has successfully conducted five scientist/educator partnerships, developed six web-based education modules, and conducted several teacher workshops. All four Hampton University student participants completed their M.S. degrees, and one is completing her Ph.D. degree in 2021. In addition to training students, over 3000 Virginia Aquarium guest were impacted via the COSEE Coastal Trends Weekend event.

The NSF-funded **Communicating Ocean Sciences to Informal Audiences (COSIA)** is a continuing collaboration between Hampton University and the Virginia Aquarium. Each spring semester, a formal course and workshops are conducted to provide college students with the communication skills required to present their research to informal audiences. This program has impacted over 100 students and 3000 Coastal Trends Weekend visitors.

The NOAA-funded Living Marine Resources Cooperative Science Center has been funded since 2001 and prepares a diverse student body for careers in marine and fisheries sciences through exemplary academic and research collaborations. This program is in partnership with University of Maryland Eastern Shore, Savannah State University, Delaware State University, Hampton University, Oregon State University, University of Miami, and the Institute of Marine and Environmental Technology to support and train over 100 students of color, either funding undergraduate or graduate student research. With the vision to be a national center of excellence for the development of a diverse body of professionals in marine and fisheries sciences through education and research of the highest quality, our main goal as it relates to student training is to prepare the future workforce for marine and fisheries science. Because of this goal, and the longevity of this program, the NOAA LMRCSC program is responsible for graduating the most number of underrepresented scientists in the Marine Sciences, in the country.

Conclusion

Becoming one of the first female African American Ph.D.'s in the ocean sciences was a struggle that provided a wonderful reward. My special experiences and perspectives enabled me to go beyond being a pathfinder. I also note that my mentorship isn't reserved strictly for POC. I was able to use what I learned to facilitate the success of many students, particularly students of color, on their way to participation in the New Blue Economy. Educating in the marine sciences means building a diverse workforce, one that has cross-cultural competency and allows POC to see themselves as vital members in the NBE.

Deidre's biography

Dr. Deidre Gibson is the Chair of the Department of Marine and Environmental Science at Hampton University. She earned her B.S. in Oceanography from the University of Washington, Ph.D. in Marine Science from the University of Georgia/Skidaway Institute of Oceanography. She is a broadly trained biological oceanographer with research interests centered on the trophic ecology, reproductive biology, and population dynamics of zooplankton, but more specifically, gelatinous zooplankton, and currently oyster restoration. While at Hampton University, she has served as PI on several NSF and NOAA grants that continue to train the next generation of African American marine scientists.

CHAPTER 22

Filling the pipeline: Northwestern Michigan College building a new generation marine technologist for tomorrow's workforce—the blue technician

Hans Van Sumeren
Northwestern Michigan College, Traverse City, MI, United States

Introduction

Addressing the technical educational needs of the new blue economy (NBE) requires a transformative perspective toward preparation and training of tomorrow's workforce. Delivery must be competency-driven, agile, and able to adapt to industry needs in a global setting. Specifically, supplying the technical marine workforce must include preparation across multiple skills sets. The blue technician of the NBE will require emphasis in areas including:

- Technology troubleshooting
- Programming and sensor integration
- Applied engineering—systems approach
- Geospatial technologies including data collection/processing/synthesis

Project management/planning

The relevance among these particular areas changes over time. Needs and requirements from decades past are significantly different from the skills of today and the rate of these changes is accelerating. In the next decade, the NBE blue technology workforce will provide support for managing the world's ocean and waterways **at scale and in real time**. Decision-making will no longer be isolated with worldwide Internet access unlimited and comprehensive in coverage. Remote or autonomous technologies will be the norm with personnel located both in the field and in shore-based control offices. The information highway will allow immediate connectivity to

Preparing a Workforce for the New Blue Economy
ISBN 978-0-12-821431-2
https://doi.org/10.1016/B978-0-12-821431-2.00010-X

critical data supporting cohesive management decisions that benefit interconnected global communities of need versus isolated areas of focus. Tomorrow's blue technician will be responsible for the accurate collection of these data from platforms or delivery mechanisms not yet developed.

The COVID-19 pandemic exposed the enormous gaps in available Internet access and capacity in rural and remote areas throughout the world. "SpaceX has been launching batches of satellites to add to its Starlink constellation, with the goal of ultimately providing low-cost, globe-spanning high-bandwidth internet connectivity. SpaceX is ultimately aiming to launch as many as 25,000 satellites in total based on current regulatory filings." (Etherington, 2020). Low-cost automated systems with real-time data delivery will drive the future of ocean monitoring and exploration. NOAA currently generates more than 10 terabytes of data per day (National Oceanic & Atmospheric Administration, [NOAA], n.d.) and unlimited global coverage will drive data collection to levels unthinkable in today's world. Given the enormity of collected data, tomorrow's blue technologist will be heavily utilized supporting these data collection and processing needs.

Historically, the typical "technician" for the marine industry came from trade-focused skills including electronics and hydraulics or 2-year associates degrees from community colleges. Beginning in the early 1980s and exploding in the 2000s, a prominent role of the marine technician was supporting the offshore oil & gas (O&G) sector as a remotely operated vehicle (ROV) technician. Around the early/mid 2010s, the global workclass ROV fleet was estimated at 800+ systems with a new ROV being built nearly every week (Robotics Business Review, 2013). A single ROV system required 9—12 personnel for 24 h operations, which meant 100s of new personnel per year who varied from entry level to supervisor level. Entry into this workforce required no formal degree but rather a "demonstrated" background in electronics and hydraulics. Given limited schools supplying this need, ensuring a competent workforce was the burden of the hiring company often completed with in-house training programs. Additional qualifications were more "lifestyle" tolerance than academic preparation, including the ability to live and work offshore in less than comfortable situations. Once a priority for entry to the industry, this lifestyle criteria has been overtaken by the need for competent technicians who can also manage data collection and processing needs. Fluctuation in oil prices continually reshapes the marketplace, and employment in this sector remains cyclical. However, developments in renewable energies markets have shifted workforce needs for these technicians as offshore leases for wind farms continues to grow (see Brown, this volume). Offshore ROV pilot technician roles remain, though the numbers of required personnel have declined and will likely never rebound to the highs previously seen. Additional developments in "resident" ROV systems living on the seafloor will further reduce the need for onsite pilot technicians. Stationed at a single platform or mobilized in an autonomous system that can transit between worksites, remote ROV pilots can live and work near the coastline anywhere, potentially removing many lifestyle obstacles (Trslić et al., 2020). While these resident systems reduce the stresses of remote relocation, the supporting blue technician workforce must deliver advanced troubleshooting techniques that are inherent with autonomous systems. Core competencies and preparation needs will change significantly as these systems become more commonplace.

Marine technician support for the O&G and renewables markets has seen a shift toward the use of autonomous systems. The core skills for autonomous underwater vehicle (AUV) operations continue to utilize many ROV competencies with additional focus toward networking of sensors, positioning systems, and data collection. As seen with the majority of Northwestern Michigan College (NMC) graduates working in this field, the typical operational duties include mobilization and deployment of systems, integration of sensor packages for geophysical data collection, and data processing. In this author's opinion, the addition of data collection and processing has moved the "new" blue technician from an associate's level degree to a bachelor-level education. There is simply no room in a 2-year degree to accomplish this level of training. The best data processors have spent time mobilizing, collecting, and troubleshooting systems in the field while also working with their own collected data to generate a final product. The bachelor training is the new norm for success at the entry level and it remains comprehensive in nature. The B.S. degree also represents an avenue for blue technicians to promote through company channels where bachelor degree attainment is often a requirement. This can be seen in marine industry sectors including the dredging and hydrography markets where bachelor degrees are hard requirements for employment.

Building the agile workforce for today and tomorrow

The Bachelor of Science in Marine Technology (BSMT) at Northwestern Michigan College has addressed these priorities through a blend of innovative programming, leveraging of partnerships, and substantial investment in people, facilities, and assets. These efforts have resulted in 100% employment of all graduates into the marine industry across multiple industry sectors.

When NMC first started to develop a bachelor degree for the marine industry, there was no roadmap to completely model a comprehensive *technical* program in the United States. Major talent pipelines existed for marine science, marine biology, hydrography, engineering, etc., but none that addressed the core needs for today's marine technician. The "ideal" technician has some components found in all these disciplines. In 2012, NMC launched the first half of this program with an Associates degree in Marine Technology. This 2-year degree was built around the demonstrated needs of the industry as outlined from multiple interactions with employers, advisory boards, and company training programs. There was no **D**evelop **A** **Cu**rricul**um** (DACum) exercise that provided this information but rather conversations with industry and listening to employers. Our research and outreach identified specific needs and competencies for personnel who could mobilize, calibrate, deploy, operate, and troubleshoot marine equipment for the purpose of collecting, processing, and delivering data. Building on these findings, professional development of NMC faculty and staff occurred through direct participation in industry in-house training and certification programs located throughout the United States. Identifying content in these programs allowed course emphasis and learning outcomes to be focused toward applications common to industry needs.

Areas of emphasis included electronics, fluid power, programming, Geographic Information Systems (GIS), and marine-related courses in ROV/SONAR technologies, sonar theory, and marine data processing. Graduates from this 2-year degree had experience operating numerous ROV systems, multiple sonar kits (multibeam, side scan, scanning), and were competent in the use of GIS and Hypack data processing software. Early program graduates found opportunities in the O&G and dredging industries including several students hired directly in multiple marine technology roles without completing their degree.

The core competencies of the associate-level program remain highly desirable skill sets for employers, but areas of emphasis continually change. Most notably, the role of low-cost, reliable, off-the-shelf components such as Arduino-based microcontrollers have provided academia a mechanism to better teach from a systems approach. These skills are coveted by employers who now routinely integrate these cost-effective technologies in daily operations. Job postings also routinely require geospatial competencies and programming languages such as Python and computer-aided design packages.

Developed in 2014, the NMC Marine Technology bachelor's degree serves as an example that there is sound logic in allowing community colleges to provide baccalaureate credentials in fields that leverage their individual expertise and assets. Community colleges do not normally grant bachelor's degrees and must rely on university partnerships (Fein, 2013). These relationships work well when numerous universities exist with laddering degree options. However, when collaborative opportunities for articulation do not exist, there is no mechanism to best serve the students and the industry. Historically, articulation agreements are typically presented as an "us and them" approach: 2 years at a community college and then completion at a 4-year university. Courses transfer but often competencies are left out. More often financially driven, true articulations require highly connected advisory boards, significant compromise between two institutions, and champions from both institutions driving the relationship.

NMC's new BSMT included 11 new courses along with additional emphasis on mathematics and physics. The intent was to further advance the comprehensive nature of key components of the associate degree. These included advanced studies in various marine platforms, continued training in data processing software, a required work/internship experience, and additional coursework in project management. These courses were designed and presented for internal approval within 8 months and had students enrolling just 4 months later, the equivalent of the speed of light in the academic world. Completing the bachelor's curriculum allowed NMC to bridge the gap between the associate's level training and the advanced competencies required in the NBE. To date, **EVERY** student who has completed the program (and all students who partially completed) has obtained a job in the marine industry across all sectors. Most employers are surprised at the breadth of knowledge and training completed and simply did not know that a college was addressing the industry needs so comprehensively. BSMT graduates have job titles including hydrographer, data processor, field engineer, ocean engineer, bathymetrist, AUV technician, marine technician, and others. To the best knowledge of this author, all graduates remain employed and there remains an extremely high need for more persons with these competencies. There is a stigma associated with the "Technician" job classification that suggests wrench-turning, low-paying, and undesirable work duties. Direct comparison of BSMT starting salaries to engineering graduate average starting salaries shows that these are very similar and employment opportunities are found globally.

Professional networks are key

NMC's Great Lakes Campus is home to the Great Lakes Water Studies Institute, the Great Lakes Maritime Academy, and our own private harbor on the shores of Lake Michigan. However, even with direct access to Lake Michigan and a dedicated campus for field operations, very few colleges or universities have all the necessary talent, equipment, and capital to create and sustain a comprehensive marine technology program. For colleges that do not support significant research programs, the acquisition of these key assets is substantially harder. NMC continually leverages multiple industry partnerships supporting programming and education.

A significant strength of the marine technology program at NMC lies in the numerous partnerships and relationships with industry. These relationships have allowed academic coursework to utilize the latest technologies and supporting talent at little or no cost to the program. As an example, Kongsberg Underwater Technology, Inc. has supported Northwestern Michigan College every year since 2009 with the no cost use of multibeam echo sounders. More importantly, they have also provided technical support for the proper mobilization, calibration, and use of the equipment. Handing off a piece of advanced survey equipment has little value to any program without the accompanying training. Troubleshooting connections and integrating components are core to the required competencies of a successful technician for the NBE. Today we host multiple industry experts and their equipment as part of our operations. Many are graduates or advisory board members of our program. Every new delivery of equipment starts the same way; unpacking, mobilizing, calibrating, and finally collecting. While sounding simple, there are many stories of students at colleges and universities who simply board a ship and see the finished product. These students never get the opportunity to understand the nuances of sensor integration and hence never gain a full understanding of how to troubleshoot or dissect a system. These networks of expertise and support are critical for the preparation of our graduates. Support includes numerous MBES systems, software, and other hardware packages that are new to the industry. Education is a business and as with any business, these networks need to be cultivated. Early in the development of the program, significant time and money were invested in creating networks of value. This included attendance at key industry conferences, trade shows, and training sessions.

Even with the ideal location, an international industry network of support, key faculty in specialty areas, a proactive college administration, and a sympathetic state government, many hurdles still remained. Facilities were adequate, but not excellent. Some marine technology equipment was owned, but not enough to sustain a program. In early 2015, the college was the recipient of a nearly $2.8 million infusion of support for equipment purchases in technical programs. The Community College Skilled Trades Equipment Program (CCSTEP) program was funded by the State of Michigan to help colleges provide up-to-date training in skilled trades and career technical education (Michigan, 2015). Nearly 2/3 of that amount was utilized to improve areas directly supporting the Marine Technology program including electronics laboratories, fluid power laboratories, ROV platforms, acoustic sonars and an indoor water tank training facility. This infusion of capital accelerated the program capabilities overnight. These dollars supported needs across the breadth of the curriculum and helped close the gap where deficiencies existed.

As stated earlier, education is a business and technology programs are expensive to build, maintain, and continually evolve. A good business plan must prioritize financial implications addressing these priorities. Programs must be creative in their approaches to long-term sustainability and numerous strategies must be employed to generate revenue to meet these ongoing demands. The majority of laboratory courses in the BSMT have user fees in addition to the tuition. These fees support a portion of yearly maintenance requirements for equipment, operations, and replacement cycles. Additionally, applied learning experiences and community connected projects have led to significant philanthropic support from individuals and foundations. NMC has developed a strong relationship with Consumers Energy, a Michigan-based utilities provider and the fifth largest utility in the United States. This partnership has allowed the delivery of advanced laboratory projects where BSMT students are performing complete survey inspections of hydroelectric facilities throughout Michigan. Final reports from these projects are presented and delivered back to the company as would occur in any company-client contract. The utility had been so impressed with the students' results that NMC was encouraged to write a grant request to the Consumers Energy Foundation from which a $50,000 grant was received to purchase a high-resolution scanning sonar. It was the student presentation to the foundation board that sealed the opportunity.

Marine Technology at Northwestern Michigan College has also engaged in global partnerships that have resulted in several tangible results and benefits to the college and our industry networks. Beginning in 2012 with a visit from six Chinese college presidents/vice presidents as part of a Vocational Education Leadership Training (VELT) (n.d.) exchange, NMC developed an MOU with the Yellow River Conservancy Technical Institute (YRCTI) located in Kaifeng, China. VELT is designed to enhance the leadership capacity of Chinese higher vocational colleges through partnerships of mutual interest with foreign institutions (VELT, n.d.). Our MOU focused on the delivery of marine-technology-specific courses to students in the surveying program at YRCTI. The course delivery was conducted in English, through translation, to a cohort of 40 students each year. This challenging environment not only strengthened NMC instructors' ability to clearly deliver curriculum (both in China and at home), but it also allowed NMC to connect multiple industry partners to our sister school for the purchase of multiple technologies. This stewardship of industry partner support resulted in $100,000 in equipment sales.

NMC has built a strong relationship with Monando State Polytechnic located in Indonesia. This relationship has led to multiple, comprehensive, student-designed class projects studying environmental impacts to coral reef systems at Bunaken National Park. Specifically, this project has connected two BSMT courses—Project Management and Marine Technology Capstone, and three NMC programs—BSMT, Freshwater Studies, and Unmanned Aerial Systems. Upon completion of the first-year investigation, the students presented their report to the NMC Foundation Board. An anonymous board member was so impressed, they completely funded the second-year expedition at a cost of approximately $50,000. It is critical to tell your story as an additional support mechanism. These student experiences are highly valued on a resume and employers embrace the "real life" laboratories.

While it has been stated that the BSMT is the new "norm" for success in the NBE, the value of microcredentials, certifications, and stackable training programs will provide opportunity for personnel in the industry to fill gaps in training and broaden opportunities (see MacKenzie and Hall, this volume). The development of the Geospatial Applications Center at NMC provides a mechanism for short course, professional-focused, comprehensive training solutions for the surveying and remote sensing industries. Relying on both internal and external expertise, this center operates in tandem with academic offerings. A significant focus of the center is marine applications and allows NMC to package academic curriculum into a professional-based delivery scheme in support of marine businesses and industries. This model serves to promote our programs, our graduates, and also provides a mechanism for engaging with ephemeral faculty who bring new expertise to the center and to NMC. Additional revenues can be realized with this approach, supporting faculty, staff, and equipment utilized in the BSMT program.

Short course modules through the Geospatial Applications Center will be more in line with future technical education models. Degree pathways may include "packages" of courses delivered through multiple entities and centered on areas of expertise. This disruptive approach to marine technology education will successfully serve the NBE but will take time for traditional academic models to embrace. Delivery of marine technology education must prioritize competencies and content, taking some risk with numerous unknowns, and focus toward success through a set of competencies as opposed to access and completion.

Meeting the needs but not the numbers

The program has more than doubled in enrollment over the first 4 years. Approximately 20% of these students come from out of state with the remainder coming from across Michigan and our local region. Approximately 30% of the student body is female, a significant increase from where it started. Graduates average ~50 days aboard a research vessel in Lake Michigan performing laboratory activities and class projects. All students receive training on multiple ROV systems, sonar systems, data collection and processing software packages, and other support equipment. Graduates find immediate employment across all sectors of the industry. Average starting wages are now $75,000 per year (2019).

The networks for employment opportunities continue to grow with program graduates returning to NMC recruiting for their companies. Several dozens of jobs are consistently available for BSMT graduates and the employment opportunities span multiple industries, locations, sizes, and lifestyles. A single company has recruited and hired nearly an entire graduating class from NMC. Postgraduation feedback from students and employers in the workforce on the strengths of the program includes:

- Breadth of the coursework
- Offshore operations and significant time utilizing equipment in applied setting
- Troubleshooting skills and processes
- GIS and data processing software
- Programming, sensor communication
- Project management skills and teamwork
- Report writing and presentation skills

Building and sustaining the technical workforce of tomorrow

Preparing tomorrow's blue technologist must continue to focus on the applied competencies required for success in the industry. Nearly every industry sector is seeing change through the development of autonomous systems, machine learning algorithms, and data collection schemes. This shift in operations suggests blue technicians will support programming and networking of sensor packages into autonomous systems at a much higher level. These autonomous systems will network with each other and drive data collection at faster speeds and broader coverage. Cyber security issues with autonomous data collection will continue to grow and NBE blue technicians will need competencies in understanding these threats in the workplace. The use of low-cost sensors and open-source programming languages such as Python that are both easy to utilize and have the power to perform (Masetti et al., 2019) will continue to grow for hydrographic surveying applications and for applications in machine-based learning algorithms used in marine instruments. Globally connected data collection and near-real-time processing will be further emphasized in the role of blue technologist. Efforts supporting the mapping of the world's oceans in the coming decades will heavily rely on blue technologists supporting all levels of operations and new approaches toward autonomous technologies. Successful integration into the NBE will not only require core technical skills and a broad understanding of data collection and synthesis, but also leadership of small teams and abilities in decision-making. A blue technologist for the new blue economy will maintain a suite of skills that blend "traditional" core technical competencies that allow effective troubleshooting skills on a variety of sensor platforms. Necessary competencies will be more focused toward programming and sensor integration of networked autonomous systems. NBE blue technicians will be increasingly involved in all aspects of data collection, processing, and synthesis and will direct or lead teams of personnel operating in the marine space. Lifestyle changes will occur including shorter deployments and more "shoreside" opportunities. From an institutional perspective, access and delivery of education and training will evolve and traditional education models will change toward competency-based, time-independent delivery methods.

The new blue economy will continue to require technicians, programmers, operators, field engineers, data processors, project managers for all aspects of the marine industry.

References

Etherington, D., February 6, 2020. SpaceX 'likely to Spin Out' and Pursue an IPO for Starlink Satellite Internet Business. Retrieved from: https://techcrunch.com/2020/02/06/spacex-likely-to-spin-out-and-pursue-an-ipo-for-starlink-satellite-internet-business/.

Fein, P., January 22, 2013. Evolution or Mission Creep? Retrieved April 18, 2020, from: https://www.insidehighered.com/news/2013/01/22/michigan-lets-community-colleges-issue-four-year-degrees-amid-controversy.

Masetti, G., Faulkes, T., Kastrisios, C., March 2019. Hydrographic survey validation and chart adequacy assessment using automated solutions. In: Paper Presented at US Hydro 2019, Biloxi, MS. Retrieved from: https://ccom.unh.edu/sites/default/files/publications/USHydro2019_Masetti_et_al_Hydrographic_Survey_Validation_and_Chart_Adequacy_Assessment_0.pdf.

Michigan Gov. Rick Snyder, February 24, 2015. $50 Million Grant Program Will Help Community Colleges Boost Skilled Trades Instruction. Retrieved from: https://www.michigan.gov/formergovernors/0,4584,7-212-96477_90815_57657-348250−,00.html.

NOAA, n.d. Big Data Program. Retrieved April 18, 2020, from: https://www.noaa.gov/organization/information-technology/big-data-program.

ROV Sales to Global Oil & Gas Worth $2.5B in 2013, February 2013. Retrieved from Robotics Business Review website: https://www.roboticsbusinessreview.com/energy-mining/rov_sales_to_global_oil_gas_worth_2-5b_in_2013/.

Trslić, P., Weir, A., Riordan, J., Omerdic, E., Toal, D., Dooly, G., 2020. Vision-based localization system suited to resident underwater vehicles. Sensors 20 (2), 529. https://doi.org/10.3390/s20020529.

Vocational Leadership Training Program, [VELT], n.d. Retrieved from: http://en.ceaie.edu.cn/velt_program?columnid=79&sub=80.

OP/ED: Big ocean data

Randy Kochevar

Whether we are talking about scientific exploration, medical research, or business management, our understanding of the world comes from ever-increasing volumes of data. This is creating an unprecedented demand for workers expert in working with data. In a report from March 2018, Silicone Republic reported an estimate that demand for data science workers would increase 20,000-fold from 2000 to 2020; and this is considered one of the hottest careers to get into (https://www.siliconrepublic.com/careers/data-science-skills-gap). Nowhere is this more true than in the oceans, where Earth-orbiting weather satellites and remote subsurface sensors bring us terabytes of new data each day—a torrent that will only grow as the New Blue Economy expands.

However, our education system is not currently set up to prepare students for data-intensive careers. As stated by Juan LaVista, Principal Data Scientist at Bing/Microsoft, "Basic skills in working with data that every person should have are not being taught in K-16 schools. Thus they are lacking at the highest levels in the broad array of professions that are becoming increasingly data-driven."

The Education Development Center established the Oceans of Data Institute in 2013, to help address the growing need for students to acquire data literacy skills as an integral part of their education; and for those who are so inclined, to prepare them for success in data-intensive careers. Our vision was—and is—to help build a world where everyone has the skills and knowledge to make informed decisions and achieve new insights and understandings using data.

From the outset, we have worked closely with data scientists and analysts from a broad spectrum of industries, including astrophysics and oceanography, database management and engineering, law enforcement, journalism, health, and agriculture, to understand the key knowledge, skills, and behaviors required to work effectively with data. We have used this information as a compass, helping us to align our efforts, both in k-12 classrooms and community college and university workforce readiness programs, to prepare students for success in an increasingly data-intensive world.

In our k-12 education programs, we set out to do this by developing curriculum, designing data interfaces, and working with educators to integrate data literacy skills into existing courses, focused on topics and concepts that are fundamental aspects of biology, ecology, or Earth science, in school across the United States and around the world.

Our first foray into this work was *Ocean Tracks*. Funded by the National Science Foundation's DRK-12 program, Ocean Tracks is a web-based interface that allows students to explore and interact with complex data from tagged open ocean animals, coupled with concurrently collected oceanographic data such as sea surface temperature and productivity, which can be applied as overlays. This interface was tested in high school science classrooms, using a series of lessons to explore topics such as predation and productivity, biodiversity hotspots, human impacts on marine life, and science-based marine spatial planning.

The Ocean Tracks curriculum was pilot tested in seven high school classrooms with 195 students. We wanted to learn about the usability of the interface; the level of engagement students felt for the interface and supporting materials; and the types of questions, reasoning, and claims students generated while using Ocean Tracks. In order to do this, we used classroom observations, screen capture recordings of students using the interface, student surveys and focus groups, reviews of student work, and teacher journals.

We found that students were readily able to navigate through the major features of the interface, including the base map (based on the Google Earth API), as well as the data overlays. However, lack of sufficient bandwidth and limited screen size on classroom laptops proved to be challenging.

Nevertheless, most students (75% on survey responses) said they enjoyed learning about marine animals and learning about their migration patterns. As one student said, "I found it most interesting that you guys actually had the power to track an animal and know exactly where they are every minute."

Some of the challenges we found included trying to find the appropriate balance between giving the students "not enough" and "too much" instruction in curriculum modules. Our goal was to provide sufficient scaffolding to enable student inquiry, and too many detailed instructions made the modules feel tedious and repetitive. Teachers also reported that they struggled to find the right balance for their students between online activities and offline discussions and group work.

Despite the challenges, we found that Ocean Tracks was successful in helping students to incorporate authentic data into their own scientific practices and thinking. They were able to take measurements, describe patterns, and generate hypotheses:

> *I noticed that the seal track follows the area of higher chlorophyll levels. There are more nutrients in the water where there is more chlorophyll so that is why they are attracted there because there is probably a greater chance of finding suitable food there… The concentration is between 10-12 where it was lingering.*
>
> ***Student work.***

And Ocean Tracks prompted thoughtful questions:

> *I thought it was cool to watch the animals to see how not only just one animal went in this certain route, but how all the animals of that species and even some animals of different species go in that particular route…*
>
> *It was also interesting how within the same species, certain animals will completely separate from the normal predicted path and go do something that's miles away. It's interesting to think about what that one specific animal is doing that's different from the rest of its species.*
>
> ***Student focus group.***

The Ocean Tracks curriculum was further developed and refined through an NSF Improving Undergraduate STEM Education (IUSE) grant ("Ocean Tracks-College Edition" or "OT-CE"), which allowed us to work with undergraduate students and instructors, who helped us to focus the curriculum more specifically on the data skills and learning outcomes we sought. At the end of the 3-year project, we found some overarching lessons that may be useful to anyone trying to teach science using authentic data:

- In presenting data analysis tasks, there is a need to strike a balance between simple and complex visualizations. For OT-CE, this entailed developing an approach that would be intuitive for a novice, yet functional enough to sustain engagement as one became more proficient. Specifically, this was operationalized in each module through a data-informed, revised learning cycle (Engage, Explore, Synthesize), with additional support in some modules in the form of a Pre-Lab Assignment and/or an Extend activity.
- There is a tension in attempting to integrate the stories scientists are trying to tell with data into a platform being built for novice users. In doing the translation, authentic stories that involve scientific data are revealed to be complex and nonlinear, even as the novice or less-than-expert learner tends to look for a clean, linear narrative. In response, the finalized OT- CE curriculum modules intentionally include built-in scaffolds related to helping learners navigate issues such as ambiguity and anomalies in data.
- There are challenges in the very essence of building a data literacy project incorporating critical thinking; there is a need to model strong process skills for conducting an investigation while simultaneously maintaining room for open-ended individual practice with exploring processes of deliberation and critical habits of mind. By the end of the project, OT-CE's approach to facilitating this involved providing parallel examples of key concepts (e.g., developing a hypothesis) and sequenced exploratory prompts designed to challenge users to develop their own ideas while maintaining a logical order of operations.

As we were wrapping up the Ocean Tracks work, we launched other projects aimed at younger audiences. We are currently in our fifth year of collaborating with the Gulf of Maine Research Institute on a NASA-funded project to build learning experiences for fifth and sixth grade students, focused on the impacts of climate change on Maine's habitats. Using funding from NSF's DRK-12 program as well as a grant from GM, we've used a modified version of EDC's *Zoom In!* learning platform to help develop students' skills in writing arguments from evidence on topics in Earth science, biology, and urban mobility. And in another DRK-12-funded project, which just launched in summer of 2019, we are working with third, fourth, and fifth graders to learn how they make meaning from different kinds of data and how they transition from small data sets that they collect, to larger, professionally collected data.

As we have expanded our programs across the k-12 landscape, we have also been working with community colleges and universities to help create pathways to careers in data-intensive fields. We recognize that as the need for workers with data skills expands—driven in part by the growing New Blue Economy—a significant portion of those jobs will not be for Ph.D.-level data scientists; rather, they will be for middle-skills workers who can carry out the routine tasks of data acquisition, cleaning and quality control, archiving, etc. The skills required to carry out these tasks can be taught through stackable credential programs or 2-year degrees, accessible to a variety of students—including not only those matriculating from high school, but also those already in the workforce seeking new career opportunities.

Since Oceans of Data Institute was launched, the demand for data skills in the professional world has exploded. The power of data to inform—and disinform—has been a center-stage issue, changing forever our political landscape. And as the role of artificial intelligence and machine learning continues to evolve and grow, the need for critical and analytical thinking skills around data will only become greater—touching all aspects of our lives. As a community, we need to ensure that these skills are an integral part of every students' education.

Author biography

Randy Kochevar trained as a deep-sea biologist at the University of California, Santa Barbara, studying the physiology of hydrothermal vent and hydrocarbon seep animals. The discovery of similar animal communities in the Monterey Submarine Canyon brought Randy up the coast, initially to the Monterey Bay Aquarium Research Institute (MBARI), and then to the Monterey Bay Aquarium, where he was involved in exhibit development, website development and print publishing, and media relations. From 2008 to 2015, Randy worked at Stanford University's Hopkins Marine Station, collaborating with Dr. Barbara Block on electronic tagging of marine apex predators including tunas, billfish, and sharks. In 2015, Randy was invited to take over as director of the Oceans of Data Institute at the Education Development Center (EDC), where he oversees a portfolio of projects to build data literacy skills in K-16 students and to help build pathways for data science and analytics careers. In addition to his role at EDC, Randy teaches Environmental Science and Environmental Science Lab at Monterey Peninsula College.

OP/ED: How to prepare the workforce for the new blue economy?

Louisa Koch

By 2030, the global ocean economy will DOUBLE in value to $3 trillion according to a report on the Ocean Economy published by the Organisation for Economic Cooperation and Development (OECD, http://www.oecd.org/sti/inno/the-ocean-economy-in-2030-9789264251724-en.htm). With the second largest Exclusive Economic Zone in the world, the United States has an abundance of natural resources in its territorial waters, but the New Blue Economy will be driven by much more than the extraction of resources; it will be driven by using data and information to address societal challenges and develop innovative solutions for them.

The New Blue Ocean Economy will be driven by advancements in many disciplines, and the most important ocean discoveries will take place where disciplines converge. Convergence requires the deep integration of knowledge, techniques, and expertise from multiple fields. Success in the New Blue Ocean Economy will depend on having a workforce with sufficient mastery of a broad range of science, technology, engineering, and math (STEM) skills to drive advancements.

So what skills are needed? To excel in science, technology, and engineering, students need to have significant understanding of math. For many, math is a challenge, one that can limit access to a science career. Finding ways to make math more accessible will increase the pool of talent available. This is a major challenge. Despite making math a national priority, the United States has not seen a significant improvement in student math skills. A 2018 international assessment of 79 countries found that US students scored below average in math (https://nces.ed.gov/surveys/pisa/pisa2018/#/math/intlcompare). So this is an area that needs continued focus.

The future ocean workforce will need a strong foundation in ocean science at its core. Physical oceanography, biological oceanography, chemical oceanography, marine biology, and ocean engineering all have important roles to play as do emerging fields such as unmanned systems, artificial intelligence, and omics.

Digital literacy will, likewise, be a key skill set for the future ocean workforce. Digital devices and the internet are transforming society and driving the New Blue Economy. Echosounders that speak whale, aquatic drones probing the ocean depths, and swarms of mini robots are some of the devices being deployed to help us better understand and benefit from the ocean. Programs such as the Marine Advanced Technology Education Program (MATE) (https://materovcompetition.org/) provide students with opportunities to learn about the use of remotely operated vehicles and to construct their own vehicle.

Digital literacy will be sufficient for some, but for others a deeper and broader understanding of computers and how to use them will be required. Computational thinking is an important higher-level skill, which allows for the formulation of problems and the development of solutions in ways that take advantage of the incredible processing power of computing and machine learning. Computational thinking involves defining problems in ways that allow them to be broken down into component parts that are easier to resolve or gain insight into, then bringing the parts back together to reassess and iterating on that process. Offerings such as the MIT open course in computational ocean acoustics (https://ocw.mit.edu/courses/mechanical-engineering/2-068-computational-ocean-acoustics-13-853-spring-2003/syllabus/) give more advanced students a way to hone these approaches.

Data fluency will also play an important role. The amount of data being generated every day is astounding. NOAA alone generates tens of terabytes of data a day from satellites, radars, ships, weather models, and other sources. Collecting this data is challenging, but being able to organize, explore, and visualize it to get the right information at the right time is an even larger challenge. Partnerships such as NOAA's Big Data Project (https://www.noaa.gov/organization/information-technology/big-data-program) make these data available for free providing more access for educators to build big data into their teaching and research.

The United States has a rich history of innovation. To continue that tradition, we need to ensure that tomorrow's workforce is equipped with the creativity needed to generate breakthrough ideas and then translate them into new products and processes. Competitions such as the Regeneron International Science and Engineering Fair (https://www.societyforscience.org/isef/) challenge students to pursue innovation and discovery. Ocean focused awards, like those given by the Consortium for Ocean Leadership and NOAA, highlight and propel ocean innovation.

No one person can master all of the skills needed to drive the New Blue Economy, which makes effective teamwork essential. Skills such as communication, collaboration, and leadership are being introduced into learning environments in many ways. These skills need to be studied and practiced to grow strong. The National Ocean Science Bowl (http://nosb.org/), which is driven by high-performing teams, is a great way to do this. While the first place prize is awarded to the team that gets the most right answers, the Sportsmanship Award, which promotes teamwork, is also highly valued.

To motivate this rich array of learning, teachers are challenging their students by focusing on real-world problems. Problems that are relevant to people's lives energize and engage a wide range of students to learn. The ocean offers an abundance of rich opportunities for real-world problem-solving. Programs such as Aquaculture in Action (https://www.mdsg.umd.edu/topics/k-12-aquaculture-education/k-12-aquaculture-education), the Bay Watershed Education and Training Program (https://www.noaa.gov/office-education/bwet), and the Ocean Guardian School Program (https://sanctuaries.noaa.gov/education/ocean_guardian/) give kids great opportunities to build their STEM skills as they investigate local and global environmental issues and undertake actions to address them.

It takes amazing teachers to provide rich learning opportunities. These teachers need support on many levels. Programs such as Project Ocean (https://www.ametsoc.org/index.cfm/ams/education-careers/education-program/k-12-teachers/project-ocean/) and the Teacher at Sea program (https://teacheratsea.noaa.gov/#/home/) help educators strengthen their ocean science while exposing them to cutting-edge research and providing field experiences that bring ocean science more deeply into their lives. Much more of this support is needed. While many teachers would benefit from these enriching opportunities, few are able to partake.

Preparing the future workforce for the New Blue Economy is a daunting task, but there are promising signs. By 2020, Millennials will make up half of the American workforce (Inc, 2019) and Millennials are tech-savvy, team-oriented, and like to be challenged. The next generation, GenZ, is starting to enter the workforce. GenZers have grown up in a digital world. They are known for being tech-savvy and entrepreneurial, skills which are in high demand. The workforce is getting more diverse; by 2032, the US workforce is expected to become majority–minority (EPI, 2016). This more diverse workforce will bring a wider set of skills and perspectives to their tasks. Finding ways to overcome barriers to entry and be more welcoming of this diverse workforce is a worthy challenge. Programs such as NOAA's Educational Partnership Program with Minority Serving Institutions (https://www.noaa.gov/office-education/epp-msi) are showing us the way.

Investing in education today is the best way to ensure the strength of our economy in the future. There is a rich array of partners investing in formal and informal ocean education. We need to continue to work together and amplify our efforts. Your support for education is key to getting us where we need to go.

Author biography

Louisa Koch, NOAA's Director of Education, educates and inspires the public and future workforce about the Earth System. She works with NOAA's amazing array of people, partners, places, and information. Louisa also serves as lead for NOAA's National Team for Regional Collaboration, helping to address complex, place-based challenges that require interdisciplinary approaches and regionally tailored solutions. Ms. Koch served as NOAA's acting Deputy Under Secretary and Deputy Assistant Administrator for Research. Prior to joining NOAA, Ms. Koch worked at the Office of Management and Budget, the Department of Defense, and the Joint Economic Committee of the US Congress. Ms. Koch earned a Master's in Electrical Engineering from the Massachusetts Institute of Technology and a Bachelor's in Physics from Middlebury College.

References

Inc., 2019. The (Millennial) Workplace of the Future Is Almost Here – These 3 Things Are About to Change Big Time. Retrieved from: https://www.inc.com/peter-economy/the-millennial-workplace-of-future-is-almost-here-these-3-things-are-about-to-change-big-time.html#:~:text=By%202020%2C%20Millennials%20(those%20born,of%20their%20entire%20employee%20base.

Economic Policy Institute, 2016. People of Color Will Be a Majority of the American Working Class in 2032. Retrieved from: https://www.epi.org/publication/the-changing-demographics-of-americas-working-class/.

Interview: Early career ocean professionals

Alfredo Giron-Nava

What are the biggest challenges faced by Early Career Ocean Professionals (ECOPs)?

I think this is very sector-dependent, but I'll start with academia, which is the sector I belong to now. I think the biggest challenge is to learn the ropes of that never-ending loop of getting funding, moving to your next position, and even once you get to a more stable position, you still have a long way to pursue tenure track. I also think the incentive structure of getting funding and publishing all the time really narrows down the impact and partnerships that scientists could have outside of academia, for example, with industry, policymakers, and international organizations.

Once you finish your education and you're looking for that first job, it's almost the same time when you are looking to transition in a personal way. Maybe have a family, find a place to buy a house. That's a big challenge that needs a lot of energy.

Finally, there is also a risk of students not knowing what their marketable skills are. I think that universities and societies have been working on this issue and teaching students that what they learn at school is useful for the job market. A lot of places have done a great job on that, however, through our global surveys, we have identified that people still don't know what their marketable skills are, which limits their job options and career paths.

How different are the challenges faced by ECOPs around the world?

There are huge disparities in access to training, funding, and even networks. One concrete example is that most global networks and funding opportunities are in English. Even though many people speak or at least understand English, it certainly is a limitation for some who cannot express their full contributions as they would do in their native languages. Through the global surveys, we have also identified that in the Global South, ECOPs feel that they need better equipped laboratories to learn more marketable skills. I think especially in these countries, universities and professional societies could partner more tightly to work on developing those curricula and attend to these limitations.

How should we change the training/education of professionals in ocean science?

I think one of the things that is often missing is a continued conversation between employers and educators about the specific skills we should be teaching in school at different levels.

Again, I think this is a call to action for societies such as MTS (Marine Technology Society) and other professional networks. I see a much heavier role of these partners in fostering conversations with employers, mentorship programs, and job boards, etc. For example, what if we go one step ahead and get employers to create and regularly update a list of desirable skills that universities could use to update their curricula?

What levels of education/training are most needed in the workforce of the future (high school, community college, college, advanced degrees, etc.)?

I think we will need a lot of skilled technicians from marine-oriented community colleges and colleges. More and more we see these programs popping up and these are the people that are working in offshore operations, coastal protection, and alternative power generation. We need those highly skilled technicians. We also need innovative engineers and scientists from college and graduate programs. So, I don't see this as a question of what level of degrees we need, but more of how do we tailor current degrees to address the gaps and needs of the blue economy?

Again, I think that professional societies need to step up in here to help make those connections and help universities tailor their programs to respond to the needs of the blue economy.

How important is professional certification (e.g., certified marine technologist) for the workforce of the future in your field?

I think it depends on the kind of jobs that you are thinking about. For example, as a fishery scientist, I don't think that professional certifications will take off in the short term because having a Ph.D. or a Master's is my credential. But if we are talking about those highly skilled technicians for the blue economy, I think it is the most straightforward way for both employers and future potential employees to connect and streamline the hiring process.

How might activities associated with the UN Decade of Ocean Science for Sustainable Development affect the development of the new blue economy?

The UN Decade of Ocean Sciences for Sustainable Development is as much about the science as the development, and in this sense I think that one of the pillars of the decade will be the blue economy and the New Blue Economy. How do we use science to create jobs? To guide coastal development? To preserve the services that the ocean provides? How do we use the vast amount of data collected in the ocean to generate jobs?

I think that the conversation about the blue economy should be at the forefront of the implementation plan of the Decade, and it should encourage partners and contributors to specifically mention how their actions are embedded into the framework of the blue economy. The Decade is not only about how my actions contribute to the scientific goals, but about how my actions contribute to sustainable development.

Focusing on the workforce associated with the blue economy, what do you think will be fundamentally different in this workforce in the year 2050 from what it looks like right now?

The first thing that comes to my mind is that people will need to have a better understanding of what the ocean provides and how the technology that we normally use on land needs to get adapted or just transferred to applications in the ocean. The ocean is such a complex system, it needs people to understand a little bit of oceanography, a little bit of climate science, a little bit of engineering, a little bit of human sciences. I feel that slowly and steadily scientific programs related to the ocean are moving in that direction. Everyone is getting a little bit of formation about everything, but there's a risk, we might end up with professionals who know a little bit of everything rather than having skills to advance the frontiers of specific fields. Still, I think that the interdisciplinary transformation of the workforce needs to happen.

In summary, I think that the workforce will be highly interdisciplinary and group-oriented, even more than now. Big problems need people that can communicate across disciplines and who know how to work together.

Alfredo biography

Alfredo Giron-Nava is a Mexican scholar and current André Hoffmann Fellow at the Stanford Center for Ocean Solutions and the World Economic Forum Centre for the Fourth Industrial Revolution. In 2019, he obtained his Ph.D. from Scripps Institution of Oceanography, where he developed quantitative methods to develop dynamic management strategies for fisheries that are heavily affected by fishing pressure and environmental variability. In May 2019, Alfredo participated in the first Global Planning Meeting for the UN Decade of Ocean Sciences for Sustainable Development, where along with a group of Early Career Ocean Professionals (ECOP), he started an ongoing effort to consider ECOP contributions to every aspect of the Decade. Since then, Alfredo has been a strong advocate to include ECOP in global, regional, and national committees and events related to the Decade and broader ocean sustainability issues, often serving as a point of contact to hear and synthesize the views of ECOP around the world. Alfredo is also the first recipient of the Walter Munk Scholar Award, awarded by MTS and the Walter Munk Foundation for the Oceans to a young scientist pioneering activities of exploration and discovery.

Interview: Traditional ecological knowledge

Noelani Puniwai

Please share your background

I was born and raised on Hawaii island in the district of Puna. I have an undergraduate degree in marine ecology and PhD degree in natural resource management. I studied a lot of coastal ecosystems. For my Master's work, I worked off the Washington and Oregon coasts on continental shelf marine invertebrates. I came back to Hawaii in the early 2000s when GIS (geographic information systems) was taking off, so became involved with mapping. I did the first analysis for Hawaii on siting protected areas and collecting both scientific and traditional knowledge of those areas. For example, areas that are convenient to dive such as Kaneohe Bay off of Coconut Island, have lots of data and research, but outside of that area, not so much; most of the information about marine environment resides in the people who live there.

So my Ph.D. work was trying to understand how to bring different knowledge systems into management. I did some work on climate change looking at what kind of information surfers knew about the climate because a surfer completely depends on oceanographic and atmospheric conditions to be able to surf, and to me they're the best coastal ocean scientists out there. I interviewed an ocean waterman to see if we could map ocean conditions because as you probably know we don't have a lot of really good current maps in Hawaii or any of our islands, because it's a complex topic with so many variables, yet the watermen have successfully navigated and used the currents for generations.

And so how do I bring in that knowledge? My dissertation work wasn't really with traditional knowledge per se as it was in the knowledge that local people had, but it led me down that path so that when I was done with my work, the Hawaiian Studies department at University of Hawaii Mānoa had an opening for mālama ʻāina faculty, and that was a perfect fit.

I am a natural resource manager, I worked in different contract roles for NOAA and private agencies and never saw myself in an ethnic studies department, but I really enjoy it here because I get to talk about the people who love these places and who want to preserve them and have a really deep—not just genealogical—but passionate love for this islands and these ecosystems and give them the skills they need to manage them.

How do you use local traditional knowledge to inform people about baselines and changes in our ocean environments?

I work at the University of Hawaii, Mānoa campus, in the Hawai'inuiākea School of Hawaiian Knowledge, with the mission to pursue, perpetuate, research, and revitalize all areas and forms of Hawaiian knowledge. The school is the only college of indigenous knowledge in a Research I institution in the United States.

I teach at the Kamakakūokalani Center for Hawaiian studies where we teach a variety of topics and address how Hawaiian studies overlap with other disciplines such as politics, history, arts, and resource management. The other faculty in my department have backgrounds in archaeology, geography, and law. And I have a background in science.

The established courses were about how people are connected to place and how we use that knowledge to manage and restore and increase sustainability of those places. I have added to the curriculum courses that are based around our connection to the ocean.

I introduce them to the Hawaiian world view of the ocean. All the gods that we have in Hawaii are environmental processes that occur. So the ocean, its movements, its moods, its changes, that is the god Kanaloa to us. It's not a personification, but it's the energies that are represented by the ocean. We named those energies Kanaloa. And so when I teach about

Kanaloa, I teach how he moves and what influences him. My last lecture, for example, was on the heavens, and that's Kāne, our god in the heaven, or the sun. Our Hawaiian traditions describe Kanaloa as a companion of Kāne. They travel together around the Hawaiian Islands opening freshwater springs.

And so when you talk about how the sun and the ocean create Earth's atmosphere, ocean movements, our trade winds, our weather patterns, everything is because of the movement of the earth, the sun, and the ocean. It's an amazing way of teaching about climate change and teaching about the ocean because you're connecting these to Kanaloa and Kāne, you can see why they are paired up in all of our traditional knowledge.

At the start of the course, students may know Kanaloa as the god of the ocean and Kāne as the god of the sun. But they soon have a deeper understanding that Kāne is the sun and he is rain. It is because of the sun that we have evaporation and rain, and it is the interaction between Kāne and Kanaloa that causes the atmospheric and oceanic patterns. This deeper understanding of the gods provides an understanding of how the world works.

When you talk about changes in our environment, we learn about them through our moʻolelo (our stories) and we have thousands written down in Hawaii. It's amazing to hear and see where people navigated in these ancient stories. Which beaches did they leave from in their canoes, where did they go. Did they always go North or did they always head West, if so you can probably tell that's the normal ocean conditions in those areas. I have my students start exploring those things and what the moʻolelo can tell you about your environment. I have them use Western science resources such as PacIOOS, but then also say go ask your grandpa, you can ask your uncle who's a fisherman. All of these sources are important, and they all work together to tell the stories about our places.

What would you like to see in terms of how traditional knowledge influences the way we are conducting technology development?

Right now, what we sense in the ocean is what's convenient. So we have a better understanding of larger patterns and movements offshore, but basic sensors don't overlap at the same scales where most human activities in the ocean occur. Nothing gathers data the size of a surfer but the surfers are the ones who are using our ocean the most often along our coastlines, which in general are the most difficult areas to sense and they're such dynamic areas.

What's the percentage of your students familiar with any of this information or the language?

I have a mix of students, but the marine biology students are definitely interested as well as the Hawaiian studies students that come with a deep love for their ʻāina, their home. During the summer, I worked with incoming students, transfer students from community colleges, and nontraditional students. They come with a love for the ocean and that's pretty much the common denominator. So the first third of the course is learning stories, and a baseline of how Hawaiians understood the environment and then we build on top of that. I go into some hardcore oceanography, meteorology, climatology cool stuff. We learn through both these sciences, and then we bring that together to talk about how it applies to us today.

We also have a graduate class that I cowrote with a lawyer, and we focus on ocean law and management in Hawaiʻi. We discuss how we codified our management system based on our environmental interactions and that many of our Hawaiian state laws are based on our traditional concepts as we were brought into the United States, but there's a lot of conflict with management at the federal level.

Can you say a little bit more about the application of traditional knowledge to inform the state laws?

Our culture is rooted in creating relationships with our deities as they all have different roles and relationships and representations. For example, Kanaloa is the ocean environment, the ocean entity. He's the ocean, and he also has specific creatures that represent him. The whale, the turtle, the octopus, the shark are all bodily representations of Kanaloa. Traditionally we had interactions with these creatures; however; those interactions are now at odds with the Marine Mammal Protection Act and conservation efforts by the EPA and NOAA. As indigenous people, we have no representation and we have no ability to actually interact as we have traditionally. NOAA is trying to work with us within the sphere of federal law, but it does not allow us to practice our traditional culture with these species because they're federally protected.

Are there success stories from other indigenous groups in working with federal law makers?

It's been interesting as we are able to connect with indigenous people around the world about how much our stories overlap. We've been talking a lot with the Maori in New Zealand about their interactions because they're Polynesian like us, so they have very similar perspectives of some of these creatures and understanding.

It comes down to a conflict in ethics. International cetacean ethics declarations, which people around the world have signed, aren't aligned with our traditional ethics. Our interactions are very different, and I feel like that is the same with other traditional groups. So how do I teach my students about Kanaloa if they have no connection with him, if they've never seen a whale, never heard one or been around one?

For example, when a whale beaches itself, NOAA mandates that the carcass is pushed out to sea or buried inland because of the smell, because tourists do not want to see a whale decomposing on the beach. And they don't want someone stealing something and that's against the law. So recently I worked with NOAA to allow a beached whale to naturally decompose on a beach. It was a tough, 2 month long process, but it was a bit easier because I billed myself as a traditional scientist. I was a scientist they could justify to their higher-ups so they could justify my monitoring of this whale. But I did it in my traditional sense, and this is the practices and the methods that we were using to monitor this whale decomposing on the beach.

Are you suggesting that the fact that you are a traditional knowledge expert carried less credence than the fact that you are a credentialed marine scientist?

Definitely. The other people that I work with don't have those scientific credentials, even though they've been fighting for a long time and they commune spiritually with these animals on a more sacred level than I can. I know that that's not my gift. They fight harder for their place at the table.

So how would someone know that you are the one to go to with respect to issues related to traditional knowledge? How do you establish that credential?

I'm just a Storyteller. I'm the one who connects all these things, but I probably am not the specific knowledge holder. I walk in all these multiple worlds, but I don't own any of them. Calling myself a practitioner has taken a long time because for me a practitioner is someone who is really well versed in particular things. I'm more just a Storyteller of all the things broadly and traditional knowledge keeping.

For example, this one uncle fisherman that I knew, he was 80 years old, and I loved him. He was an amazing fisherman, but he could only talk through stories. It took me days to truly process things he told me. I realized he was trying to tell me that he had to go 10 miles above the spot he normally fishes because he knows what the current was doing, and therefore he dropped his gear much deeper at that point knowing that was where he would find fish down current. His understanding was so complex, but his communication of that went over my head. It was just his way of communicating the information.

How would you suggest we bridge that local knowledge?

You just need to break that barrier and start talking to people as a scientist. You need to go down to the beach and start having a relationship with the people that you're working with if you're working locally. It has to be people to people.

And nurture and stress the importance of indigenous people's pride of their culture. A cultural revival happened in Hawaii in the 1970s and through that cultural revival, acknowledging our seafaring, our understanding of the ocean and that we could navigate. We had all this knowledge in our culture that brought forth a pride and desire for deeper understanding and cultural understanding, awakening our desire to want to care for our ocean and Islands.

Can you share a story about the impact that your training and education of the students has had on a particular student how has it influenced either their career choice or how they approached their career?

I'm on a Ph.D. committee for a student in the marine biology program right now and his other advisors are molecular biologists and experts in genomics in the ocean and it's been amazing serving on the committee with them. He was my first student. He was in sixth grade and I was an undergraduate. After graduating with his undergraduate degree, he became an education outreach specialist at NOAA in the Papahānaumokuākea Marine National Monument. He was using that time to decide on his graduate research topic, as he wanted to choose a topic with cultural significance. As a result, his topic is really important genetically, commercially, as well as culturally and spiritually to him.

Another example, I always try to engage my family members in my research. As a result, my sister's husband's younger brother was part of everything we did, and he became a NOAA Fisheries Observer. Not everyone has to go through college to work in the field. He's been an Observer for over 15 years now and that is an amazing job for him. He was a perfect person to serve on the long line boats going into the South Pacific because he could go handle the longer trips and was familiar with many Asian cultures.

Can you share a story about the impact of your training in traditional knowledge on your life?

Everyone knows Hawaii has the Aloha and what we feel is Aloha and to me that's the methodology that I use when I teach my students. But it's really hard to communicate that concept and get other people to understand how the methodology of Aloha is the most rigorous science that you can do.

You love your grandmother. Your love for your grandmother will make you care for her, and that love that you give to her keeps her alive. It's a reciprocal thing. Our Islands here are genealogically connected to us through our stories. We feel we are from these islands, we have love for these islands, and we're going to care for them. You don't have that same love for a random elder stranger that you meet on the side of the road. And so if you treat the lands or the ocean as that stranger elder person that you meet on the side of the road, it's that same sort of connection you're going to have with it. That's why our environments are not thriving right now, because we're not caring for them as we would care for our grandmother. You understand they get sick, and they get old and they go through these changes in these cycles and maybe they pass away but you're there through that.

I used that passing away analogy when the lava came and took a lot of shoreline communities 2 years ago. That was an area where my great-grandmother was raised and we had a beach house. It's a very different environment now, and people asked if I was sad, but I said, well, that's just the life cycle, right? You know, I loved her as I loved my grandmother and now she's passed and so while there is a different relationship with her, we do go visit her now that we can and she's 2 years old. You know, that beach, she's a baby now. It's amazing that beach came from that bottom of the ocean. So it's just your relationships that change through time, but those stories still continue.

In a sort of summary what message have you not conveyed that you'd want to convey?

The term I use right now is momona. It's a concept that means abundance. I try not to use sustainability or conservation in a lot of the work I do because if you are looking for stability, you are happy with things to be sustained. But what we really want is for things to thrive. So, through history, if you were wealthy, that was the only way to be fat, so being fat was a status symbol.

The same thing goes for environments. If an environment is rich with all its resources, it's healthy and abundant. I am committed now to sharing with people the idea of wealth and abundance and richness and knowing what that looks like.

I'm not quite sure if this is going on the economist route, but I have some colleagues who talk about the circular economies that we traditional societies had and that's what we need because a circular economy makes everybody rich. It uses the resources that you have but everybody is benefiting from it.

I want to see a rich, fat, beautiful ocean.

Author biography

Noelani Puniwai is passionate about cultivating the next generation of students to mālama 'āina. She has been trained academically (Ph.D. in Natural Resources and Environmental Management, UH Mānoa; MSc. in Environmental Science, Washington State Univ; B.A. in Marine Science, UH Hilo) to practice malama 'āina/kai. Yet through her experiences in her culture, science training, and student mentorship, she's recognized that feeling aloha 'āina, he alo a he alo, face to face must also be practiced, moving beyond exclusively intellectual pursuits. Her research interests include coastal ecosystems, cultural geography, knowledge coproduction, and seascapes. Noe believes that we can use the rigor and methodologies of pono science, the foundational wisdom of our kūpuna, and our experiential daily practice of aloha 'āina to awaken responsible action for the future of our Hawai'i.

CHAPTER 23

Building an inclusive and equitable new blue economy

Benjamin Cuker, Deidre Gibson
Department of Marine and Environmental Science, Hampton University, Hampton, VA, United States

The role of exploitation and racism in the old blue economy

To visualize an inclusive and equitable new blue economy (NBE) requires first understanding something of the old blue economy (OBE) and its failings in this regard. The OBE began when humans first began to extract resources from the ocean's edge and learned to use the seas for navigation. The evolution of coastal societies meant more extensive use of marine resources and more intensive ways of obtaining them. This engendered a mix of new technologies, such as advances in fishing gear and vessels, as well as industrialization of the expanding blue enterprise. As developed societies evolved through feudalism and into capitalism, the nature of blue economic labor relations changed. The owners of capitol, such as ships and the rights to exploit particular areas, employed fishers and sailors to do the work, and the owners garnered the wealth. This differed from the earlier artisanal model of families or villages working together to harvest resources for more or less equal shares of the take.

Preindustrial fishing cultures coevolved with the resource they were exploiting. The development of cultural restraints and taboos that prevented overharvesting took many forms, including restrictive guilds, areas of forbidden take, or seasonal closures of the extraction. Such successful systems still exist (Nunoo, 2018). The point of these traditional approaches is to feed the local population and perhaps support some trade with occasional surplus.

Enter industrialization and orientation toward profit. The industrialized model proved so effective in exploitation that it often created scarcity of the resource from overharvesting.

Industrialization of the fishing and other extractive marine enterprises wasn't about feeding the local people, but instead generating a commodity for trade. This transformed the role of fishers and seafarers into industrial

Preparing a Workforce for the New Blue Economy
ISBN 978-0-12-821431-2
https://doi.org/10.1016/B978-0-12-821431-2.00046-9

workers whose utility and worth were measured in the ratio of the cost to pay them and the wealth they generated. The devastating impacts of industrial fisheries on the ecology of the sea are well documented (Roberts, 2007), but the effects on the workers in the industry receive less attention (Rajaratnam, 2016).

Colonization, by Europeans, of Africa, India, the New World, and other places disrupted local traditional blue economies. Local peoples were either pushed off their coastal lands or saw their labor pressed into the new industrial system. Since such colonized people were invariably people of color (POC), race and racism emerged as unifying forces that enabled the colonial industrial takeover of the indigenous blue economy.

Owners of capitol strove to maximize profit, which meant minimizing pay. Such exploitation generates inherent tension between the workers and their bosses. Racism proved a useful tool to sustain exploitation of workers for maximum profit. This took different forms, for example, the most obvious was race-based slavery, where maritime workers were either owned by their captains or rented to them by an onshore slaveholder. Often the enslaved worker labored alongside free workers. The availability of the free labor from slaves diminished the ability of their free compatriots to demand better wages (Bolster, 1997). While the role of slavery in the blue economy formally ended in the United States with the defeat of the Confederacy in 1865, this institution lives on today in various fisheries associated with South East Asia (EJF, 2020).

At the end of the Civil War in the United States, newly emancipated freedmen flocked to the shores of the southern states, seeking employment in the burgeoning industrialized fisheries for oysters, blue crabs, menhaden, and other lucrative species. This was natural, as many of these women and men had worked in the OBE as slaves or sometimes as freedmen. To this day, descendants of these watermen and women continue to play an important role in the coastal economy, however, not at the level seen during Reconstruction. The end of reconstruction, in 1877, meant a reemergence of the white slave-holding class. They used Jim Crow laws to marginalize all aspects of African American life, including that associated with the blue economy. The best maritime jobs, such as well-paid harbor pilots, were reserved for white men. Guilds developed to control access to these jobs.

The rise of Jim Crow on the coastal waters of the South coincided with the technological transformation from the power of sail to mechanization. Although steam power appeared early in the 19th century, it wasn't until the 1890s that smaller engines, particularly those using internal combustion, found their way on to smaller vessels used for fishing and coastal commerce. Such mechanization was expensive and required the ability to borrow money. This shut out many African American watermen, as there were no banks to lend them the capital to motorize their boats. Engines also meant that fewer hands were needed to do the work. Therefore, the availability of maritime jobs diminished, further excluding POC. The combination of Jim Crow laws and the changing technology of exploiting the sea removed POC from the best jobs in the blue economy (Bolster, 1997).

However, industrialization of fisheries did and still does require lots of difficult hand labor for the processing of catch destined for a retail market for whole marine foods. While relatively little labor is needed to turn menhaden and similar fish into meal and oil, that's not true for delicacies such as oysters, blue crab meat, and fillet of numerous species of fish. Such fish-house work requires skill and the ability to withstand long hours of uncomfortable labor. Unsurprisingly, POC often filled these difficult and poorly compensated jobs. In the United States, this often meant local populations of African Americans and Latinx workers. The end of Jim Crow in the 1970s brought new opportunities for employment to POC. While still disadvantaged in many ways, such workers did find their way in new professions. In response, the fish processors turned to migrant laborers, primarily from Mexico and Central America. As such, race continues to inform the blue economy (Cuker, 2020a).

From aforementioned, it's clear that the OBE is part of the heritage of POC in the United States and elsewhere. Yet their numbers remain disproportionally low in various aspects of the endeavor. In particular, POC are underrepresented on the intellectual side of the OBE and particularly in the scientific fields that seek to understand the marine realm (Table 23.1).

This means that the aquatic sciences suffer from the loss of talent and perspectives engendered in its low diversity. It also means one more opportunity for POC to enjoy a vocation that brings them social equity (Cuker, 2020b). Building strong science to support a strong NBE requires a diverse set of perspectives and experiences.

Changes needed to build a diverse, equitable, and inclusive intellectual workforce for the new blue economy

Building a diverse intellectual workforce for the NBE requires a multidimensional approach. Traditionally those of us seeking to diversify the pool of students looking for employment in the OBE developed educational programs to prepare them for success.

Table 23.1 Current and projected demography of the United States and the distribution of degrees in science. Data on demography is from the Census (2020). Information on new degrees is for 2016 and comes from the NSF (2019).

	White	Asian/ Pacific Islander	African American	Hispanic	Native American
% US population in 2020	63.9	4.9	12.3	16.3	0.7
% US population projected 2050	47.0	9.0	13.0	29.0	1.0
% BS degrees awarded in all sciences	58.0	9.5	9.5	14.1	0.5
% BS ocean sciences degrees awarded	70.4	3.5	3.6	10.3	1.8
% MS ocean sciences degrees awarded	70.6	3.9	0	7.8	0
% PhD ocean sciences degrees awarded	59.0	5.8	2.9	2.2	1.2

The NSF, the National Oceanic and Atmospheric Administration (NOAA), other federal entities, and various private foundations have supported a variety of efforts to address diversity and equity in the aquatic sciences since 1990 (Cuker, 2005, 2007). Following are examples of such programs.

The NSF-funded **ASLO Multicultural Program** was created to address the lack of diversity in the aquatic sciences in 1990 (Cuker, 2001; Cuker et al., 2016). It brings about 80 diverse students from across the nation to the annual meetings of the Association for the Sciences of Limnology and Oceanography, providing full support, a variety of enrichment activities, and the attention of meeting mentors. Over 1200 different undergraduate and graduate students participated thus far (34% Male, 66% Female, 55% African American, 32% Hispanic, 7% Native American, 4% Pacific Islander). Many participants completed their Ph.D. and now serve as meeting mentors (Dr. Gibson is a product of the program, see Gibson, this volume). The program invented the concept of a "meeting mentor" that now informs similar efforts created by other scientific societies.

MAST (Multicultural students At Sea Together) ran between 2000 and 2008 and was funded by a PEW Fellowship in Marine Conservation award to Cuker (2003) and two grants from NOAA. Each year the program brought a dozen diverse undergraduate and graduate students to the Chesapeake Bay to complete a cruise on a sailing vessel. The intensive program incorporated marine science, policy, the heritage of African Americans and Native Americans under sail and on the Bay, as well as seamanship. MAST visited all of the important laboratories and museums on the Bay and engaged the students directly in scientific sampling to study eutrophication.

The NSF and NOAA-funded **The Hall-Bonner Program for Minority Doctoral Scholars in the Ocean Sciences**, which ran from 2006 to 2012. This was a joint effort between Hampton University (Drs. B. Cuker and D. Gibson), the College of William and Mary's Virginia Institute of Marine Science (Dr. Linda Schaftner), and Old Dominion University's Department of Oceanography (Dr. Greg Cutter). The Program's goal was to increase the diversity of students earning advance degrees in the ocean sciences. The students earned degrees from Virginia Institute of Marine Sciences, Old Dominion University, and Hampton University (M.S. Only). The program structure produced a critical mass of students of color in a community of scholars.

The program funded the tuition, stipend, travel to meetings, and special activities for students and faculty. Thirty students participated in the program with most earning an M.S. and six being awarded doctoral degrees.

As noted earlier, these programmatic approaches proved successful in producing students prepared for success in the OBE. However, building diversity and equity in the NBE requires more.

The NBE must value diversity, inclusion, and equity

It's insufficient to just train diverse students in the intellectual aspects of the NBE. The managers and future colleagues of these students must learn to celebrate the value that diverse students and employees bring to the enterprise. This means abandoning old ways of thinking and disrupting old models of organization. Systemic racism and other modes of social repression (based on gender, sexual orientation, national origin, etc.) pervade all aspects of our society. However, we are in a new era of people across of the globe calling out racism, colonialism, and other forms of structural oppression. The *entire* workforce of the NBE must be educated in critical race theory and the intellectual basis of all forms of oppression. Intellectual understanding of such oppression must be paired with building cross-cultural competencies. A person might be able to pass a test of knowledge about racism, yet still not understand how to operate and communicate in a nonracist way. Cross-cultural competency informs more effective communication and more collegial interactions free of unintended microaggressions.

Fully valuing diversity also requires an appreciation for contributions made by POC and other oppressed groups. This means learning some history. For example, it's important for the *entire* intellectual workforce to know about POC scientists such as Ernest Evert Just (20th century marine biologist) and Daniel Pauly (preeminent fisheries scientist). While students of color, at minority serving institutions, may learn about these giants of the aquatic sciences as critical pathfinders, it's just as important that non-POC workers in the NBE be as versed in the accomplishments of such folks. Learning about Just and Pauly helps free people from preconceived notions of who is eligible to do outstanding science. While the number of eminent diverse aquatic scientists (POC, women, and others) remains painfully small, they do exists and some have built exemplary careers with notable achievements (see, for example, perspectives of Gibson and Puniwai, this volume). The NBE workforce needs to learn about their stories.

Opening the minds of the NBE workforce also requires valuing alternative ways of knowing that emanate from traditional indigenous people coevolved with aquatic resources (see Puniwai, this volume). This valuing is best done, by incorporating indigenous people directly into the leadership of the NBE. This is critical as coastal indigenous cultures face problems of climate change and infringement on traditional resources. And, it is essential that traditional ecological knowledge informs the broader activities of the NBE; for example, leaving behind single species management based upon maximum sustainable yield theory for more expansive ecosystem-based management solutions (Cuker and Balazik, 2020).

The NBE must embrace alternative and more egalitarian modes of organization

Materialism understands that the nature of the system dictates the behavior of workers. The Western-masculine approach permeates the hierarchical organization of the OBE. Workers labor under a chain of command with a titular boss and lesser supervisors below, often arranged in some sort of branching organizational chart. Advancement means out-competing colleagues for a position at the next step up the chain. This sets up a noncooperative atmosphere. These systems at best operate as a meritocracy.

But often other factors dictate advancement and usually to the disadvantage of POC, women, and other oppressed groups. *Even functioning as a perfect meritocracy, such organization means devaluing cooperation in favor of internal competition.* Workers and managers of the NBE must be educated in alternative organizational structures. This begins with consideration of the *purpose* of an NBE organization. Traditional corporate organization focuses on producing profit for the owners (or shareholders) of the company. All else is secondary and defined as such in US law. However, companies can alternatively organize as B-corps (benefit corporations) that define corporate goals in addition to profit. In the NBE the alternative *primary* goal of a company might be to produce high-resolution benthic charts or to provide wind-powered electricity to reduce carbon emissions, rather than to simply maximize profit. It could also be to provide employment and empowerment to underserved communities.

Other alternative organizational systems exist and still others wait to be invented. The NBE companies can be employee-owned, with a democratic structure of governance. Others could be organized along the lines of cooperative endeavors, sharing capital expenses and labor.

All of these new more equitable ways of organization for NBE enterprises still must function in the larger social context. This means addressing specific national limitations to equability. For example, NBE corporations based in the United States and other underdeveloped nations in this regard must provide healthcare insurance for their workers.

Ethics and ecological knowledge must inform preparation of workers and managers in the NBE

It's possible to produce an NBE company that celebrates diversity and treats its workers well, yet functions to harm the greater society. Consider the case of a company focused on extraction of deep-sea petroleum reserves. What we know about climate change, ocean acidification, and dangers of oils spills means such an enterprise is inherently unethical. Any worker or manager educated in ethics would decline employment at such a company, given the existence of an alternative place to work with equal compensation.

Aquaculture is a rapidly growing enterprise, now accounting for half of the seafood produced globally (FAO, 2020). On the surface this may seem like a good idea, as it should reduce pressure on wild stocks. Yet the educated NBE worker should understand that much aquaculture is simply a protein transformation scheme built on converting the biomass of less valuable species into that of pricy animals destined for the luxury market. The endeavor diminishes populations of forage fish and pollutes the seas. Local fisheries and traditional aquaculture are an important and sustainable part of coastal indigenous cultures. Industrial scale fishing and fed-aquaculture systems offer no rational or ethical basis, as they exist as profit centers that provide unnecessary protein to already overfed populations in developed nations. Sufficient ecological knowledge should empower NBE workers to seek alternative employment.

Educating to understand the ocean as a commons

Workers in the NBE must understand that the ocean is a commons. Despite the designation of EEZs (Exclusive Economic Zones) and various schemes of leasing bottom area and the like, the ocean belongs to all the inhabitants (human and otherwise) of the planet. Its functioning is vital to the health of the biosphere. Hence NBE activities must be conducted with that in mind (see Spring, this volume). Consider a shipping company wanting to buy carbon credits to allow it to pollute the air with dirty fuel in a cap-and-trade system. Should that company have the right to pollute the commons of the air with CO_2 and ultimately warm and acidify the ocean commons? Is there another way to conduct human activity without monetization of labor, knowledge, and the commons?

Conclusion

The inherent inequity and lack of diversity in the OBE suggest that intentional efforts are required to correct the situation when building the NBE. Diversity is essential to bringing the benefits of the blue economy to the entirety of society. It is also essential for developing the best science to inform the NBE. This requires fully resourced institutional efforts. Yet, institutions are made of individuals, and said persons must make it their business to build diversity and equity through their everyday actions. Effective change begins from the bottom. We must hold ourselves accountable as well as our institutions as we build the NBE.

References

Bolster, W., 1997. Black Jacks. Harvard University Press, Cambridge.

Census, 2020. Demographic Turning Points for the United States. Retrieved 24 August 2020, from: https://www.census.gov/library/publications/2020/demo/p25-1144.html.

Cuker, B., 2001. Steps to increasing minority participation in the aquatic sciences: catching up with shifting demographics. ASLO Bulletin 10 (2), 17–21. https://doi.org/10.1002/lob.200110217.
Cuker, B., 2003. Minorities at sea together (MAST): a model interdisciplinary program for minority college students. Curr. J. Mar. Educ. 18, 45–51.
Cuker, B., 2005. Programmatic approaches to building diversity in the ocean sciences. Mar. Technol. Soc. J. 39 (4), 8–11. https://doi.org/10.4031/002533205787465869.
Cuker, B., 2007. Programs for building ethnic diversity in the aquatic sciences. Limnol. Oceanogr. Bull. 16 (2), 42–45. https://doi.org/10.1002/lob.200716242.
Cuker, B., 2020. Diet for a Sustainable Ecosystem. Springer International Publishing AG, Cham.
Cuker, B., 2020. George floyd and aquatic science. Limnol. Oceanogr. Bull. 29 (3), 73–74. https://doi.org/10.1002/lob.10392.
Cuker, B., Balazik, M., 2020. A fishing trip: exploiting and managing the commons of the Chesapeake Bay. In: Diet for A Sustainable Ecosystem. Springer Nature, Cham, pp. 83–86.
Cuker, B., Haxton, C., Martinez, C., 2016. How a scientific society built multicultural diversity: a 25-year-long journey. Bioscience 66 (3), 238–244. https://doi.org/10.1093/biosci/biw001.
EJF, 2020. Blood and Water. Retrieved 24 August 2020, from: https://ejfoundation.org/reports/blood-and-water.
FAO, 2020. The State of World Fisheries and Aquaculture. [online]. Available at: Fao.org. (Accessed 28 September 2020) http://www.fao.org/3/ca9229en/ca9229en.pdf.
NSF, 2019. Women, Minorities, and Persons with Disabilities in Science and Engineering: 2019. NSF - National Science Foundation. Retrieved 27 August 2020, from: https://ncses.nsf.gov/pubs/nsf19304/data.
Nunoo, F., 2018. Achieving sustainable fisheries management: a critical look at traditional fisheries management in the marine artisanal fisheries of Ghana, West Africa. J. Energy Nat. Res. Manage. 2 (1), 15–23. https://doi.org/10.26796/jenrm.v2i0.40.
Rajaratnam, S., 2016. Labour Exploitation in the Fishing Industry. Retrieved 24 August 2020, from:https://www.rsis.edu.sg/rsis-publication/nts/nts-bulletin-may-2016/.
Roberts, C., 2007. The Unnatural History of the Sea. Island Press/Shearwater Books, Washington, DC.

OP/ED: The new blue workforce: A case for diversity, equality, and inclusion by design

Kathryn D. Sullivan

I came of age professionally back in the days when women were mainly unwanted oddities in geology and oceanography. This was especially true out in field camps and aboard research ships. The bosun on one of my early cruises painted a fat red line across the fantail and announced that "no women go aft of this line on my deck." Margaret Leinen (currently Director, Scripps Institution of Oceanography) and I, both then second-year grad students, were responsible for the expedition's dredging operations, so we simply ignored this edict and went about our business. "There's no place for you to go to the bathroom" was another commonly cited and equally ridiculous so-called obstacle to including women. Did they imagine we couldn't duck behind a tree like they did, or that the ship's toilets would only work for men?

Standing your ground to be included in an expedition was just the first hurdle. The next one was being treated equitably as a full, contributing expedition scientist and given responsibilities in line with your capabilities. On the "red line" cruise mentioned earlier, the senior scientists relegated Margaret and me to routine administrative tasks such as sample logging, while treating the male grad students aboard like the competent, budding scientists we all were. Near the end of the cruise, as we steamed into port for some VIP political events, the captain's wife informed me that she expected me to serve hors d'oeuvres to the attending dignitaries. I had personally installed all the equipment in one of the major laboratories and was the sole representative of my university aboard ship, yet she saw me merely as available wait staff. Needless to say, I refused and spent the party showing the visitors through my lab rather than serving them snacks.

In some ways, it seems we've come a long way since those days. Women have now run major funding agencies and oceanographic institutions, led expeditions, and piloted submersibles to the deep seafloor. A closer look at the data reveals a much less rosy picture, however. Virtually every technical field important to our understanding and wise use of the oceans remains predominantly white and male, as are most policymaking government agencies and most ocean enterprise corporations. The number of women in positions of significant authority remains pitifully small, and the figures are even more dismal for people of color. Worse, the participation and advancement rates for all non-white male groups have shown little or no increase through time.

These statistics should alarm anyone who cares about the opportunities and challenges of the New Blue Economy. Why? For at least three reasons. First, because we now know from an abundance of research and data that diverse groups outperform homogeneous ones on multiple fronts. Diverse perspectives foster richer creativity, keener market insights, and wiser risk management, all of which lead to better product, market and social outcomes. Second, because our times demand it. Shareholders and financial sponsors in both the public and private sector are, increasingly, demanding that enterprises work to become more diverse and ensure that their internal culture and operating practices are inclusive and equitable. Third, our future requires it. The challenges of wise ocean stewardship and opportunities of the New Blue Economy cannot be met if we continue to tap just the talents of a small segment of humanity. We need the widest possible range of talents and perspectives to be in the "rooms where it happens," to be genuinely included in the work and equitably afforded opportunities and rewards.

Which reminds me of another story. Years ago, a senior colleague named John told me the story of his awakening to how culture affects participation and performance. When we talked, he was an established, highly respected researcher and dean of a premier graduate program. He was an easy-going and soft-spoken man; a gentle soul. The one thing John always hated about being an oceanographer, he told me, was the *mano a mano* combat approach to cruise planning, which pitted each scientist against the others in a fierce competition for each bit of ship time. He didn't see everyone as an adversary and detested the shouting, interruptions, and cutting insults that came with this approach. He was proud of the outstanding science that this produced, though, and so accepted the vicious competitive ethos as the price of excellence. Not long before our conversation, he went to sea on an expedition led by a female chief scientist. For the first time in his professional life, there was no shouting during the daily replanning meeting. "Everyone got to complete their sentences," he said. "They were really listening to each other and learning from each other. And they were producing the best science at sea I had ever seen." His punchline was this: Before this experience, if you had told him to make the planning less combative, he would have felt you were asking him to lower his performance standards. John was committed to excellence and had only seen one method produce it, so changing methods meant, *de facto*, settling for less. Now that he had seen an intentionally inclusive and equitable process in action, he not only knew this was wrong, but also knew how much greater the upside could be.

It is high time we become strategic about diversity, equity, and inclusion, and that we synthesize the lessons the past decades have taught us and build these imperatives into talent strategies from the outset. We must be deliberate and much more facilitative about diversity, equity, and inclusion. Drawing the red line on the fantail was an intentional act … removing it, and preventing another line from being drawn must also be done by design, and with clear intent. The emergence of the New Blue Economy presents us with an ideal opportunity to do this: A *tabula rasa* in many respects; a development of immense economic potential; a development of such critical consequence for the health of our oceans and sustainability of lives and livelihoods that we must bring our very best to the challenge. All of our very best.

Author biography

Dr. Kathryn Sullivan has a long career as a distinguished scientist, astronaut, and executive. She was one of the first six women to join the NASA astronaut corps in 1978 and holds the distinction of being the first American woman to walk in space. Her submersible dive to the Challenger Deep in June of 2020 made her the first person to both orbit the planet and reach its deepest point, as well as the first woman to dive to full ocean depth. Sullivan has held a variety of senior executive positions since leaving NASA, including Presidential appointments to the National Science Board and as Chief Scientist, Deputy Administrator and Administrator of the National Oceanic and Atmospheric Administration (NOAA). She currently serves on the boards of International Paper, Accenture Federal Services, the National Audubon Society and Terra Alpha Investments, and is a Senior Fellow at the Potomac Institute for Policy Studies. Dr. Sullivan earned a Bachelor of Science degree in Earth Sciences from University of California at Santa Cruz and a Ph.D. in Geology from Dalhousie University, Nova Scotia. Dr. Sullivan is a member of the National Academy of Engineering and the American Academy of Arts and Sciences and has been elected into the Astronaut Hall of Fame, Women Aviators Hall of Fame, Women Divers Hall of Fame, and Ohio Veterans Hall of Fame. She was recognized as one of the 46 distinguished First Women by Time Magazine (2017), the 15 Women Changing the World by the World Economic Forum (2015), and Time's 100 Most Influential People (2014). She has been awarded the Explorers Club Medal, the Rachel Carson Award, an Emmy, and nine honorary degrees. She is the author of the children's book To the Stars! (Charlesbridge Press, 2016) and Handprints on Hubble, An Astronaut's Story of Invention (MIT Press, 2019).

Interview: Veteran workforce opportunities

Bob Coniglione

Please share a brief summary of your military experience

I joined the Navy in 1976 with a GED in my pocket, no rank and no promises. Over the course of my military career, I served in a number of capacities including:

- Ships company: USS CORONADO 1977, USS MIDWAY 1978–80, USS NIMITZ 1991–94; while also serving
- Mobile Environmental Teams ("weather man in a box"): 1986–88. Deployed for 1 year with the Commander, Destroyer Squadron 21 in USS MERRILL, participated in operations aboard USS ARKANSAS, USS ANTIETAM, USS SALT LAKE CITY, HMCS KOUTNEY (w/Canadian Destroyer Squadron 2), M/V Jolly Roger.
- Shore Duty: Naval Air Station Cecil Field FL 1980–83, Naval Oceanography Command/Joint Typhoon Warning Center Guam 1984–85, Oceanographer of the Navy Staff 1988–90.

- Navy Schools: Aerographer's Mate "A" School 1978 (weather observer), Aerographer's Mate "C" school 1980 (weather forecaster), various technical classes for Satellite Interpretation, Anti-Submarine Warfare and personnel management.
- 1991 Graduate of the Navy's Senior Enlisted Academy.

In May of 1999, I retired as a Command Master Chief, Naval Meteorology and Oceanography Command, after 22 years/10 months active service.

Please share your postmilitary career path

I joined Analysis and Technology (A&T) in 1999 as a METOC (meteorology/oceanography) training Subject Matter Expert (SME). Then transitioned with A&T over to Anteon Corporation as a project manager in 2001. Then transitioned with Anteon over to General Dynamics Information Technology as a Senior Program Manager in 2006, then Director of Operations in 2009. After that, I joined PAE in 2012 as the Mission Control Center Manager at the National Data Buoy Center. I am currently the Contract Deputy Program Manager, Mission Control Manager, and IT Manager at the Buoy Center.

What were your expectations for employment when leaving military service?

At the end of my career, I was petrified about employment. At 43 years old, I had never looked for employment, written a resume, and most importantly had no clue what I was qualified to do in the civilian world.

What assistance or suggestions for postservice employment did you receive?

The Navy handed us off to something called the Transition and Assistance Program (TAP). Depending on your rank/position, there were different versions you could attend. In the end, they were basically the same thing. It may have been very well intentioned but, it was really a 3 or 4-day "information-by-firehose check in the box" for the Department of Defense (DoD). Ironically, I ended up a guest speaker at the TAP classes in Gulfport for several years. My 1 h was increased to 2 h and my time (class) was billed as the realities of leaving the military. I spoke about:

- What do you want to do—easily the biggest long-term issue.
- Learning a new language—English.
- How much are you worth

- "Resumes for Aunt Sadie"—i.e., how to write a resume in an understandable manner.
- The 12-year-old suit in your closet is not a wardrobe.

I'd like to think the Military is much better now, but I still talk to people facing military retirement all the time who have no clue what to do.

A smart officer is not going to have any problems finding another career. But enlisted people do have a much different fight. There are some who do it well, but there are some who do it terribly and when I used to teach the TAP class, one of the facts that I threw out was it takes enlisted people between three and five jobs before they settle in. That problem comes from the fact that they don't know what they want to do. And when enlisted folks get to that point, they're so concerned about getting a job.

I was very lucky. I had no less than 10 people approach me when I put in my papers to retire and say "when you get out there will be a job waiting for you." I was walking into a false sense of security when I retired. Out of the 10, only one was still standing.

When I did teach the TAP class, the folks in the class were "deer in headlights" and didn't know what to expect. I would love to think that the Navy and the military are better at it today, but I don't know that they are. Regarding my comment about checking a box, they do they put this TAP thing out there and they give you books and they tell you stuff but you need to learn a new language, English. And if you're around for 20 years or more you don't even realize that you can't communicate with someone who's never been in the military.

What—if anything—is unique to the military METOC community regarding postservice employment?

Nothing. Successful METOC officers (senior) are going to be successful civilians, not necessarily in METOC. They have postgrad science degrees and/or command experience. Those go a long way in the civilian world (especially government contracting), but they're not unique to METOC officers. While enlisted METOC people are actually at a disadvantage. Without a degree in meteorology, they won't even get looked at by the National Weather Service. And in today's National Weather Service, you need a Master's degree or a Bachelor's with a ton of experience to get hired. There are jobs in industry but the pay, hours, and benefits are no way to spend your life. If anything, the uniqueness of the technical training is a disadvantage.

I would note that currently I hire 2–3 Meteorologists a year and I'd say the number of applicants with military experience is about 5%.

Would you change anything in your experience to make the transition out of the military smoother or more efficient?

I'd certainly change my mental approach from I'm "retiring from the military" to "I'm transitioning to my next career." Too many people approach the end of a military career as a retirement only to wake up one day and realize they have to start over.

In retrospect, what was the most valuable aspect of your military career, in terms of preparing you for postservice employment?

In my case it was the network. I was lucky since I was adopted by a couple of people early in my career who put me in a great position at every turn, including retirement. But my case was the exception and not the rule.

Do you feel there is enough guidance or awareness provided to transition enlisted veterans into careers utilizing technology skills assimilated during service?

This is a double-edged sword. The military transition programs have guidance and charts and graphs that make all the technical skills sound like an easy transition to a civilian career. There are many enlisted career fields that translate well and are rewarding but there are many that don't and require transitional training and a reality check.

Do you feel there is enough support and guidance provided to enlisted veterans interested in pursuing a degree?

There have always been programs to encourage veterans to further their education. This really falls on the veteran and not the system. Whether or not an individual has the dedication and resources is the question.

I studied meteorology in the Navy and I have almost 2000 classroom hours of meteorology but I don't have a degree, and that's important to the civilian-minded guys. I see it all the time from military people. The commercial world has to understand that a Navy sonar technician can do electronic maintenance, can read graphs, and operate ROVs, gliders, and things like that. Those are the people you want behind the screen when you have a glider in the water. They have been trained to troubleshoot problems. There's a lot of smart sailors out there. They have the aptitude to apply theory and visualize things.

Do you think a microcredentialing system (crosswalk included) would be an attractive option?

For many enlisted people leaving the service after a single enlistment (4–6 years), this is a great option and the process works well with industry. Unfortunately, when it comes to government employment, I think there are obstacles. The government puts people in jobs that require specific training; often because they're too busy checking blocks that have nothing to do with the job skill required.

What's the downside of hiring military?

They get bored easily. You take a kid who's been on a ship for 3 years working 16 h a day doing six different things. And then you take them out of there and you put them in an office environment and tell him to sell copy paper or something. You want them to do the one thing every day, every hour. It becomes mind-numbing when you're used to being on a flight deck or when you're used to chasing down a problem.

How do we change the mindset so that an employer says I want to look for somebody who came out of the Navy and actually had operational experience working and has the chops to be able to get into the business world?

Your book is a great way to do that. Especially if you can get the audience to listen. We can't assume all these kids are going to go through and decide they want to get a Master's or a Ph.D. and become professors. So let's provide opportunities for them to have a broader set of skills. I would never expect the military to say, in addition to sending you to A school and C school, we'd like you to take some business classes. Microcredentials are a good way if there are crosswalks; for example, what does a Navy sonar technician do that will translate? But there's another side you have to consider. When I hire somebody right out of the military, chances are I have somebody who's coming to work every day, who understands discipline, who I can rely on. That's one of the reasons I have always looked at hiring military people, because I know half of my fear is gone. You know whether or not somebody can do the job. I can probably weed that out in the resume. What I can't weed out in the resume is their reliability and so I think part of educating the commercial world is pointing out you're getting a proven commodity and half your problem is solved.

Interview: Professional certifications

Bev MacKenzie

How can certifications help to build the workforce needed for the NBE?

The list of issues facing the ocean continues to grow. Microplastics and ocean acidification now join harmful atmospheric emissions, the spread of invasive species, and overfishing as problems that must be grappled with by the sector's regulators and other stakeholders. Neither can climate change and rising sea levels be ignored. With no respect for national boundaries, they threaten small island developing states and major ports across the developed world alike. Meanwhile, the commercial marine sector is being buffeted by the waves of the digital revolution and the networked economy. Ships are becoming ever more sophisticated and automated; supply chains are becoming tighter, yet more fragile; offshore resource extraction once limited to hydrocarbons now encompasses renewable energy and possibly soon deep sea mining. These are exciting changes that should not be feared but seized as opportunities. However, to be realized they will require new workforce with new competencies, skills, knowledge, and an ethos of safe guarding the sea and above all the innovation and the courage to push the boundaries of possibility.

Working in the marine sector is a rewarding but challenging career path. We need to ensure that the very best professionals are working with, on, or in the sea and that these professionals are also working to safeguard the seas for future generations. It's also critical that the workforce in the NBE has parity of esteem with other professions so we can attract and retain people to the sector. Professionals can have their experience and competence formally recognized, by gaining certification, which is a way of ensuring this parity of esteem.

Professional certification is considered to be a mark of excellence. It shows the individuals work to a high standard and consider how to operate safely. It demonstrates thinking about environmental sustainability and that you actively stay up to date with technological advances. All which are critical functions of working in the marine sector.

It can also be a prestigious accolade that is recognized and respected around the world. It communicates to people that competencies have been independently assessed and that you adhere to a professional Code of Conduct.

Finally, we expect to see the marine sector transition through greater digitization and automation and through a change in how we explore the ocean and use its resources. Technology will begin to play an increasingly large and interwoven component in our lives and we will also see a greater appreciation of the ocean and society. With this comes rise to the new enterprises such as autonomous shipping and we will likely see a huge demand for individuals trained in statistics, computer programming, design, and engineering. What is important to realize is that existing skills won't disappear overnight—there will still be a need for traditional seafaring skills, for example, and for skills in marine science research but training and upskilling will be of vital importance. So this skills "shift" will require two-way communication between workers—both an aging and a younger workforce that will require knowledge sharing and mentoring to teach each other different skills—the certification program supports this through the accompanying Continuing Professional Development (CPD) program.

How could certification stacking and microcredentials serve as a substitute for a degree (or can't—define your position)?

Existing certification programs across the science and engineering sector are normally based on a minimum academic standard (usually a degree or master's), but there is an increasing realization that "on the job learning" can provide a similar level of further learning. So many certification programs offer different routes with one being the undertaking of a career appraisal. These are designed to verify that over the course of your career, you have obtained an equivalent level of knowledge as a formal academic qualification with the process revolving around a report prepared that charts career progression and describes the knowledge and understanding picked up along the way and how it's been assessed.

It should contain evidence of know-how related to specific disciplines and how they have been applied in a practical setting. For example, an individual might mention your knowledge of certain theoretical models or methodologies and how they were applied in solving real-life problems as well as an understanding of their limitations.

With this in mind, certification establishes a level playing field. It recognizes that the value of a professional is more than an academic qualification, the value that it is a combination of background knowledge and professional competence.

Is there a value to certification for the academic workforce? If so, what is that value?

We believe there is a value of certification for all professionals. The overarching remit it is to bring together people united by a common goal. For example, to use ocean resources safely and sustainably and to ensure people abide by a code of conduct and ethics. Particularly given the uncertain political period in which we find ourselves and the rise in climate strikes calling on governments to take action, particularly from the younger generation, it is crucial that any action taken by decision-makers is predicated on robust and scientifically informed legislation. Professional certification means that governments can be sure that the advice they receive is coming not just from someone with a strong academic grounding but also an appreciation of the other professional competencies.

While the metrics of academic success remain peer review, funding success, etc., there is an increasing realization of the need for communication to policymakers and the public, the need to work collaboratively across industry and with society. Professional certification recognizes formally these attributes of someone working in academia. We believe there is an increase of recognition, compared to when first explorations of a certification program for marine science professions were undertaken in the United States around 10 years ago.

Finally, and perhaps most importantly, many students, graduates, post-graduate students will not be able to or chose to stay in academia. For them arguably, professional certification will be an important part of their professional life. With that in mind the academic workforce should have the ability to act as role models and as mentors for the future workforce.

Is certification a regionally or nationally specific activity, or can it be done globally?

Working on the principles of shared values, goals, and code of ethics ideally the certification should be recognized globally. This doesn't necessarily mean that the process of administering and awarding certification needs to be done globally but that if national or regional schemes exist that multilateral agreement between bodies responsible exist to enable mutual recognition and to assist the mobility of professional marine scientists, technologists, and engineers. This is particularly important should the certification extend to the "accreditation or recognition" of academic courses as contributing to an agreed academic base requirement for certification. The development and recognition of good practice in marine science, engineering, and technology education need to be done with a consistent benchmark and standard.

How important is continuing education to the certification process?

Ongoing education and training are the ultimate reason that we promote certification. Working toward certification requires Continuing Professional Development (CPD) and becoming practiced at noticing when you're doing an activity that teaches you something new. And after gaining certification, you are automatically committed to doing CPD for the rest of your working life. And that's why certification is so valuable to the NBE.

CPD is a personal responsibility of professionals to keep their knowledge and skills current so that they can perform their job in a way that safeguards the public and meets the expectations of customers and the requirements of their profession.

A number of recent tragedies have sadly put a focus on the importance of competence of professionals. The Grenfell Tower Fire, in London in 2017, led to a loss of 72 lives and put a spotlight on the competence of professionals in the building and other engineering sectors with the official report into the tragedy making strong recommendations of the importance of meaningful continuing professional development. The recommendations can be seen as being of equally relevant in the marine sector with accidents such as the loss of the Deepwater Horizon and the Costa Concordia demonstrating that those working at sea also need to be up-to-date and competent.

The formality of Continuing Education may take but one part of Professional development. Professional development naturally takes place as experience is gained in the workplace, and as an individual progresses through their career, they acquire greater independence and responsibility for projects and for managing people. Reflective analysis of this development is the focus rather than constantly attending training courses. This is not a new concept and is the most important part of CPD reporting. Reflecting on what has been done makes an individual think of the value of that activity for them self, colleagues, employer, and the wider community.

It is focused around asking key questions:

- What did you get out of undertaking the activity?
- What specifically have you learned?
- How did you learn it?
- How will you apply it in practice?
- How will it change your behavior or impact those around you?
- Has it made you aware of any additional CPD you might wish to undertake?

What benefits will the activity have for the wider community, for your colleagues, or the general public, for example?

The scope and complexity and scope of work often increase. This and any in-house training or assessment processes (such as performance appraisals) all count as Professional Development activities, but it's what you learn from them that is important and we will introduce the importance of reflection later.

How will we know if/when certification has succeeded?

The premise behind certification and the CPD commitment that goes with it is that it contributes to improved protection and quality of life, the environment, sustainability, property, and the economy. With a competent and professional workforce who demonstrate ethical behavior, we should see an improvement in the management of the ocean—moving us closer to the achievement of SDG-14. It's a lofty ambition but at the heart of sustainability lie people and their behavior.

Author biography

Dr. Bev Mackenzie has a degree in chemical oceanography and a Ph.D. in physical oceanography from the University of Liverpool, UK. On graduation she learnt the ropes in volunteer engagement at the University of Oxford, she moved to the Institute of Marine Engineering, Science & Technology (IMarEST)—a professional membership body, and learned society and International NGO as a Technical Coordinator in 2004. Until the end of January 2021 she was the Director of Policy with her role looking at strategies that provide opportunities for the exchange of ideas and practices and opportunities to influence the development of marine policy and legislation at local, regional, and global levels in support of the UN Sustainable Development Goals—in particular Goal 14 (Life Below Water). in February 2021 she joined BIMCO, the world's largest shipping association, as the representative to the International Maritime Organization (IMO) and as a member of the Marine Environment Department.

Dr. Mackenzie has overseen contributions to the work of a number of Intergovernmental Organizations leading on representation at the International Maritime Organization, the Intergovernmental Oceanographic Commission of UNESCO, the Intergovernmental Panel on Climate Change, the Informal Consultative Process on Oceans and the Law of the Sea, and the London Convention/London Protocol.

She is a Fellow of the IMarEST and of the Marine Biological Association and is a Chartered Marine Scientist and Chartered Scientist. She has been a member of the Board of the Science Council (UK) and currently sits on the Board of Plymouth Marine Laboratory.

OP/ED: Role of certifications in education programs

Steve Hall

In an era where the public, policymakers, financiers, insurers, and even industry specialists are faced with a bewildering choice of options when someone claims to be an "expert," how do we quality-control that claim, particularly when we are talking about individuals rather than corporations? With the New Blue Economy at the point of delivering a range of new products and services to market including materials mined from the ocean floor, new kinds of aquaculture, and hydrogen electrolyzed from inexhaustible supplies of ocean water as well as commerce, defense, and offshore survey by robotic ship or submarine, and many other innovative activities, how can governments, consumers, decision—makers, and licensing authorities have any level of trust that the individuals they are working with are actually competent, safe, and certified?

Clearly part of the answer is through the provision of excellent, globally recognized, baseline education, be that through schools, apprenticeships, technical training programs, and university and postgraduate education, but as we begin to enter new areas of technology and activity where there isn't a well-established standard with which to judge the competence of staff, is there a way ahead? Particularly when working in the global ocean sector where individuals from a variety of different nations, with varying levels of formal, well-regulated education standards, are competing for jobs and contacts?

Fortunately, there is a long tradition in many parts of the world of having a method of peer review of competence and ability through various means of "accrediting" or professional registration and certification of specialists in wide range of fields. It goes back a long way, to the medieval guilds who certified ability as metalworkers, ship-builders, armorers, clock-makers, apothecaries, and many other skilled trades where, in an age before the widespread availability of university education or specialist technical colleges, you had to rely on peer assessment of ability.

Just like today's world, a young person entering a trade was expected to study under skilled masters, learn by practical experience, and eventually pass a set of tests and a period of practice under supervision before they were granted full membership of their guild.

Fast forward to the 21st century and we see that certain professions have long required that practitioners not only have formal academic qualifications, but supplementary professional registration and proven skill sets, perhaps demonstrated through a probationary period of practice, as well. Examples—at least in Western jurisdictions—include medicine, accounting, pharmacy, law, and many branches of engineering. This author is based in the United Kingdom where we still have the ancient guilds, but more modern examples are the network of Learned Societies and especially the Professional Bodies who are able to certify the competence of their Associates, Members, and Fellows.

Many of these organizations came into being during the rapid learning curve that came into play at the start of the industrial revolution—engineers and scientists were sometimes self-taught, were discovering new compounds and elements, taking the first steps in using iron and steel for construction and ship-building, building bridges and coastal defenses, and inevitably mistakes were made, sometimes leading to serious loss of property and life, so by establishing these bodies, publishing peer-reviewed journals, holding meetings, and sharing best practice, the new discoveries (and how to avoid the pitfalls!) could be shared nationally, later globally, and safe methods of work established.

A key point that is relevant to the New Blue Economy is that Professional Bodies became good at using their accumulated experience and knowledge to make sensible recommendations and choices about the regulation and safe working techniques to be applied to innovative activities. The "precautionary principle" has been a feature of the way that Professional Bodies, often quite conservative (with a small "c") organizations, like to react to change—they'll happily embrace innovation and new technologies, but don't push too fast or too far—do it slowly and carefully, with peer review, and do it as well as it can be done. It might mean that things happen more slowly than a "free market" approach might provide and tends to lock out accident-prone mavericks, but does have the benefit of enhancing safety and the overall risk of harm.

Where there are several Professional Bodies working in a related area, such as the variety of different specializations of Engineer (e.g., aeronautical, naval, civil, electrical, mechanical, etc.), there may be an oversight Council, such as the UK's Engineering Council, that ensures the subcategories are operating at a sufficiently high level and are vigorous in their assessment of new entrants to the profession.

Examples of Professional Bodies whose members are active in the New Blue Economy include the Institute of Marine Engineering, Science and Technology (https://www.imarest.org/, see MacKenzie, this volume), the Institute of Civil Engineers (https://www.ice.org.uk), the Institute of Mechanical Engineers (https://www.imeche.org), The Marine Biological Association (http://www.mba.ac.uk), the Royal Institution of Naval Architects (https://www.rina.org.uk), Engineers Australia (https://www.engineersaustralia.org.au/), Institute of Engineering & Technology India (https://india.theiet.org), Verein Deutscher Ingenieure (https://www.vdi.de), Institute of Professional Engineers of Japan (https://www.engineer.or.jp/sub09/), Engineering Council of South Africa (https://www.ecsa.co.za/), and many others, with some sectors overseen by international coordination bodies such as the European Federation of Professional Engineers FEANI (https://www.feani.org).

What are the different levels of professional registration? Most bodies offer a range of levels depending upon the educational achievement, experience gained, and age of the applicant—for example, see the guidelines for the IMarEST system at https://www.imarest.org/membership/membership-registration/apply-for-professional-registration—IMarEST start with "Technician" level, go via "Registered," and eventually to "Chartered" by the time the individual is considered an experienced, fully rounded practitioner.

At each stage, individuals are assessed by experienced professionals, need to maintain a record of continued professional development, and for most organizations there is an annual fee for staying registered and other fees for one's assessment. Many employers will reimburse the fees as having a staff member go through the process is an excellent method of ensuring good training and progression, and some organizations embed the professional schemes into their internal training and staff development process.

On achievement of a given level of certification, the individual gains the legal right to use internationally recognized "postnominal" titles after their name—for example, this author can add CMarSci (short for Chartered Marine Scientist). Depending on the organization and level of certification, one may retain the status for life, or one might need to demonstrate continued professional development training and active engagement in the field to retain the title.

Professional Bodies will often work closely with educational institutions to ensure that courses are designed to supply new entrants to their profession, teach content that is accurate, up to date, and well-aligned with the requirements of the industry. Students who graduate with these sort of "accredited degrees" or technical qualifications may be exempt from having to complete certain professional examinations, or join professional development pathways a step or two up the ladder from students who studied elsewhere. A student faced with the hard decision of where to study is well advised to look first the courses that are marketed as accredited by a Professional Body—(s)he'll soon find that when it's time to look for a job, the candidates who've studied a course that is recognized in the industry will have an advantage over those that haven't.

An organization can't just set up shop on a Monday and claim to be a Professional Body—the countries hosting such bodies have rigorous procedures in place to assess the standards and protect professional reputation and status. In the United Kingdom, an application to become a recognized Professional Body has to be approved by the Privy Council (prestigious advisors to the Government) and the relevant oversight Council (if there is one) and is tightly regulated. Similar systems are in place across the world wherever true Professional Bodies are registered, so that when you go to the doctors, or want someone to build you a nuclear power station, an airliner, ship, or a high-speed railway line, you know that you're dealing with people who know what they are doing and can ask to see documentary evidence of their Professional Status.

Certified individuals, often referred to as "**Chartered**," will usually also be able to secure professional indemnity insurance, which can offer an extra level of reassurance to a customer or client that if the expert they have contracted does make a mistake, it will be possible to claim damages. Insurers aren't in the habit of throwing their money away and know that a Chartered Individual is a low insurance risk compared with someone who seeks indemnity cover without recognized credentials.

One step below the Professional Body is the "**Learned Society**," though most often a Professional Body is both. Learned Societies share many of the same criteria as a Professional Body, except that they are not permitted to confer "Professional" or "Chartered" status to their members, or strictly speaking to confer the honor of postnominal initials. They may be able to declare that their members are "Registered" or "Certified" in some way, with an assessment process in place to check on their quality, which may be almost as high as the standards used by the Professional Bodies, but they are not allowed to go that extra mile.

Examples on the New Blue Economy include the Marine Technology Society (https://www.mtsociety.org/), Challenger Society for Marine Science (https://www.challenger-society.org.uk), the Oceanography Society (https://tos.org), and Society for Underwater Technology (https://www.sut.org)—all are organizations that have highly capable and experienced members, many with Doctorates, but these Learned Societies are not authorized to issue their own professional certification in most Western countries. There will be local variations across the world, and some nations, particularly in the developing economies, will have the domestic equivalent of Learned Societies who have a high level of regulatory control over professional activities carried out within their jurisdiction. Typically this might be the licensing of activities such as professional diving, underwater construction and maintenance, leisure, aquaculture, fisheries, and even offshore energy, though for activities where there is high inherent risk such as oil and gas production, almost all nations expect to see global professional standards applied.

Learned Societies can enter into an agreement with one or more of the Professional Bodies to accredit the Learned Society's members in a scheme that is licensed from a Body able to do so, examples are SUT in Australia having a license from Engineers Australia to accredit pipeline engineers, and MTS and SUT have licensed the right to accredit Chartered Marine

Technologists from the Institute of Marine Engineering Science & Technology. Both Learned Societies have been given permission to explore new professional accreditation pathways in future, such as targeted registration for specialists in the use of deep-sea robotics, or deep-sea mining equipment—but the ultimate authority that would issue Chartered Status remains the Professional Body, who themselves must comply with the standards insisted on by the overarching Engineering Council so that quality and reputation are maintained.

There are also organizations that are "Trade Bodies" or industry partnerships such as IMCA, the International Marine Contractors Association (https://www.imca-int.com) that can accredit or certify the activities of their member companies in various ways, but usually these organizations are aimed at certifying companies/corporations rather than individual practitioners, with some being able to award International Organization for Standardization (ISO, see https://www.iso.org/home.html) certification for safe working practises, compliance with environmental protection and other worthwhile achievements.

The development of internationally recognized ISO standards for safe, clean, and sustainable Blue Economy activities is an area where the global community of marine professionals has much to offer—already there's a great deal of guidance on areas such as water quality, environmental protection, wastes, noise impacts, and more (see https://www.iso.org/ics/13/x/) but there are many gaps, and the UN Decade of Ocean Science for Sustainable Development (https://en.unesco.org/ocean-decade, see Ryabinin, this volume) is one of several new pathways that will offer an opportunity for Professional Bodies and Learned Societies to work with governments, regulators, industry, and individuals to plug these gaps and work together on an international basis to ensure safe, clean, productive, and sustainable use of the ocean and ocean resources to protect and enhance for future generations.

As we enter the Ocean Decade, license new kinds of activity, and bring to market the new bounty of the ocean, qualified Marine Professionals, certified by respected and competent marine Professional Bodies and licensed Learned Societies, will have a key role to play in ensuring that humankind can interact with the ocean as good stewards, responding quickly to change and innovation through their accumulated knowledge and experience to introduce new kinds of training, mentorship, and capacity-building expertise to look after the global ocean.

My hope is that employers, governments, regulatory authorities, and licensing agencies will embrace cooperation with Professional Bodies and Learned Societies as international providers of expert knowledge and advice, raising standards and establishing sound working practices across a broad swathe of ocean activities.

Author biography

Steve Hall is CEO of the Society for Underwater Technology, an international learned society for marine engineering, science, technology, and policy headquartered in London, with branches in 10 countries and members in 40, between 2017–20. He is now CEO of the Pembrokeshire Coastal Forum and Marine Energy Wales. Before his SUT role, he served 27 years in the UK Natural Environment Research Council, based mainly at the National Oceanography Centre, initially conducting tracer chemistry observations at sea, administering the UK contribution to the World Ocean Circulation Experiment, and managing the early "Autosub" autonomous underwater vehicle science missions. After 2006, Steve moved into a policy role, advising governments and stakeholders on a wide range of ocean policy issues, eventually becoming Head of the UK delegation to the Intergovernmental Oceanographic Commission of UNESCO, where he was elected as Vice Chair 2015–17. He was also responsible for tsunami warning systems for UK overseas territories. Steve has a long-standing interest in education and training for marine scientists and was one of the first cohort of Chartered Marine Scientists. He is a Fellow of IMarEST & SUT and a member of the Marine Technology Society.

Further reading

Chartered Professional Status, August 23, 2020. Wikipedia article. Retrieved from: https://en.wikipedia.org/wiki/Chartered_ (professional).

Engineers Australia Chartered Status, August 21, 2020. Retrieved from: https://www.engineersaustralia.org.au/start_your_journey.

European Professional Qualification Directives, August 23, 2020. Wikipedia article. Retrieved from: https://en.wikipedia.org/wiki/European_professional_qualification_directives.

Hall, S.P., 1999. Accreditation for Oceanographers? Ocean Chall. 9 (3), 7.

ISO Standards, August 24, 2020. Retrieved from: https://www.iso.org/standards-catalogue/browse-by-ics.html.

IMarEST Professional Registration, August 21, 2020. Retrieved from: https://www.imarest.org/membership/membership-registration/apply-for-professional-registration.

IMarEST Chartered Marine Scientist, August 21, 2020. Retrieved from: https://www.imarest.org/membership/membership-registration/apply-for-professional-registration/chartered-scientist-csci-chartered-marine-scientist-cmarsci.

Institution of Chemical Engineers article on 'Why Become Chartered?', 26 January, 2017. Retrieved from: https://ichemeblog.org/2017/01/26/10-reasons-to-get-chartered/.
MTS Chartered Marine Technologist, August 24, 2020. Retrieved from: https://www.mtsociety.org/cmartech-certification/.
Ocean Decade, August 24, 2020. Retrieved from: https://en.unesco.org/ocean-decade.
Professional Certification, August 23, 2020. Wikipedia article. Retrieved from: https://en.wikipedia.org/wiki/Professional_certification.
SUT Chartered Marine Technologist, August 24, 2020. Retrieved from: https://www.sut.org/sut-membership/chartered-marine-technologist-cmartech/.

SECTION VI

Policies that facilitate

When done well, public policies can be important stimuli for economic development. From a legislative perspective, US Senator **Sheldon Whitehouse** discusses some past policy successes and presents opportunities for future development. **Craig McLean** discusses the implementation of policies in the form of programs that can stimulate the blue economy and help develop the necessary workforce. Internationally, the value of efforts such as the UN Decade of Ocean Science for Sustainable Development will be critical to build out the New Blue Economy. **Vladimir Ryabinin** provides an important intergovernmental perspective. **Margaret Spring** emphasizes the importance of economic development in the context of maintaining heathy oceans. The balance between the management of resources and economic development are expressed in the context of sustainability. And **Norm Augustine** provides a very relevant cautionary note regarding risk tolerance; his op-ed is important guidance to those who are leading the establishment and oversight of workforce development programs throughout the world.

CHAPTER 24

Domestic opportunities to advance blue economy priorities

Sheldon Whitehouse
United States Senate, Washington, DC, United States

Federal policy and our oceans

Oceans have been at the center of America's economic success since our founding, when colonial cities relied on their ports for shipbuilding, fishing, and trade. Rhode Island for centuries has had a vibrant maritime economy. Today, almost half of all Americans live in coastal counties—counties that have over five times the population density of inland regions (U.S. Census, 2010; NOAA, 2020). Maritime activities in these regions contribute heavily to our nation's economy. For instance, in 2016, commercial and recreational fishing industries supported 1.7 million US jobs, generating more than $212 billion in sales and contributing $100 billion to gross domestic product (NMFS, 2016).

Recognizing oceans as a powerful economic engine, federal policymakers have strived to formalize ocean policies and governance. Congress directed the creation of the first modern agency dedicated to ocean governance in 1966 with the enactment of the Marine Resources and Engineering Development Act (P.L. 89−454). Emphasizing the economic benefits of developing oceanographic programs, the Act declared that the United States would:

> *"develop, encourage, and maintain a coordinated, comprehensive, and long-range national program in marine science for the benefit of mankind, to assist in protection of health and property, enhancement of commerce, transportation, and national security, rehabilitation of our commercial fisheries, and increased utilization of these and other resources."*

The Act established a presidentially appointed Commission on Marine Science, Engineering, and Resources to deliver on this goal. Chaired by Julius Stratton of the Ford Foundation, the "Stratton Commission," as it became known, worked diligently to present recommendations to the President and Congress for the country's marine institutions and policies.

Preparing a Workforce for the New Blue Economy
ISBN 978-0-12-821431-2
https://doi.org/10.1016/B978-0-12-821431-2.00021-4

The Commission's final report in January 1969 fulfilled Congress's intent by recommending the creation of the National Oceanic and Atmospheric Agency (NOAA) to manage federal marine programs. With strong support from Rhode Island Senator Claiborne Pell and leading ocean scientist John Knauss, Congress established NOAA in 1970 as an agency within the US Department of Commerce—a decision underscoring the significance of our oceans to the national economy (NOAA, 2006).

Congress followed the Stratton Commission and the creation of NOAA with a slew of bills to protect our oceans and coasts. The 1970s brought the Marine Mammal Protection Act; the Marine Protection, Research, and Sanctuaries Act; the Coastal Zone Management Act; the Endangered Species Act; and the Fishery Conservation and Management Act (NMFS, 2020).

The 106th Congress attempted to revive the success of the Stratton Commission by enacting legislation to establish a US Commission on Ocean Policy (P.L. 106–256) in 2000. In 2003, the Pew Commission released a widely anticipated report titled, "America's Living Oceans: Charting a Course for Sea Change" (Pew Oceans Commission, 2003), setting a national agenda for protecting our oceans (CRS, 2005).

Executive action has played a central role in ocean policy in recent years. The George W. Bush and Obama administrations created and then enlarged national marine monuments, established national marine sanctuaries, and cracked down on illegal fishing. Expanding on those ambitions, the Obama administration established an Interagency Ocean Policy Task Force to develop recommendations for a comprehensive national ocean policy. President Obama later signed an Executive Order adopting the task force's recommendations and establishing a National Ocean Policy for Stewardship of the Ocean, Our Coasts, and the Great Lakes. The Executive Order emphasized the importance of the ocean as an essential economic driver:

> *"The ocean, our coasts, and the Great Lakes provide jobs, food, energy resources, ecological services, recreation, and tourism opportunities, and play critical roles in our Nation's transportation, economy, and trade, as well as the global mobility of our Armed Forces and the maintenance of international peace and security." (Exec. Order No. 13547, 2010)*

Not all federal ocean policy has been productive. In 2018, President Trump issued an executive order reorienting President Obama's National Ocean Policy from conservation to exploitation of ocean resources. The

Trump order failed to recognize the vast economic value of healthy oceans and put coasts at greater risk from extreme weather and other effects of climate change.

Bipartisanship at work: the Senate Oceans Caucus

When I joined the Senate in 2007, Democrats and Republicans were grappling with a common set of ocean issues: illegal, unreported, and unregulated (IUU) fishing threatening the health of our wild fisheries; the glut of marine debris in our seas; and a dearth of ocean data, research, and monitoring.

Republican Lisa Murkowski shared my interest in organizing a group in the Senate to focus on these issues in bipartisan fashion. Alaska's economy relies heavily on its fisheries, which are responsible for nearly two-thirds of the nation's seafood harvest (McDowell Group, 2020). Rhode Island is similarly reliant on the ocean and coast to support our hugely valuable fishing, tourism, and port industries.

Senators Whitehouse and Murkowski announce the formation of the Senate Oceans Caucus. September 13, 2011. *(Source: Senator Whitehouse Website.)*

Together, Senator Murkowski and I determined to make the Senate Oceans Caucus a true working caucus to chart a bipartisan path forward on ocean policy. Since 2011, the Caucus has grown to 41 bipartisan members, each of whom sees the importance of protecting our oceans and coasts for future generations (ICCF, 2020). Together, we have succeeded in advancing legislation in all three of the core areas we identified at the Caucus's founding.

Our first legislative mission was to combat IUU fishing through Senate approval of four international fisheries treaties and swift passage of legislation to enable those treaties. Each year, illegal fishing practices cost the global economy nearly $23 billion, representing roughly 20% of the global seafood catch (Pew, 2013). As a major producer and consumer of seafood, the United States must ensure that imported fish are caught legally, both to maintain consumer confidence in the product and to protect our domestic fishing industry from predatory competition.

Next, we set our sights on marine debris—a priority for ensuring that marine ecosystems are healthy and tourism continues to thrive. Plastic pollution is a growing global menace, with 8 million metric tons of plastic entering the ocean each year, adding to the estimated 150 million metric tons already afloat in our oceans (Ocean Conservancy, 2020). Honorary Oceans Caucus Co-Chair Dan Sullivan (R-AK) joined me in introducing the Save Our Seas Act to tackle the marine debris crisis. Save Our Seas reauthorized NOAA's Marine Debris Program and planted the seeds for broader administrative action, including emphasizing marine debris remediation in US trade negotiations. After extensive bipartisan cooperation, President Trump signed Save Our Seas into law in October 2018.

Building on that success, in June 2019, Senator Sullivan and I introduced the Save Our Seas 2.0 Act, to improve our domestic and international waste management systems, spur innovation in plastics and other materials that end up in the sea, and take further steps to reduce the millions of tons of plastic waste that enter the oceans each year.

The Caucus has also produced legislation to improve ocean data and monitoring—a sector integral to the new blue economy that promises to advance our collective understanding of our oceans. Senator Murkowski and I have introduced the BLUE GLOBE Act to measure the value of the blue economy; train the future marine workforce; and lay the groundwork for an Advanced Research Projects Agency-Oceans, or ARPA-O, to foster the development of groundbreaking marine technology. Everyone from our fishing industry to our naval operations will benefit from these efforts to improve the wealth and accuracy of our ocean data. We have a long way to go in improving federal ocean policy, but the steady bipartisan work of the Senate Oceans Caucus shows what Congress is capable of doing to protect our oceans and coast for future generations.

Advanced Naval Technology Exercise at the US Naval Undersea Warfare Center in Newport, RI, showcasing new ocean monitoring technologies. August 29, 2019. *(Source: Courtesy of Senator Sheldon Whitehouse.)*

Guiding principles for successful ocean policy

The challenges our oceans face are serious enough that we must look beyond allegiance to political party or philosophy. Democrats and Republicans must recognize those challenges and resolve to create a cleaner, safer, and better-understood ocean. That is why a spirit of collaboration, clear communication across partisan and ideological lines, and steadfast trust in science should be guiding principles. Our bipartisan Senate Oceans Caucus has opened a pathway for this sort of success in the Senate.

In practice, these principles mean actively soliciting input from subject matter experts and from those affected by our work, as well as heeding sound research and warnings. The Oceans Caucus stays in regular contact with relevant agencies, industries, and environmental organizations to seek feedback and ideas. We ensure that our work is anchored in solid science and on-the-ground knowledge. Keeping stakeholders in the loop at every turn and listening to the science earns trust and good will, which builds our capacity for more ambitious, effective legislation.

The results are multiple fisheries treaties approved, the Port States enforcement bill passed into law, the first-ever marine plastics bill passed into law, and two important bills in process for passage: a bigger, better ocean plastics bill (already through the Senate) and a comprehensive marine data bill. Each success paves the way for future successes.

Remember Voltaire's wise words: "don't let the perfect be the enemy of the good." In a body as politically divided as our current Congress, steady forward progress deserves celebration. Continuing to move the ball forward—one successful play after another—brings legislative wins, and it opens the aperture for bigger, future progress. That is a proven strategy for improving ocean policy.

Conclusion

Our oceans are our most productive and valuable resource, yet oceans face dire threats. Seas are warming and acidifying due to carbon pollution. Valuable fisheries are fleeing traditional fishing grounds in search of cooler waters. Worsening sea level rise and storm surge threaten coastal homes, businesses, and infrastructure. Indeed the US mortgage giant Freddie Mac has warned of a crash in coastal property values based on these hazards (Freddie Mac, 2016). At the Our Ocean Conference in Oslo last year, I saw the world look to the United States for leadership in advancing ocean initiatives. American leadership will be crucial for protecting a productive blue economy. And the data-driven new blue economy will be especially important for sustainable growth. With good faith, and trust in our science, we can build upon the bipartisan success that has taken root in the Senate in recent years to advance that goal.

References

Congressional Research Service, 2005. Updated July 20, 2010. Ocean Commissions: Ocean Policy Review and Outlook. Retrieved from. https://crsreports.congress.gov/product/pdf/RL/RL33603.

Exec. Order No. 13547, 3 C.F.R., 2010, vol. 75, (140), 43021–43027. (National Policy for the Stewardship of the Ocean, Our Coasts, and the Great Lakes). Retrieved from: https://www.federalregister.gov/d/2010-18169.

Freddie Mac, 2016. Life's a Beach. Retrieved from. http://www.freddiemac.com/research/insight/20160426_lifes_a_beach.html.

International Conservation Caucus Foundation, 2020. Oceans Caucus. Retrieved from. https://www.internationalconservation.org/united-states/oc.

McDowell Group, 2020. The Economic Value of Alaska's Seafood Industry. Alaska Seafood Marketing Institute. Retrieved from. https://uploads.alaskaseafood.org/2020/01/McDowell-Group_ASMI-Economic-Impacts-Report-JAN-2020.pdf.

National Marine Fisheries Service (2018), 2016. Fisheries Economics of the United States. U.S. Dept. of Commerce, NOAA Tech. Memo. NMFS-F/SPO-187a, p. 243. Retrieved from. https://www.fisheries.noaa.gov/content/fisheries-economics-united-states-2016.

National Marine Fisheries Service, 2020 (NMFS), Retrieved from. https://www.fisheries.noaa.gov/topic/laws-policies.

National Oceanic and Atmospheric Administration, 2006. A History of NOAA. Retrieved from. https://www.history.noaa.gov/legacy/noaahistory_3.html.

National Oceanic and Atmospheric Administration, 2020. What Percentage of the American Population Lives Near the Coast? Retrieved from. https://oceanservice.noaa.gov/facts/population.html.

Ocean Conservancy, 2020. The Problem with Plastics. Retrieved from. https://oceanconservancy.org/trash-free-seas/plastics-in-the-ocean/.

Pew Charitable Trusts, 2013. FAQ: Illegal, Unreported, and Unregulated Fishing. Retrieved from. https://www.pewtrusts.org/en/research-and-analysis/fact-sheets/2013/08/27/faq-illegal-unreported-and-unregulated-fishing.

Pew Oceans Commission, 2003. America's Living Oceans: Charting a Course for Sea Change. A Report to the Nation. May 2003. Pew Oceans Commission, Arlington, Virginia.

U.S. Census Bureau, 2010. Last Revised: July 23, 2020. Coastal Areas. Retrieved from: https://www.census.gov/topics/preparedness/about/coastal-areas.html.

CHAPTER 25

Government action and the new blue economy

Craig McLean
National Oceanic and Atmospheric Administration (NOAA), Washington, DC United States

Introduction

The attractiveness of the new blue economy to many nations is rooted in local opportunity and global necessity. As addressed elsewhere in this book, the current levels of technological capability and integration of technical disciplines find this current period to be enriched with the technology to deliver epochal advances in how the 7.8 billion in the world today could live and to prepare for how the 9.7 billion people will live by 2050 (World Population Prospects, 2019). As the global economy has used the ocean to produce wealth, the lack of attention to the consequences of wealth extraction has given rise to the Sustainable Development Goals of 2030 (see, for example, contribution from Ryabinin, this volume). This realization is occurring around the world, resulting in the UN forum taking notice and action. Finding sustainable practices, in which production is designed to eliminate waste and to constructively use rather than discard byproducts, has attracted the nations of the world with the opportunities of the new blue economy. Information is power, valuable, and can guide a society to better preserve long-term prosperity. New technologies and methods are ripe for development and deployment, through circular, not linear, business models and to replace current infrastructure and technologies that are consumptive and generate unused waste products and thereby not truly sustainable. Those countries or economies that master this renewed vision to discover, expand, and employ new knowledge of the ocean will bring remarkable prosperity to their own citizens and the citizens of the world. This is a responsibility for governments to inspire and to fund. This is the realization and motivation for the new blue economy, in which knowledge is wealth upon which better decisions are enabled, and the ocean as a resource is realized in sustainable, not consumptive terms. This is a policy choice and likely an economic survival tool as the most productive,

Preparing a Workforce for the New Blue Economy
ISBN 978-0-12-821431-2
https://doi.org/10.1016/B978-0-12-821431-2.00060-3

successful, and competitive nations will surely employ ocean knowledge to optimize a nation's relationship with the ocean.

In the United States, the governmental action is funded by the work of the Congress for which the complex and successful realities are described within this volume (see Whitehouse, this volume). The Congress provides the Executive Branch with resources and direction of broad scale and depends upon the federal agencies to implement these directions with employed expertise and with Congressional oversight. Collectively, it is the work of government—federal, state, tribal, and local, to work in the best interests of the people. Increasingly, governments are learning that the long-term view, particularly noting the known reality and consequences of climate change and overexploitation of renewable resources, is more important than the short-term aggregation of immediate yet temporary wealth. If there is a theme in governmental action, it is this realization. It is expressed nation by nation and in the international communities as well.

Common global challenges

Multiple conferences and international meetings of governments and ocean constituents are demonstrating a common view of what is impeding the attainment of a sustainable global ocean future. The UN Ocean Conference (Report of the United Nations Conference, 2017), Our Ocean Conferences since 2014 (Our Ocean Conference, 2015, 2016, 2017, 2018, 2019), and other world ocean forums have shown the repeated areas of interest and initiatives of many governments to identify and address the degrading condition of the global ocean.

While the subjects of global ocean concern are in common across many governments, the challenges that confront humanity and that must be overcome in order to achieve a sustainable future are quite commonly cited in the agendas of each of the global conferences. Among the earliest to bring focus on these global issues limiting a sustainable ocean future was the Our Ocean Conference first initiated in 2014 by then US Secretary of State, John Kerry. The conference has continued since its inception and continues to focus on priority areas including: sustainable fisheries, marine pollution, ocean acidification, climate change, marine-protected areas, maritime security, and the sustainable blue economy. In this series and other similar global conferences, the more detailed subjects of plastics pollution, global coral demise, Illegal, Unreported, and Unregulated (IUU) fishing, are prominent. As this global discussion of challenges emerges, so too are solutions including commitments of remedial environmental measures and financial instruments for the blue economy. Over the course of the several years since the inception of Our Ocean and others, the encouragement of announcing commitments from nations and industries to address and deliver solutions to these challenges has been a major part of the conference, leading to a major motivation of nations and businesses around the world. From the earlier sessions in which more emotional calls to action appeared to dominate the sessions, definitive industry and government-based solutions are emerging. From 30 commitments valued at $1.8B in the first Our Ocean Conference in 2014, to the 300-plus commitments of $107B and $63B in 2018 and 2019, respectively, the activism and direction of government involvement in these solutions have clearly emerged. The majority of these commitments address the protection and conservation of the ocean through government programs, conservation measures, and pollution mitigation measures. Perhaps most striking for the new blue economy are the components that comprise increased scientific study and understanding of the ocean, targeted to reveal new information and knowledge to guide more prosperous and sustainable decision-making. The emergence of a blue economy model of sustainability, environmental concern, circular economies, and transferring the risk to the ocean to the creators of that risk is apparent. Today, commercial companies and governments are committing to the calls that two decades ago were more commonly heard from nongovernmental organizations, conservation societies, and what might have been considered as green groups. The conversation is changing. Science has informed the public as to the condition of the oceans and the need to address the current trend of degradation. Increasingly and expansively, components of global public constituencies and governments are responding to this reality.

The United States

The United States began its early chapters in blue commerce as a young nation desiring trade with other countries. The need for accurate nautical charts to enable ships to safely transit the marine corridors of commerce, fairways, and channels of the young country gave rise to the nation's first science agency, the Coast Survey, which produced those nautical charts. This was government action to increase the welfare and prosperity of the nation. There was born the linkage between science and technology and the prosperity of more than the maritime portion of a nation's economy.

The mariners and surveyors of the time did not see it in such a light, but today, those services are provided by the government through a shared role with the private sector and enable the safe navigation of ships enabling contributions to the economy and economic opportunity, providing tools for commerce, jobs, and enabling countless revenue streams (https://oceanservice.noaa.gov/facts/oceaneconomy.html).

For fisheries, early America was celebrated as a place of abundance, but in such circumstances the sustainability of a highly utilized resource is rarely in question. From the false confidence of abundance and deference to delaying hard decisions, the once endless cod resources of the United States began to decline, and through the appearance of excessive pollution saw once abundant shellfish beds in lower New York Harbor closed due to pollution before overfishing. Market failures such as the collapse of fish stocks or pollution of common waterways prevail in the historic record without the guidance of market intervention such as rigid science-based decision-making, regulation, and solutions from government such as sound fishing quotas and pollution discharge regulations. Governments do intervene in the blue economy and are now focusing on the opportunity to achieve sustainable results. But these decisions have to be based on current and valid information. The currency is knowledge and the role of government includes providing this public good.

Governments provide the information, incentive, and when necessary the controls to guide the preservation and appropriate development of ocean resources and utilization. In the new blue economy, the government is largely funding the valuable information upon which renewed economies and industries are being based. Governments provide accurate maps and charts, measure the status and forecasts of fishery stocks, measure and determine the state of the climate change indicators in oceans and the atmosphere, and measure the state of the ocean economy. In 2020, the US Government generated a comprehensive measure of the US Ocean Economy for the first time (Nicolls et al., 2020). What may surprise many policymakers and people who are outside of the marine or marine-dependent industries is that the ocean economy constituted $372.8B of the nation's gross domestic product (GDP) in 2018. This is more than the arts, entertainment, and nonocean recreation, the mining sector, utilities sector, and the combined agriculture, forestry, hunting, and fishing sectors. Leading within the ocean economy computation are tourism and recreation, at about $124B. While that may be quite surprising, consider the latent industries of offshore renewable energy and aquaculture, which are certain to grow in the coming decade as they have in Europe and other nations. These computations are a government responsibility and will serve to increase awareness for policymakers as to how large the contribution of the existing ocean economy is to the United States as a nation and to consider the opportunities of pursuing further investments, incentives, and policies.

Currently, the United States is pursuing the blue economy through a strategy defined by several initiatives and existing programs. In the world of the new blue economy, and the value of the knowledge generated by government investments in science, the United States provides about half the world's ocean observing capability through leadership in oceanographic shipboard campaigns, in situ observing systems of buoys, floats, gliders, and from satellites (ioos.noaa.gov/community/global/). The United States has been hosting and performing the stewardship of global oceanographic data as a service to international oceanographic organizations including the Intergovernmental Oceanographic Commission (IOC), the International Hydrographic Commission (IHO), and as weather and ocean observations interrelate, the World Meteorological Organization (WMO). Intergovernmental organizations perform a valuable coordinating and prioritization function in global science, raising new tools and discovered methodologies to the awareness of Member States, and designing international cooperative science programs that fuel the new blue economy such as the Global Ocean Observing System, Tsunami Warning System, Global Climate Observation System, and Global Ocean Mapping initiatives (http://ioc-unesco.org/). Many of the opportunities of the new blue economy are generated from the coordinated science planning of the global community from within the IOC, where priorities are discussed by government representatives and delegates and where the subject of science constitutes the agenda. From the floor of the IOC have come the vision and implementation of the UN Decade of Ocean Science for Sustainable Development. Consider the "Sustainable Development" of oceans to be the blue economy, and the well-designed construct of the UN Decade of Ocean Science to be targeting, through codevelopment among nations and constituencies, the proper determination of knowledge for the purposeful outcome of ocean and societal benefit. This is a global effort to serve the new blue economy (See, Ryabinin, this volume).

Current ocean science policy

Presently the policy guiding the direction of science and technology for the United States is found in the document, *Science and Technology for America's*

Oceans: A Decadal Vision (Science and Technology, 2018). This policy vision is the latest in a series of federally stated science priorities for the oceans, informed by public input and led by the federal agencies of the United States that have an ocean portfolio. There are 17 such federal agencies. This number compels a need for formal coordination and that is accomplished through the National Science and Technology Council, Subcommittee on Ocean Science and Technology. This body has produced coordinated priorities for ocean science in the United States since the inaugural version in 2007 (after rigorous work accomplished in 2005–06), *Charting the Course for Ocean Science for the United States for the Next Decade: An Ocean Research Priorities Plan and Implementation Strategy* (2007), and thereafter updated in 2013 *with Science for an Ocean Nation: An Update of the Ocean Research Priorities Plan.*

Returning to the current plan, *A Decadal Vision*, there are five goals and subordinate priorities thereunder. **Goal I** is to Understand the Ocean in the Earth System, under which are objectives to: modernize research and development (R&D) infrastructure for the nation; harness big data; develop improved models of the Earth System; facilitate the transition of research products into useful operational performance. **Goal II** is to Promote Economic Prosperity, under which are objectives to: expand domestic seafood production; explore potential energy sources including renewables; assess critical marine minerals; balance economic and ecological benefits; and to promote the blue workforce. **Goal III** is to Ensure Maritime Security, under which are objectives to: improve maritime situational awareness; understand the rapidly changing Arctic; and to maintain and enhance marine transportation. **Goal IV** is to Safeguard Human Health, under which are objectives to: prevent and reduce plastic pollution; improve forecasts of marine contaminants and pathogens; address to prevent Harmful Algal Blooms (HABs); and seek to discover natural products, which complements ocean exploration objectives and biopharmaceuticals. **Goal V** is to Develop Resilient Coastal Communities, under which are objectives to: increase preparation for natural disasters and weather events; reduce risk and vulnerabilities to coastal communities; and to empower local and regional decision-making.

These goals and objectives do support science underpinning the blue economy vision and guide the investments and direction of the federal agencies of the US federal government. With this set of goals are reinforcements of the new blue economy, including increasing the observational capabilities of marine monitoring systems. New technology and systems will generate more information of higher resolution and precision. Building a blue workforce that starts with the recognition of STEM and early interest in the K-12 domain should raise both awareness in the public discussion and the ambition to recruit a diverse national body of students interested in rewarding and important careers. Big Data, where the integration of singular focused domains of information is blended with others to enable greater discovery, is the multiplier of the information currency in the new blue economy. The result will be that some ocean discoveries will be made without having to go to sea. In an increasingly noticeable way, the objectives that the United States has announced are very similar to priorities that are emerging from other nations, as will be mentioned shortly.

Federal agency actions

In addition to this national ocean science agenda, implementing efforts to expand the ocean economy in the United States are found in the multiple agencies of the state and federal governments.

NOAA

An example at the federal level is at the National Oceanic and Atmospheric Administration (NOAA), which has adopted five pillars constituting the current effort toward a blue economy initiative (NOAA, Blue Economy). These are: seafood production, tourism and recreation, ocean exploration, marine transportation, and coastal resilience.

Seafood production

The United States is a net importer of seafood and holds the world's largest Exclusive Economic Zone (EEZ). The ambition is to triple the US domestic marine aquaculture production and to maximize the production of sustainable open water fishing. Regulatory streamlining is necessary to enable these objectives, particularly in the aquaculture component of industry. The lack of clarity and ease of the federal regulatory framework for ocean use and planning have found the industry slow to grow and discouraged domestic investment. An Executive Order signed in May of 2020 sets a direction to address these concerns. (Executive Order No. 13921).

Tourism and recreation

The coastline of the United States offers diverse ecosystems and experiences for domestic and foreign visitors and today contributes $124 billion to the

GDP of the nation's economy, annually. This area will grow through a focus on sustainable tourism and recreation in the nation's coastal areas including the Great Lakes. The United States has highly valued marine-protected areas including underwater and coastal national parks and national marine sanctuaries, estuarine research reserves, and state parks. Focusing on expanding and promoting these networks can increase the economic return in local communities and states. The component of sustainability in pursuing expansion requires attention to science-based decision-making and addressing, for example, the demise of coral species and coastal erosion, sea level rise, and land loss. These raise the importance of restoring and protecting coasts, wetlands, estuaries, coral reefs, and sea life, and the scientific information and knowledge necessary to wisely manage these valuable resources.

Ocean exploration, mapping, and characterization

The United States is one of the few countries that has a dedicated national program to explore the world's oceans. In order to utilize the area and resources of the EEZ and to know the status of the ocean areas beyond, a competent inventory and assessment of the marine areas is necessary. This mapping will include the physical, biological, chemical, archaeological, and geological attributes of the ocean. The policy direction stresses partnerships both domestic and international and will deliver information related to offshore and marine energy sources, minerals, and pharmaceutical potential. An Executive Order to include marine mineral mapping was signed in 2019 (Executive Order 13817). Additionally, a Presidential Memorandum has instructed the federal agencies of the United States to fully map the EEZ by 2030 (Presidential Memorandum, 2020). A strategy to implement this Presidential Memorandum guides all participating federal agencies (NOMEC Strategy, 2020).

Marine transportation

The marine transportation system carried and delivered more than $1.5 trillion of goods per year through US ports, both to and from trading partners. This generates $5.4 trillion in economic activity per year. Increased accuracy of navigational depth, air gap clearance between the top of a ship and the bottom of a bridge, plus knowledge of immediate oceanographic conditions promote greater security of the environment and confidence of marine shippers to maximize their efficiency and reduce shipping costs. The United States has world-class navigation services, often inspiring improvements in the international community through the International Maritime Organization (IMO) and the International Hydrographic Organization, which will likely continue.

Coastal resilience

Given the increased development at America's coast, the trend of increase in severity and intensity of storms, and other signatures of climate change, coastal development patterns of the past will not be useful to guide the coastal development of the future. Coastal zone planning will need to be more resilient to sustain the economies established in all coastal states and territories. New approaches, methods, and partnerships will provide improved information and planning tools to enable an optimum use and protection of the coastal environment by providing actionable information to coastal communities. This effort will provide science-based information and conclusions for ocean commerce, energy development, sustainable infrastructure, and coastal conservation.

Department of Energy

Another example of blue economy benefits from the science and data available to decision-makers in both government and industry is the initiative to establish offshore renewable wind energy, led by the US Department of Energy, Wind Energy Technology Office. Working with the US Department of Interior, Bureau of Ocean Energy Management, both agencies are leading an initiative to develop a favorable environment for an offshore wind industry in the country (see contribution by Brown, this volume). Offshore wind is abundant and broadly distributed, capable of producing more than 2000 GW. This possibility is known because of federally funded research and long-term observations (Powering the Blue Economy Report, 2019).

These collective initiatives are resonant with what is appearing in the blue economy direction of other nations, and in some cases the United States is no longer defining the leading edge of sustainable blue economy destined innovations. The recent position of the United States on climate change and greenhouse gas emission reduction leaves a continued reliance on old energy sources. This may challenge the speed of development of a new blue economy and will require a more thorough evaluation of policies and incentives to leave behind the traditional models of the marine economy and take a leadership position in pursuing ocean renewable energy, wind energy, aquaculture, biomedicine, sustainable and environmentally safe marine mineral extraction, and marine commerce.

International: examples from other nations

Looking to Asia and Europe for comparison of national objectives, Japan has maintained a leadership presence in the international community of ocean science, as has the European Union. For Japan, the nation's ocean policy is defined in the 2018 Basic Plan on Ocean Policy. In many of the objectives stated in the plan, scientific knowledge is essential for success. Japan's government seeks to promote industrial uses of the ocean in both energy and minerals including methane hydrates and rare earth minerals and polymetallic sulfides, as well as renewable energy opportunities. An important component of a marine nation is to educate the public on the importance of the ocean and to attract and train future generations for ocean-related jobs. Japan also seeks improved scientific knowledge of the ocean and has participated in international campaigns and programs, including but not limited to the TOGA-TAO program, which, in partnership with the United States, has given rise to increased accuracy of global weather forecasts, as well as fishery forecasts. Climate change and achieving a greater understanding of climate change on ocean resources, particularly fisheries, are important. Given the recent experience of geological disturbances including the Tohoku earthquake of 2011 and subsequent tsunami, Japan has maintained a focus on geological and seismic sciences in the ocean areas as well. In recognition of the geopolitical challenges in the region, Japan is also addressing maritime situational awareness and jurisdictional issues of historic islands. In order to implement effective government programs in pursuit of these objectives, intra-governmental coordination across relevant agencies in Japan has increased in recent years. The identification of science needs, programs to attain these needs, education and training of the public and a future workforce, and international cooperation are keys to the success of this governmental direction and the extension of successes already realized.

The European Union (European Commission, 2020) maintains a focus on climate change and its impacts on the ocean and coastal systems, marine pollution, and the economics of a blue economy including the natural capital of ocean systems and how these values intersect with the overall economy. Within the EU, there is a distributed focus of initiatives and activities across the Member States, though the unified approach is similar in subjects to the most progressive governments. Focus on renewable energy, food security, bioeconomy, and biotechnology are included in the priorities. The advance of offshore renewable energy across coastal Europe is a clear demonstration of an initiative underpinned with the resolve of government programs to support their launch and allow market forces to emerge from the public investment. As the EU pursues a climate-sensitive approach to the future, so might any progressive policy that invites science-based decision-making. As mentioned elsewhere in this volume (e.g., Rayner, Spinrad, and Ryabinin, this volume), the European Copernicus marine environmental service provides a highly regarded body of science information to guide decision-makers in establishing policies that are conducive to a more optimistic ocean future. This is accompanied by a well-financed research and development agenda for continued system development. (marine.copernicus.eu).

Conclusion

The role of government has been present in the generation of priorities, provision of resources, and creation of programs to generate ocean knowledge, data, and information that underpin the new blue economy. The common recognition of global challenges underscores the reality that every marine nation faces similar concerns, perhaps varied by region and intensity, but the problems are very much in common, so the opportunity will similarly be in common. It will be a progressive nation, caring for its citizens and future generations, that invests sufficiently to let science be publicly understood, trusted, and guide the policy decisions of government. The confluence of interests and advancements will surely give rise to the new blue economy, and the winners will be the leaders.

References

European Commission, 2020. The EU Blue Economy Report 2020. Publications Office of the European Union, Luxembourg. Retrieved from: ttps://op.europa.eu/en/publication-detail/-/publication/83843365-c18d-11ea-b3a4-01aa75ed71a1/language-en/format-PDF/source-143845572.

Executive Order No. 13817, December 20, 2017. A Strategy to Ensure Secure and Reliable Supplies of Critical Minerals, 82 FR 60835. Executive Office of the President. Retrieved from: https://www.federalregister.gov/documents/2017/12/26/2017-27899/a-federal-strategy-to-ensure-secure-and-reliable-supplies-of-critical-minerals.

Executive Order No. 13921, May 7, 2020. Promoting American Seafood Competitiveness and Economic Growth, 85 FR 28471. Executive Office of the President. Retrieved from: www.whitehouse.gov/presidential-actions/executive-order-promoting-american-seafood-competitiveness-economic-growth/#:~:text=2803(a)(2)%20and,economic%20zone%20of%20the%20United.

Nicolls, W., et al., June 2020. Defining and Measuring the U.S. Ocean Economy. Bureau of Economic Analysis, U.S. Department of Commerce. Retrieved from: ttps://www.bea.gov/system/files/2020-06/defining-and-measuring-the-united-states-ocean-economy.pdf.
NOMEC Strategy, June 2020. National Strategy for Mapping, Exploring, and Characterizing the United States Exclusive Economic Zone, 2020. Executive Office of the President, Ocean Science and Technology Subcommittee of the Ocean Policy Committee, Washington, DC. Retrieved from: https://www.whitehouse.gov/wp-content/uploads/2020/01/20200611-FINAL-STRATEGY-NOMEC-Sec.-2.pdf.
Our Ocean, 2014. US Department of State, Bureau of Oceans and International Environmental Scientific Affairs. Washington, DC. Retrieved from: https://2009-2017.state.gov/e/oes/ocns/opa/ourocean/2014/index.htm.
Our Ocean Conference, 2015. IISD Reporting Services, Valparaiso, Chile. Retrieved from: https://enb.iisd.org/oceans/our-ocean-conference-2015/.
Our Ocean, 2016. US Department of State, Bureau of Oceans and International Environmental Scientific Affairs. Washington, DC. Retrieved from: http://ourocean2016.org/.
Our Ocean, October 5–6, 2017. European Union, Malta. Retrieved from: https://ourocean2017.org/sites/default/files/ooc-2017-report.pdf.
Our Ocean, 2018. Ministry of Marine Affairs and Fisheries, Republic of Indonesia. Retrieved from: https://ourocean2018.org/?l=our-ocean-commitments.
Our Ocean, 2019. Norwegian Ministry of Foreign Affairs, Oslo. Retrieved from: https://ourocean2019.no/wp-content/uploads/2019/12/Report_OurOcean_Oslo_2019_UU.pdf.
Powering the Blue Economy Report, 2019. U.S. Department of Energy. Water Power Technologies Office, Washington, DC. Retrieved from: https://www.energy.gov/eere/water/downloads/powering-blue-economy-report https://www.energy.gov/sites/prod/files/2019/09/f66/73355-v2.pdf.
Presidential Memorandum, 2020. Ocean Mapping of the United States Exclusive Economic Zone and the Shoreline of Nearshore of Alaska, 84 FR 64699. Executive Office of the President. Retrieved from: https://www.whitehouse.gov/presidential-actions/memorandum-ocean-mapping-united-states-exclusive-economic-zone-shoreline-nearshore-alaska/.
A/CONF.230/14 Report of the United Nations Conference to Support the Implementation of Sustainable Development Goal 14: Conserve and Sustainably Use the Oceans, Seas and Marine Resources for Sustainable Development, June 5–9, 2017. United Nations Headquarters. Retrieved from: https://www.un.org/ga/search/view_doc.asp?symbol=A/CONF.230/14&Lang=E.
Science and Technology for America's Oceans: A Decadal Vision, November 2018. National Science and Technology Council, Subcommittee on Ocean Science and Technology Committee on the Environment. Executive Office of the President, Washington, DC. Retrieved from: https://www.whitehouse.gov/wp-content/uploads/2018/11/Science-and-Technology-for-Americas-Oceans-A-Decadal-Vision.pdf.
Third Basic Plan on Ocean Policy, 2018. Government of Japan. Retrieved from: https://www8.cao.go.jp/ocean/english/plan/pdf/plan03_e.pdf.
World Population Prospects 2019, United Nations, Department of Economic and Social Affairs, Population Dynamics. Retrieved from:: (https://population.un.org/wpp/).

Further reading

Nakahara, H. Report, Japan's First and Second Basic Plan on Ocean Policy and the Government Annual Reports; Considering the Formation of the Third Basic Plan on Ocean Policy. Hiroyuki Nakahara Managing Director Research Institute for Ocean Economics, Japan as in Oceans/TechnoOceans 2018.

NOAA, Blue Economy, 2018. Retrieved from: https://www.noaa.gov/sites/default/files/atoms/files/Leadership%20in%20Powering%20the%20American%20Blue%20Economy.pdf.

Report of the Sustainable Blue Economy Conference, 2018. United Nations Environmental Programme, Nairobi, Kenya. Retrieved from: https://www.unenvironment.org/nairobiconvention/news/story/report-sustainable-blue-economy-conference https://enb.iisd.org/oceans/blueeconomy/2018/html/enbplus208num31e.html.

Ryabinin, V., Barbière, J., Haugan, P., Kullenberg, G., Smith, N., McLean, C., Fischer, A., Aricò, S., Aarup, T., Pissierssens, P., Visbeck, M., Enevoldsen, H.E., Rigaud, J., 2019. The UN decade of ocean science for sustainable development. Front. Mar. Sci. **6**, Article 470. https://doi.org/10.3389/fmars.2019.00470, 31 July 2019.

World Ocean Summit, 2019. World Ocean Initiative, The Economist Group. Retrieved from: https://www.woi.economist.com/world-ocean-summit-2019/.

CHAPTER 26

UN decade perspective

Vladimir Ryabinin
Intergovernmental Oceanographic Commission of UNESCO, Paris, Île-de-France, France

Autarky—a country's self-sufficiency—is a feature of the past. Climate change, global loss of biodiversity, and the recent COVID-19 pandemic are some striking examples of heightened global interdependence of countries. Science and knowledge have become the main enabling factors for identifying and addressing problems and for capitalizing on opportunities. The United Nations has provided the main governmental platform for addressing the issues on an international level. However, the role of science in shaping and guiding international policy development still needs to be further strengthened.

One very well-known example illustrating this urgent need was the lack of action on climate change at the United Nations Framework Convention on Climate Change (UNFCCC) Conference of Parties in 2009 in Copenhagen, despite the fact that the clear science guidance had been in place. This failure led to a dedicated study by the University College London (Rapley et al., 2014) showing that distant and somewhat theoretical threats, such as climate change, are much less motivating for action by individuals and, hence, for governments in comparison with immediate and visible threats. That is why such global challenges and the science-based means of addressing them tend to remain "second priority." Once there is an understanding of a challenge of major significance, scientific consensus on its causes and means of resolving the issue, effective and very active communication to the general public and decision-makers, motivating action and explaining the consequences of inaction, becomes the most important step, to be followed by a call for international policies resulting in commitments of adhering nations. Only through a very intensive campaign did it become possible to unite the world in 2015 under the UNFCCC Paris Agreement, but the precious time was nevertheless lost.

Meteorological prediction shows a positive example of how the existence of international policies is instrumental for global international science-based cooperation. A skillful forecast for ranges longer than 2 days

Preparing a Workforce for the New Blue Economy
ISBN 978-0-12-821431-2
https://doi.org/10.1016/B978-0-12-821431-2.00027-5

requires global exchange of observations. Since the 1960s, with the advent of World Weather Watch, the first world wide web, all prognostic centers have had access to globally distributed synoptic and radiosonde data, usually within less than 1 h after the standard observing time. The current World Meteorological Organization (WMO) Information System is much more encompassing than in the 1960s and long ago started to use Internet links to exchange data and rely on satellite retrievals. Now the whole world readily benefits from meaningful guidance provided through numerical weekly outlook weather predictions. Why was this possible? Because of the existence, since 1947, of the legally binding international policy, the UN Convention on the WMO Organization (WMO, 2019) that makes national contributions to meteorological and related observations mandatory.

This is not yet the case for ocean science. Most monitoring activities in the ocean remain funded through research grants. As a result, as was recently shown by an analysis conducted by Copernicus Program Services (Buch et al., 2019), only 28% of ocean observing networks are assessed as sustainably funded in a 3–5-year timeframe. For meteorological networks, this percentage is 68%. However, this situation may be changing. Recently (UNGA, 2018), the United Nations resolved to develop an International Legally Binding Instrument (ILBI) for conservation and sustainable use of marine biodiversity beyond areas of national jurisdiction (BBNJ). The negotiation process in the United Nations is cognizant of the role of science. Development of science capacity in nations therefore becomes one of the four key components of this new emerging ILBI under the UN Convention on the Law of the Sea (UNCLOS, 1982). Requirements for ocean observations and developing capacity of ocean science to support implementation of ILBI on BBNJ should facilitate increased sustainability of ocean monitoring and research.

The development of a sustainable ocean economy is another motivating factor for the systematic strengthening of ocean science as a key condition for sustainability. The ocean economy is growing faster than its terrestrial counterpart and is expanding in directions, which are more environmentally friendly. According to OECD (2016), the ocean economy is expected to generate an annual value of the order of US $3 trillion depending on the climate change scenario. This dwarfs the global annual investment into ocean research and observations (Valdez, 2017), which is on the order of US $ billions— approximately thousand times less than the economic value generated in the ocean. This deficiency is further exacerbated by the weak funding to international coordination of ocean science and observations, which is a global investment of the order of US $ millions, again a thousand times smaller than the value of ocean research activities. As a result, we underinvest in ocean science and observation, and we insufficiently optimize the use of limited available scientific resources to keep the ocean healthy and support ocean industries.

According to the first UN world ocean assessment (UN, 2017), humankind is running out of time to start managing the ocean sustainably. The urgent need for sustainability is already reflected in many international frameworks, conventions, and protocols. The 2030 Agenda (UNGA, 2015) contains a standalone Goal 14 on the Ocean. The need to protect ocean ecosystems is mentioned in the preamble to the Paris Agreement of the UNFCCC (2015). The UN Sendai Framework for Disaster Risk Reduction considers coastal hazards (UNDRR, 2015). Conventions related to marine pollution (IMO, 1973, MARPOL, with its updates and six annexes), the Code of Conduct for Responsible Fisheries (FAO, 1995), the Regional Seas Conventions and Plans in 18 ocean regions associated with the UN Environment Program launched in 1974, and the closely connected Global Program of Action for the Protection of the Marine Environment from Land-based Activity created in 1995 (see, e.g., UNEP, 2000), the UNESCO Convention on Underwater Cultural Heritage of 2001 (UNESCO, 2001), and many other regional and global frameworks are now in place. However, the need for nations to systematically conduct ocean observations and research as the key source of guidance and policies for objective and impartial decision-making on ocean matters is only starting to be understood.

The existing frameworks pursue important and large-scale societal goals and associated targets. However, means and conditions for achieving these targets are usually not specified. For example, the 10 targets of the Ocean Goal 14 require an observing system to monitor the situation in the ocean, a predictive system to see the trends and opportunities, and a number of political decisions and implementing actions at global, national, and local level. The set of targets is ambitious but remains aspirational and has weaknesses in its enforcement.

The Intergovernmental Oceanographic Commission (IOC) of UNESCO is a body of the UN System that facilitates the development of ocean science as an unbiased service to its 150 Member States. Bringing governments and the ocean science community toward ecosystem-based management of the ocean is a relatively new area of governance. So far, national ministries of the sea or for ocean matters remain a rare exception rather than a rule. Many countries are developing their policies for using and protecting the ocean, and more and more often, they include provisions for basing such policies on best science. The question is how to accelerate this positive development. For that it is essential to make ocean science able not only to diagnose problems in the environment but to move toward offering solutions to existing problems. With this vision in mind, in 2017, the IOC proposed to the United Nations holding a UN Decade of Ocean Science for Sustainable Development, toward "the ocean we need for the future we want" (the "Decade", Ryabinin et al., 2019). That same year, the UN General Assembly proclaimed the Decade for the years 2021–30 (UN General Assembly Resolution RES/72/73, 2017) and called on the IOC to prepare an implementation plan for this major worldwide campaign, in collaboration with all relevant stakeholders: countries, UN agencies, scientific communities, private sector, and the general public.

The Implementation Plan for the Decade (https://www.oceandecade.org/) aims to offer the world an opportunity to achieve an ocean, which is clean, healthy, productive, and more sustainable and resilient. Our society will be safer with regard to ocean-related hazards and will benefit from better access to ocean knowledge, data, and information characterizing the past, current, and future predicted state of the ocean (i.e., the new blue economy). These qualities of the ocean and of our society are the desired societal outcomes of the Decade. The idea is to bring the ocean closer to people and trigger a positive behavior change toward the ocean.

The Decade is a platform for bringing together many stakeholders toward progress in ocean research and technology development. For the first time in history, we will strive to generate an adequate-resolution map of the ocean depth and a digital atlas of the whole ocean, consistent with the concept of the new ocean economy. We will have to upgrade the capacity of the ocean observing system, particularly with respect to biogeochemical, biological, and ecological essential ocean variables, embracing polar oceans, the deep layers of the ocean, and the abyssal ocean. The knowledge of ocean ecosystems will receive a boost. Much more efficient multi-hazard early warning systems for managing ocean-related risks will be built. Ocean data and information systems will be cardinally upgraded in the spirit of "big data" and with extended interoperability. Ocean science will strengthen modeling and predicting capabilities benefitting from cutting-edge Earth System science. Capacity development and transfer of marine technology will be enabled and accelerated. Through education, awareness creation, and communication, society will become more "ocean literate," overcoming the emotional deficit stemming from the remoteness of the ocean to most people. This major scope of activities should not only facilitate informed decision-making with regard to the ocean but also trigger further development of ocean-related policies eventually leading to the sustainable growth of the ocean economy.

One of the main concepts of the Decade is that the future ocean will be a managed system (Winther et al., 2020). Modern means of ocean management facilitate a sustainable ocean economy through adequate, science-based, governmentally stimulated policies and approaches of using ocean space and resources. The scope of those approaches is broad and includes integrated Coastal Zone Management, Marine (Maritime) Spatial Planning, management of fisheries and aquaculture, various climate services including adaptation to climate change and its mitigation, disaster risk reduction activities, and establishment of Marine Protected Areas (MPAs). A two-prong approach, with proactively designed scientific research and its subsequent pickup in ocean management practices, should lead to achieving the Decade societal outcomes. Moving toward "the ocean we need for the future we want" will also facilitate maintenance of peace, international cooperation, and multilateralism.

The proclamation in 2017 by the United Nations of the Decade of Ocean Science for Sustainable Development indicates that at present society and the governments of the world understand the key role of science in maintaining a healthy ocean. In the form of the Decade, we have a framework for action for 10 years to come. Apart from many tangible improvements in ocean health and sustainability, the coordinated research targeted on science-based management of the ocean space and resources will entrain and catalyze the development of ocean technology and a sustainable ocean economy. More objective and robust understanding of ocean ecosystem functioning and the existence of science-based plans for area-based management of the ocean will become a driving force behind the development of new ocean policies, aiming them at long-term protection and sustainable management of ocean space and resources.

References

Buch, E., Fernández, V., Srzic, I., Vermeulen, A., 2019. Sustainability Survey. https://insitu.copernicus.eu/library/reports/Sustainabilitysurveyupdatedreportfinal.pdf.

FAO, 1995. Code of Conduct for Responsible Fisheries. http://www.fao.org/3/v9878e/v9878e00.htm.

IMO, 1973. International Convention for the Prevention of Pollution from Ships, (As Modified by the Protocol of 1978 Relating Thereto and by the Protocol of 1997). http://www.imo.org/en/About/Conventions/ListOfConventions/Pages/International-Convention-for-the-Prevention-of-Pollution-from-Ships-(MARPOL).aspx.

OECD, 2016. The Ocean Economy in 2030. OECD Publishing, Paris. https://doi.org/10.1787/9789264251724-en.

Rapley, C.G., De Meyer, K., Carney, J., Clarke, R., Howarth, C., Smith, N., Stilgoe, J., Youngs, S., Brierley, C., Haugvaldstad, A., Lotto, B., Michie, S., Shipworth, M., Tuckett, D., 2014. Time for Change? Climate Science Reconsidered: Report of the UCL Policy Commission on Communicating Climate Science, 2014. UCL Policy Commission on Communicating Climate Science. University College London, London, UK. https://discovery.ucl.ac.uk/id/eprint/1462114.

Ryabinin, V., Barbière, J., Haugan, P., Kullenberg, G., Smith, N., McLean, C., Fischer, A., Aricò, S., Aarup, T., Pissierssens, P., Visbeck, M., Enevoldsen, H.E., Rigaud, J., 2019. The UN decade of ocean science for sustainable development. Front. Mar. Sci. 6 (470). https://doi.org/10.3389/fmars.2019.00470.

UN Framework Convention on Climate Change, 2015. Paris Agreement. https://unfccc.int/files/essential_background/convention/application/pdf/english_paris_agreement.pdf.

UN General Assembly, December 10 , 1982. Convention on the Law of the Sea. https://www.un.org/depts/los/convention_agreements/texts/unclos/unclos_e.pdf.

UN General Assembly Resolution RES/70/1, 2015. Transforming Our World: The 2030 Agenda for Sustainable Development. https://undocs.org/en/a/res/70/1.

UN General Assembly Resolution RES/72/73, 2017. Oceans and the Law of the Sea. https://undocs.org/en/a/res/72/73.

UN General Assembly Resolution RES/72/249, 2018. International Legally Binding Instrument under the United Nations Convention on the Law of the Sea on the Conservation and Sustainable Use of Marine Biological Diversity of Areas beyond National Jurisdiction. https://undocs.org/en/a/res/72/249.

UNDRR, 2015. Sendai Framework for Disaster Risk Reduction 2015–2030. https://www.preventionweb.net/files/43291_sendaiframeworkfordrren.pdf.

UNEP, 2000. Regional Seas: A Survival Strategy for Our Oceans and Coasts. https://unep.ch/conventions/info/seas/RSbooklet-E.pdf.

UNESCO, 2001. In: Records of the General Conference, 31st Session, Paris, v. 1: Resolutions, pp. 50–61. https://unesdoc.unesco.org/ark:/48223/pf0000124687.page=56.

United Nations, 2017. In: The First Global Integrated Marine Assessment: World Ocean Assessment I. Cambridge University Press, Cambridge. https://doi.org/10.1017/9781108186148. In: https://www.un.org/regularprocess/content/first-world-ocean-assessment.

Valdez, Louis, et al., 2017. Global Ocean Science Report. IOC-UNESCO.

Winther, J.-G., Dai, M., Douvere, F., Fernandes, L., Halpin, P., Hoel, A.H., Juinio-Meñez, M.A., Li, Y., Morrissey, K., Rist, T., Rubio Scarano, F., Trice, A., Unger, S., Whitehouse, S., 2020. Integrated Ocean Management. Available online at: www.oceanpanel.org/blue-papers/integrated-ocean-management.

WMO, 2019. Basic Texts, vol. 15. WMO Publications, ISBN 978-92-63-10015-3. https://library.wmo.int/doc_num.php?explnum_id=10076.

CHAPTER 27

A sustainable new blue economy

Margaret Spring
Monterey Bay Aquarium, Monterey, California, United States

Introduction

The ocean is critical to human life. Among the myriad services it provides: generating almost two-thirds of the oxygen we breathe (Burdige, 2005), regulating Earth's climate (IUCN, n.d.), and providing the seafood that sustains billions of people (FAO, 2018). The ocean also supports economic activity, from energy generation to global transportation of people and goods, as well as cultural services, including recreational and spiritual uses.

But major global powers have historically prioritized the ocean's profit potential over its capacity to sustain people's long-term needs. When 18th-century, Western decision-makers did consider the sea, it was too often from a perspective of economic exploitation, based on the false presumption that the ocean was so large and bountiful, its resources were limitless (Martin et al., 2016).

Valuing the ocean for its capacity to offer national security and provide direct economic returns led to the attitude that increased extraction would result in increased revenue, without negative consequences. This failure to recognize the ocean's central role in supporting human well-being has been coined "sea blindness" (Government Office for Science, 2018).

By the turn of the 21st century, however, industry's chronic and damaging impacts to ocean health—including declining numbers of commercially fished and hunted marine species, rising seawater temperatures and acidity levels, and increasing loads of pollutants including plastic—made it clear that the ocean's resources are not limitless. This prompted economists, policy experts, and decision-makers to begin considering the importance of the marine environment and economy to long-term national wealth.

In the United States, two reports provided comprehensive documentation of the ocean's contributions to both the US gross domestic product and societal health and well-being, making a compelling case for greater support, through funding and policies, for our nation's ocean and coasts (Pew Oceans Commission, 2003; US Commission on Ocean Policy, 2004).

Preparing a Workforce for the New Blue Economy
ISBN 978-0-12-821431-2
https://doi.org/10.1016/B978-0-12-821431-2.00022-6

More recently, thought leaders have coalesced behind a more expansive way to frame and quantify the social and economic value of a healthy, functioning ocean by acknowledging the "natural capital" it generates for humanity. Science has revealed that the ocean, and all nature, is embedded within global economies, which has led to an increasing call to fundamentally transform our approach to economic decision-making (Dasgupta, 2021; United Nations Environment Programme, 2021). In addition to providing critical nutrition for more than 3 billion people and directly supporting nearly 60 million jobs, the world's ocean buffers coastal communities from natural disasters and erosion, plays a pivotal role in global carbon sequestration and oxygen production, and provides intrinsic cultural value (Friedman et al., 2020). Efforts to quantify the economic value of these services are ongoing (Lipton et al., 2014 as well as Rayner and Kildow, this volume), but it's clear that eroding the ocean's capacity to provide them would incur staggering costs.

This "blue economy" concept, a marine counterpart to the terrestrial "green economy," expresses the economic imperative to use ocean resources sustainably and to prioritize activities that restore and protect ocean health. It is "an organizing principle that captures the interplay between the economic, social, and ecological sustainability of the ocean" (US Department of Energy [DOE], 2019). Proponents of this view acknowledge that failing to ensure sustainable ocean practices undermines consistent economic prosperity and can result in massive economic loss (European Commission, 2018; Organization for Economic Co-operation and Development [OECD], 2016; DOE, 2019).

A "blue economy" that considers the economic returns of a healthy and resilient ocean is an upgrade from the Western framework of earlier centuries, which considered only the profit from extractive uses. A blue economy supports the use of ocean resources in a way that both preserves ocean health and benefits human communities, so that we can continue using these resources sustainably into the future.

Yet even this concept is overdue for a radical revision.

The new blue economy is defined as "a knowledge-based economy, looking to the sea not just for extraction of material goods, but for data and information to address societal challenges and inspire their solutions" (Spinrad, this volume). To be truly sustainable, the new blue economy must incorporate another filter: the need to protect and restore the ocean in order to build its resilience in the face of climate change. As climate impacts intensify—and are compounded by global human population growth and intensifying market demands—we must manage the ocean in a way that preserves its capacity to sustain society into the future.

At stake is nothing less than humanity's long-term survival on our planet.

Prioritizing the ocean and acting against its biggest threats

Humanity's awareness of ocean threats and health has risen gradually over the last 50 years. In the United States, environmental activism came to public prominence on the first Earth Day in 1970, in response to disasters such as the Santa Barbara oil spill and reports of hazardous waste dumping around the nation. The ocean gained formal national attention when the United States created the first modern ocean agency in 1970, the National Oceanic and Atmospheric Administration (1970), at the recommendation of an expert panel led by Julius Stratton of Massachusetts Institute of Technology (84 Stat. 2090; 5 U.S.C §§ 906, 1970; Stratton and Baird, 1969).

Since that time, the United States has made demonstrable progress to improve fishery sustainability and marine protection, but overfishing at a global scale remains one of the most significant threats to ocean health, along with habitat degradation and pollution. Dwarfing all of these significant challenges—and compounding them—is the existential peril of climate change, which is fundamentally altering our ocean's ability to support life as we know it.

Human-caused emissions of carbon dioxide and other heat-trapping gases are making our global ocean warmer, more acidic, and more depleted of oxygen (Cullen and David, 2019). We must build up the ocean's ability to withstand these changes by reducing other stressors and accelerate international action to reduce global greenhouse gas emissions and transition to a renewable energy society.

In 2019, the Intergovernmental Panel on Climate Change issued its first global report on the impacts of climate change on the ocean and the cryosphere or frozen layer. The report confirms that the ocean, the heart of Earth's climate system, is on the front lines of our emissions assault (IPCC, 2019), and that climate-driven changes in the ocean are coming to shore as serious threats to human communities (see, for example, Englander, this volume). Threats include not only disruptions to important commercial fisheries, but also sea-level rise and coastal flooding that put entire communities at risk. Changes caused by excess carbon and heat in the ocean are rippling through marine food webs and triggering other negative impacts, including coral bleaching, shifts in the abundance and ranges of species, and harm to shell-building animals from zooplankton to crabs. Ocean climate changes are also connected with a dramatic rise in extreme weather events around the world—including more frequent and more intense droughts, wildfires, floods, and storms (IPCC, 2019). These impacts will be amplified by human population growth (United Nations, 2017) and market demands.

As climate change accelerates, the global community must affirmatively value the services of the natural world, especially the ocean, and ramp up action to develop renewable energy sources, implement innovative carbon reduction policies, and restore habitats that sequester carbon, among other steps. Better ocean data and information will enable more effective conservation strategies on a warming planet.

Although the ocean is fundamentally threatened by climate change, it also has the potential to deliver some of the highest-impact climate solutions (Hoegh-Guldberg et al., 2019). Ocean and coastal habitats such as kelp forests, estuaries, and coral reefs are not only critical feeding, and breeding grounds for marine life; they also buffer our coastal communities from intensifying storms, floods, and erosion, and support food systems. Wetlands are powerful carbon sinks—with some wetland ecosystems sequestering up to five times more carbon per acre than terrestrial forests—yet the IPCC reports that more than half have been lost over the past century (Cullen and David, 2019). Protecting and restoring "blue carbon" ecosystems are among the most ecologically and economically effective climate actions we can take.

The highest levels of government are seeking expert advice to navigate these "blue" opportunities and advance the UN Sustainable Development Goals (see Ryabinin, this volume). One such effort is codified in a series of "blue papers" and reports from the High-Level Panel for a Sustainable Ocean Economy (World Resources Institute, 2020). As we look to the future, we will need greater scientific and technical expertise, innovative tools, and a rising workforce equipped to face these global and complex challenges.

A next-generation strategy for ocean conservation

There's no time to waste: Large-scale forces set in motion by human activities are moving the dial backward for the global ocean. The problems facing the global ocean, as well as their solutions, will need to be more visible, relevant, and actionable to the public. That's especially critical when it comes to young adults, the largest voting group in history, whose generation stands to suffer the steepest costs of accelerating climate change and other drivers of ocean degradation. This generation is increasingly vocal and is successfully engaging in social and environmental change movements (Jarrin-Thomas, 2020).

Community science can achieve greater public awareness, link people to real-world problems and solutions, and create a heightened urgency to act. This is where the next generation can add value to policy and solution development as scientists, informed practitioners, and everyday community members, and link with the rising public trust in science (Pew Research Center, 2019).

We must communicate both the problems and the solutions in simple, practical terms, backed up by transparent and understandable data. We can power this next generation of data, products, and information—i.e., the new blue economy—in a number of ways. Key opportunities are emerging that use cutting-edge technology to generate robust data and to analyze those data faster, more efficiently, and more cost-effectively than ever before.

Ocean status, value, and risk

The foundation of a new blue economy is comprehensive ocean observation and measurement at an unprecedented scale. Other chapters of this book explore the potential to fill the enormous data gaps with innovative approaches to data collection, from satellites to robotics. Innovation in data collection is important, especially as we become more adept at measuring proxies for ocean conditions in the ocean itself—as opposed to in labs on land.

A recent paper identified eight "urgent, fundamental and simultaneous steps" that must be taken to better leverage science and technology to preserve ocean health (Laffoley et al., 2019). Among these top priorities: increasing ocean exploration, mapping, and observation; ensuring sufficient, long-term funding for time-series observations; expanding the Argo float network (platforms that use sensors to record properties such as temperature, salinity, and currents below the surface, see Watson-Wright, as well as Gouldman, this volume); and using technology to reconcile data from multiple sources sensibly and systematically. The authors also emphasize the need for better transparency and data sharing, asserting that "the only [ocean] data that should be regarded as confidential are those relating to national security or which are of immediate commercial sensitivity" (Laffoley et al., 2019).

Publicly available information can be hugely influential in making a strong economic case for ocean reforms. One prime example is a 2009 World Bank report that highlighted the economic losses resulting from unsustainable fishing practices such as overfishing and illegal and unreported fishing. This report estimated that global society forgoes $50 billion of potential economic benefits from marine fisheries each year and that improved governance would enable the recovery of a substantial portion of this economic loss (World Bank, 2009). Within 8 years of the report's publication, implementation of those recommendations was yielding positive results (World Bank, 2017). This economic argument for reform was supported by another study, which confirmed that for the top 25 fishing nations, the economic benefits of fishery management reform outweigh the associated costs (Mangin et al., 2018).

An additional example of how public data can have impact is the Ocean Economy joint project between the United States Bureau of Economic Analysis (BEA) (n.d.) and the National Oceanic and Atmospheric Administration (NOAA, see Kildow, this volume). This project provides statistics that capture the contributions of ocean-dependent activities to the US economy, which can inform investment decisions as well as support federal and state policymakers (BEA, n.d.). The ready availability of global fisheries and market data made these economically motivating analyses possible.

But for ocean data, we rely on a limited network of ocean buoys and satellites. These networks are expensive to maintain and operate, and their operations are at risk in tight budgetary times. As a result, we lack the necessary tools to improve forecasts of ocean conditions. Yet these projections are critical for short-term safety advisories and weather forecasting, as well as for long-term climate risk and adaptation models.

In a constrained budgetary environment, where resources can't support conventional data collection methods, we need to fill gaps, such as with regional or community science, and also develop simple indices, drawing on techniques from traditionally different fields. Such indices can be used to identify biodiversity "hot spots," estimate trends in species abundance, and trigger intervention or adaptive management. Some of these applications can be developed quite quickly and be tested and adapted in times of crisis.

For example, in 2010, faced with the unprecedented Deepwater Horizon disaster, oil spill modelers worked with fishery scientists and regulators to adapt oil spill dispersion models that enabled them to determine fishery closures on a daily—rather than seasonal or annual—basis (Lubchenco et al., 2012). This more nimble approach served not only the public, by keeping contaminated seafood out of the supply chain, but also fishermen, by keeping unaffected areas open to commercial fishing. Similarly, a transparent spatial forecasting tool, NOAA's Environmental Response Management Application, was adapted to guide recovery as well as restoration decision-making. It was made available to the public in real time, helping to address questions while guiding government action (Lubchenco et al., 2012).

The lessons learned from that all-hands-on-deck effort have driven ocean science and restoration knowledge forward. Today, we have access to technological tools we couldn't have imagined at the turn of the 21st century. The fields of genomics, data science, and optics are evolving quickly and being put to use in the ocean environment—including by the aquarium and MBARI—even as threats to ocean health mount in real time (MBARI, 2017). We can use these tools to characterize habitats, estimate species abundance, and understand or intervene when disease threatens marine life (NOAA, 2020a). There is more we can do with these tools to help us better understand ocean systems.

Today, we routinely employ indices to identify risk areas for attention or measure the success of our efforts. The federal government measures progress through NOAA Fisheries' quarterly Fish Stock Sustainability Index (FSSI), which uses weighted criteria to measure performance of certain US fish stocks against legal requirements to end overfishing and rebuild populations (NOAA, 2020b). In a similar way, the Aquarium's Seafood Watch program has, using peer-reviewed scientific data, synthesized publicly available information into "status indicators" of seafood sustainability based on existing data streams (Seafood Watch, 2020).

Over two decades, NOAA's FSSI and Seafood Watch ratings have helped reform government rules and changed consumer and business choices. They now both confirm the recovery of US fish stocks in response to strong, science-based federal management and identify major challenges on the high seas and in the waters of other nations. These data are now being used to advise global experts on areas for management intervention (Certificates and Rating Collaboration, n.d.).

Expanding this approach to use new data streams, modern tools and other metrics would enable us to continuously assess the risks to our ocean planet—and assign a cost to *not* taking action. Such indices also can identify human health related ocean risks (Sandifer, Colwell, this volume). Harnessing existing data streams to track the pulse of the ocean, would mirror how the stock market assesses economic health. Health and risk indices show promise to advance both public understanding and, potentially, more informed decision-making. These include a customizable tool called the Ocean Health Index, as well as efforts to develop ocean risk indices that link to the more mature fields of climate and financial risk evaluation (Niehörster and Murnane, 2018; Ocean Health Index, n.d.).

Public safety and environmental security require bold action in both the public and private sectors. Data limitations should not stand in the way of making informed decisions in a rapidly changing world. We don't insist on perfection in financial forecasting, insurance forecasting, or other models that support business planning in volatile environments—and we must apply that tolerance to ocean forecasting, too. We know that "black swan events" related to climate change, as well as other perturbations in the natural world and in global wealth and health, will occur (Friedman et al., 2020). The priority must be to improve ocean forecasting using the best tools available and to apply these findings to climate planning without delay.

Using the precautionary principle, as well as historical data from storms, fires, and floods, experts are making a strong case to decision-makers that the benefits of ambitious action for the ocean, including efforts to mitigate and slow the pace of climate change, outweigh any economic risks. We know that climate change is increasing the frequency and intensity of extreme weather events—and the monetary losses associated with them. Insurance companies are teaming up with scientists and conservation experts to build insurance and finance products that can increase the available data on coastal systems, encouraging investment and research that will help manage risks (Smith, 2019). One example is a cell phone app that allows real-time reporting of environmental incidents to insurance providers (Adriano, 2019).

Investors are taking this problem seriously. The world's biggest asset manager, BlackRock Inc., has made climate change central to its investment strategy—a major decision that moves business beyond the typical "corporate social responsibility" efforts we've seen in the past (Beals, 2020). Under the Climate Action 100+ initiative (Climate Action 100+, n.d.), BlackRock and more than 370 global investors, with combined assets of more than $41 trillion, are pressuring companies to achieve carbon neutrality by 2050, a target consistent with the Paris Agreement (UNFCCC, 2015). Broader environmental, social and government indices of corporate sustainability performance, such as the Dow Jones Sustainability Index, can also support positive change when top performers receive favorable financial terms (The Asset, 2021).

These are encouraging developments. Now, it's time to build the ocean into the equation.

Past as prologue: the risk and cost of inaction

The only silver lining of past catastrophes is that we can learn from them. We now have ample evidence that the exploitation of natural resources can undermine nature's resilience and trigger related economic losses. We must apply the impact data from land-based practices to improve risk-and-benefit analyses for ocean uses and make this information transparent and understandable to both policymakers and the public.

Today, laws such as the 1972 amendments to the Federal Water Pollution Control Act, commonly called the Clean Water Act (EPA, n.d.), and the 1976 Resource Conservation and Recovery Act, the landmark US law governing solid and hazardous waste disposal (RCRA, 1976), are in place to prevent the more egregious discharges of the past. At the same time, people continue to release harmful substances into the ocean at a global scale.

This includes plastic, which accounts for the largest proportion of marine pollution by volume (Our Shared Seas, 2019). Researchers from the aquarium and MBARI found an abundance of tiny plastic fragments throughout the water column in Monterey Bay, as well as in the bodies of particle-feeding animals important to the marine food web (Choy et al., 2019). This is a new risk that we have been woefully slow to address. The plastic problem is already daunting (see Smail, this volume), with almost nine million US tons of plastic entering the global ocean each year—and on track to double by 2025 (Jambeck et al., 2015).

When it comes to some other threats, however, we have a chance to prevent damage before it starts. A prime example is deep-sea and seabed mining. Despite the disastrous impacts of mining on land (EPA Superfund Sites, n.d.) and oil drilling in the deep sea (Lubchenco et al., 2012), the International Seabed Authority is currently developing regulations for opening the seafloor to extract minerals such as nickel, manganese, copper, zinc, and cobalt. This activity has yet to begin on an industrial scale (Packard and Scholin, 2018). We have time to take precautions, and we need the science, technology, and data to thoroughly assess both the biological and economic impacts before moving forward. Experts of a recent analysis concluded, "Deep-sea mining will result in disturbance to an area orders of magnitude larger than for the same level of mineral production on land" (Laffoley et al., 2019).

In making decisions about ocean resources our past mistakes make it clear that we must rely on the data we have and adopt a precautionary approach that considers the paramount value of a healthy ocean. We need investors and influencers to support science-based solutions that derive from novel data and synthesis methods. And we need a new generation of professionals to develop cutting-edge technology that can accelerate the rate of change for the better.

The global public already understands the critical importance of ocean health. A six-country public opinion study found that more than 94% of people in Chile, China, Indonesia, Mexico, Japan, and the United States believe the ocean is important to their country's economy and their own well-being (Our Shared Seas, 2020). A strong majority agree that human activity is damaging the ocean, and more than nine in 10 respondents support marine-protected areas with limitations on fishing and other commercial uses.

In other words, we have the will—we just need to find the way.

Crossing disciplines with the next generation in mind

A precautionary approach requires more complex decision-making than that of the past. Such an approach evaluates risk, including the risk of nonaction, as well as the value of ecosystem services and the costs of remediation. Making these calculations requires crossing disciplines. We can apply technology and analytical tools to inform decisions that incorporate more data, more quickly and effectively, than ever before.

Cross-disciplinary research and investments are a cornerstone of the transition to a new blue economy. So is an inspired and engaged public, which can, in turn, influence policymakers and business leaders to make significant changes—and be accountable to their commitments—on behalf of ocean health. In other words, informed constituents and consumers must keep up a steady demand for ocean solutions. For that we must look to skilled communicators in a range of disciplines who serve as a critical bridge between the experts and the public.

The next generation of professionals is poised to define how decisions are made a decade from now. They're adapting to the new normal of complex problems—focusing not only on research, but also on socially relevant, cross-disciplinary engagement. But many academic institutions still compel scientists to prioritize publication in research journals over engaging with the public and policymakers.

"We need to modify the scientific reward structure and training where academics are rewarded, not punished or just tolerated, for engaging," said Dr. Jane Lubchenco when she received the 2017 National Academy of Sciences Public Welfare Medal. "Engagement should be a core responsibility of academic scientists" (NAS, 2017).

The ocean experts of today and tomorrow must value collaboration over hierarchy and exclusion, cultivating inclusion and cross-disciplinary understanding as a matter of efficiency. When scientists and engineers work to solve problems alongside lawyers, communicators, economists, and policy experts, they are able to get to solutions more quickly.

We need a workforce trained to move beyond existing ways of measuring and valuing the ocean and its benefits and to apply those findings to real-world decisions by individuals, businesses, and governments. Employers should recruit and retain candidates from diverse backgrounds who can analyze and interpret in the fields of ocean science, medicine, engineering, conservation policy, and environmental communications. Such a workforce—trained in fields such as economics, social science, and information technology—could accelerate progress in a number of focus areas by:

- **Data synthesis.** Using multiple data sources to estimate both the ecological and social value of conserving our ocean wealth, including fish stock abundance, reproductive potential, and related services. This will require us to look beyond "dockside" or "supply chain" valuation and to assess *all* fish populations—from the most prized commercial species to undiscovered stocks, habitats or species meriting protection, especially in the deep sea.
- **Technology.** Deploying new monitoring tools and technologies, both in the subsea and through remote sensing. More widespread monitoring will improve our understanding of species diversity, abundance, and other metrics of ocean health. It will also position us to prioritize the marine wildlife and habitats in the greatest need of protection, from the deep sea to the near shore.

- **Value Natural Systems.** Helping decision-makers quantify and avoid future risk and to optimize ecosystem services and social costs and benefits—especially in light of a changing climate and ocean. Resource managers can effectively mitigate some of the biggest ocean threats, including climate change, just by improving the natural function of marine and coastal habitats. Trained professionals can identify the best criteria and tools to support these nature-based solutions.
- **Transparency.** Applying accessible tools and transparent data systems—especially through visualizations—to detect and mitigate ocean threats, from both ocean and terrestrial activities. Illegal fishing activity and sustainability performance information are both available to the government and to the public (Seafood Watch, n.d.a; Global Fishing Watch, n.d.). But the deep sea is still largely unknown, and many of its threats are not well characterized. These include agriculture, mining, aquaculture, and plastic production—as well as the continuous release of carbon into the atmosphere. We can use sensors, robotics, and eDNA to better detect, monitor, and manage the impacts.

Other strategies

These cross-disciplinary efforts, of course, require both structure and funding, and government investment is unstable. Federal resources are at the mercy of both macroeconomics and politics, causing science funding to fluctuate from administration to administration. Yet we need consistent funding and structural mechanisms to produce the continuous data sets that are the foundation of a new blue economy.

To achieve that consistency, we must keep building innovative relationships with a diversity of funding bodies that are resilient in the face of fickle government budgets and political winds. Government agencies such as NOAA, facing increasing ocean management challenges in constrained budget times, are reimagining public–private science and technology partnerships that can be applied to these problems and also promote the blue economy (see Gallaudet, this volume).

In a similar way, all sectors confronting ocean and climate threats and seeking to foster a next-generation workforce can build on a variety of successful models of collaboration:

- **Multisector and bipartisan expert groups**. To build momentum for change, experts and leaders from across political and geographical boundaries must work together to make cross-sector recommendations in response to major threats such as climate change and oil spills. One model is the bipartisan 2004 US Commission on Ocean Policy. Congress nominated the members, who were ultimately appointed by President George W. Bush (US Commission on Ocean Policy, 2004.) The commission's recommendations were largely in line with the 2003 report of the US Oceans Commission, privately funded by the Pew Charitable Trusts (Pew Oceans Commission, 2003). The groups combined into the Joint Ocean Commission Initiative, whose recommendations were implemented by both the Bush and Obama administrations and are still influencing current proposals.

- **Targeted philanthropic investment**. The philanthropic sector contributed $620 million in marine-related grants in 2016 (CEA Consulting, 2019). Included in this figure are grants for cross-disciplinary research addressing specific problems and filling knowledge gaps in areas including fisheries management and ecosystem protection. In the case of MBARI, philanthropic investment targeted closing the gap in deep ocean knowledge by pairing engineers with scientists using business-tested approaches. This investment was substantial enough to insulate MBARI's research from the shifting priorities of government agencies, allowing it to be laser-focused on delivering deep sea engineering access and scientific knowledge about ocean change.
- **Independent science for policymakers**. While governmental science bodies such as the National Academies of Science, Engineering, and Medicine play a critical role, more flexible models are being tested. One example is the California Ocean Science Trust (OST), a public benefit corporation enabled by the 2000 California Ocean Resources Stewardship Act. This legislation established a credible, independent voice—within a nongovernmental organization—to leverage expertise from the broader scientific community, accelerating progress toward state ocean priorities. The Trust retains some ties to government, but its multiparty funding model enables more independent areas of inquiry. Its approach is to engage stakeholders in the development of science products, which are applied across scientific and policy boundaries. OST staff are often cross-trained in policy, science, and communication to better support state priorities (California Ocean Science Trust, 2019).
- **Pooled resources for gap-filling projects**. All scientific institutions contain areas of specific expertise, but no single institution—or funder—can meet all needs. That's why leaders have increasingly developed models that bring the flexibility and innovation of academic and

nongovernmental research to bear on government priorities. Examples include NOAA Cooperative Institutes and the National Oceanographic Partnership Program; short-term task forces created by the Pew Charitable Trust's Lenfest Ocean Program and the Alfred P. Sloan Foundation's Census of Marine Life; and the UN Decade for Ocean Science.

- **Unified action through collective impact**. A model known as "collective impact" harnesses diverse voices to identify and advance solutions on key issues. These efforts seek to unify disparate groups around important priorities, help fill capacity gaps, and share best practices (Collective Impact Forum, n.d.). For example, the Conservation Alliance for Seafood Solutions was launched in 2008 to develop a common roadmap toward environmentally sustainable fishing and aquaculture (Conservation Alliance for Seafood Solutions, n.d.). More recently, the Aquarium Conservation Partnership has enabled US aquariums to pursue collective impact on ocean and freshwater conservation, focusing first on ocean plastic pollution (ACP, 2018).
- **Consumer-focused sustainability tools**. Consumers, especially millennials, expect companies to demonstrate socially and environmentally responsible practices—and many are willing to pay more for products that reflect these values (Certus Insights, 2019). In response, businesses are teaming up with respected organizations to make large-scale public commitments showcasing corporate social responsibility. For example, some of the largest US restaurant chains, seafood retailers, and seafood suppliers have committed to using Seafood Watch's science-based standards—driving the market toward more environmentally sustainable fisheries and aquaculture operations (Seafood Watch Standards, 2020). Other tools bring together expertise in disparate areas. For example, the Seafood Slavery Risk Tool, designed for business use, indicates the likelihood that human rights violations are occurring in a specific fishery (Seafood Watch Our Collaborations, n.d.b).
- **High-impact conservation partnerships**. A more traditional kind of collective impact is made through partnerships between government agencies and nonprofits. Often these are fund-sharing arrangements, such as land trusts and government entities working together to protect special places. For example, NOAA Fisheries provides financial and technical support for community-based habitat restoration efforts that can help vulnerable fish populations rebound (NOAA, 1970). Another such partnership that achieved major conservation impact was the 1999

California Marine Life Protection Act. Under this landmark legislation, the state of California teamed up with philanthropic and nongovernmental groups to design and implement the first statewide network of marine protected areas (Kirlin et al., 2013).

The new blue economy: an investment in our future

As humanity explores new ocean frontiers and makes new discoveries, the economic opportunities associated with our global ocean are poised to spike, just as science shows unprecedented risk to our life support system. We must build a workforce equipped with the expertise, training, and big-picture thinking to navigate these opportunities in a way that sustains ocean resources for generations to come.

Ocean conservation requires that we value the ocean and its services, but does not require that we lock resources away from humanity. To the contrary—improving our understanding of the ocean helps identify and avoid potential adverse impacts and, by guiding smart decision-making, opens up a world of possibilities. More efficient, automated data collection and analysis will spur a new wave of scientific breakthroughs, technological innovations, and market enterprises with the potential to help restore ocean health and resilience.

But we need an even wider lens to see the whole picture. A new blue economy that fully incorporates the need for resilience to climate change will require a set of clear priorities that demand a full counting of risks and costs to the environment in our policy and decision-making. It will also depend on a workforce that specializes not only in the science and innovation associated with ocean data, but also in the ability to communicate with broader audiences, understand markets, and economics, turn policymakers' attention to ocean issues, and build regional economies that thrive on ocean and coastal stewardship.

A healthy ocean provides humanity with life-giving resources. Investments in science, and in science-based solutions, will enable us to protect those resources for generations to come. To achieve a *true* blue economy, we must make better-informed decisions that improve ocean and human health. Although some of these choices may be harder to make, they will help balance our relationship with the ocean as the global human population expands to 10 billion by the end of this century.

References

Adriano, L., September 10, 2019. Ironshore Environmental Launches Real-Time Incident Reporting App. Insurance Business America. https://www.insurancebusinessmag.com/us/news/environmental/ironshore-environmental-launches-realtime-incident-reporting-app-177503.aspx.

Aquarium Conservation Partnership, 2018. Aquarium Conservation Partnership. https://ourhands.org/images/ACP_Partnerships_2018.pdf.

Beals, R.K., January 14, 2020. World's Largest Asset Manager BlackRock Joins $41 Trillion Climate-Change Investing Pact. MarketWatch. https://www.marketwatch.com/story/worlds-largest-asset-manager-blackrock-joins-41-trillion-climate-change-investing-pact-2020-01-09.

Burdige, D.J., 2005. Burial of terrestrial organic matter in marine sediments: a re-assessment. Glob. Biogeochem. Cycles 19 (GB4011). https://doi.org/10.1029/2004GB002368.

Bureau of Economic Analysis, n.d. Ocean Economy. https://www.bea.gov/data/special-topics/ocean-economy.

California Ocean Science Trust, May 2019. California Ocean Science Trust: A Vision for 2020. https://www.oceansciencetrust.org/wp-content/uploads/2019/05/OST-Vision-2020-Final.pdf.

CEA Consulting, 2019. Our Shared Seas: Global Ocean Data and Trends for Informed Action and Decision-Making. https://www.evidensia.eco/resources/300/download/.

n.d. Certificates and Ratings Collaboration. https://certificationandratings.org/.

Certus Insights, September 2019. Consumers Expect the Brands They Support to Be Socially Responsible. https://certusinsights.com/wp-content/uploads/2019/10/Markstein-Social-Responsibility-_-Certus-Insights-Research-_.pdf.

Choy, C.A., Robison, B.H., Gagne, T.O., Erwin, B., Halden, R.U., Hamilton, A., Katija, K., Lisin, S.E., Rolsky, C., Van Houtan, K.S., 2019. The vertical distribution and biological transport of marine microplastics across the epipelagic and mesopelagic water column. Sci. Rep. 9 (7843) https://doi.org/10.1038/s41598-019-44117-2.

Climate Action 100+, n.d. Global Investors Driving Business Transition. http://www.climateaction100.org/.

n.d. Collective Impact Forum. https://www.collectiveimpactforum.org/what-collective-impact.

n.d. Conservation Alliance for Seafood Solutions. https://solutionsforseafood.org/.

Cullen, H., David, A., October 11, 2019. Ocean Overlooked when it Comes to Climate Change. The San Francisco Chronicle. https://www.sfchronicle.com/opinion/article/Ocean-overlooked-when-it-comes-to-climate-change-14514785.php#.

Dasgupta, P., 2021. The Economics of Biodiversity: The Dasgupta Review. https://www.gov.uk/government/publications/final-report-the-economics-of-biodiversity-the-dasgupta-review.

Environmental Protection Agency, n.d. Summary of the Clean Water Act. https://www.epa.gov/laws-regulations/summary-clean-water-act.

European Commission, 2018. The 2018 Annual Economic Report on the EU Blue Economy. https://ec.europa.eu/maritimeaffairs/sites/maritimeaffairs/files/2018-annual-economic-report-on-blue-economy_en.pdf.

Food and Agriculture Organization of the United Nations, 2018. The State of World Fisheries and Aquaculture - Meeting the Sustainable Development Goals. http://www.fao.org/3/i9540en/I9540EN.pdf.

Friedman, W.R., Halpern, B.S., McLeod, E., Beck, M.W., Duarte, C.M., Kappel, C.V., Levine, A., Sluka, R.D., Adler, S., O'Hara, C.C., Sterling, E.J., Tapia-Lewin, S., Losada, I.J., McClanahan, T.R., Pendleton, L., Spring, M., Toomey, J.P., Weiss, K.R., Possingham, H.P., Montambault, J.R., 2020. Research priorities for achieving healthy marine ecosystems and human communities in a changing climate. Front. Mar. Sci. 7 (5) https://doi.org/10.3389/fmars.2020.00005.

n.d. Global Fishing Watch. https://globalfishingwatch.org/.
Government Office for Science, 2018. Future of the Sea. https://assets.publishing.service.gov.uk/government/uploads/system/uploads/attachment. _data/file/706956/foresight-future-of-the-sea-report.pdf.
Hoegh-Guldberg, O., Caldeira, K., Chopin, T., Gaines, S., Haugan, P., Hemer, M., Howard, J., Konar, M., Krause-Jensen, D., Lindstad, E., Lovelock, C.E., Michelin, M., Nielsen, F.G., Northrop, E., Parker, R., Roy, J., Smith, T., Some, S., Tyedmers, P., 2019. The Ocean as a Solution to Climate Change: Five Opportunities for Action. World Resources Institute. https://oceanpanel.org/sites/default/files/2019-10/HLP_Report_Ocean_Solution_Climate_Change_final.pdf.
Intergovernmental Panel on Climate Change, 2019. In: Pörtner, H.-O., Roberts, D.C., Masson-Delmotte, V., Zhai, P., Tignor, M., Poloczanska, E., Mintenbeck, K., Alegría, A., Nicolai, M., Okem, A., Petzold, J., Rama, B., Weyer, N.M. (Eds.), IPCC Special Report on the Ocean and Cryosphere in a Changing Climate. https://www.ipcc.ch/site/assets/uploads/sites/3/2019/12/SROCC_FullReport_FINAL.pdf.
n.d. International union for conservation of nature. Issue Brief (Public Policy Inst. Am. Assoc. Retired Persons): Ocean & Clim. https://www.iucn.org/resources/issues-briefs/ocean-and-climate-change.
Jambeck, J.R., Geyer, R., Wilcox, C., Siegler, T.R., Perryman, M., Andrady, A., Narayan, R., Law, K.L., 2015. Plastic waste inputs from land into the ocean. Science 347 (6223), 768–771. https://doi.org/10.1126/science.1260352.
Jarrin-Thomas, S., September 4, 2020. The Sunrise Movement Continues to Make Waves. Nonprofit Quarterly. https://nonprofitquarterly.org/the-sunrise-movement-continues-to-make-waves/.
Kirlin, J., Caldwell, M., Gleason, M., Weber, M., Ugoretz, J., Fox, E., Miller-Henson, M., 2013. California's Marine Life Protection Act Initiative: supporting implementation of legislation establishing a statewide network of marine protected areas. Sci. Direct 74 (2013), 3–13. https://doi.org/10.1016/j.ocecoaman.2012.08.015.
Laffoley, D., Baxter, J.M., Amon, D.J., Currie, D.E.J., Downs, C.A., Hall-Spencer, J.M., Harden-Davies, H., Page, R., Reid, C.P., Roberts, C.M., Rogers, A., Thiele, T., Sheppard, C.R.C., Sumaila, R.U., Woodall, L.C., 2019. Eight urgent, fundamental and simultaneous steps needed to restore ocean health, and the consequences for humanity and the planet of inaction or delay. Aquat. Conserv. Mar. Freshw. Ecosyst. 30 (1), 194–208. https://doi.org/10.1002/aqc.3182.
Lipton, D., Lew, D.K., Wallmo, K., Wiley, P., Dvarskas, A., 2014. The evolution of non-market valuation of US coastal and marine resources. J. Ocean & Coast. Econ. 2014 (1) https://doi.org/10.15351/2373-8456.1011.
Lubchenco, J., McNutt, M., Dreyfus, G., Murawski, S., Kennedy, D., Anastas, P.T., Chu, S., Hunter, T., 2012. Science in support of the deepwater horizon response. Proc. Natl. Acad. Sci. U. S. A. 109 (50), 20212–20221. https://doi.org/10.1073/pnas.1204729109.
Mangin, T., Costello, C., Anderson, J., Arnason, R., Elliott, M., Gaines, S., Hilborn, R., Peterson, E., Sumaila, R., 2018. Are fishery management upgrades worth the cost? PLoS One 13 (9), e0204258. https://doi.org/10.1371/journal.pone.0204258.
Martin, J., Maris, V., Simberloff, D.S., 2016. The need to respect nature and its limits challenges society and conservation science. Proc. Natl. Acad. Sci. U. S. A 113 (22), 6105–6112. https://doi.org/10.1073/pnas.1525003113.
Monterey Bay Aquarium Research Institute, May 3, 2017. A Glimpse into the Future of Marine Research. https://www.mbari.org/a-glimpse-into-the-future-of-marine-research/.

National Academy of Sciences, May 4, 2017. NAS Public Welfare Medalist Jane Lubchenco. https://www.youtube.com/watch?v=J1fmcboTs4A&feature=youtu.be&list=PLlKst-jESy–IGs2r1xD8iGVy4HiuBXNR.

National Oceanic and Atmospheric Administration Fisheries, 1970. Current and Past Community-based Restoration Projects. https://www.govinfo.gov/content/pkg/STATUTE-84/pdf/STATUTE-84-Pg2090.pdf#page=1 (Accessed April 24, 2020).

National Oceanic Atmospheric Administration, February 18, 2020a. NOAA Finalizes Strategies for Applying Emerging Science Technology. https://www.noaa.gov/media-release/noaa-finalizes-strategies-for-applying-emerging-science-and-technology.

National Oceanic and Atmospheric Administration Fisheries, February 3, 2020b. Status of US Fisheries. https://www.fisheries.noaa.gov/national/population-assessments/status-us-fisheries#fish-stock-sustainability-index.

Niehörster, N., Murnane, R.J., 2018. Ocean Risk and the Insurance Industry. Pub. by XL Catlin Services SE, UK, May 2018, p. 40. https://www.oceanrisksummit.com/Content/press-releases/FALK-MAIN-REPORT-FINAL-LOW-RES.pdf.

n.d. Ocean Health Index. http://www.oceanhealthindex.org/.

Organization for Economic Co-operation and Development, 2016. The Ocean Economy in 2030. OECD Publishing, Paris.

Our Shared Seas, 2019. Pollution and Development. https://oursharedseas.com/2019-update/pollution-and-development/.

Our Shared Seas, March 3, 2020. New Study Finds Ocean Is an Economic Force in Need of More Protection. https://oursharedseas.com/supportforaction/.

Packard, J., Scholin, C., June 8, 2018. The Deep Sea May Soon Be up for Grabs. The New York Times. https://www.nytimes.com/2018/06/08/opinion/the-deep-sea-may-soon-be-up-for-grabs.html.

Pew Oceans Commission, 2003. America's Living Oceans: Charting a Course for Sea Change. https://www.pewtrusts.org/-/media/assets/2003/06/02/full_report.pdf.

Pew Research Center, August 2, 2019. Trust and Mistrust in Americans' Views of Scientific Experts. https://www.pewresearch.org/science/wp-content/uploads/sites/16/2019/08/PS_08.02.19_trust.in_.scientists_FULLREPORT_8.5.19.pdf.

Reorganization Plan No 4, 5 U.S.C §§ 906, 1970. https://uscode.house.gov/view.xhtml?req=granuleid:USC-prelim-title5a-node84-leaf179&num=0&edition=prelim.

Publ. L. No. 94-580, 90 Stat. 2795, 42 USC § 6901 et seq Resource Conservation and Recovery Act, 1976. https://www.govinfo.gov/content/pkg/STATUTE-90/pdf/STATUTE-90-Pg2795.pdf.

Seafood Watch, 2020. Standards. https://www.seafoodwatch.org/seafood-recommendations/standards-revision.

Seafood Watch, n.d.a. https://www.seafoodwatch.org/.

Seafood Watch, n.d.b. Our Collaborations https://www.seafoodwatch.org/businesses-and-organizations/our-collaborations.

Smith, R., September 20, 2019. Insurers Team with Governments, Environmental Groups to Fight Ocean Risk. Insurance Business America. https://www.insurancebusinessmag.com/us/news/environmental/insurers-team-with-governments-environmental-groups-to-fight-ocean-risk-178463.aspx.

Stratton, J.A., Baird, C.F., 1969. Commission on Marine Science, Engineering and Resources. Our Nation and the Sea: A Plan for National Action. U.S. Government Printing Office, Washington, D.C, p. 230.

The Asset, 2021. Thai Union secures first sustainability-linked loan. https://www.theasset.com/article-esg/42929/thai-union-secures-first-sustainability-linked-loan.

United Nations Framework Convention on Climate Change [UNFCCC], 2015. Paris agreement. In: Proceedings of the 21st Conference of the Parties of the UNFCCC. https://unfccc.int/sites/default/files/english_paris_agreement.pdf.

United Nations, 2017. World Population Prospects: The 2017 Revision, Key Findings and Advance Tables. https://reliefweb.int/sites/reliefweb.int/files/resources/WPP2017_KeyFindings.pdf.

United Nations Environment Programme, 2021. Making Peace with Nature: A scientific blueprint to tackle the climate, biodiversity and pollution emergencies. Nairobi. https://www.unep.org/resources/making-peace-nature.

US Commission on Ocean Policy, 2004. An Ocean Blueprint for the 21st Century. https://govinfo.library.unt.edu/oceancommission/documents/full_color_rpt/000_ocean_full_report.pdf.

US Department of Energy, 2019. Powering the Blue Economy. https://www.energy.gov/sites/prod/files/2019/09/f66/73355-v2.pdf.

World Bank, 2009. The Sunken Billions: The Economic Justification for Fisheries Reform. https://openknowledge.worldbank.org/handle/10986/2596.

World Bank, February 14, 2017. Global Fisheries' Sunken Billions. https://www.worldbank.org/en/news/feature/2017/02/14/global-fisheries-sunken-billions.

World Resources Institute, 2020. High Level Panel for a Sustainable Ocean Economy, Blue Papers. https://www.oceanpanel.org/blue-papers.

OP/ED: Risk tolerance

Norman R. Augustine

Risk

Virtually all progress requires change, and virtually all change entails risk. Sustaining the ocean economy, as well as the nation's economy writ large, is no exception. The point is to not take extravagant or irrational risks; rather to accept considered, judgmental risks where the gains warrant.

Fortunately, there are well-established rules that can be applied when engaging risk. The first of these is to understand, to the extent practicable, the risks that are being confronted. Next is to assure that the benefits of favorable outcomes outweigh the liabilities of adverse outcomes. Third is to take appropriate steps to ameliorate the risks, as well as to provide backup plans in case the downsides transpire. Finally, it is essential to assure that worst-case outcomes are not existential.

The macro risk regarding the ocean economy is evident: that a healthy ocean environment might not be maintained. But there is another, more subtle, risk that concerns society's ability to understand the impact of the ocean ecosystem and indeed our overall natural surroundings. This, of course, is the realm of scientific research.

America's research enterprise is currently endangered on at least two fronts. The first is a lack of investment in research itself, especially "basic" or "fundamental" research, such as is largely conducted in our nation's universities and funded by the federal government. The second is the seriously impaired pipeline that produces the nation's future researchers. Fortunately, it is within America's wherewithal to repair each of these failings...if we possess the will to do so.

Today's US budget for basic research is about 0.2% of the Gross Domestic Product. This is roughly what we spend on cigarettes; it is about three-fourths of what we as a nation spend on the purchase of illegal drugs; and it is about one-third of what we spend on alcoholic beverages. If such be our priorities, such will be their consequences. But performing research is only part of the story: the results of that research must be applied if society is to benefit.

While many other nations have been increasing their spending on research and development as a fraction of GDP, the US position has been steadily declining in recent decades—now ranking 10th in the world. In contrast, China has recently bypassed the United States in overall R&D spending, calculated at purchasing power parity. US industry has largely withdrawn from the performance of basic research in favor of investing in applications, primarily a consequence of the near-term demand for returns by corporate shareholders who now, on average, hold their shares in an individual firm for only about 4 months. The demise of the iconic Bell Labs is Exhibit One. The product of research is often primarily a public good, and as such its support must largely depend upon federal funding.

Accompanying this concern over the nation's underinvestment in research is a disconcerting trend in how funds that are in fact devoted to research are applied. Specifically, in the intense competition for support (less than 20% of proposed valid basic research projects are now being funded), the pressure to avoid failure strongly favors risk aversion when selecting projects to be pursued. This penalizes the very research that often produces the greatest gains.

Even given appropriate funding and application of research, there remains the critical challenge of providing the talent needed to conduct that research. With the existing leakage in the nation's educational pipeline, were one to seek to provide a single additional Ph.D. researcher to the scientific workforce in the year 2033, one would need to begin today with a pool in eighth grade in the US public school system of 3000 students. This challenge is exacerbated by the serious underparticipation by women and minorities in the scientific workforce (see Sullivan, this volume).

The quality of American public education, as ranked by the most prominent international test, places US 15-year-olds well below average in math and science among OECD ("developed") nations; yet America spends more per student than any other nation except Switzerland. At the same time, the nation's higher education system, widely ranked as the world's finest, has come under assault by state governments that have imposed often draconian budget reductions on their flagship universities. Meanwhile, the federal government has attacked many of the nation's most prestigious universities by imposing a tax on their endowment earnings. These funds could otherwise have been used for such purposes as funding research or scholarships and fellowships.

The United States recently ranked 76th in the world in the fraction of baccalaureate degrees that the nation awards in science and engineering, placing it just behind Mozambique. It can justifiably be argued that America's scientific enterprise today would barely function were it not for the foreign-born individuals who came to America for their higher education and remained to contribute as members of the nation's scientific and engineering community. However, recent policy and legislation are making it less attractive for foreign nationals to study in America, as well as more difficult to remain here after receiving their degrees.

While confronting some elements of risk in sustaining a viable ocean economy may not be avoidable, confronting those risks relating to the creation of knowledge and the provision of the talent that will create that knowledge are principally a question of national priority and will.

Author biography

Norman R. Augustine was raised in Colorado and attended Princeton University where he graduated magna cum laude with a BSE and an MSE in Aeronautical Engineering. He has served as Under Secretary of the Army and Acting Secretary of the Army, as Chairman and CEO of Lockheed Martin Corporation, and as Lecturer with the Rank of Professor at Princeton University.

Mr. Augustine was Chairman of the American Red Cross, the Council of the National Academy of Engineering, the Association of the United States Army, the Aerospace Industries Association, and the Defense Science Board. He is a former President of the American Institute of Aeronautics and Astronautics and the Boy Scouts of America. He was a Regent of the University System of Maryland, Trustee of Princeton and MIT and is a Trustee Emeritus of Johns Hopkins. He served on advisory boards to NASA, the Departments of Homeland Security, Energy, Defense, Commerce, Transportation, and

Health and Human Services (NIH), and served for 16 years on the President's Science Advisory Board.

He is a member of the National Academy of Sciences, the National Academy of Engineering, the American Philosophical Society, the American Association of Arts and Sciences, the Council on Foreign Relations, and a fellow of the Explorers Club.

Mr. Augustine has been presented the National Medal of Technology by the President of the United States, the Joint Chiefs of Staff Distinguished Public Service Award, and the NASA Distinguished Public Service Medal. He has five times received the Department of Defense's highest civilian decoration, the Distinguished Service Medal, and holds honorary degrees from 35 universities.

Since retiring, he has chaired or cochaired 39 committees and commissions, mostly on behalf of the government. He has authored or coauthored four books and has traveled in 129 countries and stood on both the North and South Poles of the Earth.

SECTION VII

Closing thoughts

Liesl Hotaling, Richard W. Spinrad

Imagine the world of 2030. The New Blue Economy described in this volume will have grown by design, not by default, through careful consideration of the economic opportunities, environmental concerns, and workforce requirements. Using lessons learned from communities such as the commercial weather industry, the ocean community will undertake several important steps to ensure a strategically developed New Blue Economy. We envision that:

- Societal needs are assessed with an eye to balancing environmental stewardship, protection, and sustainability, with economic opportunity;
- Scientific and technical programs are prioritized and funded appropriately to maximize success, but with a clear tolerance for balanced risk-taking;
- Dynamical structures coordinate the roles/responsibilities and sharing of knowledge between sectors of government, industry, philanthropy, and academia on a global, regional, and local basis;
- Mechanisms are in place to ensure the most effective transfer of knowledge and technology from broad sectors of research and operations and from all sources (including traditional knowledge);
- Clear measures and indicators are established for determining success and means for redirection of effort;

- The K-12 education system is aligned and equipped to produce students prepared to step into jobs or higher educational experiences with increasingly robust data literacy skills;
- An increasingly flexible higher education system is in place to accelerate credential and degree attainment and to value fully one's life experience;
- The workforce is diverse with every effort made to ensure justice, equity, diversity, and inclusion;
- Policies, organizational structures, and governance are established, based on past successful examples, and invent or establish new components as needed.

Realistic? Maybe. Doable? Absolutely.

The Organisation for Economic Co-operations and Development (OECD, 2016) reports that global "ocean industries have the potential to make an important contribution to employment growth. By 2030, they are anticipated to employ approximately 40 million full-time equivalent jobs. The fastest growth in jobs is expected in offshore wind energy, marine aquaculture, fish processing and port activities." The European Blue Economy, currently representing around 5.4 million jobs, is set to double its employment by 2030. The BlueGeneration Project aims to convert at least 2,000 young people to employment or training in the Blue Economy by 2022 in an effort to begin to fill the employment need in Europe.

Although most of these jobs will not require advanced degrees in engineering or science, the preparation of the 21st century workforce demands an understanding of marine science and other disciplines, an ability to resolve complex environmental issues, and the ability to communicate complex ideas to a broad audience. Fostering these critical abilities will require a new set of learning opportunities, as discussed in this volume. Developing and maintaining such a workforce will rely on innovative and flexible educational programs that break through the traditional "siloed" approach to education, while offering multiple and rapid pathways for degree and certification attainment.

Programs to prepare this needed workforce should emphasize opportunities for authentic, hands-on learning and collaborative teamwork and will provide engaging and practical experience in systems thinking with marine-themed activities. These programs will include stackable micro-credentials, which represent a combination of technical course completion and/or demonstrated skill sets within a competency area to broaden the opportunity to engage students, professional development for industry professionals and people seeking workforce retraining skills. Micro-credentialing schema should also provide opportunities for crosswalking life experience to college credit, credentials, and industry certifications.

Although preparing students for employment will be the primary goal, pathways will be prepared for students to opt to pursue a higher degree. It is important to note that according to the US National Science Foundation's Science and Engineering Indicators 2020, more than half of all science and engineering graduates attend community colleges at some point in their career. This trend will most likely increase as college costs continue to rise (New America, 2019).

In this volume we have provided a definition for the New Blue Economy and indicated examples of where the opportunities lie. We have only started this treatment. The reader no doubt has identified many other potential considerations in the development of the New Blue Economy. Many relevant areas merit further discussion than what is included here: the role of external technologies coming from fields such as materials science, advanced manufacturing, and robotics; the roles of social and behavioral sciences; the role of philanthropy; and strategies for investment. No doubt new training and educational approaches will ensue, via both technological development (consider the impact of the COVID-19 pandemic on creation of new tools for distance learning) and the emergence of unexpected economic opportunities. Our objective in editing this volume was to define the New Blue Economy, present compelling cases for its development, and address the critical issues associated with building the necessary workforce. Future treatments should expand from these questions of "what" and "why," into the subject of "how"; i.e., what is needed to move forward. We are excited about the prospects of that discussion. To this end, we have succeeded if we have encouraged a broader discussion on the issue of the development of the New Blue Economy.

References

New America, Fishman, R., Nguyen, S., Acosta, A., Clark, A., 2019. Varying degrees 2019. Retrieved from: newamerica.org/education-policy/reports/varying-degrees-2019/.

OECD, 2016. The ocean economy in 2030. OECD Publishing, Paris. https://doi.org/10.1787/9789264251724-en.

Index

Note: 'Page numbers followed by "*f*" indicate figures and "*t*" indicate tables.'

N

P

Q

R

FSC
www.fsc.org
MIX
Papier | Fördert gute Waldnutzung
FSC® C083411

Kolibri 360 GmbH
Ferdinand-Jühlke-Straße 7
99095 Erfurt, Deutschland
produktsicherheit@kolibri360.de